CONTROLLING TURFGRASS PESTS

Third Edition

Thomas W. Fermanian
Turfgrass Scientist

Malcolm C. Shurtleff
Plant Pathologist

Roscoe Randell
Entomologist

Henry T. Wilkinson
Turfgrass Scientist

Philip L. Nixon
Entomologist

University of Illinois at Urbana-Champaign
Urbana, Illinois

Upper Saddle River, New Jersey 07458

Library of Congress Cataloging-in-Publication Data

Controlling turfgrass pests / Thomas W. Fermanian ... [et al.].-- 3rd ed.
    p. cm.
    ISBN 0-13-098143-5
    1. Turfgrasses--Diseases and pests--Control. 2. Turf management. I. Fermanian,
Thomas W.

SB608.T87 C65 2002
    635.9'6429--dc21   2001058050

Editor-in-Chief: *Steve Helba*
Executive Acquisitions Editor: *Debbie Yarnell*
Editorial Assistant: *Sam Goffinet*
Managing Editor: *Mary Carnis*
Production Management: *Carlisle Communications, Ltd.*
Production Editor: *Bridget Lulay*
Director of Manufacturing and Production: *Bruce Johnson*
Manufacturing Buyer: *Cathleen Petersen*
Marketing Manager: *Jimmy Stephens*
Creative Director: *Cheryl Asherman*
Senior Design Coordinator: *Miguel Ortiz*
Cover Design: *Marianne Frasco*
Cover Photo: *Eyewire (royalty-free, per Getty Images.com)*

Pearson Education LTD.
Pearson Education Australia PTY, Limited
Pearson Education Singapore, Pte. Ltd.
Pearson Education North Asia, Ltd.
Pearson Education Canada, Ltd.
Pearson Educaión de Mexico, S. A. de C.V.
Pearson Education—Japan
Pearson Education Malaysia, Pte. Ltd.

10 9 8 7 6 5 4 3 2 1
ISBN 0-13-098143-5

CONTENTS

CHAPTER 3

Abiotic Turfgrass Problems   59

CHAPTER 4

Biology and Management of Weeds in Turfgrasses   *84*

CHAPTER 5

Biology and Management of Diseases in Turfgrasses   193

CHAPTER 6
Biology and Control of Insects and Related Pests of Turfgrass   396

CHAPTER 7

Application Equipment and the Safe Use of Pesticides   481

PREFACE

*T*his book is an up-to-date account of the current state of the art and science of turfgrass pest management. The third edition has been written as a basic text for college students in turfgrass management, landscape architecture, turfgrass pathology, and economic entomology. It is also designed as a comprehensive reference for professional turfgrass managers, including golf course superintendents; lawn-care-company personnel; park and cemetery officials; those responsible for maintaining turf on sports fields, at airports, and along highways; seed and turf producers; and landscapers. It should prove a valuable reference to turfgrass consultants; representatives of a variety of agribusinesses that serve the turfgrass industry; county, area, and state extension (advisory) personnel; and master gardeners involved in the diagnosis and suggestion of control measures for turfgrass pests, officers in state and federal departments of agriculture and regulatory agencies, vocational agriculture and biology teachers, garden-center personnel, garden writers, and home lawn enthusiasts. The book contains both the technical and practical information necessary for decision making and the day-to-day operation in all areas of turfgrass culture and management.

The stimulus for writing this book has come from the rapid expansion of the turfgrass industry in the past 10 to 30 years and a tremendous increase in the numbers of professional turfgrass managers responsible for managing turf as free as possible of weeds, insects, diseases, and other pests. There are a number of excellent books available on turfgrass management and culture. However, a text-reference book that concentrates on the diagnosis, fundamental biology, and control of turfgrass weeds, insects and other animal pests, and diseases is needed.

The authors humbly submit this work as such a book.

Emphasis is placed on how to identify turfgrass pests, where and why they occur, the damage that may take place, the life cycles of pests, plus cultural, chemical, and other management strategies designed to keep pest damage to a minimum.

This edition stresses integrated pest management (IPM) systems for controlling turfgrass pests. Insects and diseases often occur in combination with weeds and/or as a result of one another. For example, thinning and killing of a turf stand by diseases and/or insects allows weeds to germinate and invade the damaged area. Insects and diseases commonly weaken turfgrass plants, making them more susceptible to another type of pest. Every IPM program on a golf course, home or industry lawn, athletic field, park, cemetery, and seed or sod farm is aimed at managing the turfgrass to its competitive best by manipulating all the cultural aspects of turfgrass management, supplemented where needed with the proper application of growth regulators and pesticides. The basic steps of any IPM program are presented in sec. 2.1.

An icon appears at the beginning of each section in the book representing one of the three major IPM steps. These icons have been inserted to help the reader focus on material that will assist them in executing these important IPM principles. It is our hope that these visual cues will assist you in understanding how presented materials relate to basic IPM principles.

Emphasis is placed on minimizing the severity of all types of pest damage through properly-organized cultural management. The effectiveness of pesticide applications (to control weeds, insects, diseases, nematodes, or rodents) can only be achieved if proper cultural management programs have been followed. Underlying mismanagement, which commonly allows a pest to become a problem, must first be resolved. Specific recommendations are given for each pest, integrating all available IPM tactics. Suggestions are given for cool- and warm-season turfgrass species growing in humid, wet, dry, and arid conditions under low, medium, or high maintenance.

The enlarged chapter on integrated management blends the various strategies and general turfgrass maintenance practices into a unified whole. The recommendations made take into account the cultural aspects of turfgrass management (such as proper establishment or renovation using locally adapted resistant species and cultivars, proper irrigation, mowing height and frequency dethatching, adjustment of the nitrogen rate and source, soil modification by coring or aerification and topdressing soil pH adjustment, traffic and soil compaction, plus judicious use of pesticides, biological, and legal control measures). Successful turfgrass management, including pests, is an art based on sound scientific knowledge. Due to frequent changes in pesticides, pesticide formulations and other chemicals used in turfgrass culture, climatic variations, and the turfs grown, this book should be supplemented by current pest control programs and bulletins available in each state or country from the local extension (advisory) office or state agricultural experiment station.

Accurate diagnosis is the first step in the control of any pest. The detailed keys in chapters 5, 6, and Appendix D plus the numerous illustrations, are designed to make positive diagnosis as simple as possible. The extensive index should also be helpful.

Much of the research information concerning turfgrass culture, including the biology and control of weeds, insects and plant diseases, is scattered throughout thousands of scientific journals, trade publications, turfgrass conference proceedings, and field day programs, and is therefore not easily available to students, turfgrass managers, and others interested in controlling the wide range of turfgrass pests. Many thousands of references were reviewed during the preparation of this text. Each of the major chapters ends with a list of the more pertinent and widely available references for those wishing to study in more detail a certain pest or maintenance activity. The text is complemented by numerous line drawings, photographs, and color plates that have been carefully selected to illustrate all of the common and uncommon pests, even the rare ones. In writing this book, the authors assume the reader possesses an elementary knowledge of such basic sciences as botany, chemistry, and soil science.

Comments regarding the general usefulness of the text and/or errors or omissions in the text are always welcome. With your suggestions, future editions can become even more valuable for the diagnosis and management of turfgrass pests.

ACKNOWLEDGMENTS

The authors wish to express their gratitude to the following colleagues for adding breadth, accuracy, and value when reviewing parts or all of the manuscript: R.C. Avenius, D.J. Blasingame, T.H. Bowyer, L.L. Burpee III, G.A. Chastagner, K. Clay, P.F. Colbaugh, J.L. Dale, P.H. Dernoeden, J.M. Dipola, M. Elliott, T.L. Finnerty, R.L. Goss, C.F. Hodges, Noel Jackson, B.G. Joyner, R.T. Kane, G.M. Kozelnicky, P.O. Larson, L.T. Lucas, T.A. Melton, W.A. Meyer, H.D. Niemczyk, R. Reed, D.L. Roberts, J.L. Saladini, G.L. Schumann, Janet L. Shurtleff, G.W. Simone, R.W. Smiley, J.D. Smith, W.C. Stienstra, N.A. Tisserat, R.W. Toler, A.J. Turgeon, T. Van Hoveln, J.M. Vargas, Jr., T.B. Voigt, R.E. Wagner, J.E. Watkins, R.L. Wick, R.E. Wolf, and G.L. Worf.

Individuals, institutions, and companies who so willingly provided illustrations are given specific credit in the illustration legend.

Limitations of space make it impossible to acknowledge adequately our indebtedness to the hundreds of research scientists, extension specialists, and writers for the enormous storehouse of information from which we drew the material presented.

A special thanks goes to artists George Mayer, Earl Mayer, and Nancy Fermanian, who drew a number of the weed species, and especially to Lenore Gray and Floyd Giles, who did some of the maintenance and equipment artwork, weed species, and some of the drawings of insects, other animal pests, and plant pathogens. We thank Carol Preston for her help in handling correspondence with the editors. Finally, we thank our wives who were quietly patient and supportive during the long hours we neglected them while referencing and writing this book. Our indebtedness to all of these persons is gratefully acknowledged.

Thomas W. Fermanian

Malcolm C. Shurtleff

Roscoe Randell

Henry T. Wilkinson

Philip L. Nixon

ABOUT THE AUTHORS

THOMAS W. FERMANIAN is an Associate Professor of Turfgrass Science in the Department of Natural Resources and Environmental Sciences at the University of Illinois at Urbana-Champaign. Dr. Fermanian received a Ph.D. in turfgrass science at Oklahoma State University in 1980. He then accepted a position in the department of Horticulture at the University of Illinois at Urbana-Champaign. Dr. Fermanian has remained at the University of Illinois for 21 years and is currently an associate professor in the department of Natural Resources and Environmental Sciences.

Dr. Fermanian has developed several knowledge engineering and qualitative data analysis tools for agriculture, along with computer-based turfgrass management systems. One of these programs, AgAssistant, was used to build the weed identification key. In addition to his interests in computing systems and programming, his major work is in low-input, site-specific management of turfgrass weeds and the use of plant-growth-regulating compounds on fine turfs.

MALCOLM C. SHURTLEFF received his B.S. in biology from the University of Rhode Island in 1943 and his M.S. (1950) and Ph.D. (1953) in plant pathology from the University of Minnesota. He is now an Emeritus Professor of Plant Pathology at the University of Illinois, where he resided on the faculty from 1961 to 1992. He served in the U.S. Navy during World War II.

Dr. Shurtleff taught plant disease control and diseases of turfgrasses and ornamentals and was an extension plant pathologist in Rhode Island, Iowa, and Illinois for 42 years, handling all types of plant disease problems. He is the author of 9 books, 16 encyclopedia articles, 393 magazine articles, and 1,651 extension and research articles on plant diseases.

Dr. Shurtleff has been honored nationally by being accorded Fellow status by the American Phytopathological Society, given the Adventurers in Agricultural Science Award of Distinction by the International Congress of Plant Protection, and the Distinguished Service Award from the USDA.

ROSCOE RANDELL is an internationally recognized entomologist who recently retired after more than 35 years of service to the University of Illinois at Urbana-Champaign. Dr. Randell began his public service as a county extension adviser and continued to work directly with turf managers, diagnosing insect problems, throughout his career. In 1985, Dr. Randell received the Illinois Turfgrass Foundation's Distinguished Service Award.

HENRY T. WILKINSON has researched and taught turfgrass pathology at the University of Illinois for over 19 years. He has discovered numerous new turf diseases in both cool- and warm-season grasses and has solved pathological problems in many different parts of the world. His main pathological expertise is in soil-borne pathogens of turfgrass. He was raised in Princeton, New Jersey. He graduated from Purdue University (B.S.) and Cornell University (M.S. and Ph.D.) and studied as a postdoctoral fellow at Washington State University.

PHILIP L. NIXON is an Extension Entomologist for the University of Illinois. After being located in the Chicago area from 1980 to 1987, he moved to his present position at the Champaign-Urbana campus. He advises and teaches turfgrass professionals on proper insect pest management and conducts research on turfgrass insects. He also is a coordinator for the Pesticide Applicator Training Program at the University of Illinois. Dr. Nixon has advanced degrees from Southern Illinois University and Kansas State University.

 Introduction

1.1 HOW TO USE THIS BOOK

This book is written for anyone who enjoys beautiful turf as free as possible of weeds, insects, diseases, and other pests. It is written for students in turfgrass management, landscape architecture, plant pathology, and economic entomology classes; lawn-care professionals; home lawn enthusiasts; golf course superintendents; park and cemetery officials; managers of athletic fields; commercial seed and sod growers; turfgrass consultants; agribusiness research and sales representatives; county extension agents; officials in state and federal departments of agriculture and regulatory agencies; those in charge of maintaining turf at airports and along highways; area and state extension turfgrass specialists, entomologists, weed specialists, and plant pathologists involved in diagnosis and control of turfgrass problems; garden-store personnel; vocational agriculture and biology teachers; garden writers; and others who grow, know, and love turfgrasses. You will find this book invaluable as a text and reference and easy to use.

An attempt has been made to write in everyday language. Scientific names of weeds, insects, and disease-causing agents are included. There is a comprehensive glossary to help with technical terms. This book blends descriptive terminology with the more technical language of weed science, entomology, nematology, plant pathology, and turfgrass management to better serve the diverse levels of knowledge and skills of a wide variety of turfgrass managers and consultants. Each chapter ends with some selected reference books found in many public libraries where more detailed information may be obtained. We have also added suggested exercises to help reinforce the main concepts presented throughout the chapter.

Three questions invariably asked about a turfgrass problem are: (1) "What's wrong?" or "What is it?"; (2) "How serious is it?" or "Will it kill my turf?"; and (3) "What should I do about it?" This book helps to answer these basic questions. These three steps—identify the problem, determine its severity, and act on it—are the foundation to integrated turfgrass management. We use three icons throughout the book to identify these important principles (see figure 2–1).

In chapter 2, integrated turfgrass management is introduced, starting with a discussion of its principles. Genetic, cultural, biological, physical, and chemical aspects of pest management are addressed. The remainder of the chapter presents the topics of selecting seed or other plant materials, and establishing and maintaining turf. A healthy and vigorous turf is basic to controlling weeds, insects, diseases, and other pests.

In chapters 4, 5, and 6, weeds, insects, animal pests, and diseases (including nematodes) are described. These chapters discuss how serious weeds, insects, other pests, and diseases may become, and where and when they can be expected to be a problem in the United States and elsewhere. Symptoms of damage are given, life cycles are described, and cultural control measures that will keep damage to a minimum are outlined. For quick reference, chemical controls are presented in tables at the end of each chapter. The simplified but comprehensive keys at the beginning of the chapters, covering weeds, pests, and diseases, plus the many photographs and line drawings, will help you diagnose essentially any turf-grass problem.

More detailed information follows in each chapter, discussing the grasses most susceptible and resistant to injury and the environmental conditions that favor or check each pest. Also included is information on turf-protection chemicals, including their common and trade names and principal uses.

Chapter 3 is a new chapter to this edition of *Controlling Turfgrass Pests*. It presents many of the nonbiotic problems that occur on many turfs. These disorders range from the infrequent, relatively minor problems to systemic situations that require modified management programs.

Chapter 7 covers the selection and use of various types of equipment used for the application of fertilizers and pesticides to commercial, public, and private turf. Tips are given on how to make spray and granular applications safe and most effective.

Chapter 8 presents techniques for the calibration of various types of equipment used for the application of fertilizers and pesticides commonly employed on turf. This material was placed in a separate, new chapter in this edition because of its critical importance to the safe and efficient use of pesticides or fertilizers.

Appendices A through C provide a wealth of information on where and to whom to write in each state for additional information on weeds, insects, diseases, and other turfgrass problems. A wide variety of useful calculations and units of measurement, conversion tables for measuring dry and liquid chemicals, measurement and rates of application equivalents, a conversion table for using fertilizers and pesticides on small areas, calculations for land areas of various shapes, tractor speed conversions, an operating chart for tractor boom sprayers, and many other calculations that you will find useful are included. A complete glossary defines the technical and nontechnical terms used in the text. The information you seek should be easy to find, especially with the extensive index.

1.2   WHERE TO FIND ADDITIONAL HELP

You can get help on lawn and fine-turf problems by contacting your county extension agent or adviser (listed under the county name in the telephone directory). When local assistance is not available, contact extension or other turfgrass specialists at your state land-grant university. A list of these institutions and their addresses is given in Appendix C.

Write, e-mail, fax, or telephone the turfgrass specialist (in the department of horticulture, agronomy, plant science, or plant and soil science) regarding turfgrass management problems and weeds and their control. Contact the extension entomologist for information about insects, mites, and other animals. The extension plant pathologist can answer questions on diseases and their control.

When contacting a specialist at your land-grant university, provide as much background information as you can about the problem or pest. Include the date collected if samples are submitted; extent of the turf area affected; cultivar and species of grass(es) affected; age of the turf if known; mowing height and frequency; degree of severity; description of the problem; date when first observed; weather and soil conditions for at least the past several weeks; fertilization, watering, and pest control measures taken in the past several weeks or months; color photographs showing the problem; and other facts that you believe may aid in the diagnosis. Do not forget your name and return address! Remember that a correct diagnosis is essential before control measures can be suggested. A diagnosis can be only as good as the specimens and information you send. Mail specimens so they arrive on a weekday (Monday through Friday).

Most county extension offices have plant clinic or specimen forms available that cover much of the background information outlined above. Many offices also have mailing boxes or tubes available for mailing specimens, which will save you the trouble of finding a suitable container.

For weed identification.   Correct identification of weeds is dependent on fresh, structurally intact samples. The identification of both broadleaf and grassy weeds is largely conducted by examining the floral or seedhead portion of the plant. If specimens without flowers or seeds are submitted, positive identification can sometimes be difficult. Find one to three samples of the weed that appear to be among the most mature. Dig out the plant, retaining all portions possible (roots, rhizomes, stolons, and the like). If the weeds are extremely large, take a representative portion. If the turf is very droughty or dry, water locally around the weeds and let them absorb water before sampling (digging). Shake loose as much soil as possible from the root system and wrap the specimen in newspaper. Then place the plant loosely in an open, plastic bag. If the roots are attached to the plant, moisten a paper towel, remove the excess moisture, and wrap the root section before wrapping with newspaper. The sample and other moistened material must be damp but not wet. Mail the weed sample in a crushproof carton or mailing tube. The samples will only remain in a condition suitable for identification for a period of 2 to 3 days. Make certain that the mailing procedure chosen will deliver the samples promptly. If seeds are present and loose, they can be pasted to a piece of paper or cardboard and submitted along with the sample. It is helpful to enclose a note to explain when the weed first appeared in the turf, how long it has persisted, and whether it is in small clumps or has spread throughout a large turf area. If flowers are not present but were visible previously, provide a description of the flower and when it first appeared.

For insect identification.   Collect more than one insect specimen found in the affected area. If possible, include more than one size or stage of the insect. Place the specimens in a pill bottle, plastic vial, or other durable small container. Wrap the container with paper or enclose in a small box. Enclose a sheet of paper with your name, address, where the insect was found, type of damage observed, and how numerous the insects were in the area affected. Mail or take the package to your county extension agent, or mail to the extension entomologist at your land-grant university. Do *not* tape the insects to a sheet of paper and place inside an envelope. The insects will be flattened beyond recognition before they can be identified.

For nematode assay.   Using a 1-in (2.5-cm) soil probe, carefully collect 15 to 30 random soil cores, 4 in (10 cm) deep, of turfgrass from the suspicious area (avoid dead turf) and an equal number of cores from an adjacent healthy area. Include as many feeder roots as possible. Keep only the soil and roots. Thoroughly mix the cores from each site and save about a pint (500 cc) of soil from each for mailing. Place the soil in sturdy plastic bags and close the open ends securely. With a waterproof pen, mark the bags HEALTHY and SUSPICIOUS and place them in a *strong* container. Be sure to include with the package your name and address, turf location, species and cultivar sampled (where known), primary symptoms observed, cultural and chemical management program followed, and the approximate size of the area sampled. Mail the package express delivery as soon after collecting as possible. Do *not* let it sit overnight, except in a refrigerator. Keep it cool. Nematodes are living animals and must reach the laboratory alive.

Where plant-parasitic nematodes are a known chronic problem, collect soil-root samples at about monthly intervals throughout the growing season to provide baseline information concerning population densities.

For disease analysis.   Carefully collect two or more 4- to 6-in (10.2- to 15.25-cm) (cup-cutter size) plugs of growing turfgrass and underlying soil (up to 2 in [5 cm] deep). The plugs should be taken at the *margin* of the diseased and healthy turf and, where possible, show a range of symptoms. Wrap or pack the turf securely in a paper bag (so that soil will not spread over the grass surface) and mail in a tight, crushproof carton or mailing tube. Do *not* add moisture to the turf before mailing—otherwise, the grass will probably arrive in a badly rotted condition, making diagnosis impossible. Direct transport express or overnight shipment of the plugs is preferable to regular mail. Fresh samples that arrive in good condition are a necessity.

In the laboratory, most turf samples are incubated in a humidity chamber for 12 to 24 hours when a leaf-infecting pathogen is suspected. The isolation of fungi and bacteria on sterile agar media requires a week to several months. Cultural and chemical suggestions are usually sent that, hopefully, will prevent further damage to the turf.

All states publish turfgrass management and pest control recommendations. Many of these publications are free. A wide variety of printed information on turfgrass problems may be obtained from your county extension office or land-grant institution. Supplement this information with the suggestions and recommendations you find in the following chapters.

Maintenance of Turfgrasses

Turfgrass pests cannot be controlled or managed over long periods solely through the use of pesticides. For a healthy, vigorous turfgrass stand, pesticides should always be used in combination with sound cultural practices. This integrated pest management (IPM) system not only manages existing pests, but also helps prevent the expansion of those pests and the generation of new ones. While IPM is the original and most widely recognized label, turfgrass managers more often practice integrated turf management (ITM) with cultural practices central to their operations, as well as direct pest control.

Of all the activities related to turfgrass management, the establishment period of a turf is the most critical to its long-term success. Proper establishment has the greatest influence over the future quality of the site. The soil environment is developed largely during establishment. Since a healthy, deep root system is the principal means for turfgrass survival during periods of stress, the preparation of the soil environment will be one of the limiting factors in the long-term success of the turf. Improper establishment can nullify even the best maintenance practices.

The initial step in the establishment of a turf is the careful planning and consideration for the future use and objectives of the site. Local and regional climatic conditions, the final soil environment, and the intensity and duration of traffic expected on the turf will largely determine the maintenance requirements after establishment. A very stressful climate can limit the use of a fine turf, due to the extensive maintenance requirement.

The level of traffic on a turf can increase continuously. For planning purposes, traffic levels can be simplified to three basic groups. An area receiving minimal traffic, such as infrequent foot traffic and little vehicle traffic, would probably require low levels of maintenance. With increasing foot traffic, such as a home lawn or public park might receive, but without extensive vehicle traffic, the turf could be maintained at a medium level. For intensely trafficked areas, receiving daily or hourly foot and vehicle traffic (soccer fields, golf course greens, tees, and, to a lesser extent, fairways), the turf would require a high level of maintenance. The possibility exists to extend a low level of maintenance to an area that might receive greater traffic. In most instances, however, the quality of the turf will also diminish.

Each level of maintenance may require that the turfgrasses chosen exhibit unique attributes. Low-maintenance turf requires plants that show thriftiness with water, growth at minimal fertilization, and hardiness to extreme hot and cold temperatures. Medium-maintenance turf requires a compromise between resistance to wear or traffic and the ability to withstand low-moisture and low-fertility situations.

It is generally kept at a higher mowing height than high-maintenance turf to help provide resistance to injury and minimize overall plant stress. Turfgrasses growing under high-maintenance conditions must exhibit a rapid rate of growth to replace injured tissue. Often they must withstand low mowing heights and exhibit a prostrate growth habit. Grasses with a high resistance to shearing or tearing are often selected for high-maintenance areas.

2.1   INTEGRATED PEST MANAGEMENT

Integrated pest management (IPM) is both a concept and a philosophy. It is a broad, multidisciplinary, systematic approach to managing all pests. All necessary control and monitoring methods (biological, cultural, regulatory, physical, and chemical) are utilized. Use of IPM strategies should result in effective yet economical management of pests with a minimum effect on nontarget organisms and the environment. IPM is based on understanding the biology and ecology of the turfgrass community.

The IPM of turf might be defined as minimizing pest populations through multiple tactics. Others define IPM as an organized and comprehensive approach to the management of key pests in an ecologically sound management system. Integrated management of turfgrass pests is much more than chemical pest control, although this is an important component. IPM does not depend on a single method or tactic to control a pest.

The amount of pest damage that can be tolerated—the aesthetic injury level—depends on the value of the turf, which in turn depends on intended use of the turf. Obviously, a highway right-of-way is not as valuable per unit area as a golf green, fairway, or home lawn.

Two other basic concepts are widely used in IPM: the economic threshold and damage threshold levels. The economic threshold level (or, more appropriately for turf, the aesthetic threshold level) is a fundamental concept of IPM. It is the point at which the density of the pest infestation requires immediate, intensive control (usually chemical) to prevent unacceptable aesthetic changes. The damage threshold level is the lowest pest population density at which damage occurs. Both concepts are important when planning pest management at low, medium, or high levels of maintenance. At present, aesthetic threshold levels have not been accurately determined for most turfgrass pests.

As far as turfgrass managers are concerned, IPM involves six basic tasks:

1. *Identify the key pests (including insects and other animal pests, weeds, fungi, nematodes, bacteria, and viruses) to be managed in the turfgrass ecosystem.* What are the key pests, potential pests, and those that may migrate in or be introduced with seed, sod, sprigs, plugs, or soil? After successfully identifying key pests to be managed in the turfgrass ecosystem, an assessment of the potential damage from these pests must be made. With this knowledge, the pests can then be ranked according to their potential importance and grouped into those that can potentially cause severe or irreversible damage, those that cause occasional minor damage or injury, and a third group of pests that causes little or no damage to the turf.

2. *Define the turfgrass management unit—the turf ecosystem.* For some pests, it may be a single unit of turfgrass (an athletic field, park, lawn, golf-course fairway or green). A whole subcontinent may be involved in the case of green bugs or certain rusts that can blow hundreds of miles through the air within a relatively few hours. The size of the management unit depends on the mobility of the pest and its dispersal potential.

3. *Develop reliable monitoring techniques, a critical component of any IPM program.* Do we know how to accurately assess the turfgrass pest populations present and measure the loss caused by each? The IPM system depends heavily on monitoring or scouting turfgrass areas on a regular basis for insects, weeds, and diseases. This information, combined with data on related matters such as recent weather conditions, past cultural practices, the grass or grasses being grown, and desired turf quality, enables the turfgrass manager to select the best strategies for reducing pest damage as effectively and economically as possible.

4. *Establish aesthetic thresholds.* These thresholds vary with the level of maintenance (low, medium, or high), the turfgrasses grown, and the pests to be controlled. Much more research needs to be done in determining the aesthetic threshold levels for turfgrass pests. Other factors to consider are budget restrictions, availability of water, fertilizer, pesticides, and dethatching and coring equipment.

5. *Evolve descriptive and predictive models of what pests are most likely to occur, when they would be expected to appear or become active, and the amount of damage that may take place.* We need to identify and fill the knowledge gaps in the life cycles of many different turfgrass pests. We need to know the biology, ecology, and epidemiology of these pests.

6. *Develop an effective and economical turf management strategy.* This should involve the coordinated and integrated use of multiple tactics or management strategies. A turf management strategy should be developed for each management unit. Although these areas generally include the boundaries of a given turf area, they sometimes need to be smaller. A good example might be the landing area in a golf-course fairway, which would vary from the general fairway management program to anticipate a higher degree of use. The same might be true for a general park area where concentrated traffic is anticipated. It is important that the strategy covers a very uniform management area to minimize variation in response to the program.

These six basic tasks can be further grouped into three general IPM procedures. The first three tasks can be combined in a procedure that attempts to identify pests observed within a defined management unit. Any sections that focus on this procedure will be identified by the icon in figure 2–1a.

The fourth and fifth tasks support the general procedure of assessing the potential impact the identified pests might have on the turf. Sections within the text that provide information on the evaluation of the pest population status, potential for change, or potential impact are identified with the icon in figure 2–1b. Finally, the last principle implies a wide range of procedures that modify the management of the turf within the predetermined maintenance level to minimize any negative impacts of the pest. Discussion of these procedures are indicated with the icon shown in figure 2–1c.

FIGURE 2–1   General procedures of integrated pest management.

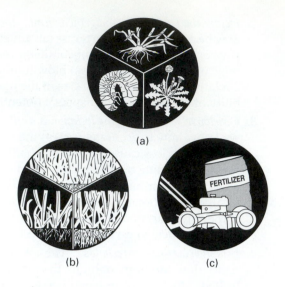

(a)

(b)   (c)

As shown in figure 2–2, a number of pieces comprise the IPM puzzle. These pieces are interlocked and must be fitted together in a certain way. This integration of techniques must be compatible with sound turf management practices. Also consider the costs of control measures over a period of years. Remember that pest control is but *one* cost in producing a vigorous and aesthetically pleasing turf. Which pest management tactics or control measures are the most cost effective over 5 or 20 years? What about more than 20 years? The management strategies selected must be economical and practical, based on the labor, equipment, and money available. When should these measures be applied? Is protective action always the best? These are tough questions that every turfgrass manager must answer sooner or later.

2.1.1   Management Tactics

The tactics used to manage turfgrass pests encompass legal or regulatory, genetic, cultural, biological, physical or mechanical, and chemical areas. Regulatory measures are normally aimed at excluding turfgrass pests from an area, state, or country. These may involve federal and state quarantines or embargoes; seed inspections; certification of seed, sod, sprigs, or plugs; regulations regarding the quality of seed or other planting material; and the planting of certain very susceptible species.

Of all the regulatory measures available, seed certification is probably the most active and covers the widest range of use. Seed certification is a program covering seed production fields and seed lots, which are inspected to ensure genetic purity of seed. Different state agencies have the responsibility for conducting the certification program, which is coordinated by the Association of Official Seed Certification Agencies (AOSCA). The certification process generally involves controlled growth of regulated seed stock in a given field, with buffer strips of similar seed stock adjacent to the field to be harvested to ensure the direct pollination of seed within the field.

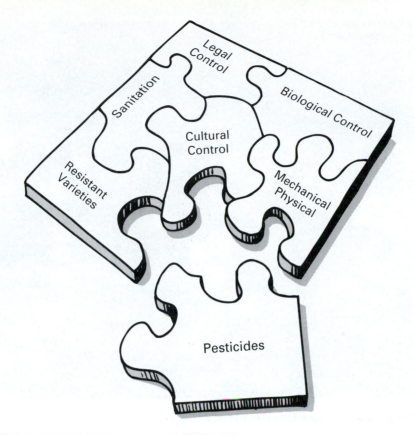

FIGURE 2–2   Integrated pest management (IPM) puzzle.

There are four classes of seed meeting or exceeding the standards established by AOSCA: breeder, foundation, registered, and certified seed. Breeder seed is the original source of all classes of certified seed. It is maintained by sponsoring plant breeders or institutions and provides a direct source of seed for foundation materials. Foundation seed is produced in fields planted with breeder seed and is tagged with white certification tags. Fields planted with foundation seed can be harvested and tagged as registered seed. These fields are used primarily to increase the supply of a particular cultivar. The registered seed tag is purple and is the source for planting certified seed. Certified seed is the seed available to the consumer. It is produced in fields planted with either registered or foundation seed. Bags of certified seed carry a blue certification tag. It is important to note that the certification program does *not* guarantee that the seed is of the highest quality as to germination, purity, or freedom from disease or pathogens; however, it does ensure that the cultivars listed on the label are true to genetic type.

Genetic.   Genetic control tactics have been widely used for generations. They involve (1) the selection, identification, and planting of adapted turfgrass species and cultivars with naturally occurring resistance to insects or other animal

FIGURE 2–3   Typical turfgrass-cultivar-evaluation study.

pests and/or diseases; or (2) the introduction of specific genes for resistance into an otherwise desirable turfgrass species or cultivar. The development of multiple disease-resistant cultivars of turfgrasses has been an important step in reducing losses from major diseases. These concepts have been demonstrated in thousands of species and cultivar evaluation studies in the United States, Canada, Europe, and elsewhere (figure 2–3).

Besides genetic selection and breeding for resistance, the identification and use of highly vigorous and competitive turfgrass species, mixes, and blends can offer successful competition with weed species, leading to the elimination of those weeds.

In many variety trials, a fast-germinating turfgrass species such as perennial ryegrass has provided superior seedling weed control when compared with slower-germinating species such as Kentucky bluegrass. A turfgrass species seeded at its optimum rate will also have significantly fewer weed problems during its establishment stage than the same turfgrass seeded at a suboptimum rate.

Cultural.   Cultural management tactics are the oldest and most widely used procedures to minimize pest damage. They include sanitation (the removal of dead or living material harboring a pest, thereby disrupting the life cycle of the pest), plus proper planting, watering, fertilizing, mowing, and cultivation practices that stimulate the vigorous growth of turfgrasses to better resist pests. The essence of cultural controls is to adjust turf-maintenance procedures to the current growth rate of the turf. This growth rate will fluctuate during the growing season, creating different levels of demand that must be met to minimize stress on the turf. The time of planting is also important in cultural control. Poor or incorrect maintenance results in a weakened turf that lacks density and vigor, thereby encouraging pest activity and weed invasion.

Water is an important factor governing the incidence and severity of a disease attack because a film of moisture on the susceptible plant surface is needed before spore germination, infection, and disease development can occur. The severity of disease can be considerably reduced with proper irrigation (for example, watering early in the morning, not during the late afternoon and evening). Any practice that shortens the time period that the grass is wet will restrict the spread and growth of practically all turfgrass pathogens.

An excess of water may increase the succulence of turfgrass plants, stress them by preventing oxygen uptake by their roots, and thus increase the plant's susceptibility to disease. Good examples of diseases accelerated by irrigation practices include Pythium blight and yellow tuft or downy mildew. The fungi causing these diseases release zoospores (reproductive structures) that swim through films of water to infect new plants. Water flowing over a grass surface rapidly transports zoospores and the spores of other fungi over wide areas, where they can infect more turfgrass plants.

Turf that is healthy requires enough water to meet its evapotranspiration and growth requirement needs. Too much or too little water, and too frequent or too infrequent irrigations, provide the conditions that favor growth and development of unwanted weed species. Poor irrigation practices can decrease turfgrass roots, grass vigor, and density, which leads to weed seed germination, growth, and survival of the adapted species. Weed colonization follows, which leads to the need for herbicides and renovation or reestablishment.

Water availability and temperature are closely related. In fact, temperature is the real driving force behind the development of insect, weed, and disease organism populations. Water availability generally plays a limiting role. Populations of most pests, along with the turf, grow rapidly as soil and air temperatures rise.

Excessive thatch and compaction must be controlled to maintain a weed-free turf. Soil that is compacted and turf that is heavily thatched restrict the entry of water, air, and nutrients into the soil, thereby reducing root growth and, eventually, the vigor of turfgrasses. The presence of knotweed, a summer annual, is a good indicator of compaction due to its tolerance and competitive advantage in compacted soils. Turf with a thick thatch commonly has high populations of shallow-rooted weeds, such as crabgrass, goosegrass, and annual bluegrass, that prefer high-moisture sites. The correct timing and frequency of core aerification and thatch removal (vertical mowing) is a relatively easy means to maintain a healthy stand of grass.

Core aerification helps to reverse the effects of compaction by allowing a better exchange of oxygen and waste gases by the roots and better water penetration. Aerification also increases a turfgrass plant's resistance to disease and insect attack by enabling it to maintain more vigorous growth through an increased root system.

An accumulation of thatch can lead to an increase in insect damage. Thatch provides a haven for insects to hide and for their populations to increase. Insecticide and fungicide activity is also inhibited because the chemicals become bound to the organic matter in the thatch. A buildup of thatch results in poor penetration of water and air, thus weakening the grass and making it more susceptible to insect damage. Many severe insect problems are associated with a thick thatch.

Mowing affects turfgrass plants by providing wounds through which pathogens (such as those causing bacterial wilt, Rhizoctonia diseases, dollar spot, red thread, and Septoria leaf spot) may infect a plant. Close mowing removes much of the photosynthetic area, thereby decreasing the stress resistance of the grass plant and its ability to recover from all types of pest damage. Why? After severe clipping, the grass plant's regenerative processes are mobilized into repairing the damage. With the need for additional resources for pest resistance, the plant definitely goes under stress.

Mowing turf at or near its optimal height for a particular use causes less depletion of food reserves, thereby allowing the plants to remain more resistant to pest attack. Mowing at the optimal height also shades out the seedlings of many weed species. Long-term close mowing results in shallow, reduced root systems, and an invasion of such weeds as annual bluegrass, oxalis, and spotted spurge.

The effects of fertilization on the development of turfgrass diseases vary. Many diseases become more damaging when grass is grown under higher or lower than normal levels of nitrogen. Brown patch and Pythium blight are examples of diseases that are more severe where nitrogen levels are excessive, while the levels of phosphorus and potassium are normal. Dollar spot, pink patch, and red thread, however, are more severe where the turf is deficient in nitrogen. With a knowledge of what diseases are most likely to occur in a given turf, under what environmental conditions these diseases occur, and their response to different fertilizer levels, the turfgrass manager can thus adjust the fertilizer program to minimize the effects of disease.

Fertilization practices also greatly affect the invasion of weeds into turf. The area covered by many broadleaf weeds often decreases as the annual amount of nitrogen increases. A fertilization program that results in a nutrient imbalance (like a low level of phosphorus, potassium, or other essential element) may also hasten weed invasion. An excellent example is the rapid development or increase in annual bluegrass populations in areas that receive a high level of phosphorus fertilization. This is most evident in closely cut turfs. Unfortunately, weeds are better able to compete in poorly fertilized turf than in turf that is well fertilized due to the greater density, vigor, and competitiveness of properly fertilized turf.

The importance of sound fertilization, watering, and cultivation programs cannot be overemphasized. Healthy, vigorous turf can often outgrow the effects of insect or weed infestations and disease attacks, while poor conditions result in more severe damage and a slower recovery. Healthy turf therefore usually requires fewer treatments of pesticides to control insects, diseases, and weeds. It also follows that turf under water, heat, compaction, thatch, fertilization, or other stresses stands a greater chance of being damaged by chemicals applied for pest control.

Biological.   Biological control can be defined as the regulation or suppression of pest populations by living organisms (antagonists, parasites, and predators). One of the more successful biological control programs for turfgrass insects at present is the use of the spore-forming bacterium *Bacillus popilliae*, or milky disease, in the control of Japanese beetle grubs. The Japanese beetle is the main host of this bacterium, although the larvae of other scarabacid insects are known to be susceptible. The bacterium is widely sold and applied as a dust that contains countless numbers of its spores. It takes several years for the milky disease to appreciably reduce beetle-grub populations. Although there are naturally occurring parasites and predators of turfgrass insects and mites, their management for reducing damaging pest populations has not been achieved. For biological control agents to be effective, pest insects must be present in the turf at high enough population levels to sustain populations of parasites and predators. Therefore, turfgrass managers must be willing to accept a certain level of damage to the turf, if the biological agent is to be effectively sustained in the turf.

An insect called the big-eyed bug is often found in turfgrass with infestations of chinch bugs. This predator attacks chinch bug eggs and nymphs and reduces their populations. Big-eyed bugs and chinch bugs are similar in appearance except for the pair of oversized eyes on the adult big-eyed bug.

Lady beetle and syrphid fly larvae commonly attack populations of greenbug in a lawn. The number of greenbug aphids builds rapidly, and rust-colored damaged areas appear (*Color plate 27*). The predator lady beetles and syrphid flies appear and reduce the numbers of greenbugs.

Other predator insects include the adults and larvae of ground beetles, which attack a number of soil insects. Noninsect predators include birds that feed on leaf-feeding caterpillars such as cutworms and sod webworms.

Insect control by disease-causing organisms other than *Bacillus popilliae* include species of *Beauvaria*. This fungus produces a disease that is quite effective in regulating chinch bug populations. Another species of *Beauvaria* attacks billbugs. Species of *Beauvaria* also have been tested as applied controls for sod webworms.

Endophytes. Endophytic fungi commonly infect and colonize many species of wild, forage, and turf grasses. The most prevalent fungal endophytes are in the genus *Acremonium*. They have been studied mostly on tall fescue, fine-leaf fescues, perennial ryegrass, and species of bluegrasses.

The asexual or anamorphic state of endophytes are species of *Acremonium*. They include *A. coenophialum*, which infects tall fescue; *A. lolii*, infecting perennial ryegrass; and *A. typhinum*, *A. uncinatum*, and other species infecting bentgrasses, some bluegrasses, fescues, ryegrasses, and others.

Biological control can be organized into two main classes: general suppression and specific suppression. General suppression is described as the biological suppression of a pest population by an indirect or nonintimate association between numerous microbes and the pest. For example, the general consumption of oxygen, water, or nutrients could limit the development of a soil-borne disease. General suppression results from multiple microbe populations living together. Therefore, complex, active ecosystems are more likely to exert general suppression on pathogens that cause disease than ecosystems that are low in microbial activity. For example, fumigated soil is very conducive to severe disease development should a pathogen find its way into the treated soil. The reason is that in fumigated soil there are no microbes to resist or compete with the pathogen. Generally, it is recommended that turf managers follow cultural practices that keep the grass growing and soil microbes active. A turf that is growing in biologically active soil is said to be "healthy," which loosely refers to the fact that the turf ecosystem is generally suppressive to pathogens and is capable of recovering from a disease once it subsides.

The second type of biological control, specific suppression, refers to the interaction between one type of organism and one pest or pathogen. In general, this type of association exists between two organisms that are naturally associated with one another in an ecosystem. Most of the efforts by scientists to develop biological controls have focused on specific suppression. This is accomplished by looking for naturally developing situations where one organism, the biological control agent, is suppressing a specific pest or pathogen. The biological control agent, usually a bacterium or fungus, is isolated and then developed for practical use by turf managers.

The use of biological control agents for the specific suppression of turf diseases has not yet delivered effective tools for turf managers. Turf diseases are mostly caused by pathogenic fungi. These fungi can exist in several different forms and often can attack several different types of plants. Numerous bacteria and some fungi have been identified that are suppressive to pathogenic fungi, but establishing them to control disease in turf has been difficult. The main reasons for this difficulty are that these organisms must be physically associated with the pathogen, and they must respond to the growth of the pathogen to be effective in specifically suppressing turf pathogens. Some have already been formulated for use in turf, but have met with limited success. For example, the bacterium *Pseudomonas aureofaciens* was found to be suppressive to the dollar spot fungus, *Sclerotinia homoeocarpa*, and to the brown patch fungus, *Rhizoctonia solani*. However, maintaining this bacterium at the required population level on growing turf leaves and crowns has met with limited success. Grass growth at the time of pathogen attach results in the generation of new and uncolonized tissue. For biological controls to be highly effective, they would have to multiply and redistribute themselves on the newly formed tissue. This can happen, but not fast enough to suppress the pathogen sufficiently. Currently, attempts to overcome this problem have included repeated applications of the bacteria with spray equipment or via injection into the irrigation system. Specific biological suppression of turf diseases does occur in nature and could be an effective tool for turf managers; however, its use will most likely need to be integrated into a management program that maintains healthy turf.

For turf, endophytes (meaning "loving, inside") are fungi that live inside (infect) the grass plant. They are generally found in the leaves and seeds of grass plants, but not in the roots. Although they infect turfgrasses, they do not cause visible harm (disease) to the grass plant as do other pathogenic fungi.

As the endophytic fungus grows in the grass plant, it produces alkaloids that are nontoxic to either it or the host plant, but are toxic to other microbes and animals.

It is unclear at this time if turfgrasses infected with endophytes appear more resistant to infection by pathogens. There are reports that some fescue turfs develop less brown patch, but it is unclear if this is due to faster recovery or true resistance to the disease. In any event, the impact of endophytes on disease severity is minor.

The shelf life of endophytes in seed is limited. Seed stored too long prior to planting can lose all viable endophytes to natural senescence. It is best to use endophyte-infected seed as soon as possible. Seed storage for years or even months can result in loss of living endophytes.

Grass infected with endophytes is toxic to animals grazing on it. Endophyte-infected grass causes fescue summer syndrome or toxicosis in cattle, ryegrass staggers in sheep, and alactia or suppressed milk development in mares. While toxic to animals, endophytes are beneficial to the pasture grass, as they protect against overgrazing. Endophytes also are beneficial to the turfgrasses they infect by having greater resistance or tolerance to:

1. Leaf-feeding insects such as sod webworm, Argentine stem weevil, billbugs, black beetle, armyworms, greenbug and other aphids, chinch bugs, and some leafhoppers
2. Some nematodes, including root lesion, root knot, and stylet or stunt
3. Environmental stresses, including heat, drought, winter hardiness, and shade
4. Weed invasion

Endophyte-infected turfgrasses may also have increased growth and tillering, greater shoot and root mass, more persistence under various stresses, and better recovery from such diseases as summer patch, necrotic ring spot, brown patch, and dollar spot (see chapter 5). Infection by endophytes also seems to enhance seed germination, seedling and spring growth, and plant vigor. Infected plants also produce more seed per plant, and the seeds are heavier.

Because considerable variation in races, biotypes, or strains of endophytic fungi exist, the types of beneficial responses to different grass-fungus associations can vary. The more useful races, strains, and species of endophytic fungi are currently being introduced into turfgrass cultivars and offer significant improvement to insect, nematode, and stress tolerances in turfgrasses. In general, the main advantage of using grass infected with endophytes is to reduce persistent insect problems. New cultivars containing beneficial *Acremonium* endophytes are available by maternal line selection and conventional breeding techniques. You can get endophyte-enhanced seed by requesting it when purchasing seed. Many turfgrass seed tags presently indicate what percent of the seed contains endophytes.

Life Cycle. Acremonium endophytes infects grass meristems, grow between the cells, and, if flowering takes place, grow into the inflorescence, ovule, and embryo. These organisms are seed transmitted. They *cannot* be spread to noninfected plants by grass clippings, mowing and other maintenance practices, or proximity to an infected plant. Turfgrass-infecting endophytes are obligate parasites that cannot naturally invade new plants, although they can be artificially introduced. The fungi do *not* naturally grow into grass roots and hence have little effect on root-feeding pests.

Species of *Acremonium* produce spores (conidia). The kidney-shaped conidia are borne singly in a head and often appear in a T-shaped configuration. The mycelia of endophytes is easily detected within leaf tissues, basal leaf sheaths, and in seed by ELISA serological assay, and by staining with rose bengal or cotton (aniline) blue in lactophenol. The specimens are cleared with methyl salicilate and viewed using bright-field microscopy.

Four classes of alkaloids—toxic to some insects, nematodes, mammals, and fungi—have been isolated from infected grasses. The chemicals occur in varying amounts and combinations in grass-fungus associations and have been identified as pyrrolizidine, peramine, lolitrem, and ergot alkaloids such as ergovaline. Some associations have one, two, or three alkaloids present; a few apparently have none.

Turfgrass pesticides, including even systemic fungicides applied in the field, have little effect on endophyte levels in turfgrasses. Long-term storage of seed, however, for several months to a year or more destroys much of the viability of these fungi, but this loss depends on the time, temperature, and moisture content of the seed.

Summary.   Endophyte-enhanced turf cultivars are yet another way for turfgrass managers to practice IPM. In so doing, managers may be producing turfgrass areas that are also more insect and stress resistant.

What Lies in the Future?   Much more research needs to be done with these beneficial turfgrass fungi. Researchers at universities and seed companies are zeroing in on selecting more effective races and strains and possibly breeding new types with additional benefits. They envision made-to-order endophytes for better handling different stresses. This new wave of fungal endophytes may be closer in the future than we believe.

Bacterial endophytes may play a role in the disease and insect resistance of turfgrasses in the future. Research is going on in several areas, including trying to introduce pesticide-producing genes into the bacteria that live within plants. There are even nitrogen-fixing bacterial endophytes. Endophytic bacteria in the future are expected to carry and transport insect-resistant and other beneficial genes into turfgrass plants.

The biological balance of fungi, bacteria, nematodes, and other microorganisms in the turf-thatch-soil environment undoubtedly plays a major role in the development of disease, but there are no outstanding examples of biological control of turfgrass diseases using organisms antagonistic or parasitic to disease-causing fungi. Biological control of nematodes, dollar spot, and possibly other diseases is enhanced by using an activated sewage sludge fertilizer. Current research in this area, however, may result in the future marketing of biological disease-control agents.

Biological control of weeds using allelopathic plants (plants that are naturally toxic to other species) in turf establishment and maintenance is in its infancy, but offers promise for the future.

Physical or mechanical.   Management tactics that use physical means of control include traps for insects, moles, or other vertebrate pests. Various noise devices that discourage injurious birds might have a place in seed-producing fields and other turfgrass areas. Providing good surface and subsurface water drainage is another valuable physical control tactic, as is having sharp mower blades, to reduce excessive wounding of the leaf tips.

Chemical.   The use of pesticides is discussed in chapters 4 to 8. This practice involves the use of chemicals applied as sprays, drenches, granules, seed treatments, and the preplant use of soil fumigants. The selective use of herbicides, insecticides, fungicides, and nematicides must remain an important part of the defense against most turfgrass pests in medium- and especially high-maintenance turfs.

Although sound turfgrass management practices can significantly reduce weed, insect, and disease activity, it does not eliminate the necessity of using pesticides when serious problems develop. In practice, sound management procedures reduce the number of pesticide applications required to control pest outbreaks when they do occur. When, by pest monitoring, you determine that the presence of a pest has equaled or exceeded the aesthetic threshold of the turfgrass, thereby requiring adjustment in the management program in terms of pesticide applications, the proper timing of these applications is crucial. With proper timing, additional applications will not be necessary, resulting in a reduction in total pesticide use.

Ideally, when a disease appears, the identity of the pathogen is determined, cultural practices are investigated to determine if they are creating or intensifying the problem, and the history of the turf is reviewed to determine if it has occurred in the past. This is where the keeping of a turfgrass diary comes in handy.

Once it is determined that cultural management practices—watering, fertilizing, mowing, dethatching, core aerification, and the pruning or removal of dense trees and shrubs bordering the turf area—cannot be modified to help suppress a disease outbreak, the use of fungicides becomes necessary. The fungicide selected should be specific for the fungus involved and be applied strictly according to label instructions. Fungicides should be regarded as only one component of an integrated pest management program.

For weed control, specific herbicides are registered for most turfgrass weeds and can be used on the common turfgrass species without injury. Of course, the weeds must be correctly identified along with the desirable turfgrass species. The most effective herbicide for a specific weed problem should be applied according to label instructions when maximum control can be obtained. The causes of the weed problem should be determined, and cultural practices should be modified to minimize the reinvasion of the weed.

To avoid turfgrass damage from weeds, insects, and diseases, regular inspections are necessary. For diseases such as Pythium blight, daily examinations of golf course greens are necessary in hot, wet weather. Insects may cause the turfgrass to die back, be stunted, or develop growth distortions. The leaves may bleach or turn yellow or brown. If such symptoms are observed, immediate action should be taken in medium- and high-maintenance turf. Insect populations can increase rapidly under the right conditions. Some insects, such as cutworms, feed only at night. Unless a special effort is made to look for them during darkness, they may go undetected for some time.

Leafhoppers, scale insects, greenbugs, sod webworm larvae, and spider mites can be detected by carefully examining the leaves, stems, and crowns of grasses. Chinch bugs are found by carefully examining the crowns and thatch. An easy method to detect many insect pests in turf is called the pyrethrum test. One tablespoon of a garden insecticide containing 1 to 2% pyrethrins is added to a gallon of water. One square yard of turf, including both damaged and healthy grass, is marked off. The gallon of pyrethrum mix is then applied with a sprinkling can as uniformly as possible over the marked area. Since pyrethrum is very irritating to most insects, this treatment will quickly bring cutworms, sod webworms, lucerne moths, skipper larvae, and other insects to the grass surface, where they can be seen and counted (if needed). Several areas within the turf should be checked to get an indication of the extent of the infestation. If pyrethrum is unavailable, you can get similar results by flooding the suspected turf area with water for 5 to 10 minutes.

White grubs and billbug larvae will not surface using the pyrethrum test or flooding. If grubs are present, they can be found by carefully digging around the roots of grass during the warm spring and fall months. If the white grub or billbug infestation is heavy, the grass roots will be eaten away, allowing you to roll the grass back like a carpet.

When a pest problem appears, consider several factors before applying any chemical. Accurate identification of the disease, insect, or weed in the turf is important; many of the insects are not pests. Of the insects that are pests, knowledge of their behavior and biology is critical in determining what chemical or chemicals should be applied for their control. Different turfgrass pests require different chemical treatments and management methods to be effective.

Such factors as temperature, humidity, soil moisture, soil type, and location influence the outbreaks of disease and buildup of insect populations. Some insect pests thrive in a warm, dry environment; others prefer a moist one. Some insect pests of turf are influenced by cycles of drought. For example, the fruit fly is more prevalent during droughts. Certain turf areas may be more susceptible to insect pest populations than others due to the grasses grown, management practices, exposure to sun, slope, and soil types. All of these factors should be considered when checking for the presence of insect pests and whether a chemical control is warranted. Remember that insect pests are commonly found first in isolated spots, most commonly at the edge of a turf. It would be unusual to find insects uniformly distributed throughout a turf area.

Turfgrass should be well watered before applying an insecticide to control pests that feed on the leaves, stems, and crowns. Apply the insecticide as soon as the turf has dried. Withhold further watering until it is necessary to water the grass to prevent wilting. This procedure allows the insecticide to remain on the turfgrass as long as possible, thereby getting maximum effectiveness from the treatment.

Sprays are the preferred application method to control turfgrass weeds, insect pests, and diseases. Granular formulations of many pesticides are available to control annual weeds, white grubs, sod webworms, chinch bugs, cutworms, billbugs, skipper larvae, slugs and snails, snow molds, leaf smuts, and other diseases.

When areas are to be treated for pests and disease-causing fungi that "operate" below ground, such as white grubs, billbug larvae, leaf smuts, summer patch and necrotic ring spot, the pesticide should be uniformly applied to the turfgrass surface and the area should be heavily irrigated (using 1 in of water or about 600 gal per 1000 sq ft) to move the chemical down into the top several inches of soil.

2.1.2   Management Procedures

No turfgrass planting is infected with every disease, weed, or insect pest to which it is susceptible. Commonly, a history of pest infestation develops in a given turf area where a single pest or a few pests occur year after year. Thorough knowledge of what

Zones of turfgrass adaptation in the United States

FIGURE 2–4   Zones of turfgrass adaptation in the United States.

weeds, insects, and diseases occur in that turfgrass area and the time of year that the pests normally appear can contribute greatly to their control. Certain states and pesticide companies have developed insect and disease calendar guides that are useful aids in turfgrass insect and disease control. They provide a turfgrass manager with a ready reference as to when to monitor for turfgrass pests and when chemical treatments are most desirable. Highly sophisticated IPM programs that involve plant growth modeling and the integration of all weed, insect pest, and disease control tactics into a cohesive system have not been developed for turfgrass managers, although several computer programs are being developed for various states. Many factors of this multidisciplinary pest control approach, however, are well understood and can lead to setting priorities for management practices based on the level of maintenance desired.

Selecting the best adapted turfgrass species and cultivars based on the climatic zones (see figure 2–4 in the next section) and the on-site use is very important. Other major considerations might include providing for a means of irrigation, an adequate soil-fertility level, and the proper soil reaction (pH). The ideal time to plan for the solution of future weed, insect, disease, and nematode problems is before the turf is planted.

The next critical stage in a pest management system is proper planting and seedling establishment. The planting of vigorous seed, sod, sprigs, or plugs free of weeds, insects, and diseases (certified and/or inspected, if possible), planted correctly to produce a uniformly dense turf, is basic to proper establishment, regardless of the level of maintenance.

Following planting and establishment, one of the most important and effective IPM practices is frequent observation of the turf for weed, insect pest, and disease problems. Close observations, assessment of pest and environmental problems, and detailed record keeping will help ensure that proper pest control action is taken at the most opportune time. Scouting combined with information written in a turfgrass pest diary can lead to adjusting or changing one or more management practices that will alleviate or moderate most pest problems.

If pesticides must be applied, and they are necessary in many situations, choose the most effective control material, while being the least toxic to nontarget plant and animal species, and the least persistent in the environment.

2.2 ZONES OF TURFGRASS ADAPTATION

Turfgrasses are adapted for optimum growth within a limited range of climatic conditions. This adaptation gives the plant a competitive edge over weeds. Climatic adaptation must be balanced with the expected maintenance level of a given turf. Turfgrasses grown under low maintenance levels must be well adapted to the prevailing local climate. Since little maintenance is provided, selected turfgrasses must be strongly competitive to dominate in the site. Zones of adaptation for a turf can often be expanded beyond their normal range when the turf is placed under high maintenance. An excellent example of this is the range of adaptation for creeping bentgrass used for golf course greens. With proper maintenance techniques, bentgrass greens can survive in Maine or Minnesota and as far south as Florida.

Temperature and precipitation have the greatest influence of climatic factors on the extent of adaptation for turfgrass species. Turfgrasses can be divided into two groups: those growing at an optimum rate in warm to hot temperatures (80 to 95°F; 27 to 35°C) and those growing at an optimum rate during cool periods of the year (60 to 75°F; 16 to 24°C). Similarly, moist, humid areas will limit the growth of some turf species, while semiarid regions might be more suitable for others. Grasses used for turf are divided into two groups: cool season and warm season. Generally, warm-season grasses are grown in the south and cool-season grasses in the north.

The United States can be split roughly into four large regions, as illustrated in figure 2–4. The upper midwest and northeastern United States, together with western Oregon and Washington, comprise an area known as the cool, humid region. The southern United States, from eastern Texas to Florida, composes the warm, humid belt. In the western United States, the area north of a line from Denver to San Francisco is referred to as the cool, semiarid region; the area south of this line often experiences a warm, arid climate. Although this artificial division of the United States can serve as an approximation for determining the adaptation of turf species, the localized environment often supersedes regional conditions. An illustration of this might be high-elevation sites in the western United States, which often support cool-season turfs, whereas only warm-season turfgrasses perform well at lower elevations. A similar situation exists in the upper warm, humid region, where a warm-season species may survive on south-facing slopes but show little or no hardiness on a north-facing site. Table 2–1 lists turfgrass species for zones of adaptation in the United States.

TABLE 2–1   ADAPTATION OF TURFGRASS SPECIES TO THE UNITED STATES

Turfgrass species	Growth habit	Cultural intensity	Leaf texture
Cool, Humid Region			
Colonial bentgrass *Agrostis tenuis*	Short stolons or rhizomes	Medium to medium high	Medium fine to fine
Creeping bentgrass *Agrostis stolonifera*	Stoloniferous	Medium high to high	Fine
Velvet bentgrass *Agrostis canina*	Stoloniferous	High	Fine
Annual bluegrass *Poa annua*	Bunch type to stoloniferous	Medium high to high	Medium to fine
Canada bluegrass *Poa compressa*	Rhizomatous	Low to medium	Medium to fine
Kentucky bluegrass *Poa pratensis*	Rhizomatous	Low to medium	Medium to fine
Rough bluegrass *Poa trivialis*	Short stolons	Low	Medium
Smooth brome *Bromus inermis*	Rhizomatous	Low	Coarse
Chewings fescue *Festuca rubra* ssp. *falax*	Bunch type	Low to medium	Very fine
Red fescue *Festuca rubra*	Short rhizomes	Low to medium	Very fine
Sheep fescue *Festuca ovina*	Bunch type	Low	Very fine
Hard fescue *Festuca longifolia* (*F. ovina* ssp. *duriuscula*)	Bunch type	Low to medium	Very fine
Tall fescue *Festuca arundinacea*	Bunch type	Low to medium	Medium to coarse
Orchardgrass *Dactylis glomerata*	Bunch type	Low	Coarse
Perennial ryegrass *Lolium perenne*	Bunch type	Medium to medium high	Medium to fine
Timothy *Phleum pratense*	Bunch type	Low	Coarse
Cool and Warm-Arid Regions			
Blue grama *Bouteloua gracilis*	Short rhizomes	Low to medium low	Fine to medium
Buffalograss *Buchloë dactyloides*	Stoloniferous	Low to medium	Fine to medium
Sideoats grama *Bouteloua curtipendula*	Short rhizomes	Low	Fine to medium
Fairway crested wheatgrass *Agropyron cristatum*	Bunch type	Low to medium	Medium to coarse
Western wheatgrass *Pascopyrus smithii*	Rhizomatous	Low to medium	Medium fine to coarse

(continued)

TABLE 2–1   ADAPTATION OF TURFGRASS SPECIES TO THE UNITED STATES (continued)

Turfgrass species	Growth habit	Cultural intensity	Leaf texture
Warm, Humid Region			

Bahiagrass
  Paspalum notatum
	Short rhizomes
	  and stolons
		Low to medium	Medium to coarse

Bermudagrasses
  Cynodon spp.
	Stoloniferous and
	  rhizomatous
		Low to high	Medium to fine

Carpetgrasses
  Axonopus spp.
	Stoloniferous	Low to medium	Medium coarse to
			  coarse

Centipedegrass
  *Eremoxhloa
  ophiuroides*
	Stoloniferous	Medium	Medium to medium
			  coarse

Kikuyugrass
  *Pennisetum
  clandestinum*
	Rhizomatous and
	  stoloniferous
		Medium	Medium to coarse

Seashore paspalum
  Paspalum vaginatum
	Stoloniferous and
	  rhizomatous
		Low to high	Medium to fine

St. Augustinegrass
  *Stenotaphrum
  secundatum*
	Stoloniferous	Medium	Coarse

Zoysiagrasses
  Zoysia spp.
	Stoloniferous and
	  rhizomatous
		Medium to high	Medium to fine

2.3   BLENDS AND MIXTURES

With few exceptions, turf sites are generally planted with more than one cultivar or species of turfgrass. Frequently, a blend of two or more cultivars of the same species of grass are utilized, or a mixture of two or more species are combined for the final plant selection. For low-maintenance areas, turfgrass mixtures are often advantageous because of their wide adaptability to various environmental conditions. Table 2–2 lists suggested species and mixtures for turf use. Several turfgrass species, such as Kentucky bluegrass, have some cultivars that are particularly adapted to low-maintenance conditions. Many cultivars of Kentucky bluegrass are particularly adapted to medium or medium-high maintenance levels. Due to the wider range of maintenance techniques, medium-maintained turf provides greater flexibility in the selection of turfgrass species and cultivars. A state turfgrass extension specialist or county extension adviser (often called a county agent) is one source for information on locally adapted turfgrass cultivars.

2.4   SEEDING OR VEGETATIVE ESTABLISHMENT

Before selecting the appropriate blend or mixture of turfgrass, it is necessary to determine which types of planting materials are available. Many turfgrass species, particularly in the warm-season group, do not produce viable seed for commercial

TABLE 2–2   SUGGESTED TURFGRASS BLENDS AND MIXTURES

Blend or mixture	Potential turf quality	Maintenance level	Adapted environment	Suggested seeding rate (1b/1000 sq ft)
Cool-Arid or Cool-Humid Regions				
Kentucky bluegrass				
Improved varieties	Good to excellent	Medium to high	Full sun	1 to 2
Common types	Fair to good	Low to medium	Full sun	1 to 2
Kentucky bluegrass*/fine fescue 50:50	Fair to good	Low to medium	Partial shade	3 to 4
Fine fescue	Poor to fair	Low	Partial shade to full sun	3 to 5
Kentucky bluegrass/perennial ryegrass 75:25	Good	Medium to high	Full sun	2 to 3
Rough bluegrass	Fair	Low	Partial shade	1 to 3
Tall fescue	Fair to good	Low to medium	Partial shade to full sun	4 to 7
Fairway crested wheatgrass	Fair	Low	Full sun	3 to 5
Cool-Arid or Cool-Humid Regions				
Bentgrass	Excellent	High	Partial shade to full sun	0.5 to 1
Zoysiagrass	Good to excellent	Medium to high	Full sun	Vegetatively established (occasionally seeded)
Transition Regions				
Kentucky bluegrass/perennial ryegrass 75:25	Good	Low to medium	Full sun	2 to 3
Tall fescue	Fair to good	Low to medium	Partial shade to full sun	4 to 7
Tall fescue/Kentucky bluegrass 90:10	Fair to good	Low to medium	Partial shade to full sun	4 to 7
Zoysiagrass	Good to excellent	Medium to high	Full sun	Vegetatively established
Warm-Arid or Warm-Humid Regions				
Bermudagrass hybrid	Excellent	Medium to high	Full sun	Vegetatively established
Common type	Good	Low to medium	Full sun	1 to 2
Tall fescue	Fair to good	Low to medium	Partial shade to full sun	5 to 7
Bahiagrass	Fair	Low	Full sun	4 to 6
Buffalograss	Fair to good	Low to medium	Full sun	3 to 6
Centipedegrass	Fair to good	Low	Full sun	¼ to ½
Zoysiagrass	Good to excellent	Medium to high	Partial shade to full sun	Vegetatively established
Seashore paspalum	Excellent	Medium to high	Partial shade to full sun	Vegetatively established
St. Augustinegrass	Fair to good	Low to medium	Partial shade to full sun	Vegetatively established

*Select Kentucky bluegrass cultivars which have been judged to be partially adapted to shade. Consult with state turfgrass extension specialist.

production. These grasses are available only as vegetative material and must be transplanted in some way. Although seed is available for some species, such as Kentucky bluegrass, vegetative planting through the use of sod can offer a much reduced time interval for establishment.

The time interval for establishment is reduced due to the fact that mature turfgrass plants are installed. It is erroneous to surmise that less site preparation is necessary than would be appropriate for seeding. When the proper sod bed is not provided, it is only a matter of time before the quality of the turf declines. While the cost of site preparation is similar for either seeding or vegetative propagation, material cost is generally less for seeding. Seeding provides a greater cultivar selection for many species. Seeding also minimizes thatch accumulation problems. Any thatch present on the site can be removed prior to seeding. Establishment is more rapid, however, through vegetative propagation, particularly with sod. The interval of development necessary for sod to provide usable turf is minimal—in many cases, only a few days.

The notion of an instant turf with sod installation is realistic, but as with all plants, a period of initial growth and rooting is necessary. This solid or continuous covering of the surface minimizes competition from germinating annual weeds and provides for lower water requirements than is necessary for seeded areas.

Along with the advantages of vegetative propagation, there are distinct disadvantages. Sodding generally represents a much higher establishment cost per square foot compared with seeding. Although properly produced sod presents a minimal thatch layer, some organic debris will be placed on the soil surface. Seeding, however, can present a susceptible host for seedling diseases and cause a greater threat of soil erosion. Table 2–3 illustrates the advantages and disadvantages for either seeding and vegetative establishment.

TABLE 2–3   COMPARISON OF SEEDING OR VEGETATIVE ESTABLISHMENT OF TURFGRASSES

Establishment method	Advantages	Disadvantages
Seeding	Less costly, greater varietal and species selection in mixture design, no initial thatch	Slow to establish, high initial water requirements, greater possibility of erosion, susceptibility to seedling pests, weeds present until cover established
Vegetative propagation	Rapid establishment, good soil erosion protection, minimized potential for competing weeds, lower water requirements than seed, assurance of a stand	High cost, heavy weight, introduction of thatch layer, potential soil-layering problems, limited varietal or species selection, generally not available in multispecies mixtures

2.5   PREPARING THE SITE

Many future problems in the care of a new turf can be avoided by following the appropriate steps in establishment for the site:

1. Control perennial grasses.
2. Rough-grade the area to be planted for the desired slope and uniformity of the surface.
3. Make soil modifications if appropriate and install drainage systems when necessary.
4. Apply pH amendments and basic fertilizer as indicated by soil tests.
5. Plow, rototill, disc, or otherwise work the soil to a depth of at least 6 in (15.25 cm).
6. Remove stones, stumps, dead roots, masonry, boards, or other debris.
7. Grade the area to a smooth, uniform surface that is free of depressions or high spots.
8. Apply starter fertilizer and rake into the soil surface.
9. Plant seeds or vegetative materials, providing firm contact with the soil.
10. Roll with a weighted roller to provide contact of the seed or vegetative material with the soil.
11. Mulch the seed bed with weed-free straw or other suitable material.
12. Irrigate to provide uniform moisture at the soil surface for proper establishment.

2.5.1   Weed Control

One of the largest causes of seedling failure is competition by weeds. Often in the renovation of an existing turf, the old turfgrass can survive if not properly controlled and become a troublesome weed in the newly established turf. Perennial weedy grasses such as bentgrass, quackgrass, and bermudagrass can persist throughout the life of the new turf. Effective chemical control measures for removing these weeds in an established turf are not available. An application of a suitable nonselective herbicide may be adequate to control most perennial grasses; however, repeat applications may be necessary. For the renovation of established turf, the dead plant material may need to be removed from the site to allow preparation of a suitable seedbed.

2.5.2   Soil Testing

Prior to disturbing the soil on the site, representative soil samples should be obtained to determine the need for soil amendments. Soil under turf should be tested routinely every 3 to 5 years by a reputable laboratory. Test results can be utilized to determine necessary fertilization needs. Changes in pH can also be detected. If the pH is found to be excessively high, then acidifying materials such as elemental sulfur, sulfuric acid, aluminum sulfate, or iron sulfate may be utilized to lower the pH. The quantity of materials required will depend on the targeted pH level and the

TABLE 2–4   STEPS IN OBTAINING AND PREPARING SOIL FOR TESTING

Step	Task
1	Collect soil samples from six to eight locations.
2	Separate any thatch from the underlying soil and submit it as a separate sample.
3	Discard roots and other debris.
4	Mix the soil samples in a nonmetallic container and remove a representative sample of soil (approximately ½ pint or 0.25 l).
5	Enclose the sample in a soil-testing sack or other nonglass container and mail the sample to a soil-testing laboratory for routine analysis.
6	Routine soil analysis usually includes tests for current levels of plant-available phosphorus, potassium, and a measurement of the soil reaction (pH).
7	Submit all pertinent information on the maintenance and care of the turf or any activities on the site over the past several seasons.

soil's resistance to the change in pH (buffering capacity). Contact your state turf-grass extension specialist for recommendations on pH amendments for your state.

Soil samples should be taken when the soil temperature is above 50°F (10°C). This will ensure an accurate potassium test, which may report an elevated level from cold soil. The steps in obtaining and processing soil samples is outlined in table 2–4. Collect a soil sample from several locations in each area in which turfgrass will be planted. Do not take samples after recent additions of lime or immediately after fertilizer applications. Samples may be collected with a soil probe or by inserting a spade deeply into the ground and taking a vertical slice of the soil, discarding roots and other debris. The vertical slice should be approximately ½ in (1.3 cm) thick and 4 to 6 in (10 to 15.25 cm) deep. Separate the thatch from the underlying soil and submit it as a separate sample. Obtain vertical slices from at least six to eight areas. These cores should be mixed in a nonmetallic container, and a representative sample of approximately ½ pint (0.25 l) of soil can be removed for submission to a soil-testing laboratory. Enclose the sample in a soil-testing sack or other nonglass container and pack in a sturdy carton. Mail the sample to a soil-testing laboratory for a routine analysis. A routine analysis of soil includes tests for current levels of soil phosphorus, potassium, and a measurement of the soil reaction (pH). Valid interpretation of the soil test results is often dependent on accurate historical information of the site. Submit all pertinent information on the maintenance and care of the turf or any activities on the site over the past several seasons. The test results may indicate whether pH adjustment is required and if fertilization is necessary. For further information about soil tests and their interpretation, consult a county extension adviser in your area or the state turfgrass extension specialist.

2.5.3   Irrigation and Drainage

To maintain a medium- or high-level turf, supplemental irrigation during drought periods will be necessary. Irrigation can range from hand-set, flexible hose irrigation to a below-ground, automated system. The major difference is in convenience and efficiency. A good system, however, should be able to uniformly deliver 1 in of water to all areas within a 24-hour period. The ability to provide moisture to the root zone can be limited largely by the rate at which the water will move into the soil.

For medium- and high-maintenance areas, it might be necessary to install a drainage system to promote adequate movement of the water through the soil. Proper drainage for turf areas is important in the control of pests and proper root function. Any drainage system needs to provide for rapid removal of excessive water from the root zone after irrigation or rain. Drainage should be provided for at both surface and subsurface areas. Surface runoff should be diverted to collecting channels and then moved away from the turf site. For athletic fields, proper crowning of the field will ensure adequate surface drainage. Most drainage systems generally consist of a perforated drainage pipe on a bed of large aggregate materials such as pea gravel at the base of a trench. Several inches of pea gravel, and a layer of sand and/or topsoil, is then added to fill the trench. The distance between drain tiles is dependent on soil type; generally, 10- to 30-ft (3- to 4.1-m) spacing is needed. To ensure that water can be moved rapidly away from the turf surface, French or open drains can be used. In this drainage system, the gravel layer is brought to the surface, which must remain open for efficient use of the system. A poorly installed drainage system can cause many future problems.

2.5.4   Modifying the Soil

Turfgrasses can survive and persist on almost any soil if adequate nutrients, water, and aeration are provided. Although turf can be grown on a heavy clay soil, a sandy loam soil is preferred, because turfgrass quality is generally better and management requirements are less. On turfs subject to heavy traffic, resistance to compaction is a highly desirable soil characteristic. Sand is resistant to compaction due to its large particle size and is used often in high-maintenance turfs. Most soils can be modified to improve their physical properties. When soils are high in clay content, organic matter (peat, rotted sawdust, and the like) or sand can be mixed with the soil to improve aeration and drainage and reduce the potential for compaction. When adding organic matter, plan to add 10 to 15% by volume of material to the soil. Muck should be avoided because it frequently contains large amounts of dispersed clay and silt that will reduce the infiltration rate. When sand is utilized to increase the total pore space in clay soils, it should be added in sufficient quantities to be effective. The mixture should contain at least 80% sand. A choice of sands is not always possible, but if available, a fine sand with particles between 0.10 and 0.35 mm in diameter should be selected. Small additions of sand may be less effective and even harmful.

For high-maintenance turfs, root zones have been designed by extensive soil modification. Most root zones incorporate the use of large quantities of sand, from 80 to 100% of the total mixture. They require careful planning and construction but will provide a wear-resistant turf with the least maintenance outlay.

Droughty, sandy soils may be improved with the addition of organic matter or finer textured soil. Four to 6 in of loam soil, low in clay content, and uniformly incorporated into the top 6 in (15.25 cm) of sand will substantially improve the water-holding capacity of the original soil. It will also provide better storage of essential plant nutrients. Any additional material added to the site should be free of perennial grass rhizomes, roots, or vegetative plant parts that might persist and increase in the newly establishing turf. Under a vigorously growing turf, soil conditions are generally improved without soil modification. This process, however, is relatively slow and may be offset by the compacting effects of severe traffic.

Droughty, sandy soils require a different fertilization program than would normally be used on a finer-textured soil. Due to the low water-retention capacity of droughty soils, frequent light applications of soluble fertilizer or the use of slow-release fertilizers is necessary for uniform nutrient release. Deeper-rooted turfgrasses (i.e., tall fescue) may be selected to obtain water from a greater volume of soil when the water-holding capacity is low.

2.5.5   Adding Basic Fertilizer Requirements

Following the recommendations based on the soil test, any fertilizers and amendments should be added to the surface prior to tilling operations. Lime or sulfur used to adjust the pH within the recommended range can be added in greater quantities than would normally be safe on the established turf. The pH amendment should be incorporated into the soil in a separate operation from the addition of basic fertilizers. Additions of phosphorus and potassium from either single or combined sources should also be thoroughly incorporated in the soil. All materials should be mixed into the upper 6 in (15.25 cm) of the soil. The use of a plow, disc, harrow, or rotary tiller will help to ensure thorough incorporation. After the addition of fertilizer materials, the rough or subsurface grade should be considered. A slope of 2 to 6% should be designed to allow surface runoff away from structures on the site. In areas with slopes greater than 25%, alternative methods such as terracing should be considered. Steep slopes are dangerous to mow. Constructing a terrace with a retaining wall, or landscaping the slope with ground covers, will provide an attractive alternative to the turf.

When extensive recontouring or modification of the soil takes place, special care is needed to ensure that the original soil level is retained around the base of existing trees. Construction of a stone-lined well or elevated terrace is necessary for the survival of trees.

After the rough slope is prepared, surface irregularities should be minimized to prevent low spots that might collect water and remain wet longer than surrounding areas. Surface irregularities can be minimized by rolling the area with a weighted roller prior to final grading.

2.5.6   Final Grade

After the incorporation of any soil amendment, a final grade should be prepared for the seed or sod bed. Final grading is similar to the subgrading, where gentle slopes of less than 6% should remain and slope away from existing structures. For seeding or vegetative plantings, a powdery surface is nonbeneficial. Remove all stones, soil clods greater than 1 in (2.5 cm) diameter, and any organic debris (sticks, twigs, or dead grass) that may interfere with the young seedlings. After the final raking of a surface, if the soil footprints are more than 1 in (2.5 cm), roll once with a weighted roller to minimize footprinting when the area is seeded or planted. The completed final grade should be 10 to 20% higher than any paved walkways or road surfaces to allow for settling. For vegetatively established sites, the surface should be moist just prior to planting. It is advisable to plan to complete the final grade immediately prior to the planting period.

For high-maintenance turfs requiring an absolutely level surface, work should proceed while standing on large sheets of plywood or other suitable material to prevent changing the contour of the surface.

2.5.7   Starter Fertilizer

A starter fertilizer can be added after the final grade is prepared to help the young seedlings through the initial stages of establishment. A complete fertilizer containing nitrogen, phosphorus, and potassium may be applied at the rate of 0.5 lb of nitrogen per 1000 sq ft (0.25 kg per 1000 sq m). Complete fertilizer rates are based on nitrogen, because phosphorus and potassium vary widely in basic fertilizers.

2.5.8   Seeding

The best time for seeding is during late summer to early fall for cool-season turfs and late spring to early summer for warm-season turfs. Soil moisture and temperature are most favorable in the fall for rapid cool-season grass establishment, and weed competition during early development of the turf is usually less severe. Early spring seeding is an alternative, but excessive soil moisture and severe competition from annual weeds can threaten successful establishment of cool-season turfs during the spring. Use of the selective preemergence herbicide siduron can minimize competition from annual grasses and some broadleaf weeds. Siduron is the only preemergence herbicide available that will selectively control summer annual grasses without injuring newly seeded turf. Midsummer plantings are frequently unsuccessful, due to high temperatures, drought, weed competition, or diseases.

When purchasing seed, the label (see figure 2–5) should be examined to ensure adequate quality. The percentage of included turfgrass species and cultivars will be clearly listed on the label, with their tested percentage of germination. The percent germination is an indication of the seeds that can be expected to germinate under optimum conditions.

Together with germination, seed purity is important in determining the necessary seeding rate. Seed purity indicates the ratio of seed to other ingredients, which are also listed on the tag. Two important "other ingredients" are the percent crop and percent weed seed in the mixture. This seed can often be a source of weed problems after establishment. These contaminants should not be tolerated in the chosen seed for medium to highly maintained turfs. Certified seed ensures genetic purity and usually good quality.

The seeding rate may have to be increased if the germination or purity percentage is low or if unfavorable growing conditions are anticipated. For rapid

PURE SEED	VARIETY/KIND	GERMINATION	ORIGIN
53.00%	Gulf annual ryegrass	85%	Oregon
17.00%	240 perennial ryegrass	85%	Oregon
9.00%	Boreal creeping red fescue	85%	Oregon
8.00%	Kenblue Kentucky bluegrass	85%	Canada
8.00%	Oasis Perennial ryegrass	85%	Oregon
2.50%	Inert matter		
2.00%	Other crop seed		
0.50%	Weed seed		

Net weight 1 pound
Noxious weed seed
81 annual bluegrass per pound; 5.1 per ounce

FIGURE 2–5   Typical seed label.

establishment with minimum weed competition, plant approximately 10 to 15 viable seeds per square inch.

For uniform growth, distribute the seed evenly across the soil surface with the recommended rate range for the blend or mixture selected. A mechanical seeder or fertilizer spreader should be used for best distribution. Hand application of seed is an alternative, but is generally less uniform. To ensure even distribution, apply half the recommended seeding rate in an east-west direction and half in a north-south direction. After seeding, rake the area lightly to cover the seed partially. For small areas, dragging the surface with the backside of a rake provides good seed/soil contact. An alternative to raking is the use of a weighted roller, metal doormat, or chain-link fence, which will ensure good seed/soil contact. Rollers using water for ballast should be checked for leaks to minimize the tracking or bunching of seed. Finally, the site should be irrigated after mulching, if necessary, to provide adequate germination.

Hydroseeding.   An alternative to dry application of seed is hydroseeding. Seeds, together with fertilizer and a pulp fiber mulch, may be suspended in water and applied to the seedbed in a high-pressure stream of water. This method is very useful for establishing difficult-to-reach sites such as slopes or rocky areas. Hydroseeding can save considerable time in establishment by combining seeding and mulching operations. Hydroseeding is most successful on sites with adequate soil moisture.

2.5.9   Vegetative Planting

Sodding.   The installation of sod is an alternative to seeding, but is generally more costly. The major advantage of this alternative is the rapid establishment of turf and/or the short interval before the site becomes usable. One can install sod any time during the growing season. A sod bed is similar to a seedbed except the surface of the latter should be moist when the sod is laid, to promote rapid rooting. A leading cause of failure with sod establishment is improper preparation of the sod bed. Laying sod on nonprepared, compacted soil prevents root extension and causes rapid desiccation during stressful periods. Lay the sod pieces with the edges snugly fitting and with ends staggered so that there are no cracks in the surface or overlapping. Stretching the sod excessively may result in shrinkage and openings in the surface during drying. Roll the sod at a 90° angle to the direction it was placed to ensure close contact with the underlying soil. This removes air pockets that would cause drying of the roots. On steep slopes, the sod should be pegged in place to prevent slippage. Water the sod thoroughly as soon as possible after laying. Daily watering during hot, dry weather, for the next 2 to 3 weeks, can provide adequate moisture during the rooting period. Less frequent watering after the sod is well knitted is required. It may take from 4 to 8 weeks for proper rooting of the newly installed sod. After that, a sodded turf may be handled as an established turf. Sods that have been produced rapidly under good growing conditions will generally root much more slowly than sods that have been grown at a slower rate under moderate fertilization.

Although sod provides the most rapid cover, other vegetative planting techniques are available. Various methods of planting are often required for turfs that form poor or loose sods or for a more economical approach to establishment. Many turfgrasses, particularly the warm-season hybrids, do not produce fertile seed, and therefore vegetative methods of establishment must be used.

Establishment by plugs.   Plugs are small pieces of sod, at least 2 in (5 cm) wide, that are usually placed 1 to 2 in (2.5 to 5 cm) in the soil and spaced 6 to 12 in (15.25 to 31 cm) apart (table 2–5).

Many warm-season and some cool-season turfgrasses are available as plugs. Plugging tools may be purchased to simplify the extraction of soil and the insertion of a new plug. After planting, soil should be packed firmly around the plugs, and the area must be extensively watered. Moderate watering every 1 to 2 days is usually adequate for proper establishment.

Establishment by sprigs.   Sprigs are individual stems, or small clusters of stems, used for vegetative establishment. They are planted in slits 2 to 3 in (5.1 to 7.5 cm) deep and 6 to 12 in (15.25 to 31 cm) apart. The sprigs should be arranged more or less in continuous lines and placed in the slit so that the upper half of the stem is above the soil level. The slit should be backfilled with soil and then rolled to ensure close soil contact with the plant material. The water requirements are essentially the same as for plugs.

Establishment by broadcast sprigging.   Sod is shredded to produce large numbers of individual pieces of lateral stems. The sod should be shredded as fine as possible. Generally, 1 sq yd of bermudagrass or zoysiagrass sod will produce 1 bu of sprigs. However, the resulting sprigs should have at least two nodes on each stem. These stems should be applied uniformly over the area at a rate of 8 to 10 bu per 1000 sq ft (3.07 l per sq m), depending on the species. Place

TABLE 2–5   ESTABLISHMENT OF TURFGRASSES BY SPRIGS OR PLUGS

Turfgrass species	Planting method	Rate per 1000 sq ft
Bermudagrass	Sprigs or plugs in rows by hand or machine, 6 to 12 in apart in 12-in rows	1 bu of sprigs or 2-in plugs on 6-in spacings, requiring 55 sq ft of sod 2-in plugs on 12-in spacings, requiring 28 sq ft of sod
Buffalograss	Seeding is usual method of establishment but can be plugged with 3- or 4-in plugs by hand or machine	3-in plugs on 12-in spacing in rows 12 in apart, requiring 63 sq ft of sod 4-in plugs on 12-in spacings in rows 12 in apart, requiring 112 sq ft of sod
Centipedegrass	Sprigs or plugs in rows 6 in apart 6-in rows	Sprigs from 8 to 10 sq ft of sod or 2-in plugs, requiring 55 sq ft of sod
Seashore paspalum	Sprigs through broadcast sprigging	5 to 14 bu
St. Augustinegrass	Sprigs or plugs, 6- to 12-in apart, 6- to 12-in row spacing	6 to 8 sq ft of sod if sprigging or 2-in plugs on 6-in spacing in rows 6 in apart, requiring 55 sq ft of sod
Zoysiagrass	Plugs 2 in square, or sprigs 6 to 12 in apart, 6- to 12-in row spacing by hand or machine	6 to 8 sq ft of sod if sprigging or 2-in plugs on 6-in spacing in rows 6 in apart, requiring 55 sq ft of sod 2-in plugs on 12-in spacing, rows 12 in apart, requiring 28 sq ft of sod

additional soil over the sprigs, or incorporate them into the existing soil by tilling or crimping to cover them partially. Whereas common bermudagrass sprigs can be buried, bermudagrass hybrid or other turfgrass sprigs must be exposed to the surface for their survival.

Rolling the area to firm the surface and ensure contact with the sprigs is essential. This method requires more water than other vegetative planting methods but provides the most rapid cover.

2.5.10   Mulching

Mulching after seeding or vegetative planting is recommended to reduce drying of the seedbed and provide a more suitable microenvironment for germination and early seedling development. Mulching also helps reduce erosion caused by wind or rain. One of the most common and effective sources of mulch is cereal straw (wheat, oats, rye, and the like). It can be spread uniformly over the seeded area at the rate of 25 to 50 lb per 1000 sq ft (12.2 to 24.4 kg per 1000 sq m). Subsequent weed problems can occur when weed seed is brought to the site in the straw. Inspect all straw to ensure that it is free of weed pests. Even in a gentle breeze, keeping the straw in place can be a problem. Watering the straw immediately after distribution and rolling can help prevent bunching or movement of straw off the site. Nylon or jute netting can be put over the straw to help tack it down. The netting usually needs to be fastened or pegged to the soil. For large sloped areas, equipment that applies asphalt or other adhesive can be utilized to help tack down the straw.

Other organic materials can be used as mulches. Wood chips or shredded bark have proven to be effective as mulches and can represent an economical choice if available near the site of establishment. Fiberous peat, ground corncobs, and sawdust can also be used as mulches but should be partially rotted to prevent the tie-up of nutrients. These materials, if dry, also compete for water, adding to the irrigation requirement normally necessary. Many synthetic mulches are commercially available. These may be shredded paper, woven paper net, burlap, fiberglass, wood pulp, or similar material. These mulches are particularly useful in hydroseeding. Their relative effectiveness in providing protection and preventing water evaporation is generally no better than that of natural materials such as straw.

2.5.11   Watering

Regardless of the method of establishment, watering is critical immediately following planting. The amount and frequency will depend on the soil type, wind, temperature, infiltration rate of the soil, and the duration and intensity of sunlight. Light daily watering for the first 2 to 3 weeks should be adequate for most times of the year. The crucial zone that must be kept damp is the upper 1 in (2.5 cm) of soil. More frequent watering may be necessary on hot, windy days to compensate for faster evaporation of water from the soil surface. The seedbed surface is extremely vulnerable to runoff or erosion during the initial period of germination. To ensure an even, light watering, use a mist nozzle to break up the water spray. Once the plant has established an extensive root system, frequent watering is no longer required.

2.6   MOWING

Start mowing a newly planted turfgrass area after the foliage has grown approximately 50% higher than the desired mowing height. This will remove not more than 33% of the leaf tissue at any one time. For example, a lawn maintained at a height of 2 in (5.1 cm) should receive its first and all subsequent mowings when it reaches 3 in (7.6 cm) in height. Approximate mowing heights for turfgrass blends and mixtures are given in table 2–6. For the first several mowings after establishment, the mower blades should be kept extremely sharp to help prevent the seedlings from being pulled out of the soil. A high-quality turf requires regular mowing at the correct height with proper equipment. Mowing is essential to the development and maintenance of a dense, uniform surface and can effectively reduce the number of weed species (chapter 4) that might invade a turfgrass stand.

TABLE 2–6   SUGGESTED MOWING HEIGHTS FOR COMMON TURFGRASSES, BLENDS, AND MIXTURES

Blend or mixture	Cutting height (in)
Bahiagrass	2 to 2.5
Bermudagrass	0.19 to 1
Buffalograss	1 to 2
Centipedegrass	1 to 1.5
Creeping bentgrass	0.125 to 0.25
Fine fescues	2 to 2.5
Kentucky bluegrass/perennial ryegrass	1 to 2
Kentucky bluegrass	
Improved varieties	1 to 2
Common types	1.5 to 2.5
Kentucky bluegrass/red fescue	1.5 to 2.5
Seashore paspalum	0.125 to 1.0
Tall fescue	1.5 to 3
Zoysiagrass	0.5 to 1

2.6.1   Cutting Height

The correct cutting height depends primarily on the turfgrass species and cultivar used, the environment (including amount of shade), and the intensity of management. Generally, the lower the cut of height, the higher the level of management required.

Cutting the grass too short weakens the turf and increases its susceptibility to weed invasions, disease and insect damage, injury from drought, and temperature extremes. Short mowing can also substantially reduce root, rhizome, and stolon development, particularly in the summer months, as well as minimize the ability of the turf to withstand stress. If the grass is cut too high, it often has a shaggy appearance that detracts from turf appearance.

2.6.2   Mowing Frequency

High, infrequent mowings can also reduce the density of most turf. Mowing frequency is dependent on the cutting height chosen and the rate of growth of the turf rather than by fixed time intervals.

Scalping, or excessive defoliation of the turf, occurs when mowing removes more than 33% of the foliage, thus reducing the amount of leaves available for photosynthesis. For optimum growth, mow the turf when it exceeds 33% of the mowing height (figure 2–6). Mowing more frequently generally has little impact on turf growth; however, extremely frequent mowing during periods of high stress can limit the accumulation of plant sugar reserves. Short mowing can also substantially reduce root development, particularly in summer months, as well as mini-

1/3

FIGURE 2–6   Mowing frequency based on growth rate.

mize the ability of the turf to withstand stress. Clippings are beneficial because they return essential plant elements and organic matter to the soil. If they are not excessively long and do not cover the turf surface after the mowing, it is unnecessary to remove clippings. They will generally sift down into the turf when dry. Mowing at the proper frequency will minimize the amount of grass clippings that remain on the turf surface. However, if there are large clumps of clippings, remove them to avoid smothering the turf.

2.6.3   Mowing Pattern

Alternating the pattern of mowing over the turf will cause the turf to grow more upright, providing a cleaner cut and enhanced appearance. A distinctive striped or checkerboard appearance will result, particularly with a reel-type mower and alternated patterns. If at all possible, 180 to 360° turns should be made off the mowing surface.

2.6.4   Mowing Equipment

Reel mowers.   There are two principal types of mowers for use on turf: reel and rotary. Reel mowers cut with a shearing action (figure 2–7). If properly sharpened and adjusted, they provide the highest-quality cut available. Improper adjustment, however, results in an uneven appearance of the turf surface, and the grass leaves appear gray to brown and/or stringy. Grass leaves may also have a similar appearance if the reel or other cutting edge (bed knife) is dull. One disadvantage of reel-type mowers is the difficulty in mowing high-cut turf. Reel mowers cannot effectively mow turf that exceeds the centerline of the reel. Mowing action is also disrupted on rough or irregular surfaces. Turfs with irregular surfaces should be smoothed through topdressing. Small pieces of debris such as stones and other hard objects can nick or mar cutting edges. Reel mowers also generally

FIGURE 2–7   Reel and bed knife of reel-type mower.

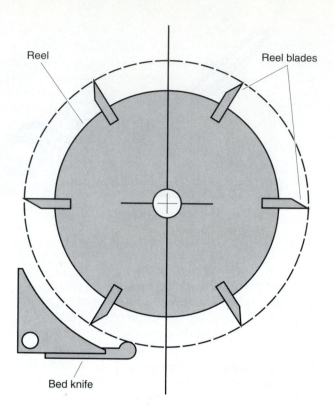

Reel

Reel blades

Bed knife

require more maintenance. Check the adjustment and sharpness of the cutting edge using the following steps:

1. Mount the mower stationary, allowing the reel to turn fully.
2. Place a strip of newspaper between the reel and the bed knife and rotate the reel slowly to cut the paper. The reel should rotate very smoothly with very little pressure, cutting the paper cleanly.
3. If the reel does not rotate or cut smoothly, adjust the movement according to the manufacturer's instructions.

Check the mowing height of the mower frequently. To ensure proper mowing heights, place the mower on a flat surface and measure the distance between the surface and the upper edge of bed knife with a small ruler. On most mowers, the cutting height may be changed by raising or lowering the castings that hold the roller at the rear of the unit. If unable to adjust mowing height, consult the instructions that were included with the mower.

Rotary mowers.   Rotary mowers are an alternative to reel-type mowers, and have become quite popular because of their lower cost and ease of maintenance. These mowers can be dangerous, however, if not used properly. The rotary motion of the blade can project stones, pieces of metal, and other debris quite forcibly for long distances, possibly injuring the operator or other persons or animals in the area. To prevent accidents, check for loose debris in the area prior to mowing; always keep fingers and feet well away from the rotary mower housing when the engine is running. A good safety measure is to detach the spark-plug wire from the spark plug after use and before any adjustments are made to the mower. Rotary mowers are more versatile than reel-type mowers. They have the ability to mow both tall turf and weeds and to mulch dry leaves. They are, however, not suitable for mowing turfs below 1 in (2.5 cm).

The cutting height of rotary mowers can be adjusted by raising or lowering the wheels. Place the mower on a flat surface, measure the height of the cutting blade from the surface, and raise or lower until the desired height is reached. To ensure a uniform and clean cut, remove rotary blades and sharpen periodically. Dull blades tend to tear the grass leaves, eventually causing deterioration of the turf.

Flail and sickle-bar mowers.   Two additional mower types are often used on low-maintenance turf. Flail mowers have a horizontally aligned shaft with vertical rotating blades. This type of mower will provide some give when coming in contact with a solid object and is used to cut tall growth up to 12 in (30 cm) above the cutting height. A sickle-bar mower provides a finer cut by way of a scissors action between a moving cutting blade and a stationary one. The movement between blades is in a horizontal plane. Sickle-bar mowers are generally used to cut tall turf, with the cutting units extended off the side of a tractor. One typical use is on highway rights-of-way.

2.6.5   Growth Retardation of Turf

Several chemical growth inhibitors of turfgrasses are currently available. These materials, when applied at the appropriate time, can help reduce the normal mowing frequency during periods of maximum growth. Many variables must be anticipated for the successful application of a growth retardant. The rate and timing of the application of a retardant is dependent on the turf species. The competitive nature of turf, through its rapid growth, provides a means for pest control and resistance to stress. Retarded turfs can require pest control and are noncompetitive during stressful periods. Table 2–7 lists currently available growth retardants and some of the necessary information for their use.

The successful application of a growth retardant is dependent on environmental conditions, both at the time of application and during the period of growth reduction. In general, the turf to be treated must be actively growing and totally free of signs of winter dormancy. In most cases, growth and development stops soon after application. For turfs in a stage of change from winter dormancy to full green-up, this will leave the dead or dying older leaves visible, providing a reduction in quality.

Turfgrass growth retardants can be grouped according to their mode of action. The first group includes those retardants that suppress both the development and growth of a plant, preventing the normal transition of vegetative growth to a mature flowering plant. The growth retardants melfluidide and maleic hydrazide are examples of this group and will reduce or eliminate the production of seedheads if applied prior to their development. It is important to apply the retardant while the turf is still in a vegetative stage of growth. If any portion of a seedhead is visible from the tip of the sheath, the effectiveness of the growth retardant in reducing seedhead production will be minimized.

TABLE 2–7   TURFGRASS GROWTH RETARDANTS AND THEIR PROPERTIES

Trade name	Common name	Mode of action	Formulation	Site of uptake in plant
Royal Slo-Glo, Retard, Liquid Growth Retardant	Malcic hydrazide	Growth and development	1.5S, 2.25S, 0.6S	Leaves
Embark	Melfluidide	Growth and development	2S	Leaves
Primo	Trinexapacethyl	Growth suppression only	1E, 25WSB	Leaves
Proxy	Ethephon	Growth suppression only	2S	Leaves
TGR Turf Enhancer, Trimmit	Paclobutrazol	Growth suppression only	2SC, several forms w/fertilizer	Roots
Cutless	Flurprimidol	Growth suppression only	50WP	Roots

*ai, active ingredient.

Labeled species	Application rate (lb ai/acre)*	Weeks of effective growth reduction	Seedhead suppression	Comments
Kentucky bluegrass, fescues, bromegrass, orehardgrass, quack-grass, quackgrass, and perennial ryegrass	2.25 to 3	4 to 8	Good	Bentgrass may be inhibited but often shows discoloration effects. Do not use on St. Augustinegrass, Not recommended as mowing reduction treatment for fine turf areas.
Kentucky and annual bluegrass, fescues, perennial ryegrass, timothy reed canary-grass, quackgrass, kikuyu grass, crested wheatgrass, orchard grass, smooth brome-grass, centipedegrass, and St. Augustinegrass	0.063 to 0.126 and 0.25 to 0.50	4 to 8	Excellent	Lower rates are for seed-head control of *Poa annua,* without growth reduction for fine, turf areas. Rates in excess of those recommended can cause discoloration or severe injury to turf. Surfactant should be added to Embark except for *Poa annua* seedhead control.
Kentucky and annual bluegrass, fescues, perennial ryegrass, bentgrass, zoysia-grass, bahiagrass, bermudagrass, centi-pedegrass, and St. Augustinegrass	0.085 to 0.17	4 to 8	Fair	
Kentucky and annual bluegrass, fescues, perennial ryegrass, and bentgrass	3.4	4 to 8	Fair	Not labeled for use on greens or any turf in California
	0.25 to 0.50	4 to 8	None	
Annual bluegrass and creeping bentgrass	0.37 to 1.5	4 to 8	None	Cutless can provide greater growth suppression of annual bluegrass rather than creeping bentgrass in mixed turfs providing for the eventual elimination of annual bluegrass. The growth suppression effects of Cutless can be minimized through the application of gibberellic acid.

The second group of growth retardants contains those that suppress growth only. This is generally achieved by reducing the production of gibberellic acid, a plant hormone necessary for shoot growth. While growth is suppressed, the plant generally matures and produces an inflorescence, but the whole plant is much reduced in size. Flurprimidol, paclobutrazol, and trinexapac-ethyl exhibit these characteristics. Flurprimidol has varying effects on different species and is most often used in mixed bentgrass and *Poa annua* turfs to provide a competitive advantage for growth of creeping bentgrass through greater growth suppression of the *Poa annua*.

Although sequential or repeat applications of growth retardants can be effective for long-term growth suppression, the possibility of injury to the turf is enhanced. Growth retardants are generally used to minimize the mowing requirements during peak periods of mowing in spring or summer. The response of the turf to an application of a growth retardant will often depend on the local environment. Consult your state turfgrass extension specialist for the optimal use of a growth retardant in your area. It is important to remember that chemical growth inhibitors or retardants will not totally substitute for mowing and should be used in integration with normal maintenance.

2.7 LEVELING THE SITE

With proper establishment of a firm, uniform seedbed, leveling should not be required. However, when irregularities do occur in the surface of a newly planted or established turf, several techniques for leveling can be utilized. For small irregularities, a weighted roller can be used to aid in removing the depressions. To avoid severe compaction, however, the ground should be moist but not wet when rolled. For large ridges or depressions, rolling is of very little value. After full establishment of the turf, the sod can be cut and removed. The area underneath can then be leveled before replacing the sod.

2.7.1 Topdressing

Topdressing, the application of a light layer of soil to the surface of the turf, can also be used to smooth irregularities. The key to efficient topdressing is through repeated light applications of soil over an extended period. The texture of materials chosen for topdressing should closely resemble the texture of the underlying soil. No more than 0.25 in (6.4 mm) of topdressing material should be applied at any time. The best rule of thumb in topdressing is "light and often."

2.8 IRRIGATION

Rainfall usually provides adequate moisture for cool-season turfgrass growth during cool spring and fall periods for many areas of the country. Extended drought periods during summer and early fall, however, may cause the grass to wilt and turn brown. In the arid and semiarid west, natural rainfall is both limited and frequently poorly distributed when it occurs. Irrigation is mandatory for much of the turf in this area. Soil moisture is crucial for active growth of warm-season turfs during the warm months of summer (April through October). Unfortunately, in most warm-season areas, rainfall is minimal during the summer months. Therefore, supplemental irrigation is usually necessary for suitable turf. Many turfs, while maintaining adequate cover under natural precipitation only, provide superior turf with supplemental irrigation. Turfgrasses such as Kentucky bluegrass may turn brown and go dormant in summer months without supplemental irrigation. These grasses generally recover and return to normal levels of quality in cooler weather. The decision as to when to irrigate will depend on the quality of the turf desired, the availability of water, the probable duration and severity of the drought conditions, and, probably most important, the budget.

2.8.1   Duration of Irrigation

To maintain a high-quality appearance throughout the growing season, an area should be irrigated as soon as the turf shows symptoms of wilting. Visual symptoms can be a rapid darkening in color or grayish appearance. Other symptoms include the lack of resilience when the turf receives traffic (that is, the appearance of footprints). Ideally, all turfs should be irrigated to saturate the root-zone area thoroughly. Following irrigation, the turf should not receive water until further signs of wilting appear. Apply enough water to moisten the soil to a depth of at least 6 in (15.25 cm). For a medium-textured soil, this might be equivalent to 1 in (2.5 cm) of water (approximately 600 gal per 1000 sq ft [2453 l per 1000 sq m]). The amount of water received by the soil surface can be measured simply by placing several coffee cans or other receptacles within the irrigation zone. The time interval required to apply the recommended quantity of water can then be used as a yardstick for future irrigations.

2.8.2   Frequency of Irrigation

Surface runoff can be a problem in soil with a low infiltration rate. Dividing the total irrigation requirement into several shorter waterings allows a slow wetting of the soil and minimizes runoff. The short irrigation cycles should be continued until the root zone is fully saturated. Light, frequent irrigations can lead to turf deterioration as a result of continuous high-moisture levels in the upper root zone. Soil saturated for extended periods places stress on the turf, which can lead to direct damage by diseases, insects, and weed competition. Damage can also result from the proliferation of roots in this restricted surface area, preventing the plant from obtaining moisture from lower soil areas during drought periods. The turf may receive irrigation any time during the day as long as the rate of application does not exceed the infiltration rate. Irrigation during periods of bright sunlight, however, is inefficient, with up to 40% water losses due to evaporation. The potential for disease is reduced by irrigating from early morning to midday to minimize the time interval during which leaf surfaces remain wet.

Soil moisture and irrigation practices also impact weed and insect populations. Excessive surface moisture favors the development of annual bluegrass, nutsedge, and other water-adapted weeds. Wet, soft turf is attractive for burrowing, nesting, and egg laying by many insects. Moisture-stressed turf is not competitive with weeds and insects and is slow to recover.

2.8.3   Water Quality

The benefits of irrigation are limited by the quality of the water supplied to the turf area. The quality requirements for irrigation water are not nearly as stringent as those for human consumption. A few basic problems can occur with low-quality water. When the water source lacks clarity due to the suspension of soil particles (silts, clays, or sand), the soil can become impervious to further watering due to the clogging of pore spaces from the introduced particles. Sand suspended in irrigation water has an abrasive effect on irrigation equipment, quickly reducing the accuracy of the delivery system.

Often, irrigation reservoirs, ponds, lakes, streams, and so on, receive runoff from surrounding agricultural areas. Pesticides in the runoff can be transported to the turf through irrigation, causing severe damage. Periodic appraisal of the irrigation source from freedom of pesticides or other toxic chemicals is a good practice. The last area of concern for the quality of irrigation water is the transmission of salts to the turfgrass site. Even dilute concentrations of salts, particularly sodium salts, added to a turf with a low infiltration rate can cause the gradual breakdown of soil structure. The analysis of salt or salt-related materials in water is a simple and inexpensive procedure.

2.8.4   Irrigation Methods

Irrigation systems come in many different shapes and sizes. The majority of components in the system can be placed either above or below ground. Second, water flow and distribution patterns can be controlled manually or automatically. Medium-maintained turf can sometimes be conveniently irrigated with above-ground manual systems. With higher maintenance requirements, below ground automatic irrigation systems are necessary to deliver the required water without restricting the use of the turf. Regardless of the system chosen, the design and installation should be left to a qualified professional.

2.9   *TURFGRASS NUTRITION*

Turfgrasses, like all plant life, require varying amounts of up to 16 common elements for optimum development and growth. When the soil pH is between 6.0 and 7.0, many, if not most, of these essential elements are supplied by the natural weathering and decomposition process of native soil. Soils in many areas of the United States, however, do not supply adequate quantities of nitrogen, phosphorus, potassium, and certain other elements, resulting in a loss of turfgrass quality.

Proper fertilization is important, therefore, for a healthy dense stand of turf that will resist weeds and recover quickly from disease or insect injury. Table 2–8

TABLE 2–8   ESSENTIAL ELEMENTS FOR TURFGRASS GROWTH

Element	Symbol	Source
Nitrogen	N	Fertilizer, soil organic matter
Phosphorus	P	Fertilizer, soil, soil organic matter
Potassium	K	Fertilizer, soil
Magnesium	Mg	Fertilizer, liming materials, soil
Calcium	Ca	Fertilizer, liming materials, soil
Sulfur	S	Fertilizer, soil, soil organic matter
Iron	Fe	Fertilizer, soil
Manganese	Mn	Generally soil only
Zinc	Zn	Generally soil only
Copper	Cu	Generally soil only
Boron	B	Generally soil only
Molybdenum	Mo	Generally soil only
Chlorine	Cl	Generally soil only
Carbon	C	Atmosphere
Hydrogen	H	Water
Oxygen	O	Water

lists the essential elements necessary for turfgrass growth. Nitrogen (N), phosphorus (P), and potassium (K) are required in the largest quantities and are the three basic components of many turf fertilizers. When all three elements are included in a material, it is said to be a complete fertilizer. Phosphorus (P) and potassium (K) requirements vary widely across the United States. The status of P and K in the soil can easily be determined through soil analysis. Soil tests will point out the relative requirements of P and K needed for optimum turfgrass growth. Consult a state turfgrass extension specialist or county extension advisor for interpretation of soil test results to determine the proper fertilization rates for phosphorus or potassium.

2.9.1   pH Amendments

The normal pH range for soil is 4.0 to 8.0. In general, turfgrasses will grow best when the pH is between 6.0 and 7.0; however, some turf species (e.g., fescues, centipedegrass), thrive well at a pH lower than 6.0. Corrective materials can be added to a soil when pH levels are outside this range. The effectiveness of pH amendments is less throughout the root zone when surface applied than when incorporated during establishment.

   If the soils are excessively acid (below pH 5.5), ground limestone or other liming materials can be added to the surface in small amounts. The material and quantities to be added to the turf will vary depending on the soil. When lime additions are required, the soil-testing laboratory will usually provide a local recommendation. If this is not available, contact the state extension turfgrass specialist. Acid soils promote the development of several common weeds, such as red sorrel and knotweed.

   For soils that are alkaline, acidifying materials such as elemental sulfur, sulfuric acid, aluminum sulfate, and iron sulfate can be used to lower the pH. These materials potentially have toxic effects on the turf if added in large quantities. The total requirements, as indicated by the soil test recommendations, should be split into smaller amounts and applied in several applications. If soil pH is amended, additional testing to ensure changes in the soil reaction is necessary. The soil pH should be tested every 6 months until the pH is within the recommended range.

Nitrogen, the element required in the largest quantity, is less stable in soil and can easily be removed through several different processes. Soil test levels of available nitrates can vary widely, even in a short period of time. Soil tests for nitrogen are therefore of very little value in determining the nitrogen requirements of an actively growing turf. Nitrogen is generally applied on a scheduled basis to provide a uniform, timely release of nitrogen.

Nitrogen can comprise up to 5% of the total weight of a turfgrass after the water is removed. This large requirement makes turfgrasses more responsive to nitrogen than to any other mineral nutrient. Nitrogen is taken up by roots and used by the plant in the nitrate (NO_3^-) or ammonium (NH_4^+) form.

2.9.2   Nitrogen Sources

Any nitrogen fertilizer applied to the turf must therefore be converted into NO_3^- and/or NH_4^- before it can be utilized. The complexity of this conversion to nitrate or ammonium will largely determine how quickly the turfgrass responds to an application of fertilizer and also how persistent this effect will be. Turfgrass fertilizers can be grouped into three broad sources, depending on the relative speed of conversion to a usable form. Table 2–9 lists the common turfgrass fertilizer sources.

Water-soluble Nitrogen.   Water-soluble or rapidly available nitrogen, which occurs in such common materials as ammonium nitrate, ammonium sulfate, and urea, provides the quickest response after application. In most cases, one of the two plant-available compounds (NO_3^- or NH_4^+) is provided by the simple dissolution of the material. Urea conversion processes take slightly longer but still occur within a few days under most conditions. These materials are also the least expensive fertilizer sources, due to their relative ease of manufacture, but have the high potential for causing chemical or fertilizer injury to turf when not properly applied.

Slowly soluble nitrogen.   Slowly soluble forms of nitrogen include natural organic materials (activated sewage sludge and animal by-products) and synthetic organic materials that are largely urea derivatives. Slowly soluble sources either have a low degree of solubility, allowing gradual dissolution over an extended period, or require some breakdown or conversion by native soil microorganisms. The latter is greatly influenced by the soil temperature. As the soil warms through the season, populations of microorganisms increase, releasing greater amounts of plant-available nitrogen.

Nitrogen sources such as IBDU and Oxamide are more efficient during cooler periods of the year because they are not as dependent on microbial activity as ureaform sources. Although these materials are more expensive per pound of nitrogen due to higher manufacturing costs, they are highly efficient as a result of minimal nitrogen losses through leaching or volatilization. An additional benefit

TABLE 2–9   COMMON TURF FERTILIZER SOURCES

Source	Usual analysis (N-P_2O_5-K_2O)
Water soluble	
Ammonium nitrate	33-0-0
Ammonium sulfate	21-0-0
Diammonium phosphate	20-50-0
Monoammonium phosphate	11-48-0
Muriate of potash	0-0-60
Potassium nitrate	13-0-44
Sulfate of potash	0-0-50
Superphosphate	0-20-0
Treble superphosphate	0-45-0
Urea	45-0-0
Slowly soluble	
IBDU (isobutylidine diurea)	31-0-0
Oxamide	32-0-0
Urea formaldehyde (UF)	38-0-0
Slow-release	
Milorganite	6-4-0
Plastic-coated urea	45-0-0
SCU (sulfur-coated urea)	32-0-0 to 36-0-0*

*Nitrogen content is dependent on the sulfur shell thickness.

of slowly soluble forms of nitrogen is the relative safety from chemical injury to the turf during application. This is a necessity when irrigation is not available.

Slow-release nitrogen.   Slow-release source of nitrogen is actually a water-soluble or rapidly available form, but it is encapsulated to prevent its rapid dissolution. Plastic and/or sulfur, the most common encapsulating materials, form a continuous membrane around a soluble form of fertilizer, allowing a gradual diffusion of fertilizer into the soil solution. If the coating material becomes cracked or split, rapid dissolution occurs, and the material responds like a water-soluble source. Careful application techniques are imperative to ensure the coating integrity of particles during application. Generally, a rotary-type spreader will minimize the breakage of these materials. Slow-release nitrogen forms are also more expensive per pound of nitrogen than water-soluble nitrogen forms; however, they generally provide a greater degree of safety with an extended period of nitrogen release.

2.9.3   Liquid Fertilizer Sources

The use of liquid fertilizer is relatively new to the turfgrass industry. Such fertilizers have been widely used for less than 20 years. Prior to 1978, liquid fertilizer solutions were combinations of urea or nitrogen salts and water, targeted for foliar uptake. As the professional lawn-care industry developed, many of the larger companies expanded nationally. Their areas of operation encompassed a wide variety of turfs and soil types. This brought about the need for greater control of the fertilizer materials and the ability to adapt application programs to a particular environment. Liquid-applied fertilizers offer both ease of handling and versatility of materials adjustments for each application. New sources of fluid fertilizer were developed to meet this demand. Materials such as Formolene, FLUF, FAN, and others were developed for the professional lawn-care market. Although most of the materials (Formolene, Folian, and the like) are a soluble source, the potential for turfgrass injury with these materials has been reduced. Formolene is a urea formaldehyde solution that offers greater safety in application than that of other soluble sources.

In addition to the soluble sources, several true liquid slow-release sources have become available. FLUF, flowable urea formaldehyde, is a stable suspension of urea-form that will provide a gradual release of nitrogen over an 8- to 10-week period. Slower-release sources such as Nutralene, with hot-water-insoluble nitrogen (HWIN), are available for liquid application in the form of a powder. These suspensions are much less stable, however, and require constant agitation for proper application.

Research has shown little difference in fertilizer efficiency between liquid or dry application. The use of a fluid-applied fertilizer is a matter of convenience and versatility. This type of formulation allows for the easy addition of other fertilizer nutrients or pesticides and is widely accepted in the professional lawn-care service industry.

2.9.4   Turfgrass Nitrogen Requirements

The total nitrogen requirement for any turf varies widely according to the turfgrasses used, maintenance level, and whether clippings are removed or allowed to remain on the site. Table 2–10 lists the range of nitrogen fertilizer needs for com-

TABLE 2–10   NITROGEN FERTILIZER RATES FOR COMMON TURFGRASSES

Turfgrass	lb N/1000 sq ft/year
Bahiagrass	0 to 1
Bermudagrass	4 to 8
Buffalograss	1 to 2
Carpetgrass	0 to 1
Centipedegrass	0 to 1
Creeping bentgrass	4 to 6
Dichondra	2 to 4
Fine-leaf fescues	1 to 2
Kentucky bluegrass	
Improved varieties	2 to 4
Common types	1 to 2
Perennial ryegrass	2 to 4
Seashore paspalum	2 to 4
St. Augustinegrass	2 to 4
Tall fescue	2 to 4
Zoysiagrass	2 to 4

mon turfgrasses. The suggested amounts reflect the soluble nitrogen necessary to support high-quality turf growth. If slowly soluble or slow-release forms are utilized, larger quantities can be added to the turf for longer persistence.

Any single application of fertilizer should deliver no more than 2 lb of soluble nitrogen per 1000 sq ft (0.98 kg per 1000 sq m) to the turf. For sensitive turfgrasses such as bentgrasses, fine fescues, and so on, this rate can be excessive, and lower rates are generally recommended. Always apply the fertilizer uniformly when the foliage is dry. To ensure even distribution, apply half of the fertilizer in a north–south direction and the other half in an east–west direction. This technique will also minimize any streaking or overlapping appearance that might otherwise occur with a single-pass application. After distribution, irrigate the area thoroughly to remove all fertilizer from the leaf tissue.

Since fertilizer elements must eventually enter the soil solution, there is little difference between liquid or dry fertilizer sources. Liquid-applied urea, however, is more susceptible to loss through volatilization if allowed to remain on the leaf blades for extended periods of time. It is advisable to irrigate immediately after the application of a liquid fertilizer. This will help to minimize volatilization losses and will also ensure a rapid turf response to the fertilizer by moving it to the root zone, where it can be absorbed. The rinsing of leaf blades will prevent damage from fertilizer injury, which can easily occur during hot days.

2.9.5   Application Schedules

Regardless of the type of nitrogen used, fertilizer applications should be applied at scheduled intervals to provide maximum benefit to the turf with minimum risk of plant injury. Generally, fertilizers are applied in the greatest quantity and frequency during periods of rapid growth. For cool-season turfs, spring and fall are the optimum times for fertilization, whereas summer fertilization is of greatest value to warm-season turf. Table 2–11 lists a generalized fertilizer schedule for most turfs. The suggested schedules are based on the use of water-soluble nitrogen and can be adjusted accordingly if a slow-release material is used. When soil test results are not available, a general ratio to follow might be a complete fertilizer with a 3:1:2 ratio of nitrogen, P_2O_5, and K_2O, respectively. Frequent light applications of fertilizer will generally result in optimum turfgrass quality.

2.9.6   Fertilizer Calculations

Converting recommended nitrogen application rates (pounds of nitrogen [N] per 1000 sq ft) into quantities of actual fertilizer is relatively simple but requires some explanation. For example, an application of 1 lb of nitrogen per 1000 sq ft of turfgrass area requires 10 lb of a 10-6-4 fertilizer, 5 lb of a 20-5-10 fertilizer, or 3 lb of 33-0-0. These amounts are determined by dividing the desired nitrogen application rate by the nitrogen percentage of the fertilizer and multiplying by 100. To apply a 23-7-7 fertilizer at the rate of 1.5 lb of nitrogen per 1000 sq ft, calculate as follows:

$$\frac{1.5}{23} \text{ lb N} \times 100 = 6.5 \text{ lb of 23-7-7 fertilizer}$$

TABLE 2–11   FERTILIZER APPLICATION SCHEDULES FOR COMMON TURFGRASSES

Turfgrass mixture or blend	Application (lb N/1000 sq ft)		
	Spring	Summer	Fall
Bahiagrass		0 to 1	
Bermudagrass		4 to 8*	
Buffalograss	0 to 1	0 to 1	
Carpetgrass		0 to 1	
Centipedegrass		0 to 1	
Creeping bentgrass	1 to 2	1 to 2	2
Dichondra		2 to 4	
Fine fescues	0 to 1		1
Kentucky bluegrass/fine fescue	0 to 1		1
Kentucky bluegrass			
Improved varieties	1	0 to 1†	1 to 2
Common types	0 to 1		1
Perennial ryegrass	1	0 to 1	1 to 2
St. Augustinegrass		2 to 4	
Tall fescue	1	0 to 1	1 to 2
Zoysiagrass	0 to 1	1 to 3	

* For high-maintenance turfs, apply no more than 1/2 N/1000 sq ft for each application in hot weather, thus reducing the possibility of injury.

†Avoid summer fertilization of cool-season turfgrass grown in the transition or warm-humid zones.

For a turf area that measures 8500 sq ft, continue the calculation as follows:

$$\frac{6.5 \text{ lb of } 23\text{-}7\text{-}7}{1000 \text{ sq ft}} \times 8500 \text{ sq ft} = 55 \text{ lb of } 23\text{-}7\text{-}7 \text{ fertilizer for the total area}$$

The same method of calculation applies to other nutrients in a fertilizer. For example, a 0-20-0 fertilizer contains 20% phosphoric acid (P_2O_5) but does not contain nitrogen or potassium. The amount of this fertilizer required to apply 1 lb of phosphoric acid per 1000 sq ft of turf is calculated as follows:

$$\frac{1 \text{ lb}}{23} \, P_2O_5 \times 100 = 5 \text{ lb of } 0\text{-}20\text{-}0 \text{ fertilizer}$$

It is important to note that the fertilizer analysis to represent phosphorus indicates the percentage of phosphoric acid that is less than half phosphorus. If fertilizer recommendations are given in pounds of phosphorus, they must be converted to the equivalent rate of P_2O_5. To do this, multiply the recommended rate of phosphorus by 2.29 to obtain the equivalent quantity of phosphoric acid.

The third number listed in a fertilizer analysis represents the total amount of potash (K_2O) contained in the fertilizer material. Like phosphorus, when potassium requirements are suggested in pounds of potassium per 1000 sq ft, the value must be multiplied by 1.20 to convert to pounds of potash. Potassium may also be found alone or in a mixed fertilizer. Muriate of potash 0-0-60 may be used when only potassium is desired. For example, to apply 1 lb of potash per 1000 sq ft, use the following calculation:

$$\frac{1 \text{ lb}}{60} \text{ K}_2\text{O} \times 100 = 1.7 \text{ lb of 0-0-60 fertilizer}$$

2.9.7   Fertilizer Spreaders

Fertilizer spreaders are of two principal types: rotary (broadcast) or drop type. The rotary spreader employs a rotating disc to distribute fertilizer well beyond the width of the spreader, thus allowing coverage of a large area in a relatively short time. The drop type, in contrast, applies fertilizer directly beneath the spreader. Accuracy of application is generally better with the drop type spreader than with the rotary type. Care should be taken to avoid overlapping or missing areas.

Both types of spreaders should be kept in good mechanical condition and should be cleaned immediately after use to avoid rusting or other deterioration of parts due to the buildup of salts from fertilizers. Periodic calibration can help to ensure uniform fertilizer distribution and point out areas of wear or mechanical breakdown. Calibration simply means determining the application rate of the spreader at a specific setting. For accurate application, the spreader must be calibrated for each fertilizer used.

2.9.8   Secondary Nutrients

Calcium (Ca), magnesium (Mg), and sulfur (S) are also necessary for optimum turf growth, but are required in lesser amounts than nitrogen, phosphorous, and potassium. These elements are generally in short supply when the pH is outside the range for optimum growth (6.0 to 7.0). When pH amendments are added to the soil, calcium, magnesium, and sulfur, which were previously unavailable, will be converted to an available form. These elements are often found in fertilizers as contaminants or impurities in the mixture. Sulfur is an added component of many complete fertilizers and may be listed on the label.

2.9.9   Minor Nutrients

Minor nutrients, also referred to as micronutrients or trace elements, are required for optimum turfgrass growth. They need to be present in smaller quantities than the major nutrients. In most medium- to fine-textured soils, there are sufficient supplies of micronutrients for available turfgrass needs. The minor nutrients, like secondary nutrients, can be in plant-unavailable forms when the pH is outside the range for optimum turf growth. In the arid and semiarid western United States, it is not uncommon to experience minor nutrient deficiencies on alkaline soils. Due to the high buffering capacity, pH adjustment in these soils may not be practical. In this situation, for high-quality turf, additions of minor elements through fertilizers is necessary. Coarse-textured soils, such as sand with its inherently low buffering capacity, also tend to exhibit extremes in pH. Coarse-textured soil may also benefit from the addition of minor-nutrient fertilizers. Golf greens with sand-based root zones may require micronutrient fertilization. Since minor-nutrient deficiencies tend to be a localized problem in many regions, it is best to consult the county extension adviser for recommendations or corrective measures.

Foliar application of iron may be used to rapidly increase the color of light green turfgrasses. Although the effect is rapid, it is often short in duration. The iron does not move far from the site of absorption in the plant and is generally removed with mowing. Foliar applications are not suitable for the long-term correction of chlorosis, because they are not intended to reach the soil. Both salt and chelated forms of iron are effective; however, salt formulations such as iron sulfate may stain building materials.

2.10 THATCH

Many turfgrasses species, when grown at rapid rates, accumulate thatch at the soil surface. Thatch is a tightly intermingled layer of living and dead stems, leaves, and roots of grasses that can develop between the actively growing vegetation and the soil surface (figure 2–8). An excessive thatch layer is undesirable because it increases the disease susceptibility of the turf; reduces tolerance to drought, cold, and heat; and can minimize the movement of air, water, fertilizers, and some pesticides through the turf into the soil. Thatch also tends to decrease the turf's capacity for growth, resulting in an all-round deterioration of turf quality. Some buildup of thatch is desirable, however, because it increases the resiliency of the turf and provides a source of nutrients through the breakdown of organic matter. An increase in thatch is directly related to the rate at which dead plant materials accumulate. If the production of dead material exceeds the ability of microorganisms to break

Soil zone

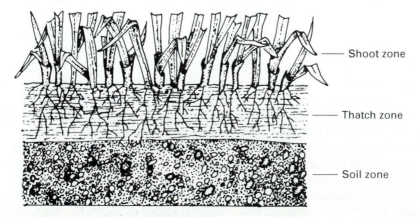

FIGURE 2–8 Profile of established turf.

down the material into its elemental components, thatch accumulates. Therefore, vigorously growing grass cultivars, heavy fertilization, excessive soil acidity (which inhibits bacterial activity), poor soil aeration and drainage, cold soils, and some pesticides are all contributors to thatch buildup.

The depth of thatch can be determined by cutting a wedge of turf and examining the profile of green vegetation, thatch, and soil (figure 2–8). If the compressed layer of dead materials just above the soil surface is greater than ½ in (1.25 cm) thick, it should be controlled.

Two basic methods are used to control excessive thatch: (1) physical removal of the thatch by various types of machinery, and (2) modification of the environment to encourage rapid bacterial activity, reduce the growth rate of the turfgrass, or both. Although effective in thatch control, physical removal of thatch can be destructive to living turf. Therefore, physical thatch removal should occur only during peak grass-growing periods. Dethatching machines are often utilized for this process. A vertical mower represents one type of dethatching machine. It contains a rotating shaft with evenly spaced, vertically oriented solid blades. A power rake, which is similar, has hinged or springlike teeth. These machines can often be obtained at garden-supply stores or equipment-rental companies.

Similar to the height-setting mechanism on a rotary mower, most dethatching machines have adjustable wheel heights to allow various depth penetrations of the turf by the blades. The blades should be set deep enough to penetrate the soil surface slightly. As much thatch as possible should be removed. Extensive removal of a thatch layer greater than ¾ in (1.9 cm) is excessively destructive to the turf, minimizing its ability for recovery. In this situation, less thatch should be removed at one time, but additional dethatching should continue during future periods of rapid growth. Because dethatching procedures can open up the soil surface for invasion by annual grassy weeds during early spring, applications of preemergence herbicides should follow this cultivation technique.

Modifying the turf environment to remove the thatch layer slowly is generally more effective for long-term control. When thatch tends to accumulate, it can slowly decrease the pH. Acidic conditions minimize the growth and reproduction of bacteria and other microorganisms, thus reducing their ability to degrade thatch. The addition of lime to raise the pH within the thatch will promote rapid degradation. Thatch tends to reduce air, water, and fertilizer movements to the soil surface. It also decreases the ability for microorganisms to multiply. Core aerification (Color plate 3) provides conduction channels for the movement of air and water to the soil throughout the thatch.

Core aerification is a process of removing small cylinders of soil and turf in a uniform pattern through the turf. Several different types of equipment are available to remove cores over large areas in a relatively short time. The majority of the machines deposit the cylinder of turf and soil onto the surface. If these cylinders are then broken up or left to disintegrate, allowing the soil to sift down into the thatch layer, an increase in thatch degradation will result. This can be attributed to the increase in microorganism activity in the soil-enriched environment.

One of the oldest practices in turfgrass management next to mowing is the application of fine layers of soil to the turf surface to provide a smooth, firm, even turf. Light, frequent topdressings also aid in the degradation of thatch by introducing soil into the thatch layer. The soil-enriched environment can now support larger populations of soil microorganisms. Topdressing also tends to minimize the rate of thatch accumulation. As mentioned previously, it is important to closely match the texture of the topdressing mixture with the underlying soil. This helps prevent any problems in the ability of the turf to absorb water.

2.11   SOIL COMPACTION

Compacted soils are characterized by poor aeration and drainage, poor surface water infiltration, shallow root growth, and reduced overall quality. Turfgrasses in compacted soils grow slowly, lack vigor, and become thin or do not grow at all. Soil compaction is more prominent on fine-textured soils subjected to concentrated foot and/or vehicle traffic. Soil compaction can be reduced by cultivation with machines that create openings extending into the underlying soil.

The two principal types of mechanical cultivators to alleviate compaction are core aerifiers and spikers. Core aerification machines remove small cores of soil, depositing them on the surface of the turf or collecting them for removal from the site. For simple alleviation of compaction, these cores can be removed or chopped or matted into the grass to provide topdressing. Spikers and slicers employ solid spines or knives, respectively, to create narrow openings in the turf. These machines, although not as effective as core aerifiers, generally are less disruptive to turf growth and can be used during periods when core aerification would otherwise cause injury. Cultivation should be carried out only during periods of rapid turf growth.

If the soil is seriously waterlogged, drainage tiles can be installed where feasible. Foot and/or vehicle traffic can be reduced or redirected by constructing walks or fences, or by planting shrubs.

Because soil is more conducive to compaction when it is very moist or wet, it is crucial to minimize traffic on a turf after a heavy rainfall or irrigation. Moderately moist or dry soils are much less prone to compaction.

2.12   INTEGRATED MANAGEMENT PROGRAMS

Turf managers can only be successful for extended periods of time when they integrate the previously described management practices into a consistent program. Inappropriate or inconsistent actions in any area of management will limit the value of the other management operations. For example, high fertility management with low or limited soil moisture will produce a poorer quality turf than one managed with low inputs of both fertility and water.

The remainder of the chapter will offer suggestions for integrating management for low, medium, or high levels of inputs. You should design a management program targeted at one of these three broad objectives based on the available resources and your desired quality goals. Typically, highway rights-of-way, large utility turfs, airport runway grounds, and similar areas are considered low-maintenance turf. Sports fields, golf course greens, tees, and fairways are considered high-maintenance turfs. The wide range of turfs in between these two extremes, such as home lawns, require a moderate level of maintenance.

2.12.1   Weed Management for Different Levels of Turfgrass Maintenance

Although only three maintenance levels may be too narrow in scope, most levels of maintenance can be adjusted to fit within one of these broad classifications. Although total weed eradication is generally not possible, higher-maintenance turfs require minimal weed populations. Similarly, low-maintenance turf will tolerate greater populations of weeds. In fact, for the most minimally maintained turfs, some weed growth is necessary to maintain a continuous living cover of reasonable density.

The initial maintenance plan or calendar should be developed solely with turfgrass requirements in mind (see table 2–12 in the next section). This maintenance schedule is rarely sufficient and must be modified to accommodate weed populations that develop on the site. An example might be in areas of poor drainage. Where soil moisture levels are at field capacity or slightly less for extended periods of time, populations of crabgrass, goosegrass, *Poa annua*, or other weeds with high moisture requirements will develop. Here, less irrigation will be necessary to hold weed populations in check.

In acidic or low pH soils, weed populations such as red sorrel will predominate. Therefore, pH amendments will be required for optimal control. Just as soil compaction can have a direct effect on turf density, the same effect influences weed populations. Many turfgrass weeds can tolerate compacted soil conditions better than do turfgrasses. To benefit the turf and minimize weed development, some alleviation from soil compaction is necessary, usually through core aerification.

Successful preemergent weed control is dependent on a continuous chemical barrier in the upper soil surface to intercept developing weed seedlings. Core aerification disrupts this barrier, allowing the emergence of weed seedlings. Therefore, preemergent herbicide applications should be delayed until after core aerification.

The cycle of applying cultural controls to benefit turf development and minimize weed encroachment should be followed by effective physical or chemical weed-management procedures. The cycle is repeated with additional necessary cultural practices, followed by direct weed-control measures. Close evaluation of turfgrass growth performance and weed populations after each step is crucial in determining necessary future operations. Again, it is necessary to keep this information in a turfgrass pest diary to follow changes in turfgrass performance as affected by weed populations. Corrective measures can then be taken to reduce weed populations. Integrated weed control is therefore a reactive procedure where the turf manager formulates new strategies based on an evaluation of previous efforts.

Low maintenance.   Low-maintenance turfs generally receive little or no direct weed-control applications. Weed populations are stabilized by maintaining adequate turf growth. In the rare event that a direct pesticide application is required, select materials that will only reduce targeted weed populations to maintain adequate ground cover.

Medium maintenance.   For medium-maintained turfs, managing most weed species is necessary. Fortunately, many weed species populations can be held at reasonable levels with cultural practices alone. Only a few species, if not occasionally treated with pesticides, will eventually threaten the existence of the turf. Weed-management strategies should be designed with these species in mind first. The program can then be modified or extended to include less critical weeds.

High maintenance.   Often, separate strategies for different weed species are required for high-maintenance turfs. In some cases, cultural practices differ between strategies. The turf manager should prioritize the individual weed-management strategies by their potential for damaging the turf. An overall plan can then be developed, not only to meet the needs of eliminating all the potential weeds but also to focus on the most bothersome species.

2.12.2   Insect Pest Management for Three Levels of Turfgrass Maintenance

Low maintenance.   Turfgrass insect pests favor high-maintenance turf for their host rather than poor-quality turf. The leaf feeders, including sod webworms, cutworms, armyworms, and other chewing insects, prefer to feed in areas of high fertility and irrigation than on low-maintenance turf such as airports, highway rights-of-way, industrial grounds, land-reclamation or waste sites, and waterway banks.

Insect control in low-maintenance areas consists of applying an insecticide to an outbreak situation. For example, a population of a general feeder, such as grasshoppers or crickets, builds up in large numbers in an industrial site; or leafhoppers attracted to lights are found in large numbers along a highway median. These turf areas could be treated if the insect population is causing severe damage or creating a nuisance to people using the area.

Only in unusual situations would areas of low maintenance need to be treated to control insects.

Medium maintenance.   Home and industrial lawns, athletic fields, parks, cemeteries, and golf course roughs are often infested with insect pests. The numbers of these pests fluctuate yearly, with visible damage appearing in periods of moisture stress. White grubs, billbugs, chinch bugs, webworms, and greenbugs are often found in these medium-maintenance areas, and their damage appears in mid to late summer. In most instances, treatment is applied when severe or not-so-severe damage is evident. Treatments, however, should be made when damage first appears. The application of an effective insecticide, combined with additional fertility and supplemental irrigation, will usually solve most insect-pest problems in medium-maintenance turfs. Where a history of repeated annual damage from a pest insect such as the Japanese beetle or bluegrass billbug occurs, treatments may need to be made to prevent damaging populations and provide season-long control.

High maintenance.   High-quality, vigorous turfgrass that is well fertilized and irrigated with frequent mowings is inviting to certain insect pests. For example, black cutworms spend repeated generations in high-quality golf greens and tees; black turfgrass ataenius grubs can appear in high numbers in fairways; and various species of white grubs may infest fairways and other fine-turf areas.

Managers of high-quality turfs soon learn which insect pests frequent these areas. They also learn when damage, if any, is most likely to appear. There are many turf areas that receive an insecticide treatment at regular intervals throughout the growing season. It is important to identify the insect pest that appears regularly in the area, note symptoms of its early damage and its potential for serious damage, and know the life cycle of the insect. The most vulnerable stage of the insect must also be understood. For example, black turfgrass ataenius grubs hatch from eggs laid on the golf course fairway by adult beetles that migrated into the area in the spring. Grubs, if a problem, will appear in early summer, and fewer than 30 to 40 grubs per square foot will usually not exhibit damage in well-managed turf.

2.12.3   Disease Management for Three Levels of Turfgrass Maintenance

Low maintenance.   Disease control is achieved most economically in low-maintenance turfs by planting mixtures or blends of locally adapted, disease-resistant grasses. In certain situations, light applications of fertilizer, mowing as high as practical, and supplemental irrigation might be feasible. Diseases are rarely unsightly enough in low-maintenance turfs to warrant other cultural or chemical measures.

Medium maintenance.   Disease control in lawns, parks, cemeteries, athletic fields, and golf course roughs should start with providing good surface and sub-surface drainage plus a well-prepared and fertile seedbed. Seed, sod, sprigs, or plugs should be of locally adapted, disease-resistant grasses and of top quality. Other cultural controls would include late spring and early fall applications of slow-release fertilizers, mowing at the suggested height and frequency, and watering deeply during droughts but only to prevent permanent injury. Dethatching and core aerification may be warranted when the thatch is over ½ in (1.3 cm) thick and compaction is a problem. Light penetration and air movement across the turf may be increased by selectively pruning or removing dense trees, shrubs, or other barriers bordering the turf areas. More light and air means that the grass will dry more quickly, and disease attacks should not be as severe. Fungicides may be called for on a preventive basis.

High maintenance.   Golf and bowling greens, tees, fairways, and high-quality athletic fields require the full spectrum of cultural and chemical disease-control practices outlined for medium-maintenance turf. Proper and timely planting, fertilization, mowing, watering, dethatching, and core aerification practices are needed to maintain turf vigor, density, and uniformity. With high maintenance will come increased disease pressure. The full spectrum of cultural controls outlined in chapter 2 and 5 in addition to chemical controls will be needed to keep damaging diseases in check. These cultural, chemical, and other management practices need to be integrated with practices to keep down damaging populations of weeds, insects, and other animal pests.

2.13   MAINTENANCE CALENDAR

Using all the technical information available for turfgrass growth still cannot ensure optimum turfgrass quality. The quality of a turfgrass site is largely the reflection of the timely application of turfgrass management techniques in a consistent and repetitive manner. Variations in climate and uses of the site can mandate great fluctuations in the scheduling of management techniques. All good turf managers, however, must work from a basic maintenance schedule to provide the consistent uniformity necessary for high-quality turf. Table 2–12 documents two basic maintenance schedules for cool-season and warm-season turfs. It is important to emphasize that these schedules

TABLE 2–12 GENERALIZED MAINTENANCE SCHEDULES FOR COOL-SEASON AND WARM-SEASON TURFGRASSES

Month	Operation	Comments
		Cool season
March–April	Rolling	Use only for correction of small surface irregularities.
April (cool-humid and arid zones only)	Dethatching	Delay dethatching until the turf has resumed active growth. Spring dethatching is not recommended for the transition and warm-humid zones.
February–March (transition and warm-humid zones)	Fertilization	One to two fertilizer applications can be made at this time. Delay first application until after the second mowing.
April–mid-June (cool-humid and cool-arid zones)		
April–mid-May	Weed control	Application of preemergence herbicide should be made in advance of summer annual grass germination. Postemergence broadleaf herbicide should be applied after the resumption of active growth. Turfgrass growth retardants should be applied after the second mowing.
February–March (transition warm-humid zones)	Reseeding	Thin areas that require reseeding should be seeded as early as possible. Seed when soil conditions permit.
mid-April–May (cool-humid and cool-arid zones)	Renovation or establishment	If perennial grass weeds are present, delay renovation or establishment until they are controlled.
June–mid-August	Mowing, irrigation, and pest control	Conduct maintenance practices only as necessary to ensure survival and desired turf quality.
mid-August–mid-September	Renovation, establishment, and reseeding	Delay renovation or establishment until perennial grass weeds have been controlled.
mid-October–November	Fertilization	Delay late-fall fertilization until after turf stops growing. Late-fall fertilization helps to promote early spring green-up.

Warm season

September–mid-October	Fertilization	One to two fertilizer applications should be made as early in fall as possible. Avoid midfall applications of fertilizer. If soils are deficient in phosphorous or potassium, incorporate with fall application.
September–mid-October	Cultivation and dethatching	Delay cultivation or dethatching until turf resumes rapid growth.
mid-September–October	Weed control	Delay the application of postemergence broadleaf herbicides until weeds are rapidly growing.
January–February	Weed control	A nonselective herbicide with no soil residual activity may be applied to existing weeds in dormant warm-season turfs.
April–August	Fertilization	Delay first application of fertilizer until after full green-up. Additional applications can be applied monthly. If phosphorous or potassium is required, spread the application over the entire season.
March–July	Cultivation and dethatching	Dethatch warm-season turfs well in advance of green-up or delay until active growth resumes. Other cultivation techniques should be applied only after resumption of growth.
March–mid-April	Weed control	Apply preemergence herbicides prior to germination of summer annual grasses. Postemergence broadleaf herbicide applications should be delayed until weeds are actively growing.
May–July	Renovation, establishment, and planting	Delay renovation or establishment of warm-season turf until similar turfs in the area have resumed full growth.
September–October	Winter overseeding	Dormant warm-season turf may be overseeded with cool-season turfgrasses to provide color through the winter months.

should represent the rough outline of a *final* maintenance calendar. Local conditions and use will justify appropriate changes in these basic schedules. Again, the state turfgrass extension specialist is an excellent source for localized maintenance scheduling. Even with these resources, maintenance schedules will change periodically. Experience is often the best source of information for accurate scheduling.

2.14   STUDY QUESTIONS

1. Define IPM.
2. What are the economic threshold level and damage threshold level for turf and how do they differ?
3. Contrast the advantages and disadvantages of sandy and clay soils.
4. What is thatch and how can it be managed?
5. How much of a 16-4-9 grade fertilizer is needed to apply the equivalent of 1 lb of nitrogen per 1000 sq ft?
6. For intensively trafficked turfs, what material(s) make the best root zone?
7. Based on table 2–12, design a maintenance calendar for your area.

2.15   SELECTED REFERENCES

Beard, H. J., J. E. Beard, and D. P. Martin. *Turfgrass Bibliography from 1672 to 1972.* East Lansing, MI: Michigan State University Press, 1977.

Beard, J. B. *Turfgrass Science and Culture.* Englewood Cliffs, NJ: Prentice-Hall, Inc., 1973.

Beard, J. B. *Turf Management for Golf Courses.* Chelsea, MI: Ann Arbor Press, 2002.

Collings, G. H. *Commercial Fertilizers,* 5th ed. New York: McGraw-Hill Book Company, 1955.

Canon R. N., D. V. Naddington, P. E. Rihii. *Turfgrass Soil Fertility and Chemical Problems.* Chelsea, MI: Ann Arbor Press, 2000.

Christian, N. *Fundamentals of Turfgrass Managers.* Chelsea, MI: Ann Arbor Press, 1998.

Daniel, W. H., and R. P. Freeborg. *Turf Managers Handbook.* New York: Harvest Publishing Company, 1979.

Dermoden, P. H. *Creeping Bentgrass Management.* Chelsea MI: Ann Arbor Press, 2000.

Etter, A. G. "How Kentucky bluegrass grows." *Annual of Missouri Botanical Gardens,* 38 (1951): 293–375.

Hanson, A. A. and F. V. Juska, eds. *Turfgrass Science.* Agronomy Monograph 14. Madison, WI: American Society of Agronomy, Inc., 1969.

Madison, J. H. *Principles of Turfgrass Culture.* New York: Van Nostrand Reinhold Company, Inc., 1971.

Sprague, H. B. *Turf Management Handbook,* rev. ed. Danville, IL: The Interstate Printers and Publishers, Inc., 1976.

Turgeon, A. J. *Turfgrass Management,* 5th ed. Englewood Cliffs, NJ: Prentice-Hall, Inc., 1999.

Vengris, J. *Lawns,* rev. ed. Fresno, CA: Thompson Publications, 1973.

CHAPTER

3

Abiotic Turfgrass Problems

Not all turf problems are due to the activities of living organisms. Some arise from turf management practices inconsistent with those presented in chapter 2. Just as likely are areas of reduced quality caused by severe injury, the result of natural phenomena that is often beyond the turf manager's control. These additional abiotic disorders, injuries, or problems are the focus of this chapter. While some of the described problems generally have little long-term impact on turf, some abiotic injuries, under the right circumstances, can be more devastating than most biotic activities. Each problem is presented separately for clarity and accessibility, but in nature most of the covered problems are found in combinations with overlapping symptoms. Many of the problems can occur throughout the growing season, so they are presented alphabetically.

3.1 IDENTIFYING THE PROBLEM

The first critical step in managing abiotic disorders or injuries is the correct identification of the sources. The key in table 3–1 will assist you.

3.2 CHLOROSIS

3.2.1 Symptoms

A general yellowing or chlorosis of small or large areas of turf can often be attributed to a nutrient deficiency. Chlorosis often reflects a lack of nitrogen. However, several turfgrass species, such as centipedegrass and St. Augustinegrass, are very susceptible to a deficiency of iron. Although not harmful on a short-term basis, chlorosis can reduce shoot growth and cause thinning when extended over a long period of time.

3.2.2 Management

Application of the appropriate fertilizer nutrient can quickly rectify chlorotic turf. However, chlorosis can persist, even when the total available nutrients are above turfgrass requirements. Saturated soils, a highly alkaline (high pH) or acidic (low pH) soil, and excessive salts in the soil can prevent the turf from absorbing the necessary nutrient

TABLE 3–1 KEY TO ABIOTIC INJURY OR PROBLEMS ON TURF

Injury or problem	Yellowing of leaves	Matted, pressed appearance	Thin turf, low leaf blade density	Black or dark, greasy appearance	Appearing uniformly across the turf	Purple to reddish-purple base of leaves	Random, circular or uniform patches or spots of dead turf	Standing water or poor infiltration	Soft, spongy turf surface	Water-soaked, limp leaves	Dark gray, wilted appearance	In shaded or low light areas
Abrasive damage		•									•	
Air pollution	•		•		•	•						
Algae			•	•								•
Black layer			•		•			•				
Buried debris							•	•				
Chemical burns and spills	•			•			•			•	•	
Chlorosis	•		•		•							
Deep thatch					•			•	•			
Dog injury	•						•			•	•	
Dull-mower injury					•							
Hail injury					•							
Heat injury	•				•	•						

Ice injury

Late-spring frosts

Lightning damage

Localized dry spot

Moss

Nutrient abnormalities

Root competition

Salt injury

Scald and heat injury

Scalping injury

Soil compaction

Summer drought or desiccation

Winter drying out or desiccation

requirements. Turfs that are chlorotic in the fall can be injured during the winter due to low food reserves and a weakened root system.

Chlorosis is generally managed through application of the appropriate nutrient in a fertilizer or through the correction of other problems. Some soils, however, such as those in arid areas, are alkaline and cannot be corrected easily. Nutrients applied to these soils may be converted to nonavailable sources. Iron, however, may be effective in correcting chlorosis when applied in the form of an iron salt (ferrous sulfate, ferrous ammonium sulfate, and so on). Iron can also be absorbed into the turfgrass plant through foliar applications of iron sulfate or iron chelate to provide a rapid greening of foliar tissue. This technique, however, is temporary at best. Should the turf not respond to the application of the corrective fertilizer nutrient, the chlorosis could be a symptom of another problem, such as insect or disease damage. Chlorosis is the most common of the listed problems. It is often considered as a symptom for many other disorders.

3.3   CHEMICAL BURNS AND SPILLS

3.3.1   Symptoms

Symptoms of pesticide damage often appear as broad swaths, narrow streaks, or other regular patterns in which the pesticide was applied. Specific symptoms might include leaf speckling, yellowing, or death soon after application—or even after several weeks during which other growth conditions are optimal.

While various petroleum and chemical products can cause direct injury to the turf, many of the symptoms and resulting injuries are similar. Generally, chemical injury is located directly on turf areas that come in contact with the material creating characteristic patterns. One of the more common chemical spills is the loss of hydraulic fluid from mowers on golf course greens or tees (figure 3–1). This damage shows up as streaks from dripping of the material. As with fertilizer burns or other salt-related injuries, the damage is due to rapid desiccation of leaf tissue from a breakdown of cell walls and death of cellular tissue.

3.3.2   Management

Fertilizers, pesticides, lime, sulfur, soaps, gasoline, hydraulic fluids, oils, and so on, can cause rapid discoloration and injury to turfgrass tissue (figure 3–2). In excess, these materials cause a rapid loss of water from leaf tissues, resulting in dehydration and death of the grass, usually in patches or streaks (figure 3–3). When applying fertilizers, herbicides, insecticides, fungicides, or other turf chemicals, strictly follow the directions and precautions on package labels. To prevent injury to the foliage, apply a fertilizer to dry foliage *only*, then water it in immediately. The major principle to consider for plant safety is the concentration of the damage-causing

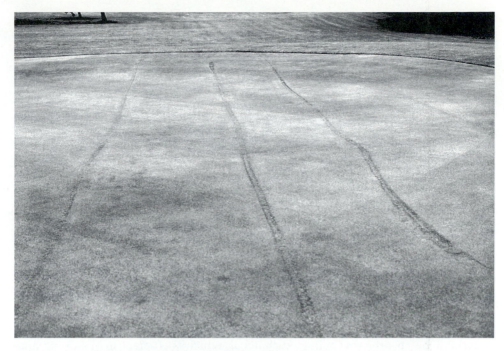

FIGURE 3–1 Damage from hydraulic fluid leaked from a mower.

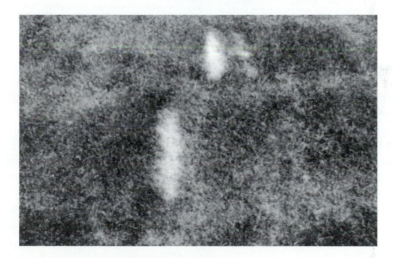

FIGURE 3–2 Injury from the overapplication of fertilizer.

material. If the safety of an application of material is in question, apply smaller amounts over an extended period of time or first test it on a small area. Higher concentrations of any material always represent an increased risk. Always calibrate spreaders and sprayers to ensure uniform and accurate application of materials (see chapter 8 and appendix B).

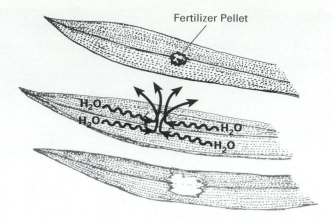

FIGURE 3–3   General mechanism of chemical or salt injury.

Treatment for chemical spills can proceed in several ways. If the spills are fairly small, treating the area with an absorbant and/or detergent or soap material can help to move the materials through the root zones or dilute their effect. For larger areas, commercially available absorbants will help to soak up or disperse the material. One absorbant is activated charcoal, which is efficient in absorbing and minimizing damage from excessive chemical materials such as hydraulic fluid, pesticides, or other detrimental substances.

For reducing the effects from spills of organic pesticides, some petroleum products, and hydraulic fluids, use 100 lb (45.4 kg) of activated charcoal to every pound (0.45 kg) of active material spilled, but no less than 2 lb per 150 sq ft (600 lb/acre; 680 kg/ha) of contaminated area. If the active material has not been diluted with water at the time of spill, apply the charcoal directly as a dry power. If the active material has been diluted with water, apply the activated charcoal in a slurry with a sprinkling can or common spray equipment. The charcoal must be incorporated into the contaminated soil, preferably to a depth of 6 in (15.24 cm). With severe spills, some contaminated soil may need removal prior to activated charcoal application.

Turf areas treated with preemergence herbicides can be reseeded earlier than normal by treating with activated charcoal. When it is desirable to terminate the effect of a preemergence crabgrass herbicide, apply charcoal slurry at a rate of 1 lb per gallon (0.11 kg/l) of water for each 150 sq ft (13.9 sq m). Water the slurry into the soil. Make sure the grass is washed free of heavy charcoal deposits. Where possible, thoroughly rake the charcoal into the soil. The area can be seeded 1 day after treatment.

3.4   BLACK LAYER

Black layer is a physical condition that has been observed most frequently in high-sand golf greens, but it can also develop in mineral soils. It became a serious problem, coincidentally, with the use of sand topdressing containing 20% or more calcium carbonate ($CaCO_3$). Black layer can also develop in overly wet mineral soil greens, and it is a natural development in wetlands. When sulfur or sulfur derivatives are applied to lower the soil pH, they are reduced to hydrogen sulfide (H_2S), an unstable gas. This sulfide combines with divalent metal ions (such as iron [Fe], magnesium [Mg], and manganese [Mn]) to form the corresponding sulfide—an insoluble dark material. A black layer develops when the dark metal sulfides form a layer in the soil, which interferes with infiltration and water movement in wet or waterlogged soil.

3.4.1   Symptoms

Turf with a black layer under it first appears yellow or bronzed, especially in low-lying, shaded, and waterlogged soils of golf greens where air circulation and internal soil drainage are poor. Eventually, the turf becomes thinned in irregular patterns, dies, and turns brown. The condition is common where sand topdressing has been layered over a heavier soil type of an older green or a thatch layer.

When plugs are taken, a coal-black layer 0.25 to 1 in (0.5 to 2.5 cm) or more in thickness can be seen in the sand profile anywhere from the thatch to about 3 in (7.5 cm) below the surface. The roots growing in this medium may be irregularly blackened and mottled. The layer of blackened sand, which occurs under anaerobic conditions, has a foul, sulfurous, rotten-egg odor, due to the production of H_2S gas by soil bacteria. The black layer interferes with downward movement of water, making the anaerobic condition worse.

Anaerobic conditions occur when oxygen is eliminated from the soil by excessive rain or irrigation, combined with compaction caused by traffic and soil layering. Excess sulfur in the soil (as elemental sulfur or as sulfate in irrigation water, fertilizers, pesticides, or amendments) replaces the oxygen as an electron acceptor for soil microbes that reduce sulfur and, as a result, releases hydrogen sulfate, which is acidic and smells like rotten eggs. The gas is toxic to grass roots, causing the grass plants to die. Blackening occurs when this unstable gas combines with Fe^- or other divalent cations to form the metal-sulfide precipitate.

3.4.2   Management

The primary management control is to eliminate, avoid, or greatly minimize soil waterlogging, which may be due to clogged soil drainage systems and leaks in irrigation lines that saturate the soil for extended periods before their discovery. Prevention includes using sand particles of the proper size when constructing greens or preparing topdressing mixes. Apply the *same* topdressing mix as used when the greens were constructed. Properly designed and installed drains are a necessity for new greens. Such subsurface drains should be inspected frequently to be sure they are clear and operating efficiently. Maintain water infiltration and drainage rates.

Where black layers have already developed, syringe greens or irrigate with care to prevent the soil from becoming waterlogged. Frequent core aerification is helpful in alleviating anaerobic soil conditions. Verti-drain and water-injection cultivation also help to ease such conditions. To increase evaporation, prune or remove dense trees, shrubs, and brush around affected greens for improved air circulation and reduced shade. Avoid the use of fertilizers and other chemicals that contain large amounts of sulfur, iron, and gypsum. Spoon-feed turf with nitrate forms of fertilizer such as potassium nitrate. Be careful when applying nitrate fertilizers to greens, as they can cause burning. Never apply more than 0.25 lb (0.12 kg) of actual nitrogen per 1,000 sq ft (100 sq m) per treatment. Water sources high in iron, magnesium, manganese, and organic matter (such as pond or sewage effluent water) with a pH over 7.0 should be avoided, if at all possible.

3.5 BURIED DEBRIS

Many different types of materials buried under turf, but near the soil surface, can reduce the quality of your turf. Buried materials can block water movement, act as heat sinks, attract fungi, or release chemicals toxic to the turf. In most cases, the effects of buried debris will be indirect and only noticed when the turf is stressed (figure 3–4). The visible symptoms of turf stressed beneath the soil by buried materials are generally found as patches characterized by one or more of the following: a reduction in growth; premature drought or heat stress; mushrooms; and death of the turf. For healthy turf, you need about 6 to 12 in (15.25 to 30.5 cm) of soil that does not have large pieces of relatively impervious material in it. For example, old tree roots; pieces of construction wood, concrete, or plaster; plastic or ceramic drain tile; and large stones can all result in local problems for turf growth. In some cases, you can determine the cause easily. For example, the presence of a clump of mushrooms is often a sign that a piece of wood or plant root is under the mushrooms. Another key to identifying buried material is that the problem does not increase in size or number in your turf, but it may show up every year in the same place! If you suspect buried debris, probe the soil to a depth of about 6 in (15.25 cm) with a slender, pointed tool (the soil should be moist). If you hit something, dig up the turf and remove the debris.

FIGURE 3–4   Buried debris (plaster) from building contractor.

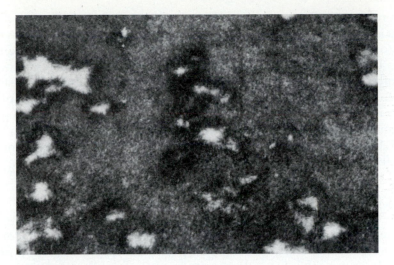

FIGURE 3–5   Injury from dog urine.

3.6 DOG INJURY

Dogs (and other large animals) and high-quality turf do not mix. Although this is unfortunate for the dog-loving segment of our society, the fact remains. Urine-soaked spots can rapidly desiccate turfgrass tissue, killing the foliage (figure 3–5). Urine, especially in the same spot and during the colder times of the year, can kill the crown and all tissues above ground. The more-or-less-circular patches are from several inches to 2 ft (0.6 m) in diameter ("big dog"). These spots may superficially resemble diseases such as fairy rings, brown patch, or dollar spot (see chapter 5). When fertility is suboptimal, the spots are often surrounded by a ring of darker green, faster-growing grass due to the release of nitrates from the urine. Although heavy irrigation and overseeding or sodding of the area can promote recovery of injured spots, the only long-lasting control is the curtailment of man's best friend's use of the turf. One compromise is to try to train the dog to use a certain area of the yard in an out-of-the-way location. Fencing the turf, or the establishment and enforcement of leash regulations, is often needed. Symptoms similar to dog injury may be caused by fertilizer spills (sec. 3.3).

3.7 MOSS

Mosses are green plants that form tangled green mats composed of a branched, threadlike growth over the turf surface. Closely mowed and open turfs with thatch are most prone to invasion by mosses. Mosses live at the soil surface and survive as small spores. When the turf becomes thin, thus allowing sunlight to reach the soil surface, and the soil is moist, mosses will grow. In the past, moss growth was reduced by the use of broad-spectrum pesticides, but many of these are no longer available. Thus, moss has again become a problem for turf managers. These plants grow in neglected turf areas low in fertility, with poor soil aeration and drainage. Soil that is highly acidic, excessively shaded, improperly watered, or heavily compacted (or suffers from a combination of these factors) encourages the growth of mosses. The best control is to keep the grass growing vigorously following the steps outlined for algae (sec. 3.8) and compacted areas (sec. 3.24). Other useful controls include raising the height of cut, controlling thatch, adjusting the pH to between 6 and 7, and establishing a more vigorous and better adapted turfgrass species or cultivar. Moss can be removed by hand raking, and the site can be made unfavorable for moss growth by treating, when first seen, with ferrous ammonium sulfate, ferrous sulfate, or ammonium sulfate.

3.8 ALGAE

Algae are a group of small, primitive, filamentous green plants that manufacture their own food. Algae and mosses are both plants, but the algae are more primitive and simple, as they do not form leaves, stems, or flowers. There are many types of algae, and they will often appear to have different coloration. Algae, like mosses, survive at the soil surface and, under conditions of light exposure, water, and moderate temperatures, they will grow. Occasionally their growth will be so rapid it appears to explode over a single night. This is known as an algal bloom, which signals the rapid growth and release of algal propagules. Under very wet surface conditions, they can form a thick, slimy, greenish-to-brownish scum on bare soil or in thin turf (figure 3–6). High fertility and a weak, thin turf encourage their growth. Algae occur in low, shaded, heavily tracked, or compacted turfgrass areas. Algal scum can dry to form a tough black crust that later cracks and peels. Algae may also induce iron-deficiency chlorosis in plants.

To prevent algal scums, establish a thick stand of healthy, vigorous turf. Provide for good surface and subsurface drainage when establishing new turf areas. Avoid frequent waterings, especially in late afternoon or early evening. Avoid overwatering and waterlogged soil. Aerify compacted turf areas using hand or power equipment. Eliminate heavily trafficked areas by providing walks, paths, or traffic guidelines such as fences or shrubs. Increase light penetration and air movement over the area, and speed drying of the turf surface by thinning dense trees or shrubs in nearby areas. When designing the landscape, space plantings to avoid too much shade. In partial shade, use shade-tolerant turfgrasses or ground covers that persist well in low-light situations. Algacides or fungicides that control algae can be applied to the area as recommended. Correct the pH if necessary. Applications of dilute solutions of sodium hypochlorite or liquid household bleach (0.01%), copper sulfate (2 to 3 oz per 1000 sq ft), or some fungicides (see table 5–5) will help control algae when combined with the cultural practices outlined above.

3.9 SUMMER DROUGHT OR DESICCATION

Dry winds during the summer months when air temperatures are above 80°F (26.7°C) can rapidly wilt and desiccate turf areas (figure 3–7). When water is applied shortly after wilting first appears, prompt recovery usually results. Common

FIGURE 3–6 Algal scum in a thin bentgrass turf.

FIGURE 3–7 Injury from summer drought or desiccation.

FIGURE 3–8   Tall fescue growing in drought-stricken Kentucky bluegrass.

turfgrasses with medium to excellent resistance to drought include buffalograss, bermudagrass, zoysiagrass, bahiagrass, most fescues, Kentucky bluegrass, and perennial ryegrass. Bentgrasses, St. Augustinegrass, centipedegrass, and fine fescue are less resistant to drought. Deep-rooted species such as tall fescue, bermudagrass, and zoysiagrass maintain their color and growth better than most other desirable grasses before summer dormancy from drought occurs (figure 3–8). These species are able to utilize water from a greater volume of soil, thus taking longer to deplete soil reserves.

For high-quality turf, a temporary reduction in the surface temperature results in reduced water transpiration, minimizing the amount of wilting. Surface temperatures can be reduced for a short period through the application of small amounts of water. This method is called syringing. The effectiveness is short-lived, however. To aid in minimizing the extent of wilt through drought, promote a deep root system through infrequent, thorough watering. Apply soluble nitrogen sparingly during stressful summer months to reduce shoot growth. Reduce compacting forces, such as foot and vehicle traffic, to minimize turf injury.

3.10   SCALD AND HEAT INJURY

Sudden and intense heat will kill turf. The time required for excessive heat to kill grass tissue can be less than a second. Scald can occur at temperatures greater than about 104°F (40°C). The amount of heat (highest temperature and length of exposure) determines the severity of the scald. For example, turf flooded with water heated above 104°F (40°C) by sunlight may take days to kill. Long before death, however, the turf will be weakened and lose regrowth potential. The extreme heat from engine exhaust can kill turf in just a few seconds. Another common scald can result from scalping the turf. If turf is mowed close to the crowns, and the sun is intense, the turf will be scalded. Sunlight heats up the crown or young leaves growing from the crown. About 3 to 7 days after the turf is scalded, white bands about ⅛–¼ in (0.3–0.6 mm) in length will appear across the leaves. These might be confused with dollar-spot lesions, but they have no darker border. These bands are leaf tissues that were killed or scalded after the turf was scalped. To prevent heat damage to turf, simply do not allow the turf to be overheated. Symptoms of mild scald can be removed with mowing, but severely scalded turf dies and needs to be replaced.

3.11 WINTER DRYING OUT OR DESICCATION

Winter desiccation (*Color plate 4*), which causes the leaves to turn white to brown in affected areas, is most severe on elevated sites and sloped areas where the grass is exposed to drying winter winds and there is little accumulation of snow. These areas also have a large surface runoff and low rates of infiltration. When air temperatures are above freezing and the soil is frozen, water is lost from foliar tissue, causing desiccation and death. In general, the severity of winter desiccation is governed by the winter air temperature, exposure of the turf during the winter, and the wind (speed and duration). The colder the temperatures, the less moisture is held in the air, and, consequently, the more the drying effect on the turf. The faster and longer the winds blow across a turf, even when it is dormant, the greater the desiccation that results. The less a turf is exposed to drying winds, the better. Snow is a natural insulator for turf during the winter. Turf under snow remains moist and may be 40°F (4.4°C), even on the coldest of days. Turf blankets, covers, and straw also protect turf from winter desiccation. The full effects of damage from winter desiccation cannot be fully determined until the turf resumes growth in the spring. All insulating materials, including snow, can also predispose turf to snow-mold disease (chapter 5).

A certain amount of winter desiccation occurs in turfs annually. This is commonly called wind burn and results in the leaves turning some shade of brown from the tip downward. The brown leaves are mowed off during the first two or three mowings in the spring, and little damage remains. However, a severe winter drought may dry out and kill the crowns and nodes of turfgrass plants, resulting in serious winter kill.

To help prevent winter desiccation, irrigate dry soils thoroughly to a depth of 6 to 8 in (15 to 20 cm) in late autumn or early winter. Help to ensure uniform infiltration on sloped areas by removing excessive thatch and alleviating compaction. Fertilize lightly during the last 6 weeks of the growing season to allow a hardening-off period to take place. Turfgrasses should enter winter dormancy at a moderate nutritional level, or snow-mold damage (see chapter 5) or other severe injury may result.

3.12   *HEAVING*

Soil saturated with water and exposed to long periods of freezing temperatures can heave. Heaving is a process that causes soil and plants to be pushed up and out of the ground by the action of freezing water, which expands when it is frozen. This can be most severe when periods of freezing are interrupted by temperatures above 32°F (0°C), which allows more free water to filter into the soil to freeze and expand. The forces generated by freezing water are tremendous and much stronger than tiny grass roots. Repeated freezing and thawing from late autumn into early spring shears off turfgrass roots and pushes the grass plants up to an inch or more above their normal position in the soil. Crowns and roots are exposed to desiccation. Seeds planted late in the fall will not produce well-rooted plants. These are very susceptible to heaving damage. To minimize injury, apply a light surface mulch of straw, and possibly erect snow fences or other barriers to promote snow accumulation. Established turfs are seldom damaged by heaving. Control by lightly rolling the area in early spring, and do not let the weakened plants become water stressed.

3.13   *ICE INJURY*

Layers of ice for prolonged periods may cause suffocation of turf, even though respiratory processes are greatly reduced during freezing weather. Most injury occurs when standing water covers the turf and then freezes. Should this happen, the turf may winter kill. Unless the ice layer remains in place for more than 3 months, removal will probably not be necessary. When possible, drain any excess water from the ice as it thaws. Fortunately, well-adapted, perennial, cool-season turfgrasses are seldom injured by ice covers.

3.14   *LATE-SPRING FROSTS*

Cool temperatures that kill new spring growth of warm-season grasses destroy food reserves stored in the plant. Several severe frosts, separated by warm periods that stimulate new growth, can completely exhaust food reserves, killing grass plants. When less severe, the root systems do not develop sufficiently to absorb water and minerals, resulting in thin, nonthrifty turf. If not too severe, the grass gradually recovers.

3.14.1   Frost Damage

Frost damage can occur in both the spring and fall, but it is more common in the fall, when the weather is cooling off. Frost is the formation of ice crystals on the surface of leaves. A related condition results from the freezing of water inside turf tissues. In both cases, the ice crystals form aggregates that are sharp and damaging to the cells that make up turf leaves. In the case of frost, the ice crystals can rip and puncture the turf, resulting in localized leaf death. For example, if there is frost on a turf, and wheels, feet, or paws move across the turf then injury occurs and appears as tan, dead areas. In the case of frozen water in superhydrated leaves, the leaves can burst and die even without external pressure. In the spring, frosts (or temperatures slightly above freezing) may not kill warm-season grasses, but the grasses may be severely weakened. Once warm-season grasses resume growth in the spring, they are using limited stored nutrients from the previous growing season. If the turf is exposed to cold conditions, it may stop growing and attempt to go into dormancy. If temperatures then increase, the plant again tries to resume growth. This stopping and starting can deplete the plant's resources to the point of death.

To reduce frost damage, do not allow traffic on frosted turf. Turf that must be used can be irrigated to warm it up and remove the frost. Do not overwater the turf in the fall. Allow the turf to acclimate to winter conditions. Do not overfertilize late in the winter, as fertilization tends to produce turf leaves with more water. Generally, frost damage is not going to result in the death of the crowns, roots, rhizomes, or stolons; thus, the turf recovers from this type of damage as the leaves grow.

3.15 AIR POLLUTION

The major worldwide source of energy is the combustion of fossil fuels. This combustion, with the release of heat, is basically the combination of oxygen and carbon in the fuel source. Along with heat, many undesirable by-products are also produced. The concentration and occurrence of air pollutants are highly dependent on the prevailing climate; thus, damage to turf is rare and nonuniform.

Several air pollutants, including ozone, peroxyacetyl nitrates (PANs), sulfur dioxide, nitrogen dioxide, fluorides, chlorine, hydrogen chloride, ethylene, and toxic dusts, can cause turfgrass leaf blades to become bleached, chlorotic, or died back from the tips or margins. Other symptoms include bands across the leaf blades, a yellow-to-brown striping of the leaves, a glossy brown discoloration before the leaves die, or reduced foliar growth without visible symptoms. Ozone can be absorbed through the leaves, causing black spots, flecking, the loss of chlorophyll, accelerated aging, and death of leaf blades. Injury from air pollutants is most common near industrial and urban areas where electric power plants, incinerators and refuse dumps, pulp and paper mills, smelters, refineries, a variety of chemical plants, and much-traveled highways are located. Turfgrass species and cultivars differ greatly in symptom expression and growth responses to the various air pollutants. Turfgrasses usually recover after a mowing or two unless there are a series of air-pollution alerts.

3.16 SALT INJURY

The application of salt (sodium chloride or calcium chloride) to roadways, walks, and drives can injure adjacent turf. With adequate drainage, most salts can be flushed through the soil profile. If salt injury persists, the use of salt-tolerant turf species can help minimize damage. One of the most salt-tolerant grasses is alkaligrass (*Puccinellia distans*). Other salt-tolerant species include bermudagrass, creeping bentgrass, and St. Augustinegrass; Kentucky bluegrass, red fescue, and colonial bentgrass are salt-intolerant species.

Where the major salt-bearing ion is sodium, it may be necessary to first deflocculate the soil before salts can be leached from the turf root-zone profile. This procedure requires the application of gypsum (calcium sulfate) to chemically exchange calcium for sodium ions. The sodium ions can then be leached from the profile, alleviating the damaging salted condition.

Three classes of salted soils are:

1. saline (EC > 4 μmhos/cm; ESP < 15% of CEC)
2. sodic (EC < 4 μmhos/cm; ESP > 15% of CEC; ph > 8.5)
3. saline-sodic (EC > 4 μmhos/cm; ESP > 15% of CEC)

Total soluble salts are estimated by measuring the electrical conductivity (EC) of the soil solution. Below 250 μmhos/cm, the scale indicates a low-salinity hazard. The exchangeable sodium percentage (ESP) is the percentage of total cation exchange capacity (CEC) consisting of sodium ions. Saline soils are corrected by leaching. Sodic soils are corrected with the addition of gypsum.

3.17   SCALPING INJURY

Severe injury can occur when large portions of the turf canopy are removed by a single mowing. This generally occurs when more than half of the green leaf tissue is removed. Radical scalping of the turf reduces its photosynthetic capacity to a level that cannot sustain the remaining tissue. Scalping injury generally appears as yellowing to death of the remaining tissue. It follows the mowing pattern and is most evident in taller areas where scalping can be more severe. In most instances, turf eventually recovers from the injury if the roots and lower crowns are kept actively growing. As with most physical injuries, scalping injury can lead to additional damage that eventually requires reestablishment.

Scalping most often occurs when correct mowing frequency is not followed. When wet weather prevents timely mowing, even at the original height, scalping injury can occur. To minimize scalping, raise the mowing height for the next mowing; then slowly lower it on subsequent mowings.

3.18   DEEP THATCH

3.18.1   Symptoms

Thatch is a tightly intermingled layer of living and dead stems, roots, rhizomes, plant crowns, and other plant parts that develop between the green actively growing foliage and the soil surface (figure 3–9). Thatch is not of uniform composition, however. Generally, the upper layers of thatch represent the most recently contributed plant materials, often with visible plant parts that are highly distinguishable. Lower levels of thatch are generally a dark-colored, spongy, humus material with no distinguishable plant parts. This lower portion of the thatch, generally called mat, can be troublesome

(a)

(b)

FIGURE 3–9 Thatch in Kentucky bluegrass: (a) close-up in clay soil; (b) thatch layer in horse-track turf.

if allowed to accumulate in layers greater than ½ in for most turfs. The thickness of thatch that can be tolerated depends on general use of the turf and the species. Finer textured turfs generally require or at least tolerate thinner layers of thatch. Coarser textured turfs can withstand slightly thicker thatch layers. Some thatch is highly desirable to provide cushioning and resilience for traffic and other compacting forces. Thatch, like other organic materials, has a relatively high cation exchange capacity and the ability to absorb moisture. These features help turf to maintain healthy growth during reduced moisture conditions and between fertilizations.

Although some thatch is desirable, an excessive layer is not conducive to deep turf root and rhizome growth. Thatch, under high summer temperatures, can rapidly dry out, creating extremely droughty conditions. If the majority of the turf roots reside within the thatch layer, a rapid turf loss can occur. Entire turfs have been lost in 1 or 2 days under these conditions. Thatch also accelerates the incidence of some destructive diseases such as necrotic ring spot and summer patch.

3.18.2   Thatch Management

 Basic management is provided by reversing conditions that promote thatch accumulation. Thatch accumulates when the forces that produce thatch material are greater than forces that promote its disappearance. This happens slowly, over an extended period of time, or rather rapidly over shorter periods. A major factor that tends to promote accumulation of thatch is the selected species and cultivar. Within each species, individual cultivars have a greater tendency for producing thatch than others. Generally this is associated with the production of materials resistant to microbial breakdown such as the lignins and hemicelluloses. Table 3–2 shows the relative tendency of common turf species for producing thatch. In addition to the genetic potential, management operations that increase plant growth such as excessive fertilization, irrigation, and so on, provide a more rapid accumulation.

Accumulation can also occur under normal production levels when forces are present to reduce the degradation of thatch material. Acidic or alkaline soils, or the

TABLE 3–2   RELATIVE TENDENCY OF TURF SPECIES TO ACCUMULATE THATCH

High	Medium	Low
Bermudagrass	Bahiagrass	Buffalograss
Creeping red fescue	Carpetgrass	Centipedegrass
Zoysiagrass	Creeping bentgrass	Perennial ryegrass
	Chewings fescue	Tall fescue
	Colonial bentgrass	
	Hard fescue	
	Kentucky bluegrass	

lack of macroorganisms (such as earthworms) or certain microorganisms, can provide an environment of low thatch degradation.

In most cases, control can be obtained through (1) increasing the opportunity for microorganisms to degrade thatch more rapidly than it accumulates; or (2) physically removing materials from the turf. Greater soil microorganism activity occurs when soil is incorporated into the thatch material. Adjusting the pH toward neutrality and removing excessive moisture also helps. All of these can be achieved through aerification and/or topdressing. Core aerification is an extremely effective mechanism for thatch reduction.

Physical thatch removal is generally provided through vertical mowing operations and subsequent removal of any thatch that has been moved to the turf surface. Although this operation is more rapid than aerification or topdressing, it does not enhance growing conditions and can be quite destructive to the turf. Vertical mowing operations should only occur when the turf is actively growing.

3.19 *HAIL INJURY*

Injury to turf by hail during a thunderstorm is generally physical and short in duration. The impact of this damage depends largely on the use of the turf and the selected species. For most turfs, hail does not represent a great threat to overall health and usefulness. For highly specialized turfs such as golf course greens, hail can severely damage playing surfaces in a short period of time.

3.19.1 Recovery

As with most physical damage, immediate repair of damaged areas minimizes the long-term impact. Hail injury on a putting green can be handled similarly to ball injuries. If not too severe, physical repair of individual hail marks will satisfactorily minimize damage. After very severe hailstorms, reseeding or refurbishing the green surface might be necessary. Whereas hail damage occurs during the storm, sand greens can often suffer desiccation if high winds prevail after a storm. Caution should be taken to prevent moisture loss.

3.20 *LIGHTNING DAMAGE*

Lightning can strike the ground directly. Simply put, lightning is intense electrical energy and, as such, produces heat, electrical current, and light, all of which can affect the living components of your turf (plants, microbes, and animals). In general, a lightning strike causes the turf to be stunted and off-color in a large circular patch (5 to 20 ft diameter [4.5 to 6 m]). There is little you can do to prevent this, and the chances of it happening are unlikely. If it does happen, continue normal maintenance, and the turf will recover.

3.21   NUTRIENT ABNORMALITIES

Insufficient or excessive quantities of available nutrients (figure 3–10) can cause abnormal changes or injury in turf. Nutrient deficiencies are generally found more often in highly maintained turf. These deficiencies appear on turfs with modified root zones such as golf course greens, tees, or sports fields. The lack of nutrient-holding capacity in a coarse soil such as sand can lead to a loss of nutrients. While only several of the 16 required plant nutrients are the cause of even a moderate number of deficiencies, most occurrences are directly related to the soil pH. Figure 3–11 illustrates the availability of nutrients across a range of soil pH values. When the soil pH is much greater or less than 3.0 to 7.0, even minor nutrients can become excessive or limiting.

The most frequent nutrient deficiency, of course, is nitrogen deficiency, which shows up as pale green, yellow, or chlorotic tissue typically in the lower, older leaves. It is most common in coarse-textured soils or after excessive leaching through rainfall or irrigation. Iron, potassium, and sulfur also present potential deficiencies, but at a much lower rate. Like nitrogen deficiency, iron deficiency appears as a chlorosis or yellowing of the plant. Iron deficiencies are also more evident in tissue between veins, leaving the veins slightly darker green. Potassium deficiency shows up as a yellowing of older leaf tissues and a loss of turgor. Leaves tend to be soft and droopy. Although nutrient deficiencies are rare, they are best determined through soil testing. It is a good idea to test coarse-textured soil on an annual basis to discover any changes in nutrient levels.

FIGURE 3–10   Injury from excessive sulphur application.

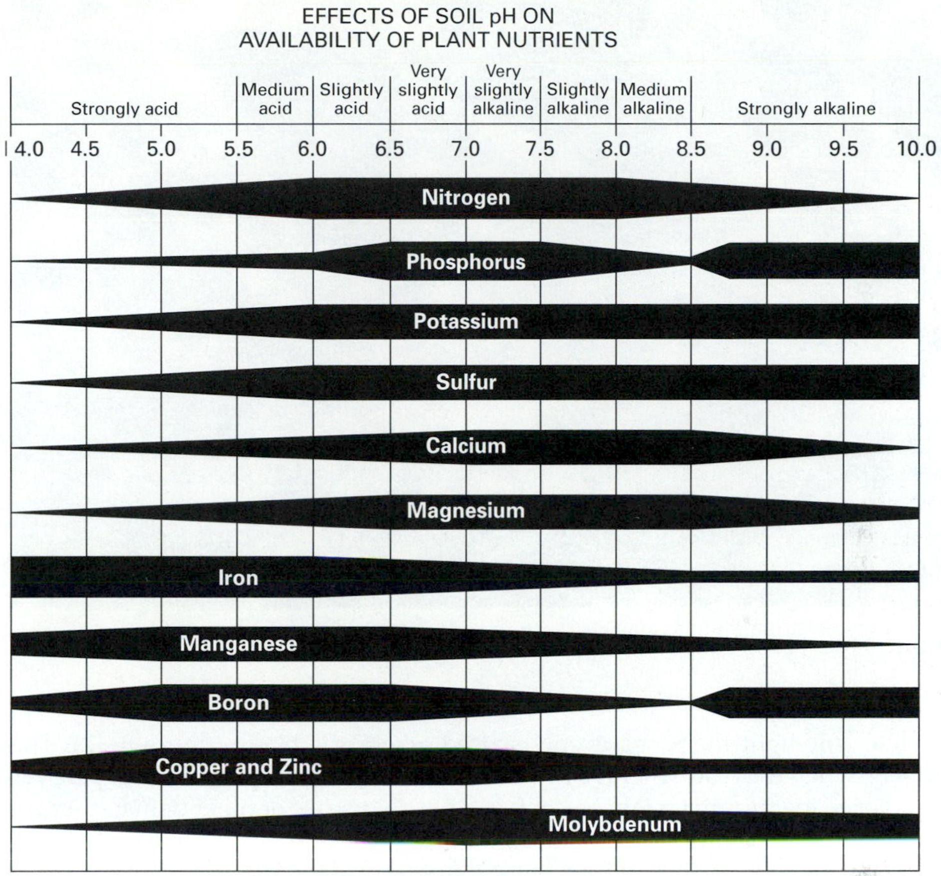

EFFECTS OF SOIL pH ON AVAILABILITY OF PLANT NUTRIENTS

FIGURE 3–11   Effects of soil pH on the availability of plant nutrients.

3.22 DULL-MOWER INJURY

When mowing equipment is out of adjustment, or cutting surfaces are dull, turf injury can occur. Dull-mower injury often appears as a light-to-fuzzy-tan or gray appearance of the turf; this is the remaining vein tissue that has not been cut. Typical damage from a reel mower with dull blades is shown in *Color plate 2*. The greater the intensity of management, the more likely the injury. Injury is more frequent on those species of turf most difficult to cut. Perennial ryegrass, tall fescue, and zoysiagrass are reasonably susceptible to mowing injury due to their clipping resistance. For these species, reel mower blades should be sharpened at frequent intervals, and equipment should operate at high enough speeds to provide a clean cut. Older cultivars, particularly perennial ryegrasses, are more susceptible to injury than newer cultivars that have been selected for ease of mowing. Once the mowers are properly sharpened and adjusted, this material can be cleanly cut.

Reel mowers present an additional problem. With improper adjustment of the reels or dull cutting surfaces, injury can occur to individual leaf blades with

FIGURE 3–12   Injury to tall fescue from an improperly adjusted reel mower.

characteristic light markings evenly spaced across the blade (figure 3–12). This happens when the blade is grabbed but not severed, creating a whitish to pale yellow color to the turf that might be confused with a disease or other damage. Again, proper adjustment and sharpening of clipping surfaces alleviates this problem.

3.23   ABRASIVE DAMAGE

Abrasion of plant tissue can occur in many ways, including dull mowing equipment. Most often, abrasion damage is the result of extreme wear from vehicle or foot traffic in small, concentrated areas. Sharp turns on delicate surfaces such as greens or tees can tear and abrade tissue. This damage generally cannot be repaired—one must wait for the eventual regrowth of the turf. Turf managers can avoid abrasion damage by making mower turns off the playing surface and avoiding sharp turns. Scuffing by players or other vehicles is often inevitable. Additional spot irrigation can help to minimize water loss from damaged surfaces.

3.24   SOIL COMPACTION

Soil compaction can predispose turf to disease problems because it interferes with root performance and general turf growth. This, in turn, causes turf to grow slower, recover from disease more slowly, and lose density. Simply put, compaction is a reduction in the number of large pores in soil. Compaction does not necessarily mean a reduction in total pore space. Soil is composed of sand, silt, and clay particles. The amounts and arrangement of these particles in soil dictate both the number and size of pores. The pores in soil are the channels for air, water, and microbes. As these change to smaller and smaller pores, the rate and availability of oxygen and moisture are reduced. Hence, clay soils, while holding the most water, make it difficult for turf to get sufficient water or oxygen. When this happens, the roots become shallow and, as such, are predisposed to more rapid changes in heat and moisture because they are near the soil surface. Deep roots do not experience the rapid climatic changes common at the soil surface. In general, compacted soils are not well suited for soil-borne microbes, either. However, turf growing in compacted soil is weak; thus, pathogens that attack in the upper root zone, thatch, crowns, and leaves do more damage. In addition, diseased turf growing in compacted soil recovers more slowly from diseases. In short, soil compaction is not good for the health of the turf or disease management.

Not all soils are susceptible to compaction. For example, sand root zones compact very little. Mineral soils are more likely to compact and, as the amount of clay is increased, the amount of compaction also increases. To prevent compaction, reduce the traffic on susceptible soils, or change the soil texture to one less susceptible to compaction. To reduce compaction, increase the amount of large soil pores. This can be done physically by creating large pores (aerification), or biologically by encouraging the activity of plants, animals (such as earthworms), and microbes.

3.25 ROOT COMPETITION

Trees and shrubs, like turf, produce roots that absorb water and minerals. Trees and shrubs differ in how roots are formed. Some are shallow, others deep. Shallow roots compete with turf for water. Sometimes the turf will win, and other times the tree or shrub will get the upper hand, but, in either case, planning and good management must be used when turf grows in close proximity to trees and shrubs. The effect of this competition on turfgrass pests is one of stress. Roots of trees and shrubs do not attract more pathogens or other pests.

Therefore, select trees and shrubs that are compatible with turf growing near them. In addition, do not stress the turf or trees and shrubs. This reduces the need for the turf to consume all the water near the soil surface and also encourages deep rooting by all plants concerned.

3.26 LOCALIZED DRY SPOT

This condition, mostly of high-sand-content root zones and mineral soil greens heavily topdressed with sand, is caused by soil-borne fungi (basidiomycetes). Localized dry spots are common in new golf greens abundantly topdressed with sand 2 or 3 years after seeding or sprigging. Soil within the patches remains very dry (and will not wet), even after rainy weather or frequent irrigation. Water does not penetrate the thatch soil layer and flows onto adjacent, unaffected turf. Hot and dry periods predispose turf to this problem.

Solid, more-or-less-circular or irregular patches of wilted, stunted, and off-color turf, first noticed when they are several feet (0.5 to 1 m) or more in diameter, are the most common symptoms. Large, irregularly shaped areas of wilted or dead turf may also occur. The turfgrass plants can eventually die from drought stress. Sometimes the presence of numerous mushrooms or fairy-ring development precedes the appearance of dry spots (see the section on fairy rings, 5.2.10).

The fungi that produce localized dry spots live in the soil and feed on dead turfgrass debris in the thatch. In this way, these fungi are beneficial to the turf. As this process continues, older mycelia are believed to die and release chemicals that coat individual sand particles with organic material, causing the sand particles to bind together, thus preventing water from passing through the thatch soil interface. This condition is normally restricted to the upper 1.5 to 2.5 in (3 to 6 cm) of soil where the fungi are found. After the fungi thoroughly colonize the thatch (which often surrounds the roots, rhizomes, and crowns), it becomes hydrophobic, much like a dried sponge. This is why hot, dry soil intensifies the severity of localized dry spots.

Practice dethatching and core cultivation to ensure water infiltration. Combine this with applications of a wetting agent, together with verti-draining and water-injection cultivation, every week or two during dry periods in the summer. Daily syringing of affected areas is also needed. Some turfgrass managers treat isolated, individual dry spots by frequent injections of water using a deep-root tree feeder. In addition, fungicides effective against basidiomycetes may be available, but using them could result in greater thatch accumulation.

3.27   STUDY QUESTIONS

1. Name five problems, biotic or abiotic, that display chlorotic symptoms or chlorosis in turf.
2. How do chemical burns occur on a leaf blade?
3. Why does tall fescue stay green in a dormant Kentucky bluegrass turf?
4. What is scalping injury? How can it be prevented?
5. How would you minimize the damage caused by spilled hydraulic fluid?
6. What adjustments in management would help to reduce the development of moss?
7. What are the three classes of salted soils, and how can they be corrected?
8. What is one of the major causes of localized dry spots?

3.28   SELECTED REFERENCES

Beard, J.B. *Turfgrass Science and Culture.* Englewood Cliffs, NJ: Prentice-Hall, Inc., 1973.

Beard, J.B. *Turf Management for Golf Course,* 2nd ed. Chelsea, MI: Ann Arbor Press, 2002.

Brunneau, A.H. (ed.). *Turfgrass Pest Management: A Guide to Major Turfgrass Pests and Turfgrasses.* Raleigh, NC: North Carolina State University Agricultural Extension Service Publication AG-348, 1985.

Christians, N. *Fundamentals of Turfgrass Management.* Chelsea, MI: Ann Arbor Press, 1998.

Daniel, W.H., and R.P. Freeborg. *Turf Managers Handbook.* New York: Harvest Publishing Company, 1979.

Hanson, A.A., and F.V. Juska (eds.). *Turfgrass Science.* Agronomy Monograph 14. Madison, WI: American Society of Agronomy, Inc., 1969.

Turgeon, A.J. *Turfgrass Management,* 5th ed. Englewood Cliffs, NJ: Prentice-Hall, Inc., 1999.

Watschke, T.L., P.H. Dernoeden, and D.J. Shetlar. *Managing Turfgrass Pests.* Boca Raton, FL: Lewis Publishers, 1994.

CHAPTER

4

Biology and Management of Weeds in Turfgrasses

The prime objective in managing a turf is to establish and maintain a uniform living plant cover for recreational use and/or aesthetic enjoyment. Uniformity of color, leaf texture, and plant density of the turfgrass stand is of the utmost importance for consistent playability and enjoyment. To achieve uniformity, a select group of plants must exist throughout the turf and grow at similar rates.

Any other plant that might grow and persist in the turf would, for the most part, deviate from the color, texture, or density of the turf. These plants are considered weeds in the turf, even if they are turfgrass species! Weeds detract from the overall appearance of the turf and compete for light, water, minerals, and space, thus reducing the vigor of the turf.

Weeds in turf, like all plants, have a minimum set of requirements for their establishment and growth. Turfgrasses are excellent competitors that, under optimum growing conditions, will often outcompete weeds for necessary resources.

4.1   TYPES OF WEEDS

Plants from many diverse families comprise the weeds that are found in turf. Turfgrass weeds can be grouped into two major categories: (1) plants that are true grasses or appear grasslike, and (b) broadleaf weeds. The term *broadleaf* can be misleading. Several weed species grouped in this classification have narrow leaves and somewhat resemble grasses. Their control, however, is similar to other broadleaf weeds.

Weeds are also categorized by their life cycle and growth habit. Annual weeds complete their life cycle in one year. Biennials complete their life cycle in two years, developing to partial maturity during the first growing season, with flowering and eventual death during the second season. Perennial plants can persist over long periods of time, usually overwintering in a dormant state. These species resume growth the following season from stored energy. They may also reproduce from seed.

Weed taxonomy and biology are used extensively in determining effective control methods. In general, the more closely a weed's growth habit resembles that of the chosen turfgrass or mixture of turfgrasses, the more difficult its control.

4.2   WEED BIOLOGY

Annual weeds can be classified by the season during which their seed normally germinates. Winter annuals germinate during late summer or fall; persist as an immature plant throughout the winter months; and produce more vegetative growth, flowers, and seed in the spring (figure 4–1). Summer annuals germinate in the spring and produce seed in summer or fall (figure 4–2). Biennial plants may have seeds that germinate in either the spring or fall (figure 4–3). Perennial plants persist over many growing seasons (figure 4–4).

WINTER ANNUAL

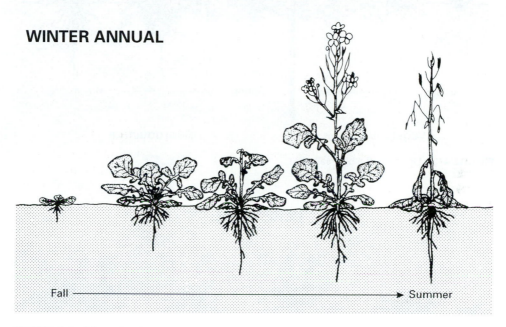

FIGURE 4–1   Life cycle of a winter annual weed.

SUMMER ANNUAL

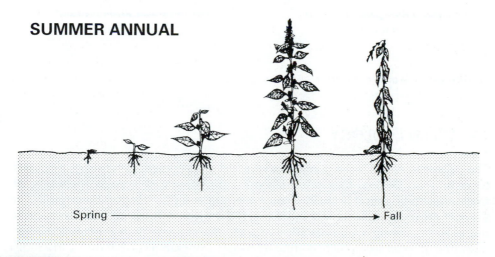

FIGURE 4–2   Life cycle of a summer annual weed.

BIENNIAL

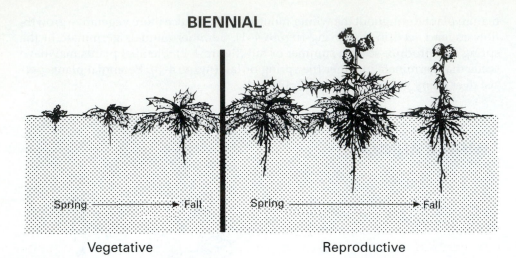

FIGURE 4–3   Life cycle of a biennial weed.

PERENNIAL

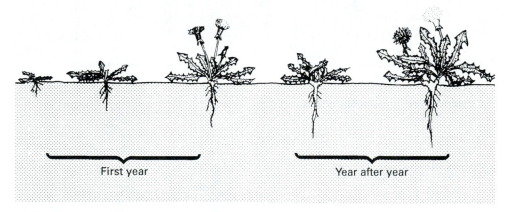

FIGURE 4–4   Life cycle of a perennial weed.

4.3   WEED ECOLOGY

The degree to which a weed adapts to the turf environment determines its persistence in a turf. Several factors influence the growth of each species, including climate, soil, and cultural or maintenance operations. Cultural operations applied to the turf during routine maintenance exerts the greatest influence on weed persistence. For example, overapplication of water during irrigation promotes the germination of weed seeds such as annual bluegrass or crabgrass that thrive in a moist habitat. The timing and intensity of aerification or vertical mowing can influence the germination of these same weeds. The turfgrass manager also has limited control over existing soil conditions. Soil fertility can affect the growth of certain weed species. Annual bluegrass responds positively to high levels of soil phosphorus.

Climatic factors also influence weed populations. Weeds and turfgrasses have specific zones of adaptation. These are largely regulated by climate. In addition, dormancy mechanisms of each weed species influences its role in the turfgrass community.

Dormancy is a physical method by which a plant persists during periods of stress. To survive stressful periods, turfgrass weeds have developed to allow the plant to resume a full growth rate when conditions return to normal. Through the production of seeds or vegetative organs such as stolons, rhizomes, tubers, corms, or bulbs, a weed can persist at a minimal metabolic rate, preserving the necessary tissue and food reserves for regrowth after the dormant period.

Annual weeds are prolific seed producers, ensuring future generations of the species. Perennial weeds, while producing some seed, may rely more on storage organs to persist from season to season. Several conditions can prevent the immediate germination of weed seeds. These include a hard seed coat that mechanically prevents the imbibition of water or expansion of the seed within the seed coat. Some species produce inhibitors that must be leached from the seed coat before germination can begin. Other turfgrass weed species require short periods of light exposure to initiate the germination process. Some species shed seed before it is fully mature, requiring an after-ripening period before germination can occur. Seeds of certain species have been shown to survive in soil for long periods of time. Lotus seeds, found in a lake bed in Manchuria, were approximately 1000 years old and still viable and capable of germinating. Several turf weeds, such as green foxtail and curly dock, have germinated after 38 years of dormancy.

4.4 INFLUENCE OF WEEDS ON TURF

The greatest influence weeds exhibit on turf is through direct competition for water, light, space, and nutrients. Weeds also influence turf in other important ways. Turf injured from insects or other animals, diseases, excessive wear, compaction, or misuse of fertilizers or pesticides results in weakened turf and thin or bare areas. Weeds often occupy these areas more rapidly than the desired turfgrass. Loss of turf cover greatly reduces overall quality and the turf's ability to compete.

If weeds persist for long periods, they can often be a good indicator of unsuitable growing conditions for the turf. A physical soil problem (compaction, poor drainage or structure, and the like) can often prevent turfgrasses from obtaining a competitive edge, allowing the further development of weeds. An example is the extensive development of prostrate knotweed in severely compacted soil. Prostrate knotweed can persist and spread under conditions of low oxygen potential, typically found in compacted soils. Most turfgrasses, however, do not grow well under these conditions. Therefore, the persistence and growth of prostrate knotweed in an area might indicate a compacted condition. The alleviation of compaction through core aerification can often reverse the balance, giving the turfgrass a new competitive edge.

Weeds, during normal development, can produce chemicals that, when exuded into the soil solution, are toxic to surrounding turfgrass plants. These allelopathic compounds can have a tremendous impact on the balance of plant species in a given area. For example, chemical compounds produced by members of the walnut family can strongly inhibit turfgrass growth. Research is being conducted to examine the effects of allelopathic compounds produced by weeds and turfgrasses that might influence turf development. Initial experiments would suggest that turfgrass species such as perennial ryegrass have the ability to inhibit the growth of other grasses. The collection and concentration of these compounds might someday prove to be beneficial for direct weed control.

4.5   WEED IDENTIFICATION

 To apply accurate and timely control methods to turfgrass weeds, it is first necessary to identify the species of weeds present. First, determine whether the weed is a grass or a nongrass broadleaf plant. All grasses have relatively narrow (longer than wide) leaves with veins running parallel to each other. Single simple leaves in alternate groups of two arise from the stem, which is generally round or slightly flattened (compressed) but never triangular. Most grasses have a fibrous or multibranching root system with all stems meeting at the crown. Nongrass monocotyledonous plants are often confused with grasses. Wild onion and wild garlic, while resembling grasses, are more closely related to other plants. The sedges and rushes, which are close relatives of the grass family, are, on occasion, controlled in a fashion similar to broadleaf weeds (application of a selective postemergence herbicide).

4.5.1   Broadleaf Weed Identification

Descriptions of broadleaf weeds without using extensive technical terms is difficult. Many broadleaf turfgrass weeds have showy, colorful flowers, which can be used as a simple means of separating these weeds into smaller groups. The identification of broadleaf weeds by floral characters is convenient. Flowers are often not present on turfgrass weeds, so the use of floral parts for identification will be minimal.

It is necessary to use a few technical terms in any basic discussion on plant structure. Just as it is impossible to describe a lawn mower without describing some of the mechanical parts (i.e., reel, bed knife, blade, motor, and so on), it is also impossible to describe a weed accurately without a minimal reference to descriptive terminology. Most of the terms used in the description of individual weeds or in the key (sec. 4.6) are included in the glossary. Some terms used extensively throughout the descriptions of broadleaf weeds help to describe the arrangement of leaves on the stem or the basic shape of the leaf. The arrangement of leaves on stems can be divided into four basic groups (figure 4–5): (1) an alternate arrangement where one leaf arises from the stem and another is attached farther up the stem but on the opposite side; (2) an opposite arrangement where two leaves appear together at the same point on opposite sides of the stem; (3) in some cases,

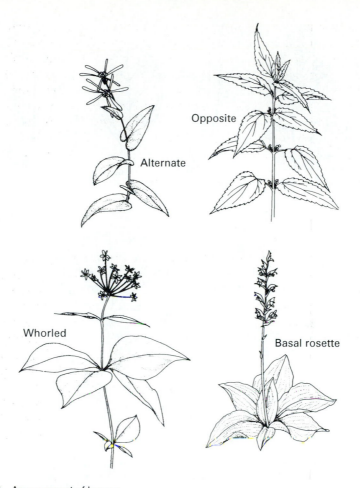

Opposite

Alternate

Whorled

Basal rosette

FIGURE 4–5 Arrangement of leaves.

leaves are found to be whorled, where three or more leaves are attached to the stem at the same site; and (4) leaves arranged in a basal rosette are all attached to the stem near ground level.

A leaf can be described as being simple (figure 4–6), in which it is not divided into smaller portions and forms one mass, or it can be compound (figure 4–7), such as that found on a locust or elm tree, in which a single leaf is the combination of many smaller leaflets. Although the diversity of leaf size and shape for broadleaf plants is great, most broadleaf weeds found in turf can be described by the leaf shapes shown in figures 4–6 and 4–7. figure 4–6 illustrates the eight basic shapes for simple leaves. Unfortunately, a species growing under different conditions can exhibit variations in leaf shape. The shapes illustrated should be used only as a guide in the identification of a species. When a leaf is described as serrate, or toothed, the indentations on the leaf margin are generally uniform and shallow, much like teeth on a handsaw. If a leaf margin is described as lobed, it is generally rounded, varying in the depth of the indentation. Pinnate leaves branch from different parts of the midvein, while

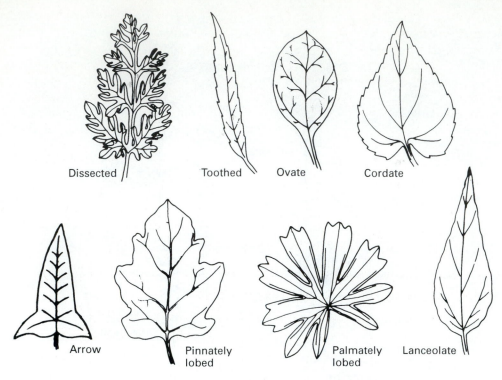

FIGURE 4–6 Simple leaf types.

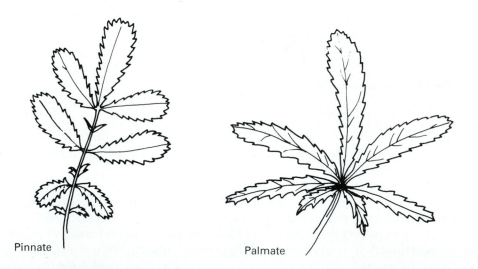

FIGURE 4–7 Compound leaf types.

palmate leaf veins all branch from a central connecting point. A compound, dissected leaf is often mistaken for a pinnately lobed, simple leaf. The difference lies in the relative size of the lobes. If the lobe spaces are as large as, or larger than, the leaf surfaces that they separate, then the leaf is considered compoundly dissected.

When comparisons of vegetative characteristics fail to identify a weed, the arrangements of the flowering portion (inflorescence) of the plant can be helpful in its identification (figure 4–8). Some weeds develop a single flower at the end of a terminal stem, also known as an apical inflorescence. When a number of flowers are attached to the terminal stem, without branching, they are called a spike. If a group of flowers are attached to the main stem by a single shorter stem, they are referred to as a raceme. A plant that exhibits some secondary branching in the flower attachment can be described as a panicle. When the branching always results in a terminal flower, either arising from the apex of the stem or from the axils of leaves, it is described as a cyme arrangement. The final arrangement in figure 4–8 is an umbel, where all branching of the inflorescence arises from the terminal end or apex of stems.

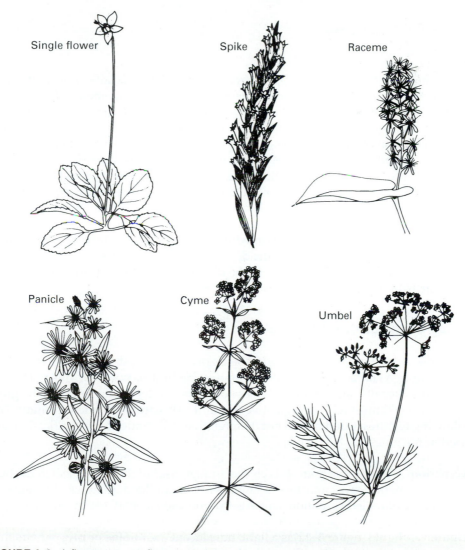

FIGURE 4–8   Inflorescences or flower arrangements.

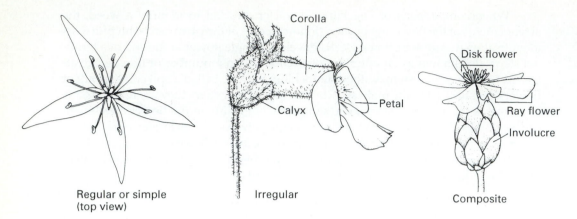

FIGURE 4–9   Parts of a flower.

The flowers themselves also provide clues to the identification of a weed (figure 4–9). If both stamens and pistils are present, the flowers are described as perfect (for instance, bedstraw and prostrate knotweed). If either structure is missing, the flower is considered imperfect (for instance, plantains and pigweed). When a portion of a flower differs in shape, the flower is considered irregular (for instance, henbit and red sorrel). When all portions of a flower are similar or can be dissected into equal halves, the flower is considered regular (for instance, chickweed and moneywort). Petal color is often useful in the identification of a weed. Composite flowers, such as the English daisy or sunflower, while resembling a single flower, actually are comprised of numerous, smaller, individual flowers. Those tightly grouped in the center are called disk flowers, while those flowers with a large petal, radiating around the outside of the disk flowers, are called ray flowers. Seed structures can also be utilized in the identification of weeds in turf; however, these structures have not been incorporated in the key.

4.5.2   Grass Identification Features

If an unknown weed is determined to be a grass, the easiest identification is often made through characteristics of the inflorescence. In turf, however, this portion of the plant is usually not available or does not persist under normal mowing heights.

Thus you must rely on vegetative (figure 4–10) features for identification. The following features, listed in order of their relative importance for identification, are used in the grass key (sec. 4.6.3).

Vernation.   Vernation (figure 4–11) is the arrangement of the young leaf in the bud shoot. In general, leaves may be classified as rolled or folded. Kentucky bluegrass is an example of folded vernation, whereas the bud shoots of tall fescue are rolled.

Ligule.   A ligule (figure 4–12) is a light, translucent membrane or ring of hairs that encloses the stem at the junction of the leaf blade and sheath, remaining fairly uni-

FIGURE 4–10   Parts of a grass plant.

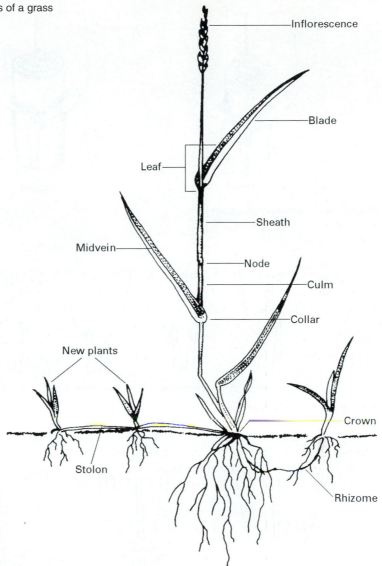

form and consistent within each species. The ligule is on the inside of the blade between the stem and blade. When the ligule is a group or ring of hairs, it is referred to as ciliate. When the ligule is a fine papery membrane, it is considered membranous and can be identified by the shape of its margin. Although the length or height of the ligule varies from one species to another, this character was not used in the key.

Collar.   The collar (figure 4–13) is an area where the leaf blade and sheath join. It is generally lighter in color than the blade. It may be divided into two sections by the midvein. The collar is also distinctively narrow or broad for each grass species.

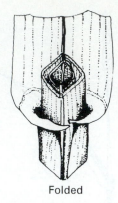

Folded   Rolled

FIGURE 4–11   Vernation of a bud shoot.

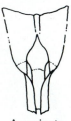
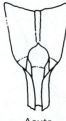
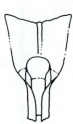

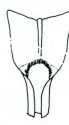
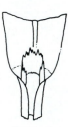

Acuminate   Acute   Rounded   Truncate   Ciliate   Toothed

FIGURE 4–12   Ligule margins.

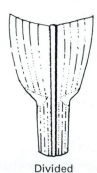

Broad   Narrow   Divided

FIGURE 4–13   Collar types.

Auricles.   Auricles (figure 4–14) are green appendages projecting from either side of the collar to the inside of the leaf blade. They range in size from small stubs to those that fully encompass the stem. Quackgrass is an example of a species with clasping auricles.

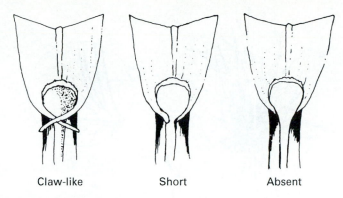

Claw-like   Short   Absent

FIGURE 4–14   Auricle arrangements of grasses.

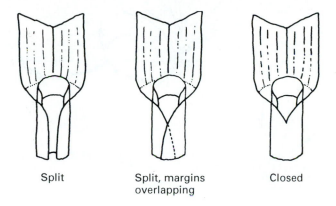

Split   Split, margins   Closed
                    overlapping

FIGURE 4–15   Sheath margins.

Sheath.   The sheath is the lower portion of the grass leaf that usually encloses the stem. It begins at the junction of the blade and collar and extends downward to the next lower node. The sheath is considered a portion of the leaf but generally adheres tightly to the stem or expanding leaves. The margin of a sheath (figure 4–15) can be split, just touching or slightly separated from the stem. More often, they are split and overlap. Rarely, the sheath is entirely fused or closed.

Blade.   The blade is the upper portion of a leaf that begins at the sheath, extending outward from the stem. The length, width, type of tip, and general roughness or smoothness of the blade can be used for identification. The grass key uses a generalized blade width to aid in identifying a grass weed. For each species, the blade can vary in width under different growing conditions (mowing height, soil moisture, age, and so on). Blade width should therefore be used cautiously and only in combination with several other identifying characteristics. The following apply to mature plants growing under optimum conditions. If the blade width averages less than 5 mm, it may be considered fine. For blade widths from 5 to 10 mm, the blade texture is considered medium, while coarse-textured blade leaves are more than 10 mm wide.

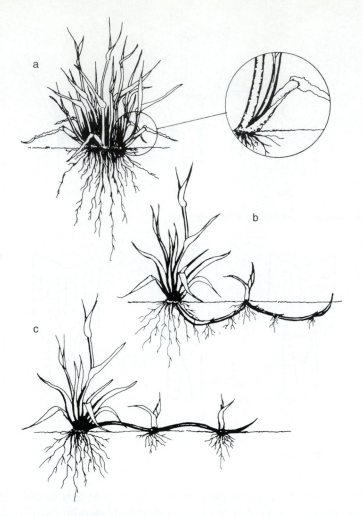

FIGURE 4–16   Growth habits of grasses: (a) bunch type with close-up of a new tiller; (b) rhizomatous growth habit; (c) stoloniferous growth habit.

Rhizomes.   A rhizome (figure 4–16b) is an underground lateral stem with the capacity to produce a new plant. Rhizomes are present or absent and can be termed strong (numerous and rapidly growing) or weak (few and often short). They are similar to aerial stems but grow laterally. In some species, they turn upward, producing new plants.

Stolons.   A stolon (figure 4–16c) is a horizontal, aboveground stem that produces roots from the nodes and provides new vertical growth for plant propagation. Stolons start from the same growing point as rhizomes but have an aboveground growth habit.

Seedhead.   The seedhead (inflorescence) is a total collection of flowers or seedbearing portions of the plant arranged in various configurations (figure 4–8). The basic flower of a grass plant is called a floret (figure 4–17a). It consists of stamens and/or pistils in association, but, unlike broadleaf plants, the surrounding modified leaves or bracts are highly specialized. The innermost bract is called the palea. Opposite the palea

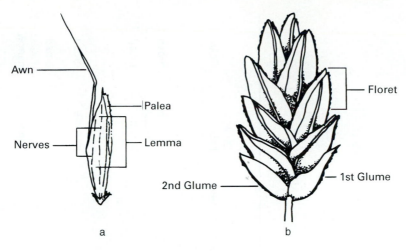

FIGURE 4–17   Grass spikelet and floret: (a) floret; (b) spikelet.

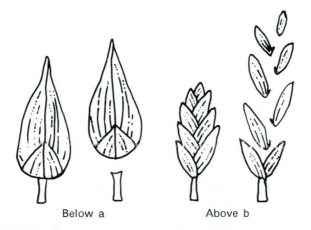

FIGURE 4–18   Disarticulation of grass florets: (a) below; (b) above.

is the lemma, which occasionally has a long, pointed tip known as an awn. When the awn is attached to the center of the lemma, dividing the lemma tip into two portions, it is considered a bifid awn. Depending on the grass species, any number of florets can be grouped together in a spikelet (figure 4–17b). The glumes, which are without other floral parts, are generally present under the lowest florets. The glumes are generally much shorter than the whole spikelet, but in some species the glumes may extend past the outer florets, encompassing the entire spikelet. There are two distinctive ways in which florets break away from the stem to distribute the enclosed seed, known as disarticulation. The disarticulation can occur either below the glumes (figure 4–18a) or above the glumes (figure 4–18b), leaving the glumes attached to the seedhead. While the grass inflorescence is not always present in grassy weeds in turf, it can be a more reliable source for identification than vegetative characteristics. Appendix D has been arranged to allow for identification by floral parts if they are present.

TABLE 4-1 KEY TO GRASS WEEDS THAT FIRST GERMINATE IN EARLY SPRING

Weed	Leaves folded in bud shoot	Leaf blade width	Auricles	Margins	Ligule		Growth habit	Collar
					Membranous	Ciliate		
Annual dropseed, *Sporobolus neglectus*		medium		round				
Bahiagrass, *Paspalum notatum*		coarse		truncate	+		rhizome & stolon	broad
Barnyardgrass, *Echinochloa crus-galli*	rolled	coarse					bunch	broad
Cogongrass, *Imperata cylindrica*		coarse		truncate	+		rhizome	
Dallisgrass, *Paspalum dilatatum*	rolled	coarse		acuminate	+		rhizome	broad
Field sandbur, *Cenchrus incertus*		medium					bunch	broad
Foxtail barley, *Hordeum jubatum*	rolled	medium		toothed	+		bunch	narrow
Fringeleaf paspalum, *Paspalum setaceum* var. *ciliatifolium*		medium		round	+		bunch	

Species								
Italian ryegrass, *Lolium multiflorum*	rolled	coarse	clawlike	round	+		bunch	broad
Johnsongrass, *Sorghum halepense*	rolled	coarse		round	+		rhizome	broad
Junglerice, *Echinochloa colonum*	rolled	medium			+		bunch	broad
Orchardgrass, *Dactylis glomerata*	folded	coarse		toothed	+		bunch	broad
Rescuegrass, *Bromus catharticus*	rolled	coarse		toothed	+		bunch	broad
Stinkgrass, *Eragrostis cilianensis*	rolled	fine				+	bunch	narrow to broad
Timothy, *Phleum pratense*	rolled	coarse		toothed	+		bunch	narrow
Torpedograss, *Panicum repens*	rolled	coarse				+	rhizome	
Velvetgrasses, *Holcus lanatus and H. mollis*	rolled	fine		round	+		bunch	narrow
Wild oats, *Avena fatua*	rolled	medium		toothed	+		bunch	broad
Witchgrass, *Panicum capillare*	rolled	medium				+	bunch	narrow

4.6 WEEDS THAT FIRST GERMINATE IN SPRING

4.6.1 Annual Grasses

Barnyardgrass, *Echinochloa crus-galli.*   Barnyardgrass (figure 4–19) is a coarse annual with broad, compressed purple sheaths. In closely mowed turfs, it lies flat on the ground and can spread out in a semicircular pattern. The absence of a ligule distinguishes barnyardgrass from many similar grasses. Barnyardgrass is more prevalent in newly established turfs and rarely persists in well-established turfs.

FIGURE 4–19   Barnyardgrass, *Echinochloa crus-galli:* (a) whole plant; (b) two views of spikelets.

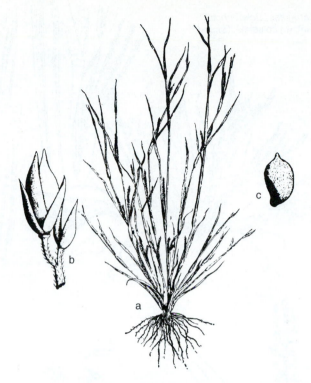

FIGURE 4–20   Annual dropseed, *Sporobolus neglectus:* (a) whole plant; (b) spikelets; (c) seed.

Annual dropseed, *Sporobolus neglectus*.   Annual dropseed (figure 4–20) is an annual grass with short stems 6 to 12 in (15.25 to 30.5 cm) high found throughout the midwestern and western states. The leaf blades are long and narrow, characterized by a tapering tip and some hair at the base. Annual dropseed is a late germinator but can persist in low-maintenance turf.

Hardgrass, *Sclerochola dura*.   Hardgrass (figure 4–21) is a cool-season annual grass that closely resembles annual bluegrass. It is most common in the west but has been identified in the midwest. This grass has short, narrow leaf blades with boat-shaped tips. The seedheads are tightly bunched racemes with awnless, three-flowered spikelets. Unlike annual bluegrass, hardgrass thrives well in sandy or coarse soils.

Junglerice, *Echinochloa colonum*.   Junglerice (figure 4–22) is an annual grass, distributed throughout the southern United States. This species is often mistaken for barnyardgrass. However, the spikelets do not have awns. The leaf blades are medium in width, and the flowers are borne on a tight panicle with tightly bunched, small spikelets. Junglerice is an occasional weed in turf and often presents a problem in newly established areas.

Wild oats, *Avena fatua*.   Wild oats (figure 4–23) is an annual grass. It is found throughout the western United States. The stems are smooth and large, reaching a height of 1 to 4 ft (0.3 to 1.2 m) where left unmowed. The leaves are coarse and long.

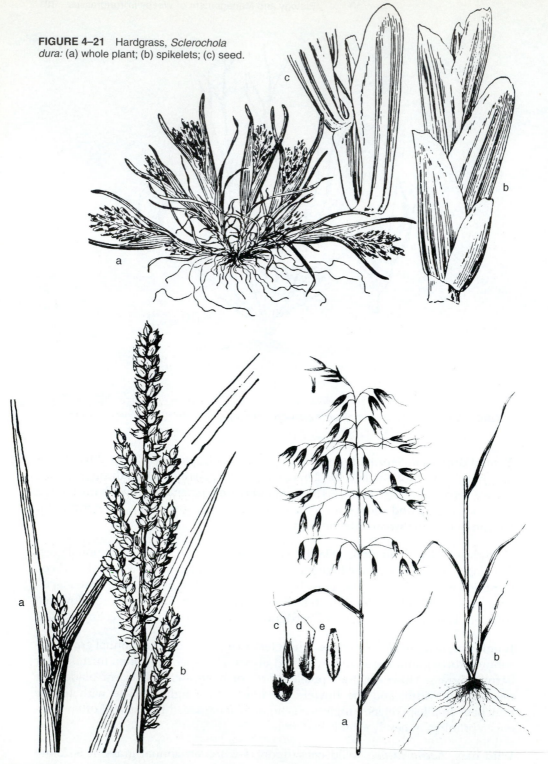

FIGURE 4–21 Hardgrass, *Sclerochola dura:* (a) whole plant; (b) spikelets; (c) seed.

FIGURE 4–22 Junglerice, *Echinochloa colonum:* (a) emerging inflorescence; (b) panicle.

FIGURE 4–23 Wild oats, *Avena fatua:* (a) inflorescence; (b) main shoot; (c) and (d) florets; (e) seed.

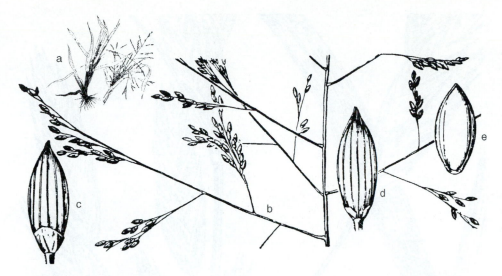

FIGURE 4–24   Fall panicum, *Panicum dichotomiflorum:* (a) whole plant; (b) branching panicle; (c) and (d) spikelets; (e) fertile floret.

The seedhead is an open, multibranched panicle with a spikelet at the end of each branch. The flowers have a distinctive long awn that is twisted. Wild oats will not long survive constant mowing and presents a problem only in newly established sites.

Fall panicum, *Panicum dichotomiflorum.*   Fall panicum (figure 4–24) is a late-germinating, cool-season annual grass. It has a short, purplish sheath with a medium-width leaf blade. The seedhead is an open panicle, with the small, awned spikelets borne on secondary branches. This species is sometimes confused with witchgrass.

Rescuegrass, *Bromus catharticus.*   Rescuegrass (figure 4–25) is an annual or biennial grass distributed throughout most of the United States, being particularly common in the warm, humid, and arid regions. Its cool-season growth habits make it particularly troublesome in dormant, warm-season grasses. It has sparingly hairy, V-shaped, medium-width leaf blades. The spikelets are tightly grouped with 6 to 12 flowers on an open panicle seedhead. Rescuegrass has little tolerance to hot weather.

Italian ryegrass, *Lolium multiflorum.*   Italian ryegrass (figure 4–26), or annual ryegrass, is a cool-season, often annual, grass. It is generally introduced to turf through seed in inexpensive seed mixtures. Italian ryegrass is a medium- to coarse-bladed grass with prominent veins. The seedhead is a true spike, with the spikelets attached directly to a single terminal stem. The flowers of annual ryegrass have prominent awns in comparison to the awnless flowers of perennial ryegrass. An additional distinctive characteristic of annual ryegrass is the long, pointed, clawlike auricles. It is found throughout the United States, where it is introduced through seeding.

FIGURE 4–25   Rescuegrass, *Bromus catharticus:* (a) leaf sheath and blade; (b) portion of panicle with spikelets.

Field sandbur, *Cenchrus incertus.*   Field sandbur (figure 4–27) is a warm-season annual grass found throughout the United States. It grows most readily in sandy or coarse-textured soils. The stems can be matlike, 6 in to 2 ft (15.25 to 61 cm) in length, with smooth, twisted leaf blades. The spikelets are borne on a terminal spike and are enclosed in a sharp, spiny burr containing one to three seeds. Field sandbur, like puncture vine, presents a hazard to humans and animals due to the painful spikes on the burr.

Stinkgrass, *Eragrostis cilianensis.*   Stinkgrass (figure 4–28) is widely distributed throughout the United States. It has many slender, smooth stems arising from the crown. The leaves have smooth sheaths that are flat with long, tapering blades. The seedhead is a multibranched and moderately spreading panicle. Each spikelet contains 20 to 40 tightly compressed flowers. Stinkgrass has a disagreeable odor. *E. pectinacea* is smaller-seeded and often occurs as a companion.

Witchgrass, *Panicum capillare.*   Witchgrass (figure 4–29) is widely distributed throughout the United States. The stems appear quite hairy. Leaf blades and

FIGURE 4–26 Italian ryegrass, *Lolium multiflorum:* (a) whole plant with inflorescence; (b) spikelet; (c) floret; (d) spikelet of perennial ryegrass, *Lolium perenne*.

FIGURE 4–27 Field sandbur, *Cenchrus incertus:* (a) whole plant; (b) modified spikelet or bur.

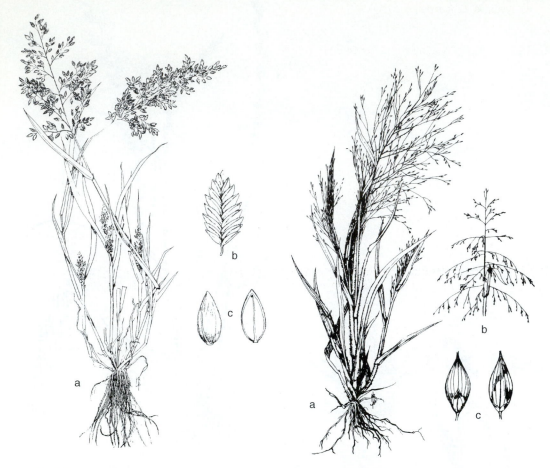

FIGURE 4–28   Stinkgrass, *Eragrostis cilianensis:* (a) whole plant; (b) spikelet; (c) seed.

FIGURE 4–29   Witchgrass, *Panicum capillare:* (a) whole plant; (b) extended panicle; (c) spikelets.

sheaths are covered with dense, soft hairs and are medium in width. The seedhead is a many-branched panicle that is open and spreading at maturity. The spikelets are awnless and contain one flower. The entire seedhead breaks from the plant at maturity and can be blown across the ground by the wind. Witchgrass is sometimes confused with fall panicum.

4.6.2   Perennial Grasses

Weeping alkaligrass, *Puccinellia distans*.   Alkaligrass (figure 4–30) is a cool-season grass species found throughout the northern United States, growing in soils with medium to high salt levels. Alkaligrass is a fine-bladed, bluish-green, bunch-type grass. The seedhead is an open panicle with four- to six-flowered spikelets.

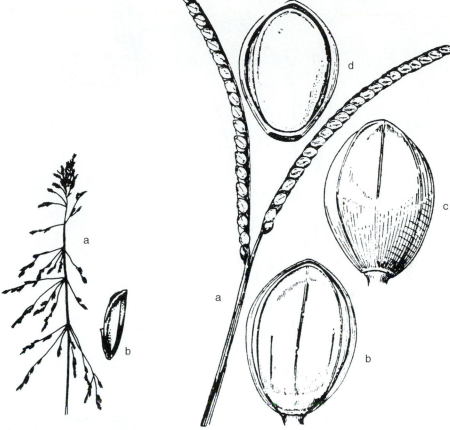

FIGURE 4–30   Weeping alkaligrass, *Puccinellia distans:* (a) inflorescence; (b) floret.

FIGURE 4–31   Bahiagrass, *Paspalum notatum:* (a) one-sided raceme; (b) and (c) spikelets; (d) floret.

Alkaligrass is often used on roadsides where salting is carried out for the purpose of snow and ice removal.

Bahiagrass, *Paspalum notatum.*   Bahiagrass (figure 4–31) is a warm-season perennial found extensively through the warm, humid region. It is a coarse-textured, wide-bladed grass, with short, heavy stolons and rhizomes. The seedhead is a raceme with solitary spikelets. Bahiagrass is extensively grown as a low-maintenance turf through the southernmost areas of the warm, humid region. It can become a troublesome weed in finer-bladed, warm-season turfs. It is well adapted to low-fertility, coarse-textured soils.

Foxtail barley, *Hordeum jubatum.*   Foxtail barley (figure 4–32) is a cool-season perennial grass found throughout the United States except in the southeast. It is a

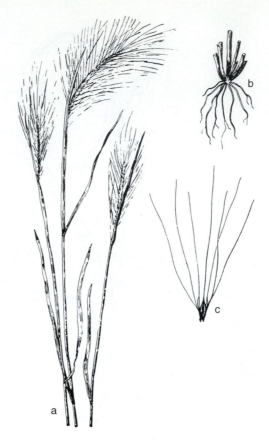

FIGURE 4–32   Foxtail barley, *Hordeum jubatum:* (a) upper portion of plant with inflorescence; (b) base of plant; (c) seed with awns.

medium-bladed, yellow-green grass with a bunch-type growth habit. The seedhead of foxtail barley is a spike with one-flowered spikelets and extremely long awns. Foxtail barley germinates early in spring and is a persistent pest in semiarid areas.

Cogongrass, *Imperata cylindrica.*   Cogongrass (figure 4–33) is a warm-season perennial grass found throughout Florida and the Gulf coast. Cogongrass, which spreads by rhizomes, has medium-width leaf blades that are long and hairy. The spikelets are formed in a loosely branched panicle, with long, silky hairs.

Dallisgrass, *Paspalum dilatatum.*   Dallisgrass (figure 4–34) is a warm-season perennial. It occurs extensively throughout the warm, humid region. Dallisgrass is one of the first grasses to begin growth in the spring. It is coarse bladed and light yellow. The seeds are borne on a tightly branched raceme seedhead with rows of four tightly grouped spikelets. Dallisgrass can be an aggressive grass in medium-maintained turfs.

FIGURE 4–33 Cogongrass, *Imperata cylindrica:* (a) whole plant with rhizome; (b) ligule and portion of sheath and blade; (c) spikelets.

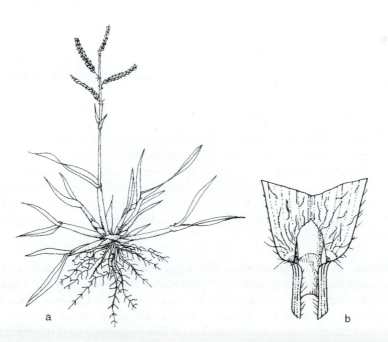

FIGURE 4–34 Dallisgrass, *Paspalum dilatatum:* (a) whole plant with inflorescence; (b) ligule.

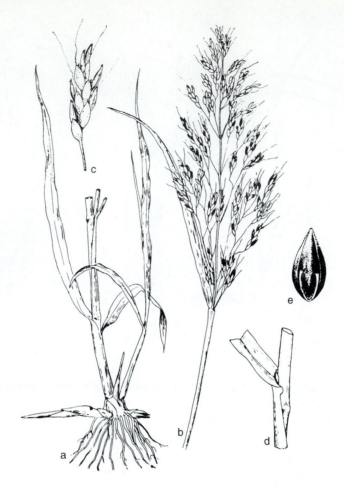

FIGURE 4–35   Johnsongrass, *Sorghum halepense:* (a) main culm with rhizomes; (b) inflorescence; (c) group of spikelets; (d) leaf blade and sheath; (e) seed.

Johnsongrass, *Sorghum halepense*.   Johnsongrass (figure 4–35) is a warm-season perennial grass found in high-cut, low-maintenance turfs. It is a pale green, coarse-bladed rhizomatous grass and a persistent pest in many crops. The seed-heads of johnsongrass are open panicles with paired spikelets, the bottom spikelet having a long, distinctive awn. Johnsongrass is a difficult-to-control weed in highway rights-of-way in the south. Its persistence can be minimized through frequent mowing at low heights.

Orchardgrass, *Dactylis glomerata*.   Orchardgrass (figure 4–36) is a cool-season, perennial grass found throughout the United States. It is principally used as a pasture or forage grass but can be introduced to turf as a contaminant in seed mixtures. It is a coarse-bladed, pale green bunch grass. The seedhead of orchardgrass is a short-branched panicle of four-flowered spikelets. In low-maintenance turf, orchardgrass is generally a problem only where it is introduced during establishment.

FIGURE 4–36   Orchardgrass, *Dactylis glomerata:* (a) whole plant with inflorescence; (b) spikelet; (c) floret.

Fringeleaf paspalum, *Paspalum setaceum* **var.** *ciliatifolium.*   Fringeleaf paspalum (figure 4–37) is a warm-season perennial grass found throughout the warm, humid region. It thrives in sandy or droughty soil, is coarse bladed, and has weak rhizomes. The seedhead of fringeleaf paspalum is a one- to three-branched raceme with two rows of spikelets on the underside. This species is extremely variable. The presence of hair or the shape of spikelets will vary even on a single plant.

Smutgrass, *Sporobolus indicus.*   Smutgrass (figure 4–38) is a warm-season perennial distributed throughout the southeastern United States. The leaf blades are narrow, long,

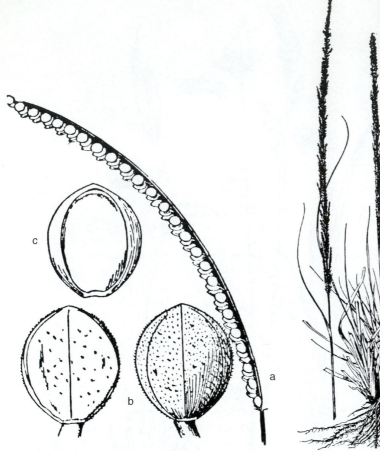

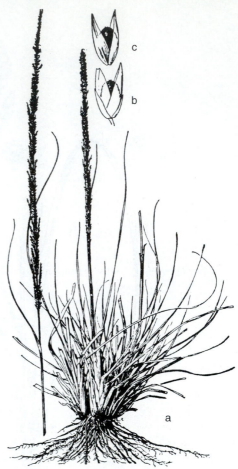

FIGURE 4–37 Fringeleaf paspalum, *Paspalum setaceum* var. *ciliatifolium:* (a) one-sided raceme; (b) spikelets; (c) floret.

FIGURE 4–38 Smutgrass, *Sporobolus indicus:* (a) whole plant with inflorescence; (b) spikelet; (c) floret.

and pointed. The seedhead is a spikelike panicle. Smutgrass gets its name because a black fungus commonly grows on the maturing seedhead.

Timothy, *Phleum pratense*.   Timothy (figure 4–39) is a cool-season perennial grass found throughout the northeastern United States. Timothy is commonly used as a forage or pasture grass and occurs as a weed in low-maintenance turf, where it is introduced in seed. It is bluish green and grows in coarse-bladed, bunch-type clumps. The seedhead of timothy is a single terminal spike with one-flowered, awned spikelets. It is easily recognized by the swollen or bulblike stems at the base of each shoot.

Torpedograss, *Panicum repens*.   Torpedograss (figure 4–40) is a warm-season perennial found along the Gulf coast. It is strongly rhizomatous, with a bladeless

FIGURE 4–39   Timothy, *Phleum pratense:* (a) whole plant with inflorescence; (b) glumes; (c) floret.

FIGURE 4–40   Torpedograss, *Panicum repens:* (a) whole plant with rhizomes; (b) spikelets; (c) floret.

sheath at the base of all shoots. Higher up on the stem are medium-width leaf blades. The seedhead is an extensively branched, open panicle, with two-flowered spikelets.

Velvetgrasses, *Holcus lanatus and H. mollis.*   Velvetgrasses (figure 4–41) are cool-season perennials found in northern United States. *H. lanatus* is fibrous rooted, and *H. mollis* is rhizomatous. They are grayish green with fine-pointed leaf blades. The seedhead is a shortly branched panicle of two-flowered spikelets in bunches. Velvetgrass survives close mowing and can be a persistent pest in moist soils.

FIGURE 4–41   Common velvetgrass, *Holcus lanatus:* (a) whole plant; (b) *Holcus mollis*, base of plant with rhizomes; (c) spikelet; (d) floret; (e) spikelet; (f) floret; (g) mature fertile floret.

4.6.3   Annual Broadleaf Weeds

Bedstraws, *Galium spp.*   Bedstraws (figure 4–42) are annuals, multibranched, with narrow, rough leaves, in a circle of six to eight on jointed stems. The flowers are small and white with four petals, borne on slender branches attached to the joints of the stem. The seeds are formed in round pods that have stiff bristles. Bedstraws can be found throughout the eastern United States and northwest coast but do not persist in closely mowed turf.

TABLE 4-2 KEY TO BROADLEAF OR SIMILARLY CONTROLLED WEEDS THAT FIRST GERMINATE IN EARLY SPRING

Weed	Flower color	Flower arrangement	Simple or compound	Leaf Margins	Leaf Shape	Leaf Arrangement	Leaf Attached by petioles
Bedstraws, *Galium* spp.	white	cyme	simple	entire	lanceolate	whorl	sessile
Buffalobur, *Solanum rostratum*	yellow	single terminal flower	simple	lobed	pinnately lobed	alternate	+
Little burclover, *Medicago polymorpha* and *M. arabica*	yellow	clusters	compound	entire	palmate	alternate	+
Carpetweed, *Mollugo verticillata*	white	single flower in leaf axils	simple	entire	lanceolate	whorl	+
Redstem filaree, *Erodium cicutarium*	red to purple	terminal umbel	simple	lobed	deeply lobed	whorl	+
Carolina geranium, *Geranium carolinianum*	pink to purple	single terminal	simple	lobed	palmately lobed	whorl	+
Prostrate knotweed, *Polygonum aviculare*	white	single flower in leaf axils	simple	entire	lanceolate	alternate	+
Kochia, *Kochia scoparia*	green	single flower or panicle in leaf axils	simple	entire	lanceolate	alternate	+
Common lambsquarters, *Chenopodium album*	green	spike	simple	entire	cordate	alternate	+
Common mallow, *Malva neglecta*	white to purple	single flower in leaf axils	simple	serrate	ovate	alternate	+
Parsley-piert, *Alchemilla arvensis*	green	panicles in leaf axils	compound	lobed	palmate	alternate	+
Prostrate pigweed, *Amaranthus blitoides*	green	single flower in leaf axils	simple	entire	ovate	alternate	+
Poorjoe, *Diodia teres*	purple	single flower in leaf axils	simple	entire	lanceolate	opposite	sessile

(continued)

TABLE 4–2   KEY TO BROADLEAF OR SIMILARLY CONTROLLED WEEDS THAT FIRST GERMINATE IN EARLY SPRING (continued)

| Weed | Flower color | Flower arrangement | Simple or compound | Leaf | | | |
				Margins	Shape	Arrangement	Attached by petioles
Puncturevine, *Tribulus terrestris*	yellow	single flower in leaf axils	compound	entire	pinnate	opposite	+
Common purslane, *Portulaca oleracea*	yellow	single flower in leaf axils	simple	entire	lanceolate	opposite	+
Annual sowthistle, *Sonchus oleraceus*	yellow	panicle	simple	lobed	pinnately lobed	alternate	+
Prostrate spurge, *Euphorbia humistrata* and *E. supina*	white	single flower in leaf axils	simple	entire	ovate	opposite	+
Spotted spurge, *Euphorbia maculata*	white to red	single flower in leaf axils	simple	entire	ovate	opposite	+
Prostrate vervain, *Verbena bracteata*	purple	spike	simple	lobed	pinnately lobed	opposite	+
Waterpod, *Ellisia nyctelea*	white to blue	single flower in leaf axils	simple	deeply dissected	palmately lobed	lower leaves alternate; upper opposite	+
Wild carrot, *Daucus carota*	white	umbel	simple	deeply dissected	palmately lobed	opposite	+
Curly dock, *Rumex crispus*	red	raceme	simple	entire	lanceolate	alternate	+
Bull thistle, *Cirsium vulgare*	red to purple	single terminal flower	simple	lobed	lanceolate	alternate	+
Creeping bellflower, *Campanula rapunculoides*	purple	spike	simple	lobed	cordate	alternate	+
Florida betony, *Stachys floridana*	purple	cyme	simple	serrate	ovate	opposite	+

Plant	Flower color	Inflorescence	Leaf type	Leaf margin	Leaf shape	Leaf arrangement	
Field bindweed, *Convolvulus arvensis*	white to pink	single flower in leaf axils	simple	entire	arrow-shaped	alternate	+
Creeping buttercup, *Ranunculus repens*	yellow and white	single terminal flower	compound	lobed	palmate	alternate	+
Virginia buttonweed, *Diodia virginica*	white to purple	single flower in leaf axils	simple	entire	lanceolate	opposite	sessile
Catnip, *Nepeta cataria*	purple	cyme	simple	entire	cordate	opposite	+
Mouseear chickweed, *Cerastium vulgatum*	white	panicle	simple	entire	lanceolate	opposite	sessile
Chicory, *Cichorium intybus*	blue	single flower in leaf axils	simple	pointed lobes	pinnately lobed	basal whorl	sessile
Silvery cinquefoil, *Potentilla argentea*	yellow	panicle	compound	lobed	palmate	alternate	+
Clovers, *Trifolium* spp.	white to pink	terminal	compound	entire	palmate	alternate	sessile
English daisy, *Bellis perennis*	yellow and white	terminal	simple	serrate	ovate	alternate	+
Oxeye daisy, *Chrysanthemum leucanthemum*	yellow and white	terminal	simple	lobed	ovate	alternate	+
Dandelion, *Taraxacum officinale*	yellow	terminal	simple	pointed lobes	pinnately lobed	basal whorl	+
Dichondra, *Dichondra repens*	white to green	terminal	simple	entire	ovate	alternate	+
Wild garlic, *Allium vineale*, and wild onion, *Allium canadense*	white to green	terminal	grasslike				
Mouseear hawkweed, *Hieracium pilosella*	yellow	single flower in leaf axils	simple	entire	lanceolate	opposite	sessile

(continued)

TABLE 4–2   KEY TO BROADLEAF OR SIMILARLY CONTROLLED WEEDS THAT FIRST GERMINATE IN EARLY SPRING (continued)

Weed	Flower color	Flower arrangement	Simple or compound	Leaf			
				Margins	Shape	Arrangement	Attached by petioles
Common lespedeza, *Lespedeza striata*	purple	raceme	compound	entire	palmate	alternate	+
Moneywort, *Lysimachia nummularia*	yellow	single flower in leaf axils	simple	entire	ovate	opposite	+
Yellow nutsedge, *Cyperus esculentus*, and purple nutsedge, *Cyperus rotundus*	yellow or purple	umbel-like	grasslike; three-sided stem				
Poison ivy, *Rhus radicans*	green	panicle in leaf axils	compound	entire	pinnate	alternate	+
Poison oak, *Rhus toxicodendron*	green	panicle in leaf axils	compound	lobed	pinnate	alternate	+
Red sorrel, *Rumex acetosella*	red	raceme	simple	entire	arrow-shaped	basal whorl	+
Speedwells, *Veronica* spp.	white to blue	spike in leaf axils	simple	entire	lanceolate	opposite	+
Little starwort, *Stellaria graminea*	white	cyme	simple	entire	lanceolate	opposite	+
Wild strawberry, *Fragaria vesca*	white	single flower in leaf axils	compound	serrate	palmate	alternate	+
Canada thistle, *Cirsium arvense*	purple	single terminal flower	simple	lobed	lanceolate	alternate	+
Hairy vetch, *Vicia villosa*	purple	raceme	simple	entire	pinnate	alternate	+
Violets, *Viola* spp.	purple	single flower in leaf axils	simple	serrate	ovate	alternate	+
Common yarrow, *Achillea millefolium*	white to yellow	umbels	simple	serrate	palmate	alternate	+

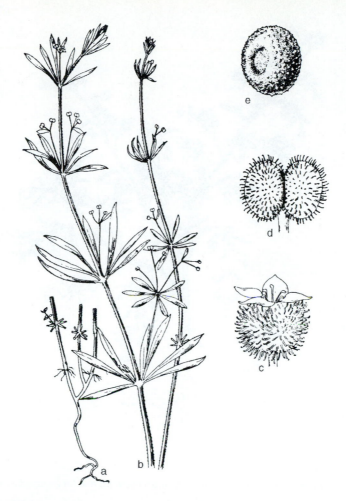

FIGURE 4–42 Bedstraw, *Galium aparine:* (a) lower section of plant; (b) upper portion of stem; (c) flower; (d) seedpod; (e) seed.

Buffalobur, *Solanum rostratum*. Buffalobur (figure 4–43) is an annual that generally occurs in low-fertility soils. The stems are multibranched, can reach heights of 24 in (61 cm), and are covered with hairy, long, stiff, yellow prickles. The leaves are long, alternate on the stem, are dense and hairy, and cut into deep, rounded lobes. The flowers are yellow with five lobes and form a spike with rough burrs that enclose the seeds. Buffalobur is found mostly in semiarid regions.

Little burclover, *Medicago polymorpha* and *M. arabica*. Little burclover (*Color plate 5*) is a winter annual weed found throughout most of the eastern and southern United States (*M. arabica*). The stems are round and smooth, with a creeping growth habit. Leaves of little burclover are compound with leaflets in groups of three, resembling those of clover or black medic. Occasionally, they have white or dark red spots across the surface. The yellow-orange flowers appear in loose clusters. The seedpods appear twisted and are covered with barbed or hooked spines.

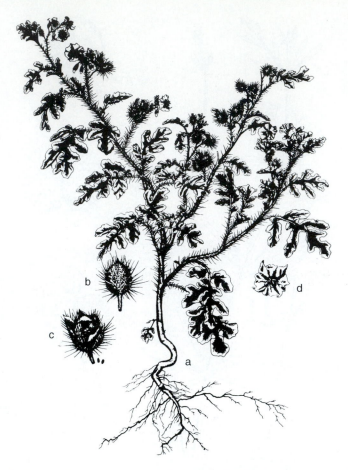

FIGURE 4–43   Buffalobur, *Solanum rostratum:* (a) whole plant; (b) and (c) seedpods; (d) flower.

These seedpods can be a nuisance in turf and can easily attach themselves to clothing or the fur of animals. Little burclover, with its prostrate growth habit, can spread even in closely mowed turfs.

Carpetweed, *Mollugo verticillata*.   Carpetweed (figure 4–44) is an annual with smooth, tonguelike leaves. The stems branch in all directions, forming flat, circular mats. Carpetweed is slow to germinate in the spring but spreads rapidly in hot weather. Its many branches radiate from a single taproot. Single leaves are arranged like wheelspokes around the nodes. Small greenish-white flowers arise from the leaf axils. Carpetweed occurs in turfs throughout the United States.

Redstem filaree, *Erodium cicutarium*.   Redstem filaree (*Color plate 11*) is a prostrate, spreading annual that forms rosettes. It is widely distributed throughout the upper half of the warm, humid region. The deeply lobed leaves are hairy, dark green, and, on average, about 1 in (2.5 cm) long. The flowers are small, with five bright petals colored rose to purple.

FIGURE 4–44   Carpetweed,
Mollugo verticillata.

FIGURE 4–45   Carolina geranium,
Geranium carolinianum: (a) whole
plant; (b) flower; (c) seedpod;
(d) seed.

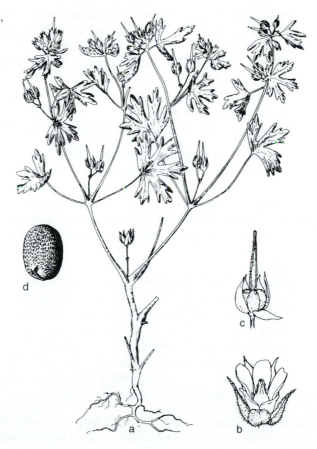

Carolina geranium, *Geranium carolinianum.*   Carolina geranium (figure 4–45) is
an annual or, sometimes, a biennial. Its root system is fibrous; stems are erect and
branch at the base, reaching heights of 20 in (51 cm). The leaves are 1 to 3 in (2.5 to
7.6 cm) across and alternate with deeply cut, fingerlike divisions. The flowers are

FIGURE 4–46 Prostrate knotweed, *Polygonum aviculare:* (a) whole plant; (b) flower on stem.

small, five petaled, colored pale pink to lavender, and generally occur singly or in loose clusters at the tips of stems and branches. Carolina geranium can be found throughout most of the United States.

Prostrate knotweed, *Polygonum aviculare.* Prostrate knotweed (figure 4–46), an annual, is low growing and appears early in spring. Its features vary depending on its maturity. Young plants have long, slender, dark green leaves that occur alternately on a tough, knotty stem. Mature plants have smaller, dull green leaves and inconspicuous white flowers. The swollen, or knotted, joints along the stem are covered with a thin, papery sheath. Knotweed persists in soil-compacted areas and is an excellent indicator of compacted soil or areas of excessive wear. It occurs throughout the United States but is particularly abundant in cool areas.

Kochia, *Kochia scoparia.* Kochia (figure 4–47) or Mexican fireweed is an annual found on dry sites. It has smooth, green, multibranched stems that grow to heights of 6 ft (1.8 m). The leaves are alternate, simple, hairy, 1 to 2 in (2.5 to 5.1 cm) long, and attached directly to the stem. The flowers are small, greenish, without petals, and borne in the junction of the stem and upper leaves. Due to its adaptation to dry areas, it is most commonly found in semiarid places.

Common lambsquarters, *Chenopodium album.* Common lambsquarters (figure 4–48) is an annual widely distributed throughout the United States. The stem arises from a taproot and is smooth, with red or light green stripes. It can obtain heights of 3 to 4 ft (0.9 to 1.2 m). The extent of stem branching varies considerably, depending on other plant competition. The leaves are alternate, 1 to 3 in (2.5 to 7.6 cm) long, with a distinct, white, mealy coating on the underside. The edges are somewhat toothed. The flowers are small, green, without petals, and borne on the ends of branches and in the axils of stems. Common lambsquarters does not persist well in mowed turf but can be troublesome in newly established sites.

Common mallow, *Malva neglecta.* Common or roundleaf mallow (figure 4–49) is an annual or short-lived perennial. It has a long taproot and rounded leaves with five distinct lobes. The leaves are opposite on the stem and closely resemble those of ground ivy, for which it is sometimes mistaken. The flowers have five pinkish-white petals and arise from the leaf axil on the main stem. Unlike ground ivy, the spreading branches of mallow do not root at spots that touch the ground. Common mallow can be found throughout the United States but is especially persistent in cool regions.

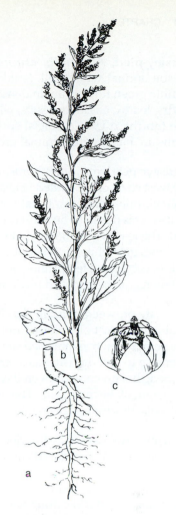

FIGURE 4–47 Kochia, *Kochia scoparia:* (a) whole plant; (b) flower.

FIGURE 4–48 Common lambsquarters, *Chenopodium album:* (a) taproot; (b) stem with inflorescence; (c) flowers.

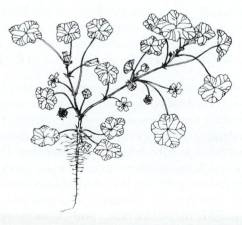

FIGURE 4–49 Common mallow, *Malva neglecta.*

Parsley-piert, *Alchemilla arvensis.*   Parsley-piert (*Color plate 6*) is a small, cool-season annual with small, fanlike leaves that appear in clusters on the stem. The minute green flowers are in dense clusters in the leaf axils. Parsley-piert is common in dry, loamy, calcareous soils, where it appears in early (southern United States) or late (northern United States) spring and throughout the summer. Both leaves and stems are hairy, with terminal growth reaching 4 to 9 in (10.2 to 23 cm).

Birdseye pearlwort, *Sagina procumbens.*   Birdseye pearlwort is found extensively on the west coast of the United States, where it can heavily infest soils of low fertility. It is a small, fine-leafed plant that spreads into a dense circular or oval patch. Birdseye pearlwort is normally an annual, but occasionally it can be found surviving as a perennial. The leaves are smooth and narrow. Flower stalks are longer than the leaves and end with tiny white flowers. Although most common turf management practices have little effect, a good fertilization program can often discourage its spread. Early small infestations can be removed physically for control.

Prostrate pigweed, *Amaranthus blitoides.*   Prostrate pigweed (figure 4–50) is an annual that is most often found in farm fields. Its prostrate growth habit can present a problem in newly established sites. Pigweed is a prolific seed producer and has good tolerance to hot, dry weather, competing well under such stress. The leaves are spear shaped, dull green, and covered with dense, coarse hairs. Red or light green stripes run the length of the main stem. The seeds develop in bushy terminal spikes and along the leaf axils. Prostrate pigweed occurs throughout the United States.

Poorjoe, *Diodia teres.*   Poorjoe (figure 4–51) is a creeping annual with long, narrow leaves, which makes it somewhat difficult to distinguish from coarse-textured grasses. The leaves are opposite and attached directly to long trailing stems covered with soft, inconspicuous hairs. Round, buttonlike seed capsules form at the leaf axils. The starlike, four-petaled purple flowers are conspicuous.

Puncturevine, *Tribulus terrestris.*   Puncturevine (figure 4–52) is an annual found in warm, humid, or arid regions. It has multibranching, hairy stems arising from simple taproots and a prostrate growth habit. The leaves are compound, oblong, opposite, and hairy. The flowers are small, yellow, five petaled, and produced in the leaf axils. The seeds mature in a pod containing five sharp burrs stiff enough to penetrate shoes or golf cart tires. Therefore, puncturevine presents a serious problem, even when it occurs in small numbers. It grows throughout the southern half of the United States.

Common purslane, *Portulaca oleracea.*   Common purslane (figure 4–53) is an annual often found in newly established sites, where it can present a major source of competition. Common purslane has a fibrous root system with thick, succulent, often reddish stems that form dense mats 1 ft or more (30.5 cm) in diameter. It thrives well in extremely hot, dry weather due to its ability to store moisture. The leaves are alternate, or clustered, and thick and fleshy, similar to those of a jade plant. The waxy coating on the leaves not only minimizes water loss but also makes common purslane difficult to control with herbicides. The flowers are small and yellow and occur in the leaf axils. The seeds may be dormant in the soil for many years. Common purslane is widely distributed and occurs throughout the United States.

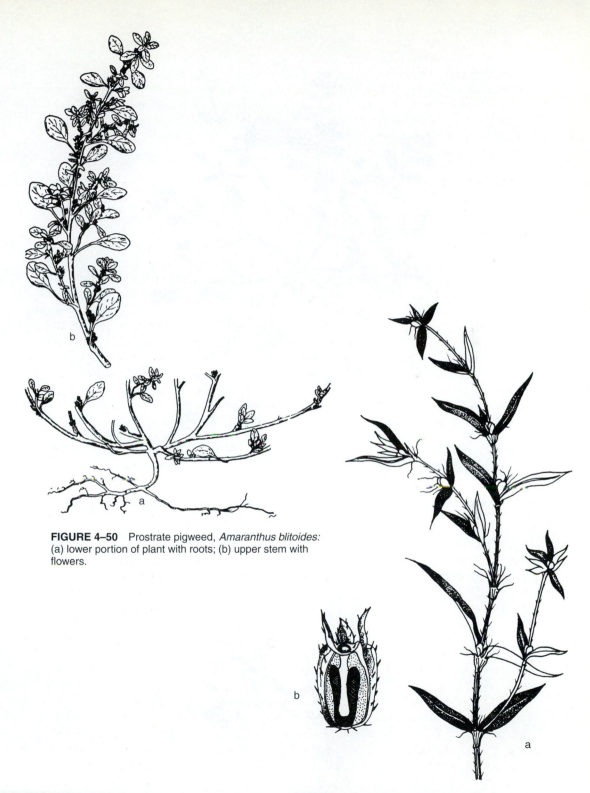

FIGURE 4–50 Prostrate pigweed, *Amaranthus blitoides:*
(a) lower portion of plant with roots; (b) upper stem with
flowers.

FIGURE 4–51 Poorjoe, *Diodia teres:* (a) whole
plant; (b) seed capsule.

FIGURE 4–52 Puncturevine, *Tribulus terrestris:* (a) whole plant; (b) seedpod with burrs.

FIGURE 4–53 Common purslane, *Portulaca oleracea.*

FIGURE 4–54   Annual sowthistle, *Sonchus oleraceus*.

Annual sowthistle, *Sonchus oleraceus*.   Annual sowthistle (figure 4–54) is an annual found in cool, humid regions. The stem is smooth, 1 to 6 ft (0.3 to 1.8 m) tall, and has a milky juice. The erect growth habit does not allow persistence under mowing conditions. Annual sowthistle is thus a problem only in newly established sites. The leaves at the base of the stem have short petioles, while leaves at the top of the stem are directly attached. The leaves are long and deeply toothed with short spines at the margin. The flowers occur in light yellow masses, similar to a daisy, and are borne on stalks at the top of the plant. Annual sowthistle can be mistaken for dandelion.

Prostrate spurge, *Euphorbia humistrata* **and** *E. supina.*   Prostrate spurge (figure 4–55) is an annual appearing in late spring. The stems are prostrate, forming a mat. The leaves are opposite and frequently have a red blotch in the center. The stem, when broken, oozes a milky sap. Prostrate spurge occurs throughout the northern United States and persists well under closely mowed turf.

Spotted spurge, *Euphorbia maculata.*   Spotted spurge (figure 4–56) is an annual similar to prostrate spurge. It germinates in late spring or early summer and has a

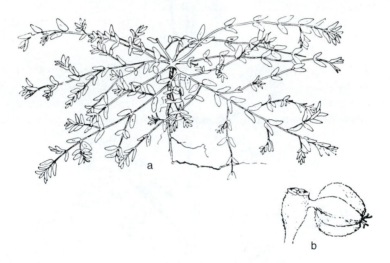

FIGURE 4–55   Prostrate spurge, *Euphorbia humistrata:* (a) whole plant; (b) seedpod.

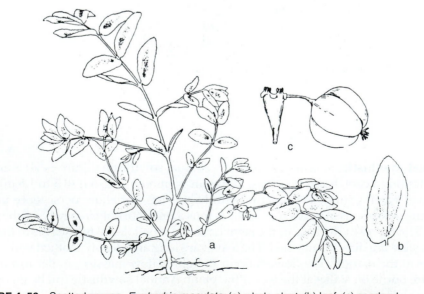

FIGURE 4–56   Spotted spurge, *Euphorbia maculata:* (a) whole plant; (b) leaf; (c) seedpod.

shallow taproot. It is more erect in its growth, however, than prostrate spurge but can spread from 6 in to 3 ft (0.15 to 0.9 m). There are generally fewer (than prostrate spurge) but larger leaves on the stem, each leaf with a conspicuous red blotch in the center. Spotted spurge is found throughout the eastern United States, being more abundant in warmer areas.

Prostrate vervain, *Verbena bracteata*.   Prostrate vervain (figure 4–57) is an annual occasionally found in turf. The hairy stems are prostrate from a few inches to more than a foot long, branching freely at the base. The numerous leaves are small, rough, opposite, and deeply lobed or toothed, with conspicuous hair. The flowers are blue or purplish, small, and in dense spikes almost hidden by the surrounding green petals. Prostrate vervain is generally found in warm, humid, or arid regions.

Waterpod, *Ellisia nyctelea*.   Waterpod (figure 4–58) is a cool-season annual found throughout the northern United States. It germinates early in spring, with flowers appearing in April or May and seeds maturing in early summer. The roots are shallow and fibrous, and the stem is succulent and branching with sparse hairs. Unmowed,

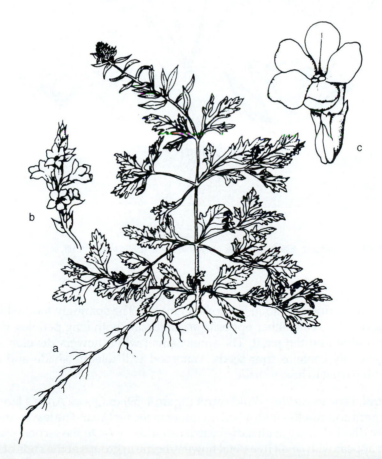

FIGURE 4–57 Prostrate vervain, *Verbena bracteata:* (a) whole plant; (b) inflorescence; (c) flower.

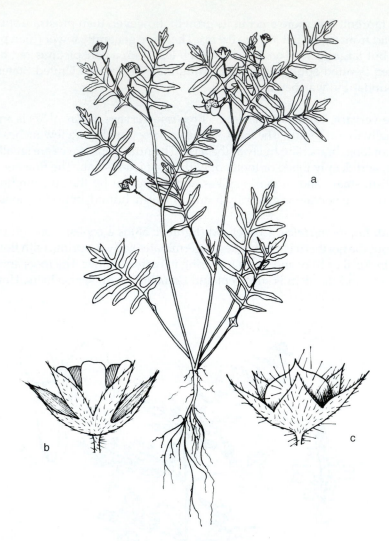

FIGURE 4–58   Waterpod, *Ellisia nyctelea:* (a) whole plant; (b) flower; (c) fruit.

the plant grows from 3 to 15 in (7.6 to 38.1 cm) tall. The sparingly toothed leaves are deeply divided and are either opposite and alternate with long petioles, depending on their position on the plant. The single, small white flowers develop into fruit, which generally contains four seeds. Waterpod will tolerate shade and occurs in lawns with partial to heavy shade.

Wild carrot, *Daucus carota*.   Wild carrot (figure 4–59) or Queen Anne's lace is a biennial that produces rosettes with a fleshy taproot in the first year. The leaves are alternate and finely divided, hairy, with a distinctive carrotlike odor. In the second year, the floral stalks appear with small five-petal flowers, borne in groups at the ends of branches. Wild carrot can survive close mowing and produces clusters of flowers that are flat and

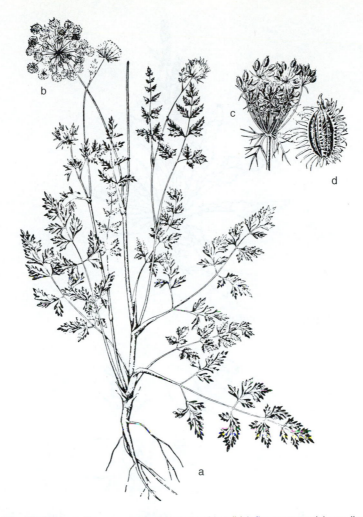

FIGURE 4–59   Wild carrot, *Daucus carota:* (a) whole plant; (b) inflorescence; (c) seedhead; (d) seed.

near the ground. It is often confused with yarrow, a similar fine-leafed perennial. Wild carrot is widely distributed throughout the eastern United States.

Curly dock, *Rumex crispus*.   Curly dock (figure 4–60) is a low-growing perennial found throughout the United States. A branched taproot forms a rosette of long leaves with wrinkled margins. The flowers form in dense clusters on erect branches that develop in unmowed situations. Curly dock is competitive during summer stress, due to food reserves in the taproot.

Bull thistle, *Cirsium vulgare*.   Bull thistle (figure 4–61) produces a rosette of deeply cut, hairy, spiny green leaves. The spines are hard and can puncture the skin. In the second year, flowering stems bear reddish-purple spiny flowers. Bull thistle does not tolerate mowing and is generally a problem only in newly established turf. It has wide distribution throughout the cool, humid region.

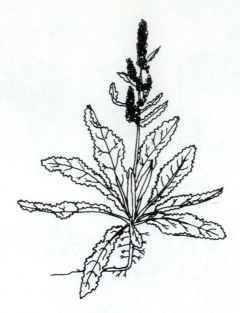

FIGURE 4–60 Curly dock, *Rumex crispus.*

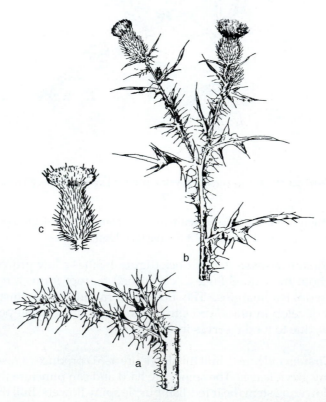

FIGURE 4–61 Bull thistle, *Cirsium vulgare:* (a) leaf and stem with spines; (b) flower arrangement; (c) flower.

4.6.4   Perennial Broadleaf Weeds

Creeping bellflower, *Campanula rapunculoides.*   Creeping bellflower (figure 4–62) is a perennial common throughout the upper cool-humid region. It spreads throughout the turf by short stolons. The stems are generally erect, reaching heights of 3 ft (0.9 m), and contain a milky juice. The basal leaves have a long petiole and are heart shaped with minimal serrate margin, whereas the upper leaves are smaller and directly attached to the stem. The flowers are numerous and bell shaped, approximately ¾ in (1.9 cm) long, with five purple teeth scattered along the upper portion of the stem. Creeping bellflower, sometimes used as an ornamental, can become a troublesome weed in turf.

Florida betony, *Stachys floridana.*   Florida betony (figure 4–63) is a cool-season weed found mostly in the southeastern United States. It generally appears in fall in cool, moist weather spreading through underground white tubers. The stems are square and not

FIGURE 4–62   Creeping bellflower, *Campanula rapunculoides:* (a) upper stem with inflorescence; (b) lower stem and roots; (c) immature plant.

FIGURE 4–63   Florida betony, *Stachys floridana.*

FIGURE 4–64   Field bindweed, *Convolvulus arvensis:* (a) whole plant with stolons and rhizomes; (b) variation in leaf shape.

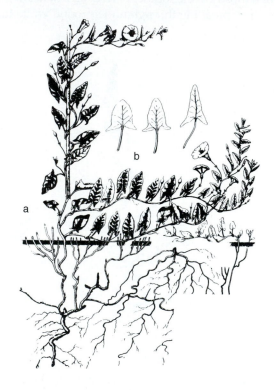

branched, with leaves opposite, broad at the base, and tapered to a round tip. The edges are serrated or sawtoothed. Florida betony has irregular, trumpet-shaped lavender flowers. It is generally dormant during the summer season.

Field bindweed, *Convolvulus arvensis.*   Field bindweed (figure 4–64) is a low-growing, creeping perennial found throughout most of the United States. It has a very extensive, deep, multibranched root system that ensures its survival through periods of stress. The stems are smooth and slender, 2 to 7 ft (0.6 to 2.1 m) long, with a vinelike growth habit that twists and covers plant material in its path. The leaves are arrow shaped but variable in shape and size. The white and pink flowers are funnel shaped, approximately 1 in (2.5 cm) across, and formed singly in the leaf axils. Field bindweed is a serious weed in agriculture and is considered a noxious weed in many states.

Creeping buttercup, *Ranunculus repens.*   Creeping buttercup (figure 4–65) is a perennial introduced from Europe, found in the northern United States. It thrives in moist, rich soils and occurs in lawns and along ditches. The stems are low and hairy, and creep, rooting at the nodes. This growth habit quickly covers the ground with a network of plants. The leaves are long petioled and alternate, three divided, and three lobed. They are hairy, dark green, and sometimes have lighter spots. Yellow, five-petaled flowers appear between May and August, producing numerous

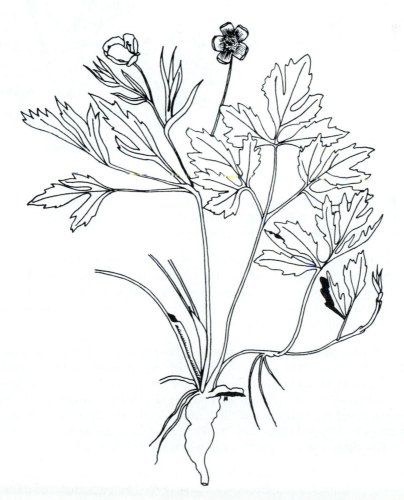

FIGURE 4–65   Creeping buttercup, *Ranunculus repens.*

FIGURE 4–66   Virginia buttonweed,
Diodia virginica: (a) whole plant;
(b) seedpods.

seeds. The perfect flowers are approximately ½ in (1.25 cm) long and borne at the end of long, terminal stalks. Improving soil drainage and lowering the mowing height aids in the control of creeping buttercup.

Virginia buttonweed, *Diodia virginica.*   Virginia buttonweed (figure 4–66) is one of the most troublesome turf pests in the southeastern United States. It is a prostrate growing perennial with somewhat fleshy lanceolate leaves. The white or purplish flowers resemble those of poorjoe but have two sepals instead of four. Virginia buttonweed is most severe in moist areas.

Catnip, *Nepeta cataria.*   Catnip (figure 4–67) is an erect-growing perennial found throughout the United States. The stems, which are covered with fine, short hairs, are 2 to 3 ft (0.6 to 0.9 m) tall, square, and light green. The heart-shaped leaves are opposite and pointed with soft-toothed margins. The leaves are darker green on the upper surface and light green or whitish underneath. The flowers, with the petals formed into a two-lipped tube, form in dense clusters at the ends of the stems and branches. Catnip is a serious pest only in newly established turfs.

Mouseear chickweed, *Cerastium vulgatum.*   Mouseear chickweed (figure 4–68) is a low-growing perennial found throughout most of the United States. The roots of this species are shallow, branched, and fibrous. The stems are generally hairy, slender, and

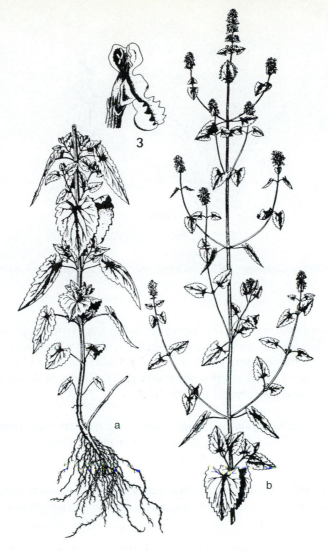

FIGURE 4–67 Catnip, *Nepeta cataria:* (a) lower portion of plant; (b) upper portion of plant and inflorescence; (c) flower.

FIGURE 4–68 Mouseear chickweed, *Cerastium vulgatum.*

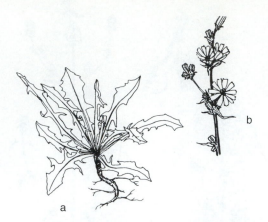

FIGURE 4–69   Chicory, *Cichorium intybus:* (a) basal whorl of leaves; (b) upper branch with flower.

spread closely over the ground. The leaves are small, very hairy, opposite, and some-what thickened or fleshy. The flowers are small, white, five petaled, and notched at the tip. The bracts surrounding the flower are also hairy. Growth of mouseear chick-weed is stimulated by close, continuous mowing. The stems of this species can root from nodes that come into contact with the soil.

Chicory, *Cichorium intybus*.   Chicory (figure 4–69) is an erect perennial found throughout the northern United States. It has a large, deep, fleshy taproot and stems that are erect, branched, and smooth with a milky sap. The basal leaves are long and deeply lobed, forming a basal rosette; leaves growing on the stems are smaller and much less lobed. Small, blue, daisylike flowers are borne on long peti-oles. Chicory is quite resistant to mowing and is more prevalent in low-maintenance or newly established turf.

Silvery cinquefoil, *Potentilla argentea*.   Silvery cinquefoil (figure 4–70) is an erect perennial found throughout the northern parts of the cool, humid region. The stems are long and prostrate, with leaves having five to seven sharply toothed leaflets. The lower leaf surface is densely covered with short hairs and is silvery in color. The yellow-petaled flowers are about ½ in (1.25 cm) across, borne on short stalks in the leaf axils. Silvery cinquefoil is an indicator of poor soil fertility. It is a tough, wiry plant, and is often confused with wild strawberry. Unlike the three-lobed strawberry leaf, silvery cinquefoil leaves are divided into five leaflets. It is seldom a serious turf pest.

Clovers, *Trifolium* spp.   Several species of clover, including burr, crimson, hop, and white clover, can often invade turf, presenting a difficult weed problem. Each species has flowers of a unique color and size (*Color plates 7 to 9*). They are generally borne on separate stems in a single, tightly bunched cluster. Clover stems are prostrate and root from nodes that touch the soil. The leaves are compound with three short, soft leaflets. The leaves of white clover, the most common turf pest, have white markings across

FIGURE 4–70   Silvery cinquefoil, *Potentilla argentea*.

FIGURE 4–71   English daisy, *Bellis perennis*.

each leaflet. Clovers are often used in low-maintenance seed mixtures, but can become dominant in a turf. They are widely distributed and common throughout the United States.

English daisy, *Bellis perennis*.   English daisy (figure 4–71) is a low-growing perennial first introduced as an ornamental. It occurs throughout the west coast. It usually grows in soil that is low in fertility. The leaves are rounded, slightly toothed, and narrow at the base, and they vary from nearly smooth to slightly hairy. The daisylike flowers are whitish to pink with bright yellow centers. English daisies can form extensive patches in turf in a short period of time.

Oxeye daisy, *Chrysanthemum leucanthemum.*   Oxeye daisy (figure 4–72) is a low-growing to upright perennial that produces rhizomes. The stems are smooth, seldom branched, and 1 to 3 ft (0.3 to 0.9 m) high, bearing simple, alternate leaves that are usually conspicuously low. The flowers are daisylike in appearance. Oxeye daisy is an indicator of low fertility, but is seldom a problem in mowed turf. It is widely distributed throughout the cool, humid region.

Dandelion, *Taraxacum officinale.*   Dandelion (figure 4–73) is a low-growing perennial common throughout the entire United States. Its roots are thick, fleshy, and often branched. Dandelion stems never elongate and produce a basal rosette of leaves. The leaves are simple but are deeply lobed, with the lobes pointing back toward the stem. The flowers are borne on single stalks with bright yellow flowers. Dandelions often regenerate from pieces of root or stem. When mature, dandelion seed develops pappi, which are transmitted through the air over large distances.

Dichondra, *Dichondra repens.*   Dichondra (*Color plate 10*) is a low-growing perennial that is well adapted to a turf site. The stems form a low, dense mass that spreads

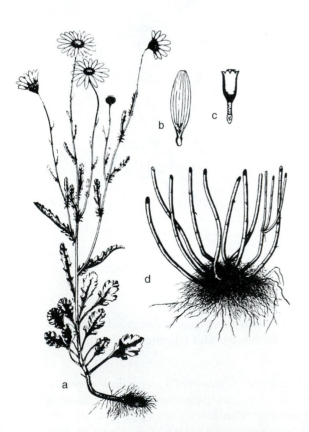

FIGURE 4–72 Oxeye daisy, *Chrysanthemum leucanthemum:* (a) whole plant with flowers; (b) ray flower; (c) disk flower; (d) lower stems and roots.

FIGURE 4–73 Dandelion, *Taraxacum officinale:* (a) whole plant; (b) mature seedhead.

across the ground. The leaves are kidney shaped and pale green. The small, pale green flowers are inconspicuous. Dichondra is often used in southern California as a turf. As a weed in turf it is highly competitive, due to similar growth requirements. It occurs throughout most of the warm, humid zone.

Wild garlic, *Allium vineale;* **and wild onion,** *Allium canadense.*   Wild garlic (figure 4–74) and wild onion (figure 4–75) are similar species of perennial, grasslike plants. Both species have a hollow stem, 1 to 3 ft (0.3 to 0.9 m) tall, that is smooth and waxy. The leaves of wild garlic are slender, hollow, nearly round, and attached to the lower half of the stem; the leaves of wild onion are flat and not hollow. These weeds have a characteristic strong odor. They are difficult to control in turf due to their waxy leaf coating that resists the penetration of most herbicides. They can reproduce in many

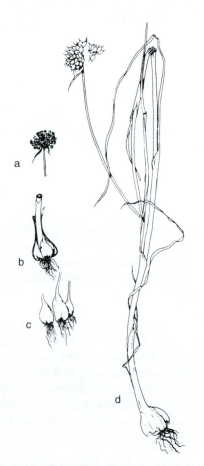

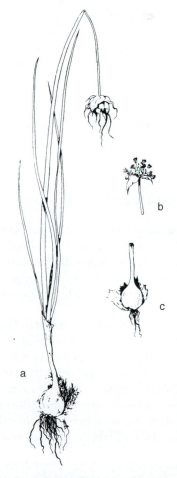

FIGURE 4–74 Wild garlic, *Allium vineale:*
(a) flower cluster; (b) old bulbs and bulblets;
(c) underground bulblets; (4) whole plant.

FIGURE 4–75 Wild onion, *Allium canadense:*
(a) whole plant; (b) flower cluster; (c) old bulb.

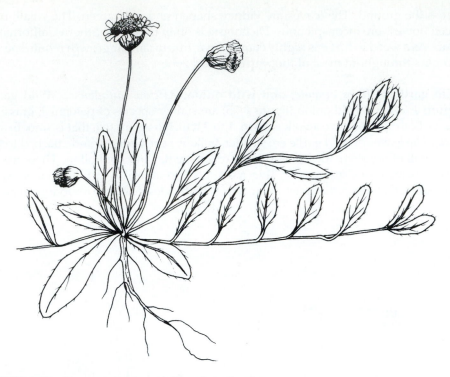

FIGURE 4–76   Mouseear hawkweed, *Hieracium pilosella*.

ways, including underground and aboveground bulblets, flowers, and a central bulb. The flowers are greenish white, small, and borne on a short stem above aerial bulblets. Both species often occur together.

Mouseear hawkweed, *Hieracium pilosella*.   Mouseear hawkweed (figure 4–76) is a perennial that varies in height from 6 in to nearly 3 ft (0.15 to 0.9 m). The stems are covered with bristlelike hairs and spread by slender rhizomes and stolons. The spatula-shaped leaves may reach 10 in (25.4 cm) long and are most often found in basal rosettes. The flowers are yellow and daisylike.

Healall, *Prunella vulgaris*.   Healall (figure 4–77) is an erect or low-growing perennial found throughout the cool, humid zone. Healall stems are long, branched, square, and hairy when young but smooth when mature. The leaves are oval, opposite, with smooth margins or very slightly notched, and 1 to 4 in (2.5 to 10.2 cm) long with long petioles. The flowers are violet or purple with two-lipped tubes and form in groups at the ends of branches. Due to healall's creeping growth habit, it can form dense patches that escape close, continuous mowing. Despite its name, healall has no known medicinal value.

Common lespedeza, *Lespedeza striata*.   Common lespedeza (figure 4–78) is a warm-season perennial found throughout the warm, humid region. It is well adapted to acidic soils of low fertility. Lespedeza is commonly grown for erosion

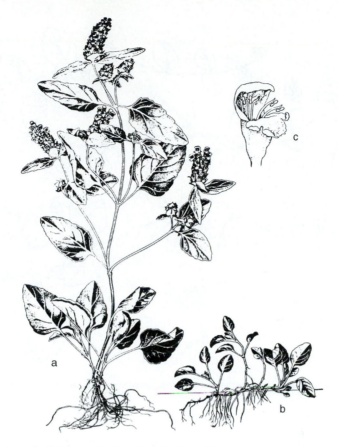

FIGURE 4–77 Healall, *Prunella vulgaris:* (a) whole plant; (b) rhizomatous growth habit; (c) flower.

FIGURE 4–78 Common lespedeza, *Lespedeza striata* (Courtesy B. J. Johnson).

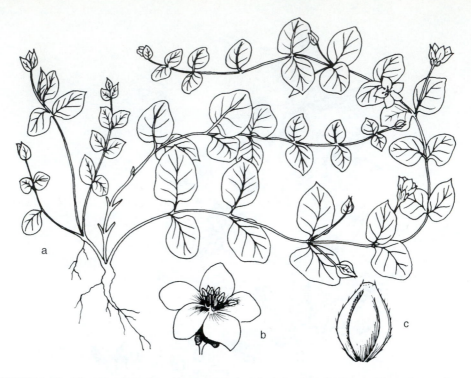

FIGURE 4–79   Moneywort, *Lysimachia nummularia:* (a) whole plant; (b) flower; (c) seed capsule.

control and soil improvement and as a forage, but it can escape and become a troublesome weed in turf. Lespedeza has small purple flowers and compound leaves with three leaflets. It is widely distributed through the southern United States and is particularly well adapted to the transition zone.

Moneywort, *Lysimachia nummularia.*   Moneywort (figure 4–79) is a prostrate perennial found throughout the eastern United States from Georgia to Canada. It is most abundant in moist, rich, shaded soils and reproduces by seed but more often through a creeping growth habit. The leaves are simple, short petioled, nearly round and opposite; they have a smooth margin and are ½ to 1 in (1.25 to 2.5 cm) in diameter. Solitary, five-petaled, yellow flowers arise from leaf axils and are borne on slender pedicels. Moneywort was introduced as an ornamental from Europe and is still used as a ground cover. Moneywort blooms occur throughout summer from June to August. It is sometimes confused with ground ivy and is also called creeping jenny and creeping charlie. Its adaptation to moist, shady sites makes it more difficult to control.

Yellow nutsedge, *Cyperus esculentus;* **and purple nutsedge,** *Cyperus rotundus.* Yellow nutsedge (figure 4–80) and purple nutsedge (figure 4–81) are low-growing perennials that resemble grasses. They occur throughout the United States, particularly in turf under a medium to high level of irrigation. Nutsedges are light

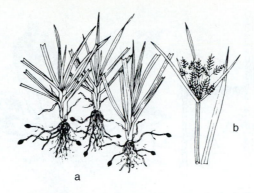

FIGURE 4–80 Yellow nutsedge, *Cyperus esculentus:* (a) whole plant with nutlets; (b) inflorescence.

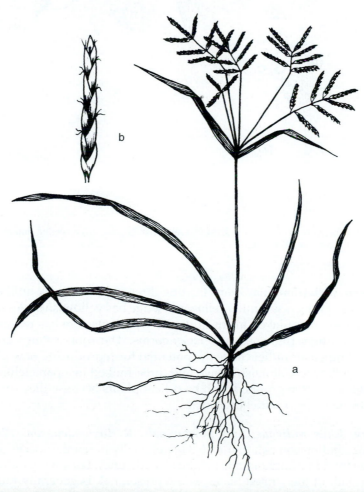

FIGURE 4–81 Purple nutsedge, *Cyperus rotundus:* (a) whole plant; (b) portion of inflorescence.

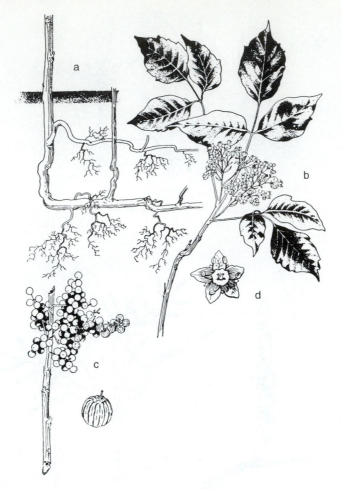

FIGURE 4–82   Poison ivy, *Rhus radicans:* (a) root and base of plant; (b) flowering branch; (3) cluster of berries and single berry; (4) flower.

yellow-green, with triangular stems bearing three-ranked leaves, unlike the two-ranked leaves of the grass family. Yellow nutsedge has yellowish scales at the base of the stem and onionlike corms or nutlets. Purple nutsedge has reddish to purple scales, produces rhizomes, and lacks corms. The root systems are fibrous. Deep-rooted tubers or nutlets have the potential for reproducing new plants. The flowers are yellow to yellowish brown and three-ranked in a paniclelike arrangement. Sedges are often grouped with broadleaf plants because they can be selectively controlled in turf with herbicides.

Poison ivy, *Rhus radicans*; and poison oak, *R. toxicodendron*.   Poison ivy (figure 4–82) and poison oak (figure 4–83) are woody perennial species within the same genus found throughout most of the United States. Poison ivy is dominant in the eastern half of the United States, while poison oak is generally found in the west. The plants may be a low-growing shrub or a vine climbing up in trees or

FIGURE 4–83 Poison oak, *Rhus toxicodendron*.   **FIGURE 4–84** Red sorrel, *Rumex acetosella*.

fence rows. The leaves have three shiny leaflets, 2 to 4 in (5.1 to 10.2 cm) long, with pointed tips. The edges of leaflets may be smooth or irregularly toothed. The flowers are small, green, five petaled, and borne in a head of 1 to 3 in (2.5 to 7.6 cm). Each flower produces a berry that is small, white, round, and hard. All parts of poison oak and poison ivy contain a material that can cause blistering of the skin. Poison ivy leaves often turn bright red in the fall. Both species vary in leaf shape, rooting habit, and the amount of hair on the leaves.

Red sorrel, *Rumex acetosella*.   Red sorrel (figure 4–84) is a low-growing, creeping perennial found throughout the eastern United States. It has an extensive, shallow-branching root system. New plants are generated along the branches. The stems are generally short but can range up to 16 in (40.6 cm) high if unmowed. Several stems may arise from a single crown. The leaves are arrow shaped and somewhat thick and fleshy, and they measure 1 to 3 in (2.5 to 7.6 cm) long. Early growth consists of a rosette of basal leaves. The flowers are yellow to red and borne on branching seed stalks at the ends of stems. Male and female flowers are borne on different plants. The male flowers are yellow to yellow-green, and female flowers are red to reddish brown. Red sorrel persists well in acidic, low-fertility soils and can be an indicator of a low pH. It can be difficult to control due to its thick, fleshy leaves and extensive root system.

Speedwells, *Veronica* spp.   Speedwells (figure 4–85) are a number of species of low-growing annual or perennial weeds that are similar in appearance. They are distributed throughout the cool-humid region. Speedwells have a fibrous root system with smooth, branching stems that are seldom over 8 in (20.3 cm) tall. The leaves are simple and narrow, with opposite leaves at the base and slightly toothed, alternate, and smooth-margined leaves on the stems. The axillary flowers are small and white to blue, giving rise to characteristic heart-shaped seed capsules. Speedwells are one of the most difficult turfgrass weeds to control.

Little starwort, *Stellaria graminea*.   Little starwort (figure 4–86) is a perennial generally found in coarse soils in the eastern United States and Canada. Its leaves are simple, opposite, narrow, and broader near the base. From May through July,

FIGURE 4–85   Speedwell, *Veronica officinalis:* (a) whole plant; (b) seed capsule.

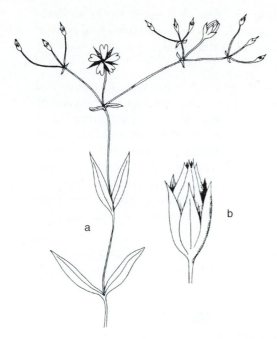

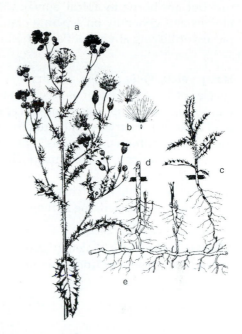

FIGURE 4–86   Little starwort, *Stellaria graminea:* (a) whole plant; (b) seed capsule.

FIGURE 4–87   Canada thistle, *Cirsium arvense:* (a) whole plant; (b) down with seed; (c) new shoot; (d) base of stem; (e) root system.

white, perfect flowers appear on long-stalked, terminal, spreading cymes. The stems of little starwort are nearly prostrate, slender, and without hair. This weed is more abundant in damp, poorly drained soils.

Wild strawberry, *Fragaria vesca*.   Wild strawberry (*Color plate 12*) is a perennial found throughout the United States. It generally resumes growth early in spring and blooms from May to July. The regular, small flowers have five rounded white petals. The compound leaves are in groups of three leaflets with sharply toothed margins. Both the stems and leaves are densely hairy. Wild strawberries have short, upright stems that produce numerous lateral stolons. American Indians used strawberry juice and water to treat inflamed eyes. Root infusions were used to treat gonorrhea and mouth sores.

Canada thistle, *Cirsium arvense*.   Canada thistle (figure 4–87) is an extensively spreading, rhizomatous perennial widely distributed through the northern United States. The multibranched roots can extend several feet deep into the soil. The stems can reach 2 to 5 ft (0.6 to 1.5 m) tall and branch at the top, and they are slightly hairy when young. The leaves have spiny margins and are somewhat lobed and smooth. Both the stems and leaf margins have hard, pointed spines that can easily penetrate the skin. The bright lavender flowers generally do not exist in mowed turf but are borne in apical bunches. Male and female flowers are in separate blooms and generally on separate plants. New plants can arise from broken root pieces, nullifying any potential physical control of this plant.

Hairy vetch, *Vicia villosa*.   Hairy vetch (*Color plate 13*) is just one of a number of species of *Vicia* that occur as weeds in turf. Hairy vetch is used as a cover crop or a green manure crop, sometimes escaping cultivation. It is often a component in seed mixtures. Hairy vetch occurs throughout the United States and can persist under normal mowing heights. The leaves are generally compound, alternate, narrow, and oblong, with terminal climbing tendrils on the stem. Blue and white bicolored, ½ in flowers form on a one-sided raceme, growing downward. They are irregular and pealike. Vetches are used extensively as a forage crop and are generally high in nitrogen due to their association with nitrogen-fixing bacteria.

Violets, *Viola* spp.   Violets (figure 4–88) are a group of low-growing perennial species that produce strong rhizomes and extensive root systems. They are gen-

FIGURE 4–88   Violet, *Viola* spp.

FIGURE 4–89   Common yarrow, *Achillea millefolium.*

erally introduced from cultivated plantings. Their leaves are small, round, and borne on a long petiole. The flowers are generally white to lavender or purple. Violets are extremely difficult to control in shaded turf and are often resistant to selective herbicides.

Common yarrow, *Achillea millefolium.*   Common yarrow (figure 4–89) is a perennial that produces rhizomes. The stems can be 1 to 2 ft (0.3 to 0.6 m) tall, and they branch at the top. The leaves are soft, covered with hair, and finely divided, and they appear fernlike. The basal leaves are longer than the leaves that rise from the stemless branches. The flowers are small, white, and in groups of 5 to 10. Yarrow can persist well under close mowing on poor soils but does not compete well in rich soils.

4.7   WEEDS THAT GERMINATE IN LATE SPRING AND SUMMER

4.7.1   Annual Grasses

Crabgrasses, *Digitaria* **spp.**   Large crabgrass, *Digitaria sanquinalis* (figure 4–90), and smooth crabgrass, *D. ischaemum* (figure 4–91) are annuals that persist well under most turf conditions. They are coarse bladed and light or apple green in color. Crabgrasses are highly competitive in turf, and their spreading growth habit tends to minimize recovery by turfgrass species. Crabgrasses can germinate throughout the entire growing season after soil temperatures have warmed in the spring. Germination occurs after each irrigation or rainy period, thus requiring persistent control.

TABLE 4–3 KEY TO GRASS WEEDS THAT FIRST GERMINATE IN LATE SPRING AND SUMMER

Weed	Leaves folded in bud shoot	Leaf blade width	Ligule			Growth habit	Collar
			Margins	Membranous	Ciliate		
Crabgrasses, *Digitaria* spp.	rolled	coarse	truncate or round	+		bunch	broad
Yellow foxtail, *Setaria glauca*	rolled	coarse	truncate		+	bunch	broad
Goosegrass, *Eleusine indica*	folded	coarse	toothed	+		bunch	broad
Bermudagrass, *Cynodon dactylon*	folded	coarse			+	rhizome and stolons	narrow
Kikuyugrass, *Pennisetum clandestinum*	folded	coarse			+	rhizome and stolons	narrow
Nimblewill, *Muhlenbergia schreberi*	rolled	fine	round	+		stolons	broad
Zoysiagrasses, *Zoysia* spp.	rolled	medium			+	rhizome and stolons	broad

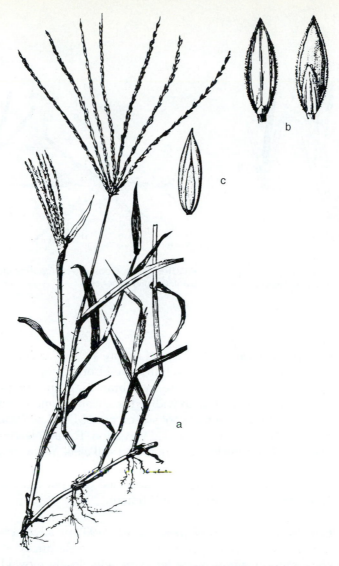

FIGURE 4–90 Large crabgrass, *Digitaria sanquinalis:* (a) whole plant; (b) spikelets; (c) floret.

FIGURE 4–91 Smooth crabgrass, *Digitaria ischaemum.*

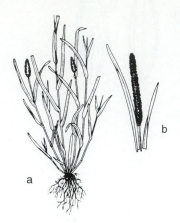

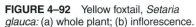

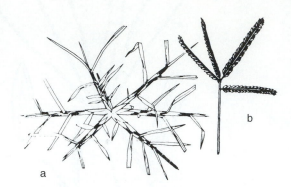

FIGURE 4–92   Yellow foxtail, *Setaria glauca:* (a) whole plant; (b) inflorescence.

FIGURE 4–93   Goosegrass, *Eleusine indica:* (a) whole plant; (b) inflorescence.

Yellow foxtail, *Setaria glauca.*   Yellow foxtail (figure 4–92) is a cool-season annual grass found throughout the United States. Introduced from Europe, it is most prevalent in thin turfs on rich soils. The stems are erect, branching, and compressed at the base. Leaf blades are flat and medium in width. The sheath is smooth and has a few long hairs at its base. Unlike most other cool-season grasses, yellow foxtail has a ring of hairs for a ligule. Young shoots have a rolled vernation. The inflorescence of yellow foxtail is a narrow spikelike panicle of one-flowered spikelets. Yellow foxtail receives its name from the yellowish appearance of the seedhead as it dries late in the season. It is generally not a serious weed in a well-established turf, but can present serious competition to seedling grasses during establishment.

Goosegrass, *Eleusine indica.*   Goosegrass (figure 4–93) or silver crabgrass is a warm-season annual that generally germinates several weeks after crabgrass in the north during the spring. It can grow under extremely closely mowed conditions, such as a putting or bowling green. Under these conditions, the lower portions of each stem are white or silvery. All spikelets on the seedhead are borne on one side of a branched panicle. Goosegrass grows abundantly in compacted or poorly drained soil. It is found throughout the United States, particularly in irrigated turf.

4.7.2   Perennial Grasses

Bermudagrass, *Cynodon dactylon.*   Bermudagrass (figure 4–94) is a warm-season perennial found throughout much of the United States. It is fine to medium bladed and low growing, producing both rhizomes and stolons. Bermudagrass is extremely aggressive and one of the most rapidly growing grasses commonly found in turf. It has a deep root system that provides tolerance to drought, but is particularly troublesome in moist soils. The seedhead of bermudagrass is a whorl of three or four racemes with small, one-flowered spikelets. Bermudagrass is the

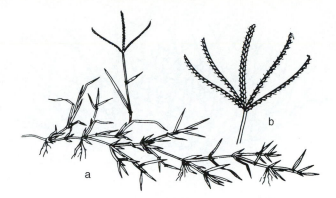

FIGURE 4–94   Bermudagrass, *Cynodon dactylon:* (a) whole plant; (b) inflorescence.

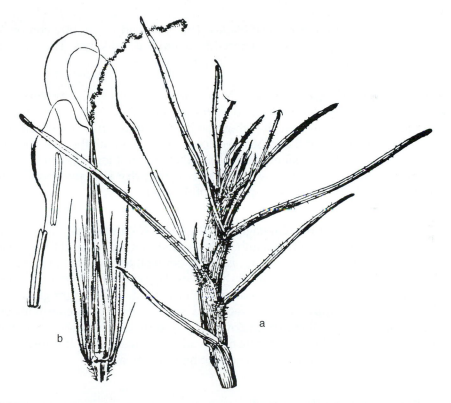

FIGURE 4–95   Kikuyugrass, *Pennisetum clandestinum:* (a) shoot with enclosed inflorescence; (b) spikelet.

principal turfgrass species used in warm, humid regions, but can become a troublesome pest when it invades other turfs.

Kikuyugrass, *Pennisetum clandestinum.* Kikuyugrass (figure 4–95) is a warm-season perennial found in southern California and Hawaii. It is low growing, with both rhizomes and stolons, and used somewhat as a turf species. The stems are

FIGURE 4–96   Nimblewill, *Muhlenbergia schreberi.*

compact and hairy with narrow leaf blades. The seedhead is a tightly compacted panicle at the tip of the flowering stem with two to four spikelets. The lemma is distinctly awned. Kikuyugrass is very difficult to control.

Nimblewill, *Muhlenbergia schreberi.*   Nimblewill (figure 4–96) is a warm-season perennial found throughout the United States with the exception of the northern border states. Nimblewill forms bluish-green patches in the turf. It has short, tapered leaf blades and stems that root at the nodes in contact with the soil. The seedhead is a spikelike panicle of single-flowered spikelets. Nimblewill is a pest in cool-season turfs, where it is generally aggressive during dry summer months and forms brown patches during cooler periods of the year.

Zoysiagrasses, *Zoysia* spp.   Zoysiagrasses (figure 4–97) are warm-season perennial grasses found throughout the southern half of the cool regions and the southern United States. Common zoysiagrass (*Z. japonica*) is a well-adapted turfgrass species for many places. It can be introduced to a nonzoysiagrass turf where it becomes a troublesome weed. Zoysiagrass is medium to fine bladed, and upright growing. It has both stolons and rhizomes. The seedhead is a compacted raceme with one-flowered spikelets resembling a spike. Zoysiagrass is difficult to control due to its aggressive growth habit, although its rate of spread is relatively slow. It is resistant to wear and can outcompete other turf species in high-use areas.

4.7.3   Perennial Broadleaf Weeds

Spotted catsear, *Hypochoeris radicata.*   Spotted catsear (figure 4–98) is a perennial that occurs throughout the northern United States, from the Pacific coast through eastern Canada. The yellow flowers of spotted catsear resemble dandelions and are generally over 1 in (2.5 cm) in diameter and borne on the end of a branch stalk nearly 1 ft (0.3 m) high. The plant is often found as a basal rosette of irregular, round-lobed, hairy leaves with a thick, milky taproot. The flowering stems are leafless and multibranched.

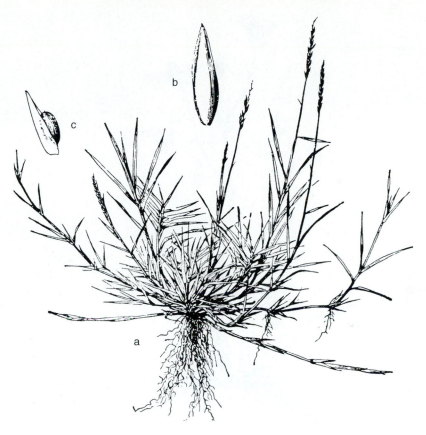

FIGURE 4–97   Zoysiagrass, *Zoysia japonica:* (a) whole plant; (b) spikelet; (c) floret.

TABLE 4–4   KEY TO BROADLEAF OR SIMILARLY CONTROLLED WEEDS THAT FIRST GERMINATE IN LATE SPRING OR SUMMER

Weed	Flower color	Flower arrangement	Leaf				
			Simple or compound	Margins	Shape	Arrangement	Attached by petioles
Spotted catsear, *Hypochoeris radicata*	yellow	single terminal flower	simple	lobed	lanceolate	whorl	+
Mugwort, *Artemisia vulgaris*	green to yellow	panicles arranged in axils	compound	serrate	pinnate	alternate	+
Lawn pennywort, *Hydrocotyle sibthorpioides*	white	clusters	simple	serrate	ovate	alternate	+

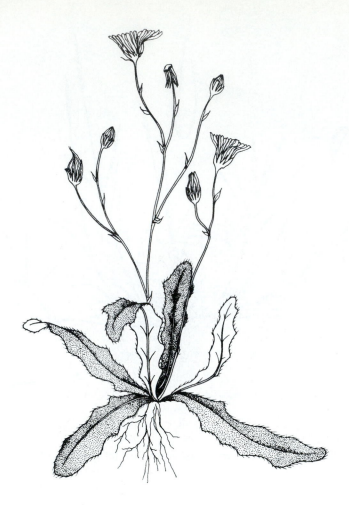

FIGURE 4–98   Spotted catsear, *Hypochoeris radicata*.

Mugwort, *Artemisia vulgaris*.   Mugwort (figure 4–99) is a perennial most often found in western and northern United States. A native of the western states, it has been introduced eastward and is represented by several forms. Mugwort seed generally germinates in July. It has erect, rigid branching stems that are often reddish and can reach 1 to 3 ft (0.3 to 0.9 m) when not mowed. The leaves are alternate and deeply lobed. These leaves have a woolly, white mass of hair on the underside and a long petiole. The flowers are numerous, small, and borne in spikelike clusters. Mugwort thrives well on calcareous soils and increases both by seed and short rootstocks.

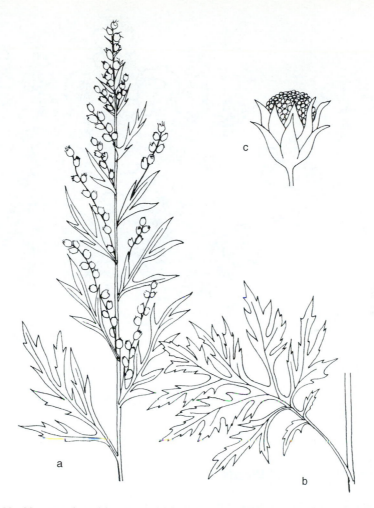

FIGURE 4–99   Mugwort, *Artemisia vulgaris:* (a) flowering shoot; (b) lower leaf; (c) flowerhead.

Lawn pennywort, *Hydrocotyle sibthorpioides*.   Lawn pennywort (figure 4–100) is a perennial most noticeable during the summer in the midwest and northeastern United States. Lawn pennywort was introduced from Asia and was used as a groundcover or ornamental plant. It increases by seed and stems rooting at nodes. Lawn pennywort can be found in medium-maintained turfs. The leaves are alternate and nearly rounded or shield shaped, ½ to 1 in (1.25 to 2.5 cm) in diameter. They are generally glossy and have a slender petiole. The stems are prostrate and root at the nodes. The flowers are white.

FIGURE 4–100   Lawn pennywort, *Hydrocotyle sibthorpioides.*

4.8   WEEDS THAT GERMINATE IN SPRING AND FALL

4.8.1   Annual Broadleaf Weeds

Black medic, *Medicago lupulina.*   Black medic (figure 4–101) or yellow trefoil is an annual, a biennial, and often a perennial found throughout the cool regions. It has a shallow taproot with slender and multibranching stems that spread from 1 to 2 ft (0.3 to 0.6 m). The leaves have three leaflets; the center one is on a short petiole, similar to clover. The stems do not take root at the nodes. The flowers are bright yellow and compressed into small clusters. Black medic can be distinguished from white clover by the lack of markings on the leaves and the distinctive yellow flowers. Black medic, a member of the legume family, is common throughout the United States.

Yellow woodsorrel, *Oxalis stricta.*   Yellow woodsorrel (figure 4–102) or oxalis is a perennial, with hairy stems weakly branched at the base. These may root at the nodes. It can be easily mistaken for a clover, with leaflets in groups of three. The middle leaflet, however, is not borne on a short stalk. The flowers are small, with five conspicuous, bright yellow petals. As the flowers mature, cucumber-shaped seedpods are formed. When dry, the seeds may scatter for several feet in all directions. Yellow woodsorrel occurs extensively throughout the United States and is a very common turf weed.

TABLE 4–5   KEY TO BROADLEAF OR SIMILARLY CONTROLLED WEEDS THAT GERMINATE IN SPRING AND FALL

Weed	Flower color	Flower arrangement	Leaf				
			Simple or compound	Margins	Shape	Arrangement	Attached by petioles
Black medic, *Medicago lupulina*	yellow	spike arranged in axils	compound	serrate	pinnate	alternate	+
Yellow woodsorrel, *Oxalis stricta*	yellow	cyme	compound	entire	palmate	opposite	+
Ground ivy, *Glechoma hederacea*	purple	clusters in leaf axils	simple	lobed	ovate	opposite	+
Broadleaf plantain, *Plantago major*	green or brown	spike	simple	entire	ovate	whorl	+
Buckhorn plantain, *Plantago lanceolata*	green or brown	spike	simple	entire	lanceolate	whorl	+

FIGURE 4–101   Black medic, *Medicago lupulina.*

FIGURE 4–102   Yellow woodsorrel, *Oxalis stricta.*

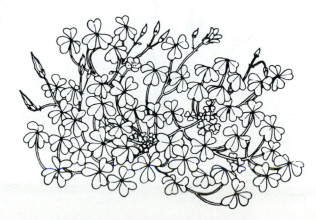

4.8.2 Perennial Broadleaf Weeds

Ground ivy, *Glechoma hederacea.* Ground ivy (figure 4–103) or creeping charlie is a low-growing perennial widely distributed throughout the United States. It forms dense patches in turf. It has shallow-rooted, creeping stems. The square stems root at the nodes that come in contact with the ground. The leaves are almost round or kidney shaped, with round-toothed edges. Leaves are generally bright to dark green, somewhat hairy, opposite, and borne on long petioles. Ground ivy is well adapted to moist, shady areas, where it can easily outcompete turfgrass. The purplish-blue flowers have the typical trumpet shape found in the mint family. Due to its extensive stolons and root system, ground ivy is difficult to control.

Plantains, *Plantago* spp. Broadleaf (*P. major*, figure 4–104), blackseed (*P. rugelli*), and buckhorn (*P. lanceolata*) plantains (figure 4–105) are low-growing perennial species found throughout much of the United States. These species form basal

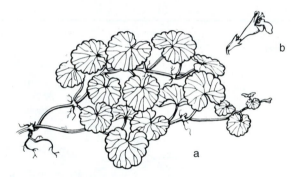

FIGURE 4–103   Ground ivy, *Glechoma hederacea:* (a) whole plant; (b) flower.

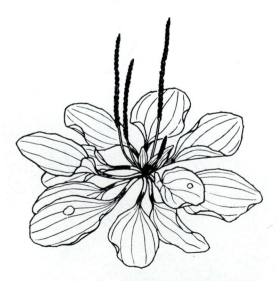

FIGURE 4–104   Broadleaf plantain, *Plantago major.*

FIGURE 4–105   Buckhorn plantain, *Plantago lanceolata.*

rosettes of leaves. Buckhorn leaves are long, narrow, hairy, and 2 to 10 in (5.1 to 25.4 cm) long. Broadleaf and blackseed plantain leaves are broad, simple, and egg shaped. Blackseed plantain has a purplish petiole. The flowers of all species are borne on a leafless stem. Buckhorn plantain flowers are in a tight spike at the end of the stem, whereas broadleaf and blackseed plantain flowers are borne on an elongated spike on the stem. Plantains are common weeds in turf.

4.8.3 Annual Grasses

Annual bluegrass, *Poa annua.*   Annual bluegrass (figure 4–106) is a winter annual or perennial that can persist in closely mowed turfs for many seasons. It is apple green in color and produces hundreds of whitish-green seedheads at all mowing

TABLE 4–6   KEY TO GRASS WEEDS THAT GERMINATE IN SPRING AND FALL

| Weed | Leaves folded in bud shoot | Leaf blade width | Auricles | Ligule | | | Growth habit | Collar |
				Margins	Membranous	Ciliate		
Annual bluegrass, *Poa annua*	rolled	fine		acute	+		bunch	narrow
Bentgrasses, *Agrostis* spp.	rolled	fine		round	+		stolons	narrow
Tall fescue, *Festuca arundinacea*	rolled	coarse	small	truncate	+		bunch	broad
Quackgrass, *Agropyron repens*	rolled	coarse	clawlike		+		rhizome	broad
Redtop, *Agrostis alba*	rolled	coarse		acute	+		rhizome	broad
Perennial ryegrass, *Lolium perenne*	folded	fine	short	truncate	+		bunch	broad

FIGURE 4–106   Annual bluegrass, *Poa annua.*

heights. Seedhead production persists, even at the extremely close mowing heights of a putting green. The ligule of annual bluegrass is acute, which readily distinguishes it from the truncate ligule of Kentucky bluegrass. It grows well on compacted soils under moist and shaded conditions and frequently occurs in dense patches. Seedheads are produced throughout the growing season but are particularly abundant during midspring. Closely mowed, perennial-type annual bluegrass is susceptible to winter damage and many diseases.

4.8.4   Perennial Grasses

Bentgrasses, *Agrostis* **spp.**   Bentgrasses (figure 4–107) are cool-season perennial grasses found throughout the United States. They are medium to dark green, medium to fine-leaved, prostrate, stoloniferous grasses. The seedhead is an open panicle of one-flowered spikelets with glumes longer than the flowers. Most bentgrass species are aggressive in closely mowed, moist soils of moderate to high fertility. In favorable environments, bentgrass can spread rapidly and outcompete most other turfgrasses.

Tall fescue, *Festuca arundinacea.*   Tall fescue (figure 4–108) is a coarse-bladed, cool-season perennial grass found throughout the United States. It is used widely in warmer areas, where it provides a good turf with some tolerance to shade, drought, and heat. Tall fescue is a medium green bunch-type grass. It is often used as a single species for low- to medium-quality turfs, but forms objectionable bunches or clumps of weeds in finer-bladed turfs. The seedhead of tall fescue is a slightly branched panicle of several flowered spikelets.

Quackgrass, *Agropyron repens.*   Quackgrass (figure 4–109) is a cool-season perennial found throughout the northern United States. Quackgrass is a medium-bladed, often blue-green rhizomatous grass that is persistent in cool-season turfs. The seedhead is a spike of multiflowered spikelets. Quackgrass has long, clawlike, though not very conspicuous auricles that clasp the stem. Once established, it is difficult to eradicate in turfs maintained at higher than ¾ in (1.9 cm).

FIGURE 4–107   Creeping bentgrass, *Agrostis stolonifera:* (a) whole plant; (b) panicle inflorescence.

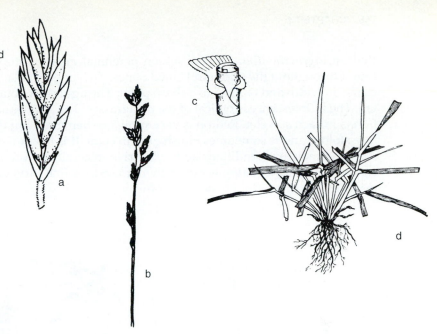

FIGURE 4–108 Tall fescue, *Festuca arundinacea:* (a) spikelet; (b) inflorescence; (c) rolled vernation; (d) whole plant.

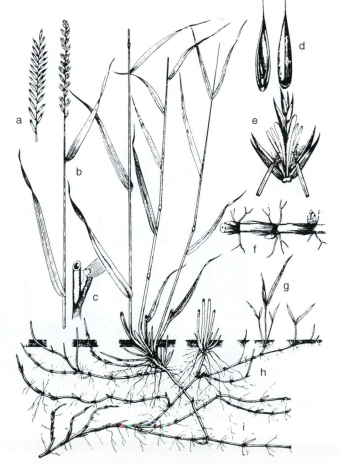

FIGURE 4–109 Quackgrass, *Agropyron repens:* (a) spike; (b) stems, leaves, and inflorescence; (c) auricle and ligule; (d) seed; (e) spikelet; (f) buds on rhizome; (g) new shoots; (h) origin of new shoots; (i) rhizome and roots.

Redtop, *Agrostis alba.*   A cool-season perennial grass, redtop (figure 4–110) is found throughout the northern United States. The long leaf blades are medium to coarse in width and tapered to a sharp point. The ligules are rounded to acute and tall. The inflorescence is typical of the bentgrasses. The glumes are longer than the florets in each spikelet. Redtop is very similar to bentgrass but does not have surface stolons and is sometimes bluish gray in color. It is tolerant of a wide range of soils and moisture conditions and will thrive in acidic or porous soils. Due to its coarse leaf texture, redtop is generally not used for fine turfs but is more appropriate for pasture or highway rights-of-way use.

FIGURE 4–110   Redtop, *Agrostis alba:* (a) main shoot and rhizomes; (b) ligule; (c) inflorescence; (d) spikelet; (e) floret.

Perennial ryegrass, *Lolium perenne*.   Perennial ryegrass (figure 4–26) is a cool-season bunchgrass found throughout the United States. It is often used as a turf. It is well adapted to medium maintenance. It is generally less tolerant to environmental extremes than Kentucky bluegrass, however, and is used principally in mixtures with other turfgrass species. As a very small percentage in a mixture, perennial ryegrass can become a serious weed. Perennial ryegrass is an excellent seedling competitor and can dominate in a mixture. While it tolerates a wide range of soil and environmental conditions, heat, cold, and drought can take their toll on perennial ryegrass stands. The leaves are narrow to medium in width, light to dark green, and generally shiny on the underside. The auricles are short and stubby. The inflorescence is a long spike of many-flowered spikelets. Perennial ryegrass can be distinguished by the presence of short or stubby auricles, unlike the long clasping auricles of annual ryegrass. Also, perennial ryegrass is folded in the bud shoot, whereas annual ryegrass is generally rolled.

4.9   WEEDS THAT GERMINATE IN THE FALL

4.9.1   Annual Grasses

Downy brome, *Bromus tectorum*.   Downy brome (figure 4–111) or downy chess is widely distributed throughout the United States, with the exception of the extreme southeast. Downy brome, a cool-season grass, is often abundant in cultivated fields and sometimes in fall-seeded turf. The medium-coarse leaf blades and sheaths are pubescent. The inflorescence consists of a dense panicle, 2 to 10 in (5.1 to 25.4 cm) long, with soft, drooping spikelets. Each spikelet forms 4 to 10 seeds with long conspicuous awns.

Smooth brome, *Bromus inermis*.   Smooth brome (figure 4–112) has smooth stems and the flowers lack awns. It produces rhizomes and has a perennial growth habit. Smooth brome is a cool-season perennial grass often used for forage and erosion control. It has an open panicle seedhead with multiflowered spikelets. The sheath is not split or overlapped.

TABLE 4–7   KEY TO GRASS WEEDS THAT GERMINATE IN THE FALL

Weed	Leaves folded in bud shoot	Leaf blade width	Ligule Margins	Ligule Membranous	Growth habit	Collar
Downy brome, *Bromus tectorum*	rolled	fine to medium	toothed	+	bunch	narrow, divided
Smooth brome, *Bromus inermis*	rolled	coarse	truncate	+	rhizomes	broad

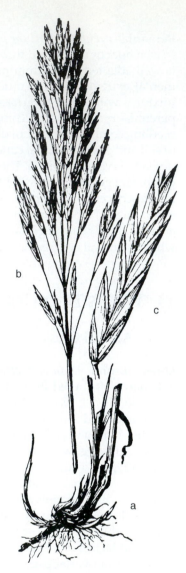

FIGURE 4–112 Smooth brome, *Bromus inermis:* (a) base of plant with rhizomes; (b) panicle inflorescence; (c) spikelet.

FIGURE 4–111   Downy brome, *Bromus tectorum:* (a) whole plant; (b) floret; (c) panicle inflorescence; (d) spikelets.

4.9.2   Annual Broadleaf Weeds

Hairy bittercress, *Cardamine hirsuta*.   Hairy bittercress (figure 4–113) is a winter annual found throughout the eastern United States. In turf, it is generally found as a basal rosette of leaves that are deeply pinnately lobed and somewhat resemble a dandelion, with the exception of hair across the surface. Small white flowers are borne on racemes from February through May. Hairy bittercress is generally a pest in newly established turfs in moist, sandy soils.

TABLE 4–8   KEY TO BROADLEAF OR SIMILARLY CONTROLLED WEEDS THAT GERMINATE IN THE FALL

Weed	Flower color	Flower arrangement	Leaf				
			Simple or compound	Margins	Shape	Arrangement	Attached by petioles
Hairy bittercress, *Cardamine hirsuta*	white	raceme	simple	lobed	deeply dissected	whorl	+
Lawn burweed, *Soliva pterosperma*	green	cyme	simple	entire	deeply dissected	opposite	+
Mayweed chamomile, *Anthemis cotula*	white to yellow	single terminal	simple	lobed	ovate	alternate	+
Common chickweed, *Stellaria media*	white	single terminal	simple	entire	ovate	opposite	+
Purple deadnettle, *Lamium purpureum*	purple	cluster of flowers in axils	simple	serrate	ovate	opposite	+
Henbit, *Lamium amplexicaule*	purple	cluster of flowers in axils	simple	lobed	ovate	opposite	+
Wild mustard, *Sinapis arvensis*	yellow	panicle	simple	lobed	pinnately lobed	alternate	+
Field pennycress, *Thlaspi arvense*	white	raceme	simple	serrate	deeply dissected	alternate	+
Field pepperweed, *Lepidium campestre*	white to green	raceme	simple	lobed	pinnately lobed	basal whorl	+
Yellow rocket, *Barbarea vulgaris*	yellow	raceme	simple	lobed	pinnately lobed	basal whorl	+
Shepherdspurse, *Capsella bursa-pastoris*	white	raceme	simple	lobed	pinnately lobed	basal whorl	+
Swinecress, *Coronopus didymus*	white	flowers in raceme in axils	simple	lobed	pinnately lobed	alternate	+

FIGURE 4–113   Hairy bittercress, *Cardamine hirsuta* (Courtesy B. J. Johnson).

Lawn burweed, *Soliva pterosperma*.   Lawn burweed (*Color plate 14*) is a low-growing winter annual found in warm, humid regions. The finely divided leaves are in narrow segments and appear twisted. Many conspicuous greenish flowers are attached directly to the spreading stems. The seeds produce a sharply pointed spine that can easily pierce the skin. Lawn burweed was first imported from South America as an ornamental, but it quickly became a troublesome weed in turf.

Mayweed chamomile, *Anthemis cotula*.   Mayweed chamomile (figure 4–114) is a winter annual producing a short, thick taproot. It is also known as dog fennel or stinking fennel. The stems are erect, slender, multibranched, nearly smooth, and can be 12 to 18 in (30.5 to 45.7 cm) tall unmowed. Its flowers, ½ to 1 in (1.25 to 2.5 cm) across, resemble those of daisies, growing singly at the ends of branches. Mayweed chamomile occurs throughout the eastern United States and closely resembles pineappleweed (figure 4–115) found in the west. It has a strong, disagreeable odor.

Common chickweed, *Stellaria media*.   Common chickweed (figure 4–116) or starwort occurs throughout the United States. It is a winter annual, germinating in the fall and producing abundant growth in early spring. In protected areas, common chickweed may persist throughout the summer. It prefers moist, shady areas, where it spreads readily, impeding normal turf growth. Its creeping growth habit

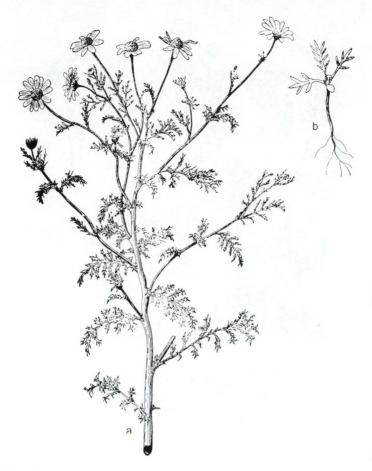

FIGURE 4–114   Mayweed chamomile, *Anthemis cotula:* (a) whole plant; (b) seedling.

allows dense patches to form, competing well with the existing turf. In dormant, warm-season turfgrasses, common chickweed can spread extensively. The leaves are a bright, shiny green, and they are rounded and tapered to a point and borne opposite each other on hairy stems. Stem nodes touching the ground take root to form new plants. Common chickweed flowers are white with five deeply notched petals. They are extremely small but conspicuous.

Purple deadnettle, *Lamium purpureum*.   Purple deadnettle (figure 4–117) is a winter annual that occurs throughout the United States. It is generally more prevalent in fertile, moist soils. Purple deadnettle is more erect than henbit and can grow from 6 to 18 in (15.25 to 45.7 cm) in height. It has square, purplish branched stems that can root from the nodes. The leaves are somewhat heart shaped and hairy. The purple flowers, which can be found from April to October, are approximately ½ in (1.25 cm) in length. These appear in whorls or bunches at the top of the stems.

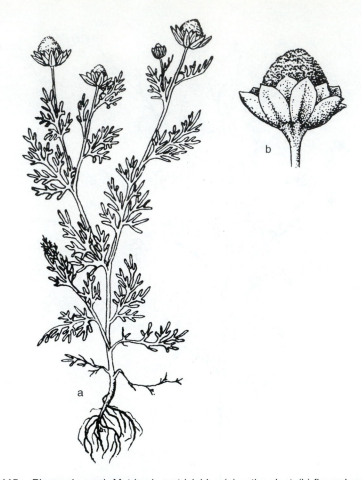

FIGURE 4–115   Pineappleweed, *Matricaria matricioides:* (a) entire plant; (b) flower head.

FIGURE 4–116   Common chickweed, *Stellaria media*.

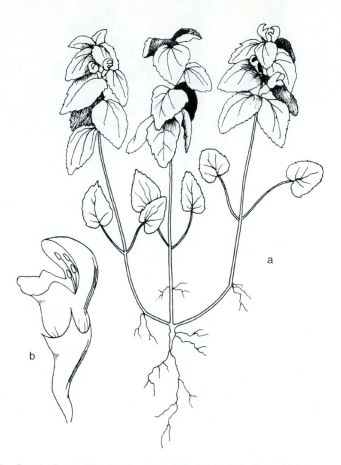

FIGURE 4–117   Purple deadnettle, *Lamium purpureum:* (a) whole plant; (b) flower.

Henbit, *Lamium amplexicaule.* Henbit (figure 4–118) is typically a winter annual that usually germinates in early fall. It also germinates to a certain extent in early spring. Henbit, a member of the mint family, has a typical square stem. The leaves are rounded, coarse toothed, hairy, deeply veined, and grow opposite each other on the main branches. The flowers are trumpet shaped and pale purple. Henbit is primarily an upright grower, but can root and vine from nodes. In warm-season turfs, henbit can be a particular early-season problem.

Wild mustard, *Sinapis arvensis.* Wild mustard (figure 4–119) is a winter annual common in the warm, humid region and occasionally found in cool, humid zones. Its leaves are narrow and decrease in size toward the tip of the stem. They are alternate and irregularly lobed. The flowers of wild mustard are conspicuous, with four yellow petals that appear in clusters at the ends of the branches. Wild mustard seeds live in the soil for many years.

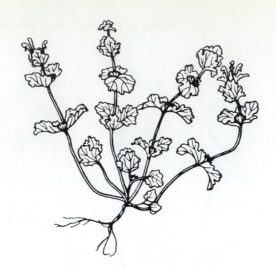

FIGURE 4–118   Henbit, *Lamium amplexicaule.*

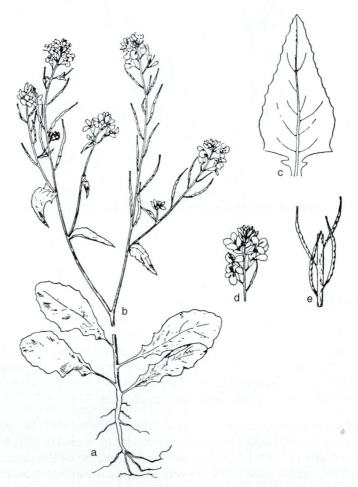

FIGURE 4–119   Wild mustard, *Sinapis arvensis:* (a) lower part of plant; (b) inflorescence; (c) leaf from upper part of stem; (d) flower cluster; (e) seedpod.

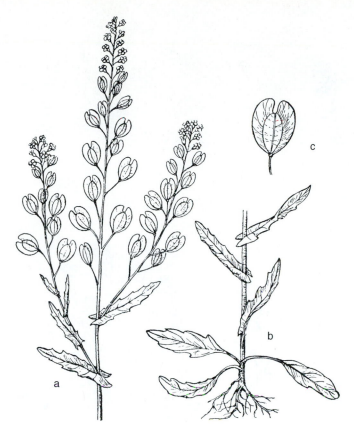

FIGURE 4–120   Field pennycress, *Thlaspi arvense:* (a) whole plant; (b) lower part of plant with roots; (c) seed.

Field pennycress, *Thlaspi arvense*.   Field pennycress (figure 4–120) is a winter annual with an erect stem and is up to 20 in (50.1 cm) tall in nonmowed areas. Its leaves are alternate, simple, toothed, and ½ to 2 in (1.25 to 5.1 cm) long. The leaves clasp the upper stem with earlike projections. The flowers are white with four petals and are borne alternately along terminal branches. Field pennycress is generally found in cool regions.

Field pepperweed, *Lepidium campestre*.   Field pepperweed (figure 4–121) is a winter annual. It resumes rapid growth in spring. Although generally a weed of open field and waste areas, it thrives well in turfs. Field pepperweed has a rosette growth habit through fall and early spring, when a single, long-flowering stem appears. As the flowering stem develops, the rosette disappears and is replaced by arrow-shaped leaves that clasp the stem. Field pepperweed is a member of the mustard family, with small, white flowers that contain four petals. The leaves are bright green with toothed margins and blunt or rounded tips. The seedpods have a distinct, strong, peppery taste typical of the mustard family. Field pepperweed is

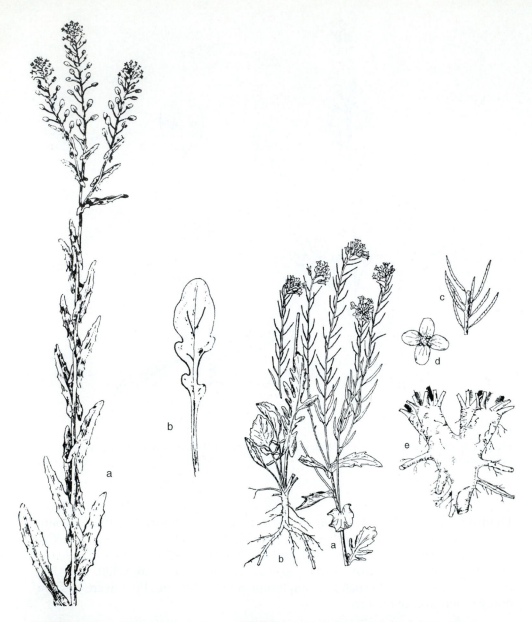

FIGURE 4–121 Field pepperweed, *Lepidium campestre:* (a) upper part of plant with inflorescence; (b) leaf from basal whorl.

FIGURE 4–122 Yellow rocket, *Barbarea vulgaris:* (a) upper part of plant with inflorescence; (b) basal rosette of leaves with roots; (c) seedpod; (d) flower; (e) root.

found throughout northern United States. *L. virginicum*, a similar species, is found throughout the southeastern United States.

Yellow rocket, *Barbarea vulgaris.* Yellow rocket (figure 4–122) is a winter annual of the mustard family. It forms a rosette of leaves in early fall and generally remains in this state due to mowing of the turf. The leaves of yellow rocket are bright, shiny green and deeply notched along the edges. They terminate in a large, rounded lobe. Its flowers are yellow, small, have four petals, and are found in clusters at the tips of the uppermost branches. Yellow rocket is found in cool regions.

Shepherdspurse, *Capsella bursa-pastoris.*   Shepherdspurse (figure 4–123) is a winter annual that forms a basal rosette during early fall. While in the rosette stage, shepherdspurse can be misidentified as a dandelion. The leaves are coarsely lobed but generally more narrow than dandelions. As the weather warms in spring, a multibranched seed stalk arises from the center of the rosette. The leaves on these stems are arrow shaped and coarsely toothed. The flowers of shepherdspurse are typical of the mustard family; that is, they are small, white, and four petaled. As the seeds mature, they form heart-shaped capsules resembling the purses used by shepherds in biblical times. Shepherdspurse can be found in cultivated areas throughout the world. It is primarily a problem in newly seeded turf.

FIGURE 4–123   Shepherdspurse, *Capsella bursa-pastoris:* (a) whole plant; (b) roots with basal rosettes; (c) flower; (d) seedpods.

Swinecress, *Coronopus didymus*.   Swinecress (*Color plate 15*) is a winter annual found throughout the southeastern United States. It forms a basal rosette of leaves that are pinnately lobed toward the apex. Small, white flowers form in racemes on stems with opposite leaves. Swinecress produces an unpleasant pungent odor when bruised or ruptured.

4.10   CULTURAL CONTROL METHODS

 Nonchemical controls for minimizing weed populations in turf are simple in concept. With few exceptions, turfgrass weeds will not flourish well under the same conditions as those optimal for turf-grass growth. Thus, the turfgrass manager must provide growing conditions that enhance the rapid growth and development of turf-grasses at the expense of weeds.

Cultural and routine maintenance operations, such as mowing, irrigation, and fertilization, should proceed on schedule at the appropriate rate and timing. Overfertilization is detrimental to turfgrass species and encourages an increase in weed populations. Underfertilization limits the competitive ability of the turf to outgrow weeds, minimizing the duration of chemical weed-control practices.

The soil environment is also critical to proper turfgrass growth and the minimization of weed populations. The pH of turf soils, when feasible, should be kept between 6.0 and 7.0. In compacted fine-textured soils, core aerification (*Color plate 3*) should be utilized to alleviate compaction and provide more oxygen for root growth. Soil moisture should be adequate but not excessive to promote healthy turfgrass-root development.

Turfgrass weeds are opportunistic and will invade damaged turf areas. It is important to ensure weed control in turf damaged from insect or disease problems. The turfgrass must be encouraged to regrow in damaged areas to gain a competitive edge on potential weeds. Many weeds can be excluded from turfgrass sites by frequent, regular mowing. Mowing practices offer the best opportunity for controlling a wide range of weed species. Most of the weeds presented in this chapter can be held to very low populations through normal, consistent mowings. If weed populations begin to increase, lower the mowing height slightly. Do not lower the mowing height below the recommended range for the turf. Increasing the mowing frequency can also help. Keep mowers properly adjusted and blades sharp to minimize stress on the turf and to cut tough weeds consistently (*Color plate 2*).

4.11   CHEMICAL CONTROLS

Herbicides are pesticides that kill one or more plant species. Only a small number of currently available herbicides are used in turfgrass weed control. They can be classified into one of three categories depending on the nature of their activity.

1. *Contact herbicide.*   Contact herbicides kill plant parts that come in contact with the chemical. They are absorbed by living tissues of a plant to a small

degree but do not move throughout the plant tissues. Since only the tissue contacted is killed, surfactants are often added to improve the coverage of the foliage. Contact herbicides have a very rapid effect and are used to quickly remove annual weeds. The injury symptoms of a plant treated with a contact herbicide are usually visible within hours. Temperature may affect the control. Cool weather may slow the effectiveness of the contact herbicide. Many contact herbicides used have little or no soil activity. For example, diquat is a contact herbicide often used to renovate or clean up dormant, warm-season turf. Contact herbicides do not provide effective control of perennial species, which may regrow from underground storage organs or roots.

2. *Systemic herbicides.*   Systemic herbicides are absorbed by external tissue of plants (leaves, stems, and roots) and transported throughout the plant to provide killing action to the entire plant. Most herbicides used on turf are systemic in nature. They may be selective or nonselective. A selective herbicide is one that has a greater degree of activity on one group of plants while not injuring others. Nonselective herbicides provide the same amount of control for most species. For example, 2,4-D is a selective systemic herbicide that has high activity on most broadleaf weeds while showing little effect on most grasses. Glyphosate (Roundup Pro or Kleenup) is an example of a nonselective systemic herbicide that kills most species of plants found in a turf landscape, including the turf itself.

3. *Soil sterilants.*   Soil sterilants are chemicals that render the soil toxic to higher plant life. The difference between a soil sterilant and a nonselective herbicide is that the chemical remaining in the soil kills seeds that germinate for extended periods of time, thereby preventing regrowth of plants. The length of time soil remains void of plants depends on the material used, rate of application, and the environmental conditions, such as light, temperature, moisture, and soil properties, which affect the rate of decomposition of the herbicide. Soil sterilants are not generally used in turfgrass management. They are used, however, in preventing plant growth under fences, near curbing, or in other areas where mowing is difficult or impossible. Some soil sterilant herbicides move laterally in the soil solution and affect areas not initially treated with the sterilant. Prudent use around the edges of turf is necessary to ensure an attractive appearance and to prevent injury to landscaping materials that have roots extending into treated areas.

4.11.1   Control Strategy

The first step in the selection of an appropriate control strategy is in the correct identification of the weed. Most herbicides will control a broad spectrum of weeds, but only within a targeted life cycle.

Turfgrass weeds can be divided into three convenient groups, based on the most effective mode of weed control. Annual grasses and many broadleaf weeds are controlled with the use of herbicides applied prior to germination of the targeted weed species. These are called preemergence herbicides. Perennial grasses, due to their growth habits that closely resemble those of the turf, provide a difficult problem as to selective control. For the most part, they are controlled with a nonselective herbicide and require renovation of the affected areas as a second step. Most broadleaf weeds and many grasslike weeds can be controlled selectively with a group of herbicides that have varying degrees of plant safety for different turfgrass species. These herbicides are usually applied to the actively growing weed and are absorbed through the foliage. They are called postemergence herbicides.

Although these broad classifications of weed control cover most weeds found in turf, there are many exceptions, particularly in different areas throughout the country. Information on local weed-control methods and selection of herbicides should be obtained from the state extension turfgrass specialist. This information will generally contain specific control methods for common weeds in your area.

Environmental conditions both at the time of application and during the period of herbicide absorption and translocation will greatly influence the ability of the herbicide to control the weeds. The greatest absorption of herbicide will take place in weeds that are actively growing. Cool, dry soil conditions will minimize the effectiveness of a herbicide application. Under dry conditions, it is advantageous to irrigate prior to a herbicide application or wait for a rainfall.

The relative stage of growth of weeds can also be an important factor in their control. Young, rapidly expanding weeds are more sensitive to injury and control from herbicides than are older, maturing plants. Applications of herbicides shortly after plant emergence are most effective when the plants are young and actively growing.

4.11.2   Preemergence Herbicides

Most annual grasses can be controlled with a wide spectrum of materials used as preemergence herbicides (table 4-9). These chemicals are for the most part safe and can be used on most turfs. However, most limit the versatility that a turfgrass manager might have in reseeding or overseeding a thin turf to provide additional plant density. With the exception of siduron and ethofumesate, preemergence herbicides are injurious to turfgrass seedlings. Siduron has selective activity on warm-season annual grasses and can be used in a cool-season turf seedbed. It can control a broad spectrum of weeds when applied at the time of seeding of perennial ryegrass turf. Application of ethofumesate should be made directly to the soil surface prior to any mulching.

The period of active control of a preemergence herbicide depends on the material used, rate of application, soil temperature and moisture, and soil texture in general. Most preemergence herbicides are degraded in the environment through their gradual breakdown by microbial organisms. Losses through leaching, runoff, or chemical decomposition are generally minimal. The breakdown rate of a preemergence herbicide is therefore accelerated when conditions enhance the growth and development of microorganisms. Warm, moist soil with adequate aeration will provide the best environment for breakdown, whereas cool, dry, or compacted soils will slow the degradation process.

TABLE 4-9   PREEMERGENCE HERBICIDES USED TO CONTROL WEEDS IN TURF*

Herbicide: common name (trade name)[†]	Weeds controlled	Tolerant turfgrasses	Rate (lb AI/acre)[‡]
Atrazine (Atrazine)	Goosegrass, crabgrasses, annual bluegrass, barnyardgrass, common purslane, corn speedwell, burclover, yellow woodsorrel, carpet bunweed, chickweed, henbit, spurweed	Centipedegrass, zoysiagrass, bermudagrass, St. Augustinegrass	1 to 2
Benefin (Balan)	Crabgrasses, barnyardgrass, fall panicum, yellow foxtail, goosegrass, junglerice, seedling johnsongrass, Italian ryegrass, field sandbur, carpetweed, prostrate knotweed, lambsquarters, prostrate pigweed, purslane, annual bluegrass, chickweed	Kentucky bluegrass, perennial ryegrass, centipedegrass, tall fescue, fine fescues, zoysiagrass, bermudagrass, St. Augustinegrass, bahiagrass	2 to 3
Benefin and trifluralin (Team)	Crabgrasses, goosegrass, yellow foxtail, barnyardgrass, annual bluegrass, spotted spurge, prostrate spurge, yellow woodsorrel	Bentgrass (high cut only), Kentucky bluegrass, perennial ryegrass, centipedegrass, tall fescue, fine fescues, centipedegrass, zoysiagrass, bermudagrass, St. Augustinegrass, bahiagrass	1.5 to 3
Bensulide (Betasan, Lescosan)	Crabgrasses, barnyardgrass, fall panicum, yellow foxtail, goosegrass, field sandbur, carpetweed, sheperdspurse, lambsquarters, prostrate pigweed, henbit, annual bluegrass	Kentucky bluegrass, zoysiagrass, rough bluegrass, creeping bentgrass, perennial ryegrass, St. Augustinegrass, tall fescue, fine fescues, redtop, bermudagrass, bahiagrass, centipedegrass	7.5 to 10
Bensulide and oxadiazon (goosegrass/crabgrass control)™	Crabgrass and goosegrass	Bermudagrass, zoysiagrass, tall fescue, bentgrass, Poa annua, or perennial ryegrass; may also be used on bermudagrass and bentgrass greens	6 + 1.5
DCPA (Dacthal)	Crabgrasses, annual bluegrass, spotted spurge, prostrate spurge, yellow foxtail, barnyardgrass, goosegrass	All turfgrasses except when grown at putting green height	10 to 15
Dithiopyr (Dimension)	Goosegrass, crabgrasses, annual bluegrass, barnyardgrass, crowfootgrass, kikuyugrass, ryegrass, smutgrass, lespedeza, fall panicum, hairy bittercress, parsley-piert, common purslane, pineappleweed, speedwell, shepherdspurse, yellow woodsorrel, spurge, chickweed, henbit, Carolina geranium	Kentucky bluegrass, zoysiagrass, creeping bentgrass, perennial ryegrass, St. Augustinegrass, tall fescue, fine fescues, redtop, bermudagrass, bahiagrass, buffalograss, carpetgrass, centipedegrass, kikuyugrass	0.25 to 0.88

(continued)

181

TABLE 4–9   PREEMERGENCE HERBICIDES USED TO CONTROL WEEDS IN TURF* (continued)

Herbicide: common name (trade name[†])	Weeds controlled	Tolerant turfgrasses	Rate (lb AI/acre)[‡]
Diphenamid (Enide)	Most grasses	Dichondra turfs only	10
Ethofumesate (Prograss)	Annual bluegrass, large crabgrass, barnyardgrass, yellow foxtail, canarygrass, burclover, common chickweed, common purslane, redroot pigweed	Perennial ryegrass, Kentucky bluegrass, creeping bentgrass, St. Augustinegrass, tall fescue, dormant bermudagrass	0.75 to 3
Isoxaben (Gallery)	Most broadleaf weeds	Kentucky bluegrass, zoysiagrass, bentgrass, perennial ryegrass, St. Augustinegrass, tall fescue, fine fescues, bermudagrass, bahiagrass, buffalograss, centipedegrass	0.5 to 1
Metolachlor (Pennant)	Annual bluegrass, crabgrass, sprangletop, yellow nutsedge	Bermudagrass, centipedegrass, St. Augustinegrass, zoysiagrass, bahiagrass	2 to 4
Monuron (Cooke Oxalis Control)	Annual grasses, chickweed, woodsorrel	Dichondra turfs only	1
Napropamide (Devrinol)	Goosegrass, crabgrasses, stinkgrass, barnyardgrass, seedling johnsongrass, carpetweed, fall panicum, ragweed, common purslane, pigweed, lambsquarters	Bermudagrass, centipedegrass, St. Augustinegrass, zoysiagrass, bahiagrass	2
Oryzalin (Surflan)	Crabgrasses, barnyardgrass, crowfootgrass, yellow foxtail, goosegrass, seedling johnsongrass, Italian ryegrass, field sandbur, carpetweed, prostrate knotweed, henbit, purslane, annual bluegrass, chickweed	Centipedegrass, buffalograss, tall fescue, zoysiagrass, bermudagrass, St. Augustinegrass, bahiagrass	2 to 3
Oryzalin + benefin (XL)	Crabgrasses, barnyardgrass, crowfootgrass, yellow foxtail, goosegrass, seedling johnsongrass, Italian ryegrass, field sandbur, carpetweed, prostrate knotweed, henbit, purslane, annual bluegrass, chickweed	Centipedegrass, buffalograss, tall fescue, zoysiagrass, bermudagrass, St. Augustinegrass, bahiagrass	4 to 6
Oxadiazon (Ronstar)	Goosegrass, crabgrasses, annual bluegrass, barnyardgrass, fall panicum, hairy bittercress, carpetweed, common purslane, lambsquarters, speedwell, spotted catsear, swinecress, yellow woodsorrel	Perennial ryegrass, Kentucky bluegrass, bermudagrass, St. Augustinegrass, tall fescue, zoysiagrass	2 to 4

Herbicide (trade name)	Weeds controlled	Tolerant turfgrasses	Rate (lb AI/acre)
Oxadiazon and benefin (Regalstar)	Crabgrass, foxtail, barnyardgrass, fall panicum	Bermudagrass, St. Augustinegrass, centipedegrass, zoysiagrass, bentgrass (fairways only), bluegrasses, ryegrasses, and tall fescue; not for use on home lawns	2 + 1
Oxadiazon and prodiamine (Regalstar II)	Crabgrass, foxtail, barnyardgrass, fall panicum	Bermudagrass, St. Augustinegrass, centipedegrass, zoysiagrass, bentgrass (fairways only), bluegrasses, ryegrasses, and tall fescue; not for use on home lawns	2 + 0.4
Pendimethalin (Scotts Weedgrass Control 60WDG, Pre M) Light Rate	Crabgrasses, foxtails, barnyardgrass, fall panicum, annual bluegrass	Kentucky bluegrass, perennial ryegrass, fescues, bermudagrass, St. Augustinegrass, centipedegrass, bahiagrass, zoysiagrass	1.5
Pendimethalin (Scotts Weedgrass Control 60WDG, Pre M) Heavy Rate	Crabgrasses, foxtails, barnyardgrass, fall panicum, annual bluegrass, prostrate spurge, yellow woodsorrel, hop clover, cudweed, evening primrose, chickweed, henbit	Bermudagrass, bahiagrass, St. Augustinegrass, centipedegrass, zoysiagrass, tall fescue	3.0
Prodiamine (Barricade)	Crabgrass, foxtail, barnyardgrass, fall panicum, and other annual grasses	Bermudagrass, St. Augustinegrass, centipedegrass, zoysiagrass, bentgrass (do not apply to greens), bluegrasses, ryegrasses, and fescues	0.5 to 1
Pronamide (Kerb)	Cool-season annual and perennial grasses	Bermudagrass, St. Augustinegrass, centipedegrass, and zoysiagrass; use on warm-season turf only	1 to 1.5
Siduron (Tupersan)	Crabgrasses, foxtails, barnyardgrass	Kentucky bluegrass, tall fescue, fine fescues, smooth brome, perennial ryegrass, orchardgrass, Seaside, Highland, Astoria, and C-7 creeping bentgrasses	6 to 12
Simazine (Princep)	Annual bluegrass, little burclover, crabgrasses, lawn burweed, chickweed, speedwell, henbit, hop clover, yellow foxtail, barnyardgrass	Bermudagrass, St. Augustinegrass, zoysiagrass, centipedegrass	1 to 2

*Refer to current product labels for full list of weeds controlled, tolerant turfgrasses, and application rate ranges.
†Trade names are only listed as examples. Contact a local turfgrass extension specialist for a current list of available products.
‡AI = active ingredient.

An additional, important consideration in the effectiveness of the preemergent herbicide is the degree of adsorption of the materials to clay particles in the soil. The greater the clay content in the soil, the greater the adsorption, leaving less herbicide available for absorption into emerging plants. Herbicide labels often instruct the user to apply more material in heavier clay soils to compensate for this process. In lighter, sandy soils, it is necessary to reduce the rate of application to minimize the possibility of injury to the turf. Second applications, often at reduced rates, are sometimes necessary to provide season-long weed control. Providing control of annual grasses is the primary use of preemergence herbicides. A secondary benefit is often realized in the control of annual broadleaf species. This minimizes the need for postemergent control of broadleaf annuals.

4.11.3   Postemergence Herbicides

Postemergence herbicides are utilized to selectively control many broadleaf and grass weeds in turf. These materials are systemic, absorbed through leaf and stem or root tissue, and transported throughout the plant to affect all parts. table 4–10 lists the selective postemergence materials for broadleaf-weed control on turfs. 2,4-D, MCPP, MCPA, and dicamba are often available in varying combinations. They may also be tank-mixed. When applied in combination, these materials exhibit a synergistic property. This enhanced activity may also cause injury to the turf. The user is cautioned to seek advice on the relative concentration of materials to be used in combinations and read the label carefully. Optimum application conditions require moist, warm soil conditions for active weed growth. Postemergence herbicides must remain on leaf and stem surfaces for a period of time after application for thorough absorption. They should not, therefore, be applied prior to an anticipated rain. Cool weather and/or dry conditions after application can minimize the effectiveness of postemergence herbicide applications.

Table 4–11 lists the relative sensitivity of broadleaf-weed species to different postemergence herbicides. Sensitivity of each weed to the listed herbicide or herbicide combination is given in a scale of sensitive to resistant. Response ratings indicate control possibilities under ideal conditions. Poorer control is generally found under less than optimal conditions. Lack of a rating for a herbicide-weed combination indicates that information was not available at the time of publication. This sensitivity can be altered by the combination of materials, which often provides a synergistic degree of control. Most postemergence herbicides remain active long enough to prevent early reseeding or renovation of a turfgrass site. Very few postemergence herbicides have minimal soil activity, so very few can be used in seedbeds to help control broadleaf weeds competing with turfgrass seedlings.

Annual grass weeds occasionally escape preemergence control. A group of postemergence herbicides is available for their control (table 4–12). These herbicides should be applied in sequential applications to provide consistent, thorough control. Once the annual grass plant reaches a state of maturity, rates of herbicides required for consistent control are injurious to the turf.

TABLE 4–10 POSTEMERGENCE HERBICIDES USED FOR SELECTIVE CONTROL OF BROADLEAF WEEDS IN TURF*

Herbicide: common name (trade name[†])	Rate (lb/AI[‡]/acre)	Remarks
Bentazon (Basagran)	1	Use to selectively control yellow nutsedge in turf; repeat applications are sometimes necessary for total control
Bentazon and atrazine (Prompt)	0.5 to 0.75	See table 4.3 for spectrum of weeds controlled
Bromoxynil (Brominal, Chipco Buctril, etc.)	0.375 to 2	May be used for broadleaf seedling control in newly seeded seeded turf or in combination with 2,4-D, MCPP, or dicamba for use on established turfs except bentgrass
Chlorsulfuron (Corsair)	0.125 to 0.25	Used to selectively control tall fescue in Kentucky bluegrass turf
Clethodim (Envoy)	0.125 to 0.25	Bermudagrass control in centipedegrass on sod farms only
Clopyralid (Lontrel)	0.09 to 0.5	See table 4.3 for spectrum of weeds controlled
2,4-D	1	See table 4.3 for spectrum of weeds controlled
2,4-D + dicamba	0.44 to 1.75	See table 4.3 for spectrum of weeds controlled
2,4-D + dichlorprop	0.9	See table 4.3 for spectrum of weeds controlled
2,4-D + MCPP + dicamba	1.44 to 1.56	See table 4.3 for spectrum of weeds controlled
2,4-D + MCPP + dichlorprop	1 to 2	See table 4.3 for spectrum of weeds controlled
MCPA + MCPP + dichlorprop	1 to 2	See table 4.3 for spectrum of weeds controlled
2,4-D + MCPP + MSMA + dicamba	1 to 1.5	See table 4.3 for spectrum of weeds controlled
2,4-D + triclopyr	0.75 to 1	Controls most weeds listed in table 4.3; provides partial control of violets
Dicamba	0.25 to 1	See table 4.3 for spectrum of weeds controlled
Fluazifop (Fusilade II)	0.03 to 0.09	Bermudagrass suppression in tall fescue or zoysiagrass
Halosulfuron (Manage)	0.031 to 0.062	Use to selectively control yellow and purple nutsedge in turf; repeat applications are sometimes necessary for total control
Imazapic (Plateau)	0.063 to 0.125	Bahiagrass, crabgrass, yellow and purple nutsedge, and annual sedge; for centipedegrass only grown as commercial turf or golf courses; may also be used on selective native grass species
Imazaquin (Image)	0.25 to 0.5	Purple and yellow nutsedge, green kyllinga; use on warm-season turf only
Metsulforon (Manor)	0.005 to 0.038	May be used on most established turfgrass species except bahiagrass and perennial ryegrass; often used to remove perennial ryegrass from overseeded warm-season turfgrass species
Metribuzin (Sencor)	0.25 to 0.5	Use on warm-season turf only
MCPP	0.5 to 1	See table 4.3 for spectrum of weeds controlled
Simazine (Princep)	1	See table 4.3 for spectrum of weeds controlled; use on warm-season turf only
Triclopyr (Turflon Ester)	0.5 to 1	See table 4.3 for spectrum of weeds controlled
Triclopyr + clopyralid (Confront)	0.38 to 0.75	See table 4.3 for spectrum of weeds controlled

*Refer to current product labels for full list of weeds controlled, tolerant turfgrasses, and application rate ranges.
[†]Trade names are only listed as examples. Contact a local turfgrass extension specialist for a current list of available products.
[‡]AI = active ingredient.

TABLE 4–11   SUSCEPTIBILITY OF BROADLEAF WEEDS IN TURF TO POSTEMERGENCE HERBICIDES*

Broadleaf weed	Atrazine/Simazine	2,4-D	MCPP	Dicamba	Triclopyr	Triclopyr + clopyralid	2,4-D + MCPP	2,4-D + dichlorprop	2,4-D + MCPP + dicamba	2,4-D + triclopyr
Bedstraw		I†		S						
Betony, Florida	S-I	I	—	S			—	I	I-S	
Bindweed, field		S-I	—	S			S-I	S	S	
Bittercress, hairy		S-I	—	S			S	S	S	S
Burweed, lawn	S-I	I	S-I	S			S-I	—	S	S
Buttercup, creeping		S-I	—	S			S	S	S	
Carpetweed	S	S	—	S			S	S	S	
Carrot, wild		S	S-I	S	S		S	S-I	S	I-R
Catnip		S		S			S	S	S	S
Catsear, spotted		S	S	S					S	
Chamomile, mayweed		I	S	—						
Chickweed										
Common	S	R	S-I	S			S	S	S	S
Mouseear	I	R	S-I	S			S	S	S	S-I
Chicory		S	S	S			S	S	S	
Cinquefoil, silvery		S-I	S-I	S-I			S-I	S-I	S-I	
Clover										
Crimson		S	S	S			S	S	S	
Hop	S	I	S	S			S	S	S	S
White	S	I	S	S		S	S	S	S	S-I
Daisy										
English	I-R	R	—	S		S	—	—	S	I-S
Oxeye		R	—	—			—	—	—	
Dandelion		S	S-I	S	S		S	S	S	
Deadnettle, red		I	—	S			S	S	S	
Dichondra	S-I	S-I	—	S-I	S	S	—	—	S	
Dock, curly		I	I-R	S	S				S	
Filaree, redstem		S-I	—	S					S-I	I

186

Weed									
Garlic, wild		—	R	S-I			S-I			S
Geranium, Carolina		S-I	S-I	S-I		S	S-I	S-I	S	S
Hawkweed, yellow		S-I	R	S		S-I	S-I	S-I	S	S
Healall	S	S	S-I	S		S-I	S	S	S	S
Henbit	S	—	—	—		—	—	—		
Ivy										
Ground		I-R	—	S-I	—	I-S		S-I	S	S
Poison		—	—	—		—		—		
Knotweed, prostrate	S	R	—	S		R	S-I	S-I	S	S
Kochia		S	S	S		S	S-I	S	S	S
Lambsquarters		S	S	S	S	S	S-I	S	S	S
Lespedeza, common	S	I-R	S	S-I		—	S-I	—	S	S
Mallow, roundleaf		I-R	—	S		—	—	S-I	S	S
Medic, black		R	—	S		S	S	S	S	S
Moneywort		S	S-I	S		S-I	S-I	—		
Mugwort	S	S-I	—	S		—	—	—		
Mustard, wild		S	R	—		S	S	S	S	S
Onion, wild		—	S	R		—	—	—		
Parsley-piert	S	R	—	S		R	S-I	S-I	S-I	S
Pennycress, field		S	S-I	—		S	S	S	S	S
Pennywort, lawn	S	S-I	S-I	S-I		S-I	S-I	S-I		
Pepperweed, field		S	S	S-I		S	S	S	S	S
Pigweed, prostrate		S	S	S		S	S	S		
Pineappleweed	—									
Plantain										
Broadleaf	I-R	S	I-R	R		S	S	S	S	S
Buckhorn	I-R	S	I-R	R		S	S	S	S	S
Poorjoe		S	S	S		S	S	S		
Puncturevine		S	R	S		S	S			
Purslane, common		—	S-I	S	S	—	—	—	S	S
Shepherdspurse		—	S-I	S		S	S-I	S-I	S	S
Sorrel, red		R	R	S		—	S-I	S	S	S

(continued)

TABLE 4-11 SUSCEPTIBILITY OF BROADLEAF WEEDS IN TURF TO POSTEMERGENCE HERBICIDES* (continued)

Broadleaf weed	Atrazine/Simazine	2,4-D	MCPP	Dicamba	Triclopyr	Triclopyr + clopyralid	2,4-D + MCPP	2,4-D + dichlorprop	2,4-D + MCPP + dicamba	2,4-D + triclopyr
Sowthistle, annual	S									
Speedwell										
Creeping	S	R	R	R			I-R	I-R	I-R	
Purslane	S	I	I	I			I	I	S	
Spurge, prostrate	S-I	I-R	I	S-I			I	S-I	S	S-I
Starwort, little		I	I	S						
Strawberry, wild		R	R	S-I			I	R	S-I	
Thistles										
Bull		S	I	S		S	S-I	S-I	S	S
Canada		I	I	S		S	S-I	S-I	S	S
Vervain, prostrate		S-I	R	S-I			S	S		
Violets		R	R	R			I-R	I	I-R	I-R
Woodsorrel, yellow	I	I	I-R	I			I-R	I-R	S	I
Yarrow, common	I	I		S			I-R	I	S	S

* Indication of relative sensitivity of weed species under optimum conditions (warm, moist soil for rapid growth); cool, dry conditions will generally reduce susceptibility.

†S = susceptible; I = intermedially susceptible; R = resistant.

188

TABLE 4–12 POSTEMERGENCE HERBICIDES USED FOR SELECTIVE CONTROL OF ANNUAL GRASS WEEDS IN TURF

Herbicide: common name (trade name)	Rate (lb AI/acre)	Tolerant turfgrasses	Weeds controlled	Remarks
Asulam (Asulox)	2	Tifway (419) Bermudagrass, St. Augustinegrass	Crabgrass, goosegrass, sandspur, smutgrass	Do not apply to freshly mowed turf or turf under stress. On bermudagrass use Tifway only
Diclofop-methyl (Hoelon, Illoxan)	0.75 to 1	Bermudagrass	Goosegrass	For use on golf courses and sod farms. Young goosegrass plants are easiest to control. The high rate is needed for more mature plants. Larger, mature goose-grass will not be adequately controlled. Do not mow 24–36 hours after applying
Dithiopyr (Dimension)	0.25 to 0.88	Kentucky bluegrass, zoysiagrass, creeping bentgrass, perennial ryegrass, St. Augustinegrass, tall fescue, fine fescues, bermudagrass, bahiagrass, buffalograss, carpetgrass, centipedegrass	Crabgrasses	For early postemergence control of crabgrass. After tillering, tank-mix with MSMA or Acclaim
DSMA (Weedone Crabgrass Killer, Scott's Summer Crabgrass Control, DSMA Liquid, etc.)	4 to 6	Consult label for formulation; do not use on St. Augustinegrass, bentgrasses, or fine fescues	Crabgrasses, dallisgrass, yellow nutsedge	High rates are needed for mature crabgrass; requires two to three resprays at 5- to 10-day intervals; apply when temperatures are below 85°F; will often result in yellow, discolored turf for a short period
Metribuzin (Sencor)	0.25 to 0.5	Bermudagrass	Goosegrass, annual bluegrass	For bermudagrass only

(continued)

189

TABLE 4-12 POSTEMERGENCE HERBICIDES USED FOR SELECTIVE CONTROL OF ANNUAL GRASS WEEDS IN TURF (continued)

Herbicide: common name (trade name)	Rate (lb AI/acre)	Tolerant turfgrasses	Weeds controlled	Remarks
Metribuzin + MSMA	0.125 to 2.0	Bermudagrass	Goosegrass, annual bluegrass, crabgrasses, dallisgrass, yellow nutsedge	For bermudagrass only
MSMA (Acme Crabgrass and Nutgrass Killer, Bueno 6, etc.)	4 to 6	Consult label for formulation; do not use on bentgrass, St. Augustinegrass, red fescue, dichondra, or zoysiagrass		Air temperature and turf type determines degree of sensitivity; apply when air temperatures are below 85°F; most turfs are generally more sensitive to MSMA than to DSMA
CSMA (Weedone Crabgrass Killer)	2 to 5	Consult label for formulation; use low rate in bentgrass, dichondra, or fine fescue turf; may injure St. Augustinegrass, fescue, and some bentgrass species	Crabgrasses, dallisgrass, yellow nutsedge	Do not use when temperatures exceed 85°F; mature weeds may require repeat applications at 5- to 10-day intervals; may temporarily discolor or yellow turf
Pronamide (Kerb)	1.5	Bermudagrass	Annual bluegrass	Can be applied both pre- and postemergence for control of annual bluegrass
Quinclorac (Drive)	0.75	Most turfgrass species, except bahiagrass, centipedegrass, St. Augustinegrass, or dichondra	Crabgrass, barnyardgrass, white and hop clover, common dandelion, and foxtail	Most effective when used with a surfactant or crop oil
Sethoxydim (Vantage)	0.3	Centipedegrass	Crabgrass, goosegrass	Safe on centipedegrass seedlings after the third mowing
Simazine (Princep)	1	Bermudagrass	Annual bluegrass	Do not use during spring green-up
Fenoxaprop ethyl (Acclaim)	0.117 to 0.250	Kentucky bluegrass, perennial ryegrass, fine fescues, tall fescue, annual bluegrass	Crabgrasses, goosegrass, barnyardgrass, foxtail spp., panicum spp., johnsongrass	Apply prior to second tiller; do not apply to Kentucky bluegrass less than 1 year old; do not tank mix with other herbicides

TABLE 4–13   NONSELECTIVE POSTEMERGENCE HERBICIDES USED TO CONTROL WEEDS IN TURF

Herbicide: common name (trade name)	Rate (lb AI/acre)*	Remarks
Cacodylic acid (Phytar 560)	6 to 8	Apply as a preplant for control of principally annual weeds
Diquat dibromide (Reward)	0.5 to 1	Contact herbicide; will not control established perennial weeds; surfactant will enhance coverage; no soil residue; may seed after treatment
Glufosinate-ammonium (Finale)	0.75 to 1.25	Has no residual activity in the soil; repeated treatments may be necessary for complete control of some weeds
Glyphosate (Roundup, Kleenup, Touchdown)	1 to 2	Glyphosate has no residual activity in the soil; repeated treatments may be necessary for complete control of some weeds

*AI = active ingredient.

4.11.4   Nonselective Herbicides

Nonselective herbicides are most often used to control perennial grasses in turf (table 4-13). For small weed populations, a spot or localized application is generally appropriate. For large populations, broadscale applications are made with a follow-up reestablishment of the turf. Nonselective herbicides are generally chosen on the basis of the time they remain active in the soil and the possibility of their causing injury to turfgrass seedlings or sod. Herbicides such as glyphosate or glufosinate-ammonium, which have limited activity in the soil, are often chosen to minimize the interval between weed control and reestablishment of the turf. It is imperative that the application of nonselective herbicides precede a period in the year optimal for turfgrass establishment. Poor timing of nonselective weed control is generally not successful, due to reinvasion of the turf with similar weeds.

Other nonselective herbicides (atrazine, bromacil, diuron, and the like) that act as soil sterilants are sometimes used to maintain weed-free areas near established turfs. Their high degree of soil activity, however, limits their use in turf. They can often cause problems to turf and desirable ornamental plants if soil movement (for instance, erosion) occurs.

4.12   STUDY QUESTIONS

1. What are the three major groups or categories of turf weeds? Describe the three major life cycles.
2. How do weeds influence turf growth? Provide some examples.
3. What are the steps in identifying turf weeds? Define the vegetative characters used in grass identification.

4. Give the normal first annual appearance (season) of the following weeds:

Fall panicum

Timothy

Carpetweed

Curly dock

Yellow nutsedge

Goosegrass

Lawn pennywort

Yellow woodsorrel

Quackgrass

Shepherdspurse

5. What are the basic objectives of cultural control methods?

6. What group of herbicides is used to control perennial grasses and why?

4.13   SELECTED REFERENCES

Anderson, W. P. *Weed Science Principles.* St. Paul, MN: West Publishing Co., 1977.

Anonymous. *Herbicide Handbook of the Weed Science Society of America*, 5th ed. Champaign, IL: Weed Science Society of *America*, 1981.

Anonymous. *Weeds of the North Central States.* College of Agriculture Bulletin 772, Urbana-Champaign IL: Agricultural Experiment Station, University of Illinois, 1981.

Ashton, F. M., and A. S. Crafts. *Mode of Action of Herbicides.* New York: John Wiley & Sons, Inc., 1981.

Fernald, M. L. *Straws Manual of Botany*, 8th ed. New York: American Book Company, 1950.

Gleason, H. A., and A. Cronquist. *Manual of Vascular Plants of Northeastern United States and Adjacent Canada.* Princeton, NJ: D. Van Nostrand Company, 1963.

Gould, F. W. *Grass Systematics.* New York: McGraw-Hill Book Company, 1968.

Hitchcock, A. S. (revised by Agnes Chase). *Manual of the Grasses of the United States.* United States Department of Agriculture Miscellaneous Publication 200, Washington, DC, 1950.

King, L. J. *Weeds of the World: Biology and Control.* Plant Science Monograph (N. Poulnin, ed.), New York: Wiley-Interscience, 1966.

McCarty, L. B., J. W. Everest, D. W. Hall, T. R. Murphy, and F. Yelverton. *Color Atlas of Turfgrass Weeds.* Chelsea, MI: Ann Arbor Press, 2001.

Watschke, T. L., P. H. Dernoeden, and D. J. Shetlar. *Managing Turfgrass Pests.* Boca Raton, FL: Lewis Publishers, 1994.

Biology and Management of Diseases in Turfgrasses

5.1 INTRODUCTION

All species of plants, both wild and cultivated, are subject to disease attack. Plant diseases should be considered as a *normal* part of nature and one of many environmental factors that help to keep the hundreds of thousands of living plants and animals in balance with each other. All pathogens that cause disease in turfgrass are also beneficial to the ecology of turf, and as such should never be eliminated. Instead, turfgrass diseases should be managed to ensure that their severity is maintained at an acceptably low level. When humans select and cultivate plants, they should recognize that diseases are one of the many expected hazards. The longer and more intensively a plant or cultivar is grown, the greater the number of diseases identified by plant pathologists. The ever growing number of diseases represents both the process of evaluation and the recognition of *new* diseases as they become important or obvious enough to be identified. As cultural practices have changed (for instance, mowing equipment and height of cut, dethatching, cultivation, fertility, pesticides, and turfgrass cultivars), so have diseases. When changes are made in the predominant cultivars and grasses grown, new diseases have appeared.

Cultivated plants, including turfgrasses, are usually *more* susceptible to disease than their wild relatives—partly because large numbers of the same species or cultivar, having a uniform genetic background, are grown close together in pure stands, sometimes over large areas. Once a disease-causing organism or agent (a *pathogen*) becomes established under these conditions, it often spreads rapidly and may become very destructive. In addition, the cultivation of any species or cultivar is another example of our constant attempt to control nature.

Plant diseases are not new. Fossils prove that plants had diseases 250 million years or more before humans appeared on earth. Plant parasites capable of causing disease undoubtedly arose and developed as plants arose and developed on earth.

5.1.1 What Is a Plant Disease?

When a plant is more or less *continuously* irritated, injured, or subjected to stress over a fairly long period of time by some harmful factor in its environment that interferes with its normal structure (appearance), growth, or functional activities, it is said to be diseased. The disease that results is expressed by characteristic symptoms and/or signs. *Symptoms* are the visible expressions of a turfgrass plant's reaction to the activities of an outside factor. *Signs* are indications of disease from direct visibility of a pathogen or its parts and products. Diseases such as a wilt or viral infection affect the entire plant; others, such as leaf spots and powdery mildews, affect only part of a plant.

Considerable knowledge of a plant's normal growth habits, cultivar characteristics, and normal variability within a species, as related to the environmental conditions under which the plant is growing, are required for recognizing disease. Symptoms of a disease may vary greatly depending on the turfgrass species, cultivar, and environmental conditions.

Plant injury, in contrast, usually results from *momentary* or *discontinuous* damage. Broadly speaking, a plant is considered diseased when it does not develop or produce normally and function to its full genetic potential. There is often no sharp distinction between a healthy and a diseased plant. Disease, according to some scientists, is merely an extreme case of poor growth.

No one quarrels that lightning, mechanical damage by a lawn mower, or a chemical spill (fertilizer or pesticide burn) results in injury. Rodent, insect, and mite feeding are generally considered to be injuries. But how about a toxin secreted by an aphid or chinch bug when it feeds on a grass plant over a period of several hours or days that results in a slow and progressive scorching, wilting, or dieback of leaves? Is this an example of disease or injury? There is no sharp distinction between the two.

Disease is often a major factor in the success or failure of a turfgrass stand. The best cultural practices may fail if a very susceptible species or cultivar is grown where a major disease is widespread. Examples include Kentucky bluegrasses susceptible to necrotic ring spot, summer patch, leaf smut, or a Helminthosporium disease;[1] spring dead spot of bermudagrass and buffalograss, in the more northern regions of the warm-season turfgrass belt; and St. Augustinegrass decline (SAD) in many warm-to-hot areas where this grass is grown.

5.1.2   Causes of Grass Diseases

Diseases of turfgrasses are conveniently divided into two principal groups:

1. Disorders caused by injury or unfavorable growing conditions (called nonparasitic, noninfectious, physiogenic, or abiotic)
2. Diseases caused by living pathogens or parasites including infectious bacteria, mollicutes (phytoplasmas and spiroplasmas) fungi, viruses, and nematodes (called infectious, biotic, or transmissible)

Pathogens are organisms or agents capable of causing disease. Most pathogens are parasites (organisms or agents that obtain food from living organisms), but there are some exceptions.

It is important that these two groups be distinguished, since management (and therefore control) measures depend on an early and accurate diagnosis of disease or injury (be it caused by insect, drought, extremes in temperature, or chemical burn). Noninfectious disorders are *not* contagious; they do not spread from diseased to

---

[1] Diseases formerly attributed to species of *Helminthosporium* are now thought to be caused by other fungi; however, by convention, the name "Helminthosporium disease" is still useful.

healthy turfgrass plants. Included here are an excess, deficiency, or unavailability of water, light, air movement, one or more of the 20+ essential soil elements (including nitrogen, phosphorus, potassium, calcium, iron, magnesium, manganese, boron, copper, molybdenum, zinc, and sulfur), unfavorable soil moisture and/or oxygen relationships for normal root development, extreme soil acidity or alkalinity (a soil pH below about 5.5 and above 7.5), leaf and crown bruises, mower shredding of leaf blades, and scalping. Other noninfectious disorders include animal urine or salts, pesticide and fertilizer injury, extremely high or low temperatures, shallow or compacted soil, buried debris causing shallow roots, excessive thatch, shading or root competition, and other injurious impurities in the air or soil. Turfgrass plants in poor health due to unfavorable growing conditions probably far outnumber plants significantly attacked by disease-causing fungi or other pathogens.

Suspect a noninfectious problem when a large turf area suddenly deteriorates, especially when other nongrass plants in a given area or environment are similarly affected. Although noninfectious agents are capable of inducing disease, they commonly cause plants to become more susceptible to attack by infectious agents. More commonly, two or more noninfectious disorders and several infectious agents act together to produce a complex or poorly defined disease or syndrome. Here it is necessary to eliminate the unfavorable growing conditions before, or simultaneously, with controls aimed at checking the growth and spread of infectious agents.

5.1.3   Parasites or Pathogens

Parasitic organisms obtain some or all of their food from living plants. If these organisms cause injury or disease, they are called pathogens. An *obligate parasite* (such as a rust or powdery mildew fungus, plant-parasitic nematode, or virus) requires living plant cells as food to complete their life cycle, whereas a *facultative parasite* (which includes most turfgrass fungi) may obtain its nutrition from either a living or dead plant. These organisms generally cause disease when unfavorable environmental conditions have predisposed the turfgrass plant (or *host*) to infection or when moisture and temperature conditions are particularly favorable for the pathogen.

Parasites that cause infectious diseases often spread readily from diseased to healthy plants by air currents; flowing or splashing water; insects, mites, or other animal life; human activities (such as mowing, dethatching, cultivating, and walking); movement of soil, thatch, seed, sod, or sprigs; and other ways.

Diseases to be discussed in detail are those caused by infectious fungi, bacteria, viruses, and nematodes. The size relationships of various pathogens are shown in figure 5–1. These pathogens get their nourishment from grass plants and cause roots, stems (crowns, tillers, rhizomes, and stolons), or leaves to grow or die abnormally.

Fungi.   Over 250,000 species of fungi have been described, probably less than one tenth of the total number. Approximately 22,000 different species of fungi are plant parasites that can cause plant diseases. Only a few of these are known to cause turfgrass diseases. Many fungi are able to attack living plants only at certain times and feed on dead organic material (*saprophytes*) in plant

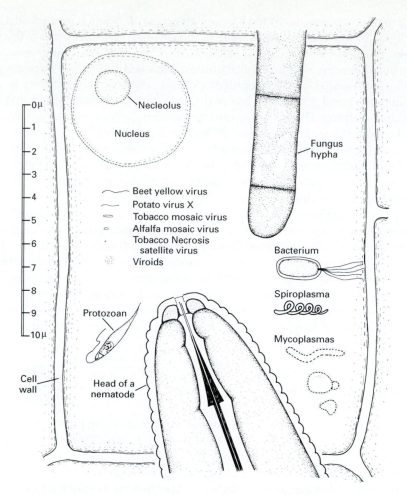

FIGURE 5–1   Schematic diagram showing the relative sizes and shapes of certain plant pathogens in relation to a single plant cell. Viroids and protozoa are not known to infect turfgrasses. The scale to the left is marked out in micrometers (μm) or microns (μ), where each unit equals about ⅟₂₅,₀₀₀ of an inch. Modified from Agrios, *Plant Pathology,* 4th ed., San Diego, CA: Academic Press, 1997.

debris or soil for the remainder of their life cycle. Most plant-pathogenic fungi are capable of growing on a variety of culture media and living or nonliving plant tissues. The great majority of turf diseases are caused by fungi.

These largely microscopic organisms lack chlorophyll and hence cannot produce their own food by the process of photosynthesis. Most fungi are multicellular, usually composed of threadlike or filamentous mycelium or hyphae (figure 5–2), with well-developed cell walls and nuclei. The great majority are microscopic; others, such as mushrooms, are clearly visible, both above and below ground.

Nourishment is obtained from dead or living plants. Most of the fungi that can be found in turf (1 lb [453 g] of topsoil contains 4.5 to 225 million fungi) feed on dead and decaying organic matter such as dead roots, stems, and leaves found in the soil

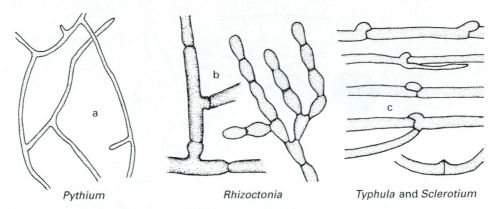

Pythium   Rhizoctonia   Typhula and Sclerotium

FIGURE 5–2   Fungus hyphae. (a) Delicate nonseptate hyphae of *Pythium;* (b) young (left) and mature (right) hyphae of *Rhizoctonia solani;* (c) hyphae typical of *Typhula, Coprinus, Limonomyces,* and *Sclerotium,* showing various types of clamp connections.

or in the thatch. These fungi are generally considered to be beneficial, since they aid in thatch decomposition. Together with bacteria, fungi break down organic matter in the thatch and soil into nutrients that can be utilized by turfgrass plants. If it were not for bacteria and fungi, our world would be piled deep with plant and animal remains.

Over 400 different fungi have been found growing in or on turfgrass plants. About half of these fungi attack living turfgrass plants under favorable conditions of temperature and moisture. Disease-causing fungi can be found in large numbers on dead and decaying leaves and stems, where grass plants are growing poorly under stress conditions. Weakened or injured plants often lose much of their natural resistance to fungi and exhibit more damage from disease than do vigorous plants. But some fungi can also cause disease on vigorously growing turf.

A typical fungus begins life as a microscopic, sexual or asexual spore that can be compared to the seed of a higher plant. The microscopic spores are of many different shapes, sizes, and colors (figure 5–3) and are borne in a variety of ways. Perhaps average fungus spores, if laid end to end like bricks, would fit 2,500 to 3,000 to an inch (2.5 cm). Some spores appear as mold growth on the surface of leaves (gray and Cercospora leaf spots of St. Augustinegrass, red thread, and pink patch, Helminthosporium [*Bipolaris, Drechslera, Exserohilum*] leaf spots, copper spot, leaf smuts, rusts, powdery mildew, and slime molds); others are borne in speck-sized, dark, fungus-fruiting bodies embedded in diseased plant tissue (*Septoria, Ascochyta, Leptosphaerulina,* and *Gaeumannomyces*). Spores play an important role in the multiplication, dissemination, and survival of fungi. Fungal spores and other parts are easily transported by air currents, splashing or flowing water, insects, mites, birds, slugs, spiders and other animals, mowers or other turf equipment, shoes, infested soil, and grass parts, including seeds, sod, sprigs, plugs, and clippings. Humans also spread fungi on their hands and clothing. Foliage that is wet, sticky, hairy, or rough traps more spores than plant surfaces that are dry, smooth, and free of hairs (trichomes). Thick-walled resting spores allow certain turfgrass fungi (such as *Pythium, Sclerophthora,* and *Fusarium*) to withstand unfavorable

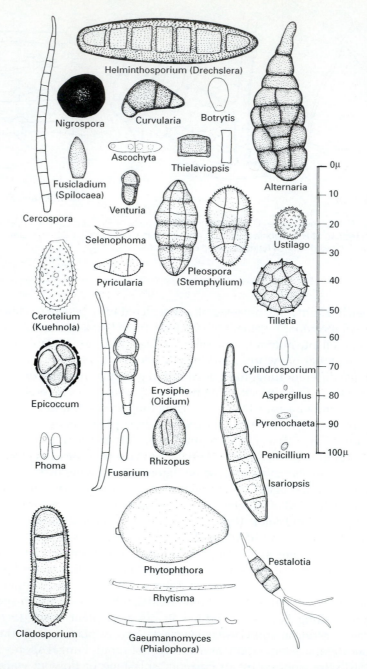

FIGURE 5–3 Representative fungus spores drawn to scale. Some of these fungi do not infect turfgrasses. The color, size, shape, and the way in which spores are borne is important in the classification of fungi. The scale to the right is marked into units of 10 micrometers (μm) or microns (μ).

growing conditions such as extreme heat or cold, drying, and flooding. Resting spores may lie dormant in soil or thatch for several years or more before germinating. Still other fungi, including species of *Rhizoctonia, Sclerotium, Typhula,* and *Coprinus,* produce unusually dark, hard, compact masses of hyphae called bulbils or sclerotia that can withstand even more unfavorable conditions than can resting spores. These structures may also lie dormant in the thatch or upper half inch (1.3 cm) of soil for several years before germinating and initiating disease. Fortunately, most fungus spores and hyphae are easily killed by adverse conditions.

Under moist conditions, the spore germinates and produces one or more branched, tubular filaments called *hyphae* that are 0.5 to 100 microns in diameter (figure 5–2). The hyphae grow and branch to form a fungus body called a *mycelium.* The mycelium forms a woolly mass, an interlacing tangle of hyphae, or even a compact solid body such as a bulbil or sclerotium. A parasitic mycelium may grow on the surface of its plant host, appearing as delicate, whitish, cobweblike threads (such as powdery mildew, *Pythium, Nigrospora,* dollar spot, and snow-mold fungi). Or the mycelium may be completely inside the host plant and not evident on the plant surface.

Fungal hyphae penetrate a plant by (1) growing into a wound made by turfgrass equipment, hail, wind, insect and nematode feeding, or previously occupying fungi; (2) growing through a natural opening (like a stoma); or (3) forcing their way directly through a plant's epidermis by a combination of pressure and enzyme action.

Fungi are often more common and damaging to plants in moist or overly fertilized turf than in dry or moderately nourished grass. Examples include snow molds, take-all patch, most leaf-spot diseases, Rhizoctonia blight or brown patch, and Pythium blight. Free moisture is essential to rapid reproduction, spread, and infection of grass plants by all fungi except powdery mildew. Rusts, dollar spot, red thread, and pink patch are examples of diseases that commonly injure slow-growing, nitrogen-deficient turf when light rains or morning dews persist.

Fungi are classified into genera and species by their type of asexual and sexual reproduction, morphological characteristics, and other features. Fungal pathogens, especially those with a sexual reproductive phase, exhibit a great deal of genetic variability, which means that populations can readily adapt to environmental or chemical changes.

Species of fungi, like other types of parasites, may be subdivided into strains or physiologic races that may differ in: (1) the species and cultivars of plants they attack; (2) virulence—ranging from weakly parasitic to highly virulent; (3) temperature range at which infection, disease development, and reproduction occur; and (4) ability to develop at lower levels of soil moisture and other conditions.

Fungal diseases are controlled through cultural practices, by growing resistant species and cultivars, preferably in blends or mixtures; and by the timely applications of fungicides to the seed, foliage, and soil.

Bacteria.   Many of the several thousands of bacterial species (including mycoplasmas and spiroplasmas, figure 5–1) are beneficial; most are harmless to humans or animals. Less than 300 types of bacteria cause disease in humans, animals, and plants. Approximately 150 species incite disease in plants.

Bacteria exhibit three fundamental shapes: spherical, spiral or curved rods, and rods (figure 5–4). They are one-celled prokaryotes (lacking a nuclear membrane)

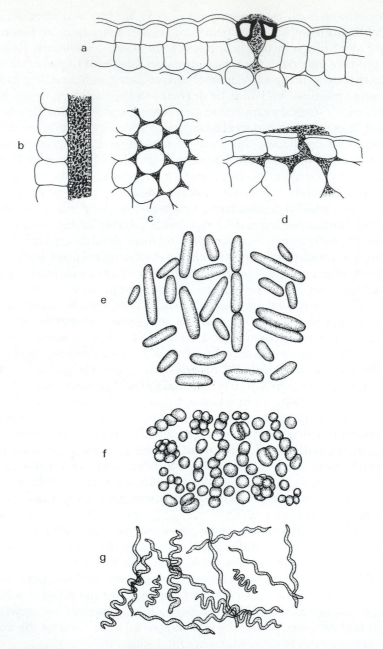

FIGURE 5–4   Bacteria. These one-celled organisms enter plants through natural openings such as stomates (a) or through wounds (d). They commonly migrate within plants by movement through water-conducting vessels (b) or between cells (c). Bacteria are of three basic forms: bacilli or rods (e), cocci or spherical (f), and spirilla or spiral (g). Practically all plant-infecting bacteria are rod shaped, being 0.6 to 3.5 μm in length and 0.5 to 1.0 μm in diameter.

found everywhere in our environment (air, water, soil, and on or in plants, animals, and humans). Bacteria that cause plant disease are microscopic, one-celled organisms that are rod-shaped and measure about 0.5×2 µm (up to about ¹⁄₁₂,₀₀₀ of an inch long). They can be seen only with a good compound light microscope at about 400 to 600 magnification. Placed end to end, it would take 12,000 to 30,000 bacteria to cover 1 in. Some species have whiplike flagella that may aid them to move in water (figures 5–1 and 5–4). Fortunately, plant-pathogenic bacteria do not form spores and hence are readily killed by heat or chemicals.

Bacteria are identified on the basis of cell and colony morphology, host specificity, physiological requirements for growth, Gram reaction (+ or −), and their response to a wide variety of biochemical and serological tests.

Bacteria, like fungi, lack chlorophyll and hence cannot manufacture their own food. Most bacteria feed on dead organic matter (*saprophytes*), but a few are pathogenic. They do not require a living grass plant for multiplication and growth.

Bacteria enter plants only through natural openings, such as stomates, or through wounds (figure 5–4) produced primarily by insects, nematodes, adverse weather, or humans with turfgrass equipment or shoes. Watersoaked and succulent tissues often predispose plants to invasion by bacteria. Free moisture and moderate-to-warm temperatures (65 to 85°F; 18.3 to 29.4°C) are general requirements for reproduction and disease development. The bacteria that infect plants are usually favored by a near-neutral (pH 6.5 to 7.5) growth medium. Fortunately, most bacteria that cause plant diseases are quickly killed by exposure to high temperatures (10 min at 125°F or 51.7°C), dry conditions, and sunlight. Many pathogenic bacteria in soil are eaten by protozoa (figure 5–1) or other minute animal. Some are inhibited by antibiotic substances liberated by other soil-inhabiting organisms, including other bacteria, actinomycetes, and fungi.

When conditions are unfavorable for growth and multiplication, bacteria remain dormant on or in plants, seeds, thatch, turf equipment, or soil. A few are known to live for several months in the bodies of insects. Most plant-pathogenic bacteria are facultative parasites; few are obligate parasites. The great majority of the species of bacteria that infect plants cannot survive in soil once host-plant residues are decayed.

Bacteria are spread through planting, cultivation, mowing, and transporting diseased plant materials. Animals (including insects, mites, and nematodes), splashing or flowing water, and windblown rain or dust are other common disseminating agents.

Bacteria reproduce by simple fission, where a mother cell divides in half to produce two identical daughter cells. Under the most favorable conditions, bacteria can reproduce at an astonishing rate. If a single bacterial cell divided in half and all its descendants did likewise every 20 min for just 12 hours, nearly 70 billion bacteria would be produced. Due to lack of food, the accumulation of toxic wastes, unfavorable temperature or moisture conditions, and other limiting factors, reproduction soon slows and finally stops.

Bacteria, like fungi, overwinter on and in thatch and other plant refuse, soil, turfgrass plants, seed, and occasionally in insects.

Most bacterial diseases are controlled by preventing injuries that allow bacteria to enter plants, starting with disease-free planting material, and growing resistant grass species and cultivars.

Infected grass plants may display a wide range of general symptoms (such as angular leaf spots, blotches, wilts, and flower and seed blights). Some bacteria produce toxins that poison the plant and produce chlorosis, watersoaking, death of plant tissues (*necrosis*), and other symptoms or signs.

The only widespread and important turfgrass disease is bacterial wilt, which is most serious on Toronto (C-15) creeping bentgrass.

Bermudagrass stunting disease in southern Florida is believed to be caused by a small, xylem-inhabiting bacterium, *Clavibacter xyli* ssp. *cynodontis*. This disease is predominantly a problem in bermudagrass golf greens, where unsightly circular, chlorotic patches of dying grass are seen. Stunting disease is worst under such environmental stresses as low light intensity.

Viruses.   The 850+ known plant-infecting viruses infect, replicate, mutate, and otherwise act like living organisms *only* when they are in living plant cells. They are truly obligate parasites. Only a handful of turfgrass diseases are caused by viruses. The only widespread and serious disease is St. Augustinegrass decline (SAD).

Over 200 viruses have been isolated from infected organisms, purified, and crystallized in the laboratory. Outside a living host cell, they are like chemical substances, yet they function more like living organisms inside infected host cells.

Virus particles (or *virions*) can be filamentous (long, rigid, or flexuous rods), roughly spherical (polyhedral or isometric), or bacilliform in shape. They are much smaller than bacteria (perhaps $\frac{1}{200,000}$ of an inch long; see figure 5–5) and can be seen only with a high-resolution electron microscope. They range in diameter from 10 to 70 nm (a nanometer [nm] is $\frac{1}{1000}$ of a micron or micrometer), and rod lengths can exceed 3 μm. With few exceptions, viruses are macromolecules composed of a core of ribonucleic acid (RNA) or deoxyribonucleic acid (DNA), surrounded by a protective protein or lipoprotein coat. Viruses, in some ways, resemble the chromosomes present in all living plant and animal cells. They cannot be cultured on artificial media, and their diagnosis requires specialized laboratory techniques.

Viruses cause disease by imposing a different set of genetic information on the biosynthetic apparatus of host-plant cells that results in drastically altered host metabolism. The nucleic-acid portion of the virus causes plant cells to reproduce large numbers of virus particles. Normal plant-growth processes are diverted to abnormal ones. The changes are expressed in various ways. The more obvious symptoms of virus infection are yellowing and loss of vigor due to destruction of chlorophyll. The most common symptom of turfgrass plants infected with a virus is a light green or yellow and dark green mosaic or mottle, stunting, and poor growth. Viruses rarely kill a plant host in a short time. Depending on environmental conditions, many turfgrass or crop plants and weeds may harbor one or more viruses but show no external symptoms, particularly at temperatures above about 85°F (29.4°C).

Because of their weakened state, virus-infected plants are often more susceptible to nematode attack, other diseases, and winter injury. Virus diseases are often confused with nutrient deficiencies or genetic abnormalities, insect-induced toxemias, pesticide or fertilizer injury, mite feeding, or other injuries. This is because the amino acids and sugars needed for healthy plant growth are siphoned preferentially to produce more virus particles at the expense of normal metabolism.

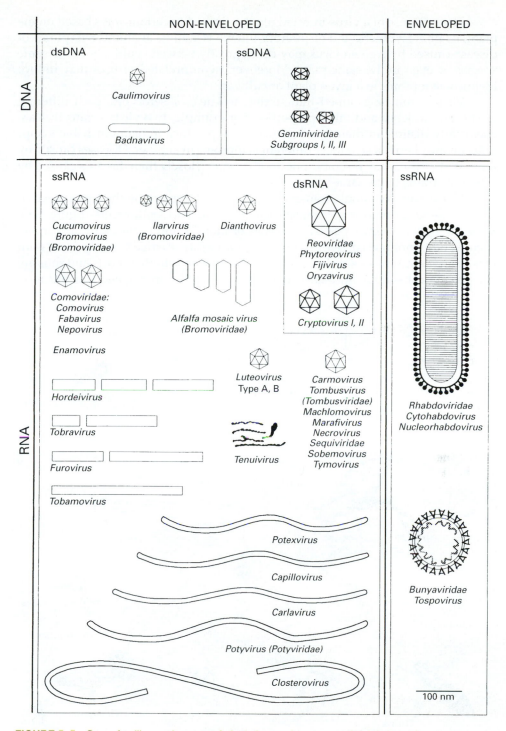

FIGURE 5–5 Some families and groups of plant viruses drawn approximately to scale.

The response of a virus-infected plant is limited in certain ways based on the plant, its photosynthetic and respiratory functions, and its anatomy. Symptoms of disease caused by a given virus may differ greatly between cultivars of the same species or even in the same plant. Moreover, two unrelated viruses may induce identical symptoms in a given plant or cultivar.

Viruses can be grouped into strains, isolates, or serotypes, each differing greatly in virulence and other properties. For example, two virus strains that are chemically distinct in their nucleoprotein may produce indistinguishable symptoms in one plant but strikingly different symptoms in another. Diseases caused by unrelated viruses may resemble one another more closely than diseases caused by different strains of the same virus.

Viruses are commonly transmitted from plant to plant by the feeding activities of insects, primarily specific species of aphids and leafhoppers (figure 5–6). Some circulative viruses persist for weeks or months and replicate in their insect vectors; other nonpersistent or stylet-borne viruses are carried for only a few minutes to an hour. Viruses are almost always systemic within their hosts and thus are generally most serious on plants that are vegetatively propagated (for instance,

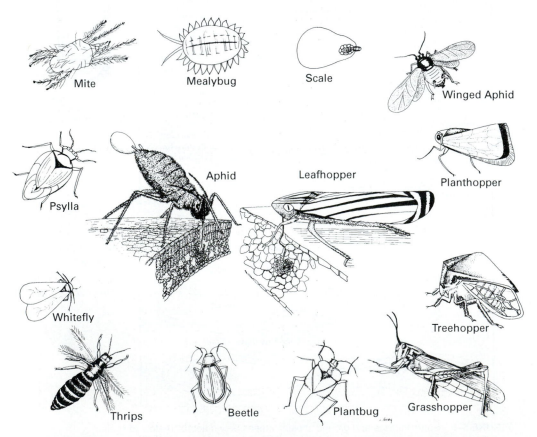

FIGURE 5–6 Insects that transmit plant viruses. The most important vectors are species of aphids and leafhoppers.

turfgrasses are infected by viruses in sod, plugs, and sprigs). Over 100 viruses are spread in seed. Viruses may also be disseminated by the transfer of infected plant sap into healthy plants. A good example is the St. Augustinegrass decline virus, which is spread by mowing first an infected turf area and then mowing healthy turf. A few viruses are spread by pollen, mites, slugs and snails, birds, mammals, primitive soil-borne fungi, nematodes, or possibly other minute fauna.

Viruses often overseason in biennial and perennial crops and weeds, in the bodies of insects, but not in dead plant parts (except tobacco mosaic or TMV). Once infected with a virus, plants normally remain so for life. Most if not all plant-infecting viruses persist in a number of different plant hosts; many of these, however, may be symptomless. The latter plants (commonly weeds or wild plants) are latent carriers and often a major source of infection to turfgrasses, crop, and ornamental plants.

Viruses are identified by (1) host specificity; (2) particle morphology under an electron microscope; (3) mode or modes of transmission; and (4) their physical, biological, and serological properties. The specific reactions of indicator host plants often aid in their identification. Plant viruses are grouped taxonomically into families by their shape and size, type of nucleic acid (RNA or DNA) contained, whether the virus is enclosed in an envelope, other physical properties, mode of transmission, and serology (figure 5–5).

Virus diseases are controlled principally by limiting their transmission in vegetative plants or plant parts, by growing resistant species or cultivars, and by controlling their vectors.

Nematodes.   More than 18,000 species of nematodes have been described. They have been called the most numerous multicellular animals on earth. About 3 to 10 billion nematodes are estimated to live in an average acre (0.4 hectare) of soil, most of which are in the top 6 in (15.2 cm). As many as 20,000 plant-parasitic nematodes of 10 species can be found in 1 pint (500 cc) of soil. Nematodes are important agents of disease in turf with many genera and species present. Fortunately, relatively few of the 60 genera and 3000 species of plant-parasitic nematodes commonly feed on the roots of grass plants and cause disease. Even when parasitic nematodes are found associated with the roots of turfgrass plants, fairly high populations of most species must be present for visible aboveground damage to occur. Nematodes seldom kill grass plants by themselves; however, they are capable of greatly curtailing the growth of plants by reducing their root systems.

Nematodes are unsegmented roundworms, sometimes called nemas or eel-worms. The great majority are mobile, slender, and invisible to the naked eye because of their transparency and small size (figure 5–7). Practically all forms fall within the range of $\frac{1}{10}$ to $\frac{1}{100}$ of an inch in length; the largest being 4 to 5 mm long. Exceptions are several sluggish ectoparasites and adult females of certain genera (such as root knot, cyst, and cystoid), the bodies of which become swollen and saclike.

Nematodes are very common, free-living in fresh and salt water, decaying organic matter, moist soils, and tissues of other living organisms. It is likely that every form of plant and animal life is fed upon by at least one type of nematode; however, most types are harmless, feeding primarily on soil microorganisms or other nematodes. Several hundred species are considered beneficial to humans

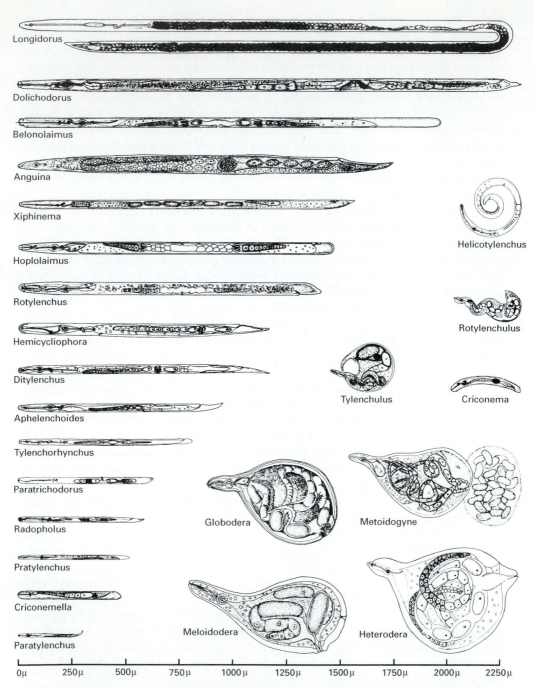

Longidorus

Dolichodorus

Belonolaimus

Anguina

Xiphinema

Helicotylenchus

Hoplolaimus

Rotylenchus

Hemicycliophora

Rotylenchulus

Ditylenchus

Tylenchulus   Criconema

Aphelenchoides

Tylenchorhynchus

Paratrichodorus

Radopholus   Globodera   Metoidogyne

Pratylenchus

Criconemella

Paratylenchus   Meloidodera   Heterodera

0μ   250μ   500μ   750μ   1000μ   1250μ   1500μ   1750μ   2000μ   2250μ

FIGURE 5–7   Morphology and relative size of 24 of the most important plant-parasitic genera of nematodes. The line across the bottom equals 2250 μm or μ, about ¹⁄₁₂,₀₀₀ of an inch long. Most nematodes are wormlike in form and transparent and thus cannot be seen without a good light microscope.

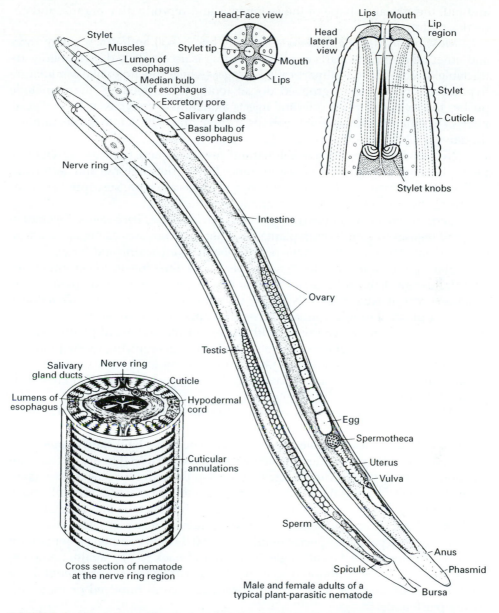

Head-Face view

Stylet
Muscles
Lumen of esophagus
Median bulb of esophagus
Excretory pore
Salivary glands
Basal bulb of esophagus

Stylet tip
Mouth
Lips

Nerve ring

Head lateral view
Lips   Mouth
Lip region
Stylet
Cuticle
Stylet knobs

Intestine

Ovary

Testis

Salivary gland ducts
Nerve ring
Lumens of esophagus
Cuticle
Hypodermal cord
Cuticular annulations

Cross section of nematode
at the nerve ring region

Egg
Spermotheca
Uterus
Vulva

Sperm

Spicule
Anus
Phasmid
Bursa

Male and female adults of a
typical plant-parasitic nematode

FIGURE 5–8 Morphology and main characteristics of typical male and female plant-parasitic nematodes. (Courtesy G. N. Agrios, as published in *Plant Pathology*, 4th ed., San Diego, CA: Academic Press, 1997.)

since they are predators or parasitic on plant-feeding nematodes, fungi, bacteria, algae, insects, protozoa, or other microscopic animal and plant life.

Nematodes have bodies differentiated for feeding, digestion, locomotion, and reproduction (figure 5–8). Some anatomical features used for identification include the size and shape of the adult female, the size and shape of the stylet and tail, the size and shape of the esophagus and reproductive organs, and cuticular patterns.

Most plant-parasitic nematodes usually live in soil and attack young roots and other belowground parts of plants. Feeding is accomplished by a hollow, retractable, needle-like mouthpart called a *stylet* (which functions like a miniature hypodermic needle) or a grooved dorsal tooth. When feeding, the nematode pushes the stylet into plant cells and injects a mixture of enzymes that predigests plant-cell contents. The liquefied contents are then drawn back into the nematode's digestive tract through the stylet.

Nematode feeding lowers the natural resistance and reduces the vigor of plants. Feeding damage affords easy entrance for a variety of fungi and bacteria as well as other nematodes. Nematode-damaged plants are more susceptible to winter injury, drought, other pathogens, and insect attack.

Certain nematode genera (*Longidorus, Paratrichodorus, Trichodorus, Xiphinema*) transmit viruses to various crop plants—but not to turfgrasses as far as we know. Other nematodes transmit or make portals of entry for bacteria and fungi.

Plant-parasitic nematodes may live part of the time free in soil around roots or in fallow gardens and fields. Parasitic nematodes tunnel inside plant tissue (*endoparasites*) or feed externally at the root surface (*ectoparasites*). Nematodes may enter plants through wounds and natural openings or by pushing their way in between and through cell walls with the aid of the stylet. All plant-parasitic nematodes (about 3000 species) are obligate parasites requiring living plant tissues for reproduction. Nematodes find plant roots by their random movement through the soil, but some are apparently attracted by sensing heat or chemicals given off by roots.

All plant-parasitic nematodes reproduce by laying eggs (figure 5–9). These are deposited in or on plant tissues, especially roots, or in soil. Some nematode eggs are retained within the dead female bodies that become cysts. Eggs hatch, sometimes after months or even years, releasing young wormlike *juveniles* (larvae) that are usually born ready to start feeding. Juveniles develop through four stages, each completed by a molt, and become adults after the final molt. Nematodes multiply much faster than higher animals but much more slowly than bacteria and fungi. Reproduction without males (parthenogenesis) is common for many species.

Most plant-parasitic species require 20 to 60 days to complete a generation from egg through four larval stages to adult and back to egg again. Some nematodes, such as the common dagger nematode (*Xiphinema americanum*), have only one generation per year, but still produce several hundred or more offspring per female.

Soil populations of plant-parasitic nematodes are affected by the following:

1. Climate, including length of the growing season, temperature, and the distribution and amount of rainfall (mild winters or heavy snow cover may result in high numbers of nematodes surviving the winter months)
2. Soil type, texture, and structure
3. Populations of antagonistic or parasitic bacteria, viruses, the 50 or more species of nematode-trapping fungi, protozoa, mites, and predatory nematodes
4. Toxic chemicals in soil or secreted by plant roots

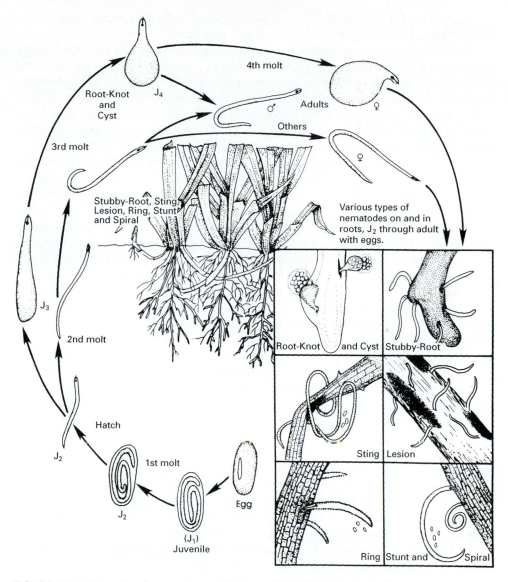

FIGURE 5–9   Life cycles of common parasitic nematodes that attack turfgrasses.

5. Crop rotation or past cropping history; and species, cultivar, age, and nutrition of currently growing plants

Certain nematodes live strictly in light, sandy soils; others build up high populations in heavier loam and clay soils. Some seem to thrive best in heavy soils. Crop damage is much more common and severe in light, sandy soils (such as golf greens) than in heavy, clay soils.

Most plant-parasitic nematodes are inactive at temperatures below 50°F (10°C) and above 95°F (35°C). The optimum temperature for most nematodes is between 60 to 90°F (15.6 to 32.2°C) but varies greatly with the species, stage of development, activity, growth of the plant host, and other factors.

Some species of nematodes, especially ectoparasitic forms, are killed easily by air-drying soil. Other types remain alive but in a dormant state under the same conditions. When dormant or as cysts and eggs, or when embedded in plant tissues, they are much more difficult to kill by chemicals (*nematicides*) or heat than when moist and actively moving. Exposure to moist heat, such as steam or hot water at 120°F (48.9°C) for 30 min, is sufficient to kill most nematodes and nematode eggs. At higher temperatures, shorter periods provide control.

Once a plant-parasitic nematode has been accidentally introduced into a turf area or field to which it is not native, several years usually pass before the nematode population builds sufficiently to cause conspicuous disease symptoms. Nematodes also move very slowly through soil under their own power—rarely more than 12 to 24 in (30.5 to 61 cm) a year. Nematodes are easily disseminated, however, by any agency that moves infested soil, plant parts, or contaminated objects. These include all types of turf equipment, vehicles, machinery carrying soil, tools, containers, flowing water, wind, land animals, birds, clothing and shoes, and movement of sod, plugs, sprigs, and the like (especially with soil around the roots).

Nematodes damage roots in two major ways. Root cells are injured or killed mechanically or chemically (apparently through the secretion or injection of enzymes emitted through the stylet). Such damage renders the roots less effective in absorbing and transporting water and nutrients (figure 5–9). In many cases, root growth ceases. Nematodes also interact with fungal or bacterial pathogens to cause disease complexes that often damage roots more severely than when either the nematode or other pathogen is present alone. Nematode damage is difficult to separate from insect, disease, or other pest injury, especially late in the season.

Diagnosis of nematode injury should *never* be made solely on plant symptoms. Symptoms of nematode damage often mimic those induced by other factors, such as low or unbalanced fertility, sunscald or frost, excessive soil moisture and poor drainage, drought, insect or mite injury, wilt or root-rot fungi, and herbicide damage. Common symptoms of nematode injury include stunting and unevenness in height, loss of green color and yellowing (often similar to nitrogen, iron, magnesium, manganese, or zinc deficiency), slow general decline, temporary wilting on hot and bright days, lack of normal response to water and fertilizer, dieback of leaves, or killing of entire plants. Root systems have a scarcity of feeder roots and often are stubby or excessively branched ("hairy-root"), discolored, shallow, light-to-dark-brown or black, and decayed (figure 5–9). Accurate diagnosis of a nematode problem requires proper sample collection, proper handling of soil and roots, and laboratory analysis by a competent nematologist (chapter 1). Only a few plant-parasitic nematodes cause unique disease symptoms. These include root-knot and cyst nematodes. Root-knot nematodes, the most easily detected species, produce galls or knots on the roots of over 2000 kinds of plants, including turfgrasses. Most root-feeding species, however, cause no specific symptoms, at least aboveground.

The first indication of nematode injury may be appearance of scattered, circular-to-oval or irregular areas of stunted plants with yellowish foliage and dieback of leaf tips. The areas are small at first and gradually enlarge (over a year or more). Turfgrass plants in the center lack vigor, gradually decline, wilt, and often die.

Being obligate parasites, plant-infecting nematodes cannot be cultured on artificial media in the laboratory. Nematodes are extracted from soil by various flotation, sieving, and centrifugation procedures and from roots by incubation techniques. Normally, mixtures of parasitic nematodes and nonparasitic nematodes occur in, on, or about plant roots. Inferences about the pathological importance of each plant-parasitic species are drawn from knowledge of their feeding habits, relationship of above- and belowground symptoms to the species, numbers present, and frequency of occurrence. Inoculation of plants with individual species is usually necessary to demonstrate parasitism and the degree of pathogenicity.

In many situations, nematodes cannot be considered as a single factor in determining a turfgrass problem, since fungi, bacteria, viruses, and a variety of environmental stresses often combine with nematodes in causing disease.

Nematodes are managed by starting with disease-free planting material and following the best cultural practices to ensure steady, vigorous growth. If the results of a nematode analysis indicate potentially serious population levels, a soil fumigant can be applied before planting, or a nematicide can be applied to established turf. Because eradication is impossible in the field, periodic checks of nematode populations by soil sampling are most desirable when troublesome types are found in large numbers. Scientists are also studying the effects of a number of chemosterilants and the possibilities of biological control measures (such as introducing parasitic bacteria and fungi) in reducing soil populations.

Other infectious agents.   Viroids, mollicutes (phytoplasmas, spiroplasma), and protozoa (figure 5–1) are known to cause diseases of plants; there are only a few mollicutes known to occur in turfgrasses.

Turfgrass diseases vary in severity from year to year and from one locale to another, depending on the environment (principally soil and air moisture conditions, temperature, humidity, and grass nutrition), the relative resistance or susceptibility of the grass plant, and the pathogen or pathogens involved.

For an infectious disease to develop, four conditions are necessary: the presence of a virulent pathogen, a susceptible host (turfgrass plant), a favorable environment, and time for disease to develop (from infection and colonization of the host plant to the presence of symptoms). If any of the first three conditions are missing, disease will not develop.

A disease usually goes through four stages: infection or invasion of the host, incubation, development of symptoms, and reproduction of the organism or agent (*inoculum production*). Infection is the process of the pathogen entering the turfgrass plant. Incubation is the time period between infection and the appearance of visible symptoms. During incubation, the pathogen is obtaining nourishment from the host plant before symptoms appear. Inoculum production is the formation of reproductive propagules (fungus spores and sclerotia, virus particles, bacteria, nematode eggs, and juveniles) for spread and survival.

Table 5–1 lists a number of cultural management practices that aid in managing turfgrass diseases and includes a listing of diseases that are at least partially controlled by each practice.

TABLE 5–1   CULTURAL MANAGEMENT PRACTICES THAT AID IN MANAGING DISEASES

Cultural management practices	Diseases partially controlled
Provide good surface and subsurface drainage when establishing a new turf area. Fill in the spots where water may stand. Before establishment, remove stumps, large roots, construction lumber, bricks, concrete, plaster, tin cans, and other debris. Uniformly mix all soil amendments (e.g., peat moss, calcined clay) into the soil. Test the soil reaction (pH) and follow the recommendations in the soil report. A pH between 6 and 7 is best for most grasses (5.5 is best for reducing pink snow mold, snow scald, take-all patch, and summer patch).	Helminthosporium diseases, Rhizoctonia blight or brown patch, yellow patch, dollar spot, red thread, pink patch, summer patch, snow molds, necrotic ring spot, Pythium diseases, fairy rings, take-all patch, downy mildew or yellow tuft, spring dead spot, seed rot and seedling blights (damping off), nematodes, Physoderma leaf spot, Cephalosporium stripe, bermudagrass decline, mushrooms, puffballs
Grow locally adapted, disease-resistant grasses in blends or mixtures, or substitute grass species. Check with your local county extension (advisory) office or turfgrass specialist for suggested grass species and cultivars to grow and for available disease resistance in turfgrass cultivars. In shaded areas, grow shade-tolerant cultivars or species.	Helminthosporium diseases, rusts, leaf smuts, Fusarium blight, summer patch, necrotic ring spot, snow molds, powdery mildew, dollar spot, anthracnose, take-all patch, Pythium blight in overseeded turf, nematodes, gray, Septoria and Stagonospora leaf spots, red thread, St. Augustinegrass decline, spring dead spot, bacterial wilt, Rhynchosporium scald, southern blight, yellow ring
Buy only top-quality certified sod, sprigs, plugs, or pathogen-free seed from a reputable dealer. When possible, plant at suggested rates when the weather is cool and dry. The seedbed should be well prepared and fertile. Avoid overwatering, especially from planting to seedling emergence or establishment.	Helminthosporium diseases, Pythium diseases, seed rot and seedling blights (damping off), leaf smuts, rusts, dollar spot, summer patch, necrotic ring spot, Fusarium blight, snow molds, yellow patch, yellow ring, nematodes, bacterial wilt, St. Augustinegrass decline
Fertilize (supply nitrogen, potassium, and phosphorus) according to local recommendations and soil tests. Recommendations will vary with the soil type, length of the growing season, grasses grown and their use, and amount of watering. In hot weather, avoid overstimulation with fertilizer, especially with a quick-release nitrogen product. High levels of potassium (potash) and phosphorus may help suppress disease development. Reduce winter injury and snow-mold damage by avoiding nitrogen fertilization after about 6 weeks prior to winter dormancy. Slowly released forms of nitrogen fertilizers are generally recommended. (Attacks of red thread, pink patch, rusts, dollar spot, anthracnose, summer patch, necrotic ring spot, some leaf spots, frost scorch, and melting out are lessened by applications of fertilizer; the opposite is generally true of diseases such as Pythium blight, brown patch, gray leaf spot, leaf smuts, snow molds, powdery mildew, yellow patch, and downy mildew or yellow tuft.)	Helminthosporium diseases, powdery mildew, rusts, brown patch, snow molds, dollar spot, fairy rings, Fusarium blight, summer patch, necrotic ring spot, superficial fairy rings, leaf smuts, rusts, red thread, pink patch, seed rot and seedling blights (damping off), Pythium diseases, anthracnose, Nigrospora blight, take-all patch, Ascochyta, Cercospora, gray, Phaeoramularia, Septoria, Stagonospora and minor leaf spots, downy mildew or yellow tuft, copper spot, southern blight, Leptosphaerulina leaf blight, spring dead spot, nematodes, bacterial wilt, St. Augustinegrass decline, zoysia patch, Rhynchosporium scald, frost scorch, bermudagrass decline, Curvularia blight, yellow ring, yellow patch
Mow frequently at the height recommended for the area, season, and grasses grown. Essentially all diseases are increased by scalping. Remove no more than one-third of the leaf height at one time. Keep the turf cut in late fall until growth stops. Keep the mower blades sharp.	Helminthosporium diseases, powdery mildew, Pythium diseases, brown patch, snow molds, dollar spot, copper spot, rusts, Fusarium blight, summer patch, necrotic ring spot, red thread, pink patch, anthracnose, Cercospora, gray, Septoria,

TABLE 5–1   CULTURAL MANAGEMENT PRACTICES THAT AID IN MANAGING DISEASES (continued)

Cultural management practices	Diseases partially controlled
Removing the clippings may help to control red thread, rusts, leaf smuts, leaf spots, brown patch, Fusarium blight, and snow molds. Avoid mowing when the grass is wet.	Stagonospora, Ascochyta, Phaeoramularia and minor leaf spots, downy mildew or yellow tuft, Curvularia blight, Leptosphaerulina leaf blight, bacterial wilt, spring dead spot, Nigrospora blight, St. Augustinegrass decline, bermudagrass decline, nematodes, frost scorch, slime molds
Water established turf thoroughly during droughts. Moisten the soil to a depth of 6 in at each irrigation. Repeat every 7 to 10 days if the weather remains dry. Water as infrequently as possible to allow gaseous exchange between soil and atmospheric air. Avoid frequent light sprinklings, especially in late afternoon or evening. Daily watering may be needed in certain cases to prevent wilting, necrotic ring spot, summer patch, and localized dry spot, or where parasitic nematode populations are high. Keep bentgrass greens on the dry side to prevent a takeover by annual bluegrass. Daily light watering is suggested on warm-season grasses.	Helminthosporium diseases, brown patch, red thread, Pythium diseases, dollar spot, pink patch, powdery mildew, summer patch, necrotic ring spot, Fusarium blight, rusts, leaf smuts, fairy rings, Nigrospora blight, nematodes, seed rot and seedling blights (damping off), take-all patch, anthracnose, Ascochyta, Cercospora, gray, Phaeoramularia, Septoria, Stagonospora and minor leaf spots, southern blight, Leptosphaerulina leaf blight, zoysia patch, Curvularia blight, downy mildew or yellow tuft, localized dry spot, white blight, slime molds, Physoderma leaf spot, bermudagrass decline
Increase light penetration and air movement to shaded turfgrass areas and speed drying of the grass surface by selectively pruning the lower limbs or removing dense trees, shrubs, hedges, etc., bordering the turf area. When landscaping, space plantings and other barriers to avoid excess shade and increase air movement across the grass.	Helminthosporium diseases, brown patch, rusts, dollar spot, red thread, pink patch, powdery mildew, Pythium blight, snow molds, seed rot and seedling blights (damping off), Ascochyta, gray, Cercospora, Phaeoramularia, Septoria, Stagonospora and minor leaf spots, slime molds
Remove excess thatch in early spring or fall when it accumulates to ½ in on high-cut grasses, ⅛ in for fine turf. Thatch control will reduce essentially all diseases as many pathogens survive in thatch between disease attacks. Use a vertical mower, power rake, or similar dethatching equipment. These machines may be rented at many garden-supply and rental stores or purchased from turf-equipment suppliers. If thatch is increasing, you are usually overgrowing the turf.	Helminthosporium diseases, brown patch, red thread, pink patch, dollar spot, rusts, snow molds, anthracnose, fairy rings, superficial fairy rings, summer patch, necrotic ring spot, Fusarium blight, Pythium blight, leaf smuts, Ascochyta, Cercospora, gray and Ramularia leaf spots, white blight, localized dry spot, yellow ring, yellow patch, southern blight, Curvularia blight, spring dead spot, take-all patch, nematodes, slime molds, etc.
Core (aerify) compacted and severely thatched turf areas one or more times each year, during cool weather, using a hand-aerified or power machine. After aerification, proper measures should be taken to prevent turf desiccation. Hydroaerification (perforation of turf by high-pressure water) can be done at most times during the growing season. It is especially useful for sand-based turf or turf with weak root systems. For chronically compacted areas, eliminate foot and vehicle traffic by putting in walks, fences, shrubbery, patios, parking areas, etc.	Helminthosporium diseases, Pythium blight and root rot, summer patch, necrotic ring spot, Fusarium blight, rusts, snow molds, dollar spot, gray leaf spot, brown patch, southern blight, anthracnose, spring dead spot, zoysia patch, bermudagrass decline, nematodes, Curvularia blight, fairy rings, superficial fairy rings, localized dry spot, yellow ring, take-all patch
Follow suggested insect and weed control programs for the area and grasses being grown. Some insects transmit disease-causing fungi, bacteria, and viruses; weeds may harbor them.	Practically all diseases. This area has not been researched extensively by plant pathologists. See chapters 4 and 6 for suggestions on managing weeds and insects.

5.2 KEYS TO INFECTIOUS TURFGRASS DISEASES

The following are four keys to most of the infectious turfgrass diseases that appear in the United States and many other areas of the world. Also included are a number of noninfectious diseases or disorders (see also chapters 2 and 3) that are com-

KEY TO WARM-WEATHER DISEASES (60 TO 75° F; 16 TO 24° C); occur in late spring or early summer or early fall	Dollar spot	Copper spot	Downy mildew or Yellow tuft	Yellow patch	Yellow ring	Take-all patch	Necrotic ring spot	Large brown patch	Fairy rings	Powdery mildew	Rusts	Slime molds	Red thread/Pink patch	Anthracnose; basal stem rot	Bermudagrass decline and root rot	Pythium crown and root rot	Gray leaf spot	Helminthosporium leaf spots	Cercospora-Phaeoramularia leaf spots	Ascochyta leaf spot or blight	Septoria-Stagonospora leaf spots	Leptosphaerulina leaf blight	Fusarium leaf spot	Minor leaf spots and streaks	St. Augustinegrass and Centipedegrass decline	Mushrooms and Puffballs
A. Mostly circular patches in turf <7 in across; often follows heavy dews or rainy weather	●	●	●																							
1. Straw-colored, somewhat sunken patches; may be covered with white mold in damp weather	●																									
a. Outer leaves with whitish to tan lesions with dark borders	●																									
2. Coppery red to salmon pink patches; found mostly in acid soils in coastal regions		●																								
3. Yellow patches or tufts up to 4 in across; plants easily pulled from turf			●																							
B. Circular patches or rings up to 2–3 feet or more across; often with green centers				●	●	●	●																			
1. Dull yellow, tan, straw- or bronze-colored rings, crescents or streaks				●																						
2. Lemon to golden yellow rings with dense white mold (mycelium) in thatch					●																					
3. Sunken, yellow to reddish or dull brown patches with blackened crowns and roots						●																				
4. Tan to straw-colored sunken patches or rings with blackened crowns and roots							●[a]																			
C. Straw-colored, mostly circular patches up to 18–20+ ft across								●[b]																		
D. Rings or arcs up to 15 ft or more across; often with outer ring of dark green grass									●																	
E. Mostly irregular patches or patterns in turf										●	●	●														
1. Powdery mold on leaves; leaf spots are not usually evident										●	●	●														
a. Milky white to gray mold; found in shade and easily wiped off										●																
b. Bright yellow, orange or reddish brown pustules on leaves											●[a]															
c. Turf first slimy to "greasy"; then covered with powdery structures												●[c]														
2. Tan or blighted patches; usually up to 15 in across, but may be larger													●	●	●	●										
a. Coral-pink to red "threads" or pinkish mold bind leaves together													●													
b. Plants yellow or tan then red, bronze, or brown with black specks on stems														●[d]												
c. Light yellow then dark patches; turf becomes thinned with blackened roots															●[e]											
d. Plants yellow to reddish brown, wilt and die; roots and crowns discolored																●[e]										

monly confused with true diseases. The keys are organized according to the air-temperature ranges at which disease symptoms are *first visible*.

5.2.1 *Anthracnose*

Anthracnose, caused by the fungus *Colletotrichum graminicola,* occurs worldwide, causing a foliar blight of nearly all turfgrasses under prolonged hot and damp or overcast and rainy weather when the grass is under stress. Annual bluegrass (*Poa annua*) fine-leaf fescues and bentgrasses are most susceptible, followed by perennial

	Dollar spot	Copper spot	Downy mildew or Yellow tuft	Yellow patch	Yellow ring	Take-all patch	Necrotic ring spot	Large brown patch	Fairy rings	Powdery mildew	Rusts	Slime molds	Red thread/Pink patch	Anthracnose; basal stem rot	Bermudagrass decline and root rot	Pythium crown and root rot	Gray leaf spot	Helminthosporium leaf spots	Cercospora-Phaeoramularia leaf spots	Ascochyta leaf spot or blight	Septoria-Stagonospora leaf spots	Leptosphaerulina leaf blight	Fusarium leaf spot	Minor leaf spots and streaks	St. Augustinegrass and Centipedegrass decline	Mushrooms and Puffballs
F. 3. Leaves spotted; turf may be thinned and weakened																	•	•	•	•	•	•	•	•		
a. Distinct oval to round, oblong, or elongated spots																	•	•	•							
(1) Blue-gray to brown spots with darker margin																	•									
(2) Tan to ash-gray spots or blotches with narrow dark border																		•								
(3) Dark brown to purple spots with tan to gray centers																			•							
b. Leaves with blotches or mottling; may die back from tip																				•	•	•	•	•		
(1) Leaf spots or blotches usually without dark borders																				•	•	•	•			
(a) Dark specks (pycnidia or perithecia) in older lesions																				•	•	•				
• Purplish to chocolate brown spots; turn tan then yellow																					•					
• Yellow to gray-green or brown blotches on leaves																						•				
• Yellow to reddish brown spots that turn white																							•			
(b) Dark specks absent in lesions																							•	•		
• Oval to irregular, dull tan spots with a dark border																							•			
• Other types of leaf spots or stripes																								•		
4. St. Augustinegrass and centipedegrass become more yellowed, then later dies																									•	
5. Mushrooms (toadstools) or puffballs pop up in turf but not in rings																										•

[a] Usually most evident in hot, droughty weather when grass is not growing.
[b] Mostly found on St. Augustinegrass and zoysiagrass.
[c] Powdery structures are easily wiped off leaves.
[d] Foliar blight phase of anthracnose occurs to stressed turf in hot weather.
[e] Disease is more dependent on soil moisture than temperature.

HOT-WEATHER DISEASES (OVER 75° F; 24° C); occur from late spring to late summer	Pythium blight	Copper spot	Dollar spot	Summer patch/ Necrotic ring spot	Fusarium crown and root rot	Brown patch or Rhizoctonia blight	Southern blight	Yellow ring	Large brown patch	Fairy ring	Anthracnose	Helminthosporium crown and root rot	Curvularia blight	Plant-parasitic nematodes	Fusarium leaf spot	Cercospora-Phaeoramularia leaf spots	Gray leaf spot	Leptosphaerulina leaf blight	Rusts	Slime molds	St. Augustinegrass and Centipedegrass decline
A. Mostly round patterns of dead grass	●	●	●	●		●	●		●												●
1. Sunken patches up to 6 in across; appear in wet weather	●	●	●																		
a. Leaves matted, slimy, then withered; wet grass covered with cottony mold	●a																				
b. Coppery red to salmon pink patches; found mostly in acid, infertile soils		●																			
c. Straw-colored, somewhat sunken patches. Light tan leaf spots with dark borders			●b																		
2. Patches up to 2 to 3 ft across; may develop green centers					●	●	●		●												
a. Straw-colored patches; centers often remain green; appear in stress periods					●c	●	●		●												
b. Turf thins out, turns straw-colored in round to irregular patches					●c																
c. Light or dull brown patches; grass is not matted. Occurs during wet periods						●d															
d. Yellow to white, enlarging patches with fluffy white mold in thatch							●e														
e. Lemon to golden yellow rings with dense white mold (mycelium) in thatch								●													
3. Straw-colored, roughly circular patches up to 18 to 20+ feet in diameter									●												
4. Rings or arcs of thin, brown and/or green grass up to 15 ft across or more										●											
B. Irregular patterns of weak, thinned, wilted, dead, or dormant grass											●	●	●	●							
1. Large areas of turf appear dry, then wilt and turn brown when under drought stress											●	●	●	●							
a. Turf is reddish brown, finally tan to brown. Black specks on older leaves											●										
b. Dry, "patchy" yellow to brown turf areas that may become thinned												●	●	●	●						
(1) Some oval or eye-shaped spots with dark margins; crown and roots rot												●c									
(2) Irregular mottled patches of yellow thinned turf; may later wilt														●							
(3) Enlarging areas of stunted, thin, yellow grass; may wilt when dry															●c						

ryegrass, and centipedegrass. Other grasses attacked include Canada and Kentucky bluegrasses, tall fescue, bermudagrass, St. Augustinegrass, bahiagrass, and zoysiagrass. The fungus commonly enters through a variety of wounds. Anthracnose usually occurs first on the oldest (outermost) leaves, where the turfgrass is weakened by other causes such as Helminthosporium diseases, Leptosphaerulina leaf blight, summer patch, extremes in soil fertility, a thick thatch, heat and

	Pythium blight	Copper spot	Dollar spot	Summer patch/Necrotic ring spot	Fusarium crown and root rot	Brown patch or Rhizoctonia blight	Southern blight	Yellow ring	Large brown patch	Fairy ring	Anthracnose	Helminthosporium crown and root rot	Curvularia blight	Plant-parasitic nematodes	Fusarium leaf spot	Cercospora-Phaeoramularia leaf spots	Gray leaf spot	Leptosphaerulina leaf blight	Rusts	Slime molds	St. Augustinegrass and Centipedegrass decline
B. 2. Leaves spotted, turn yellowish or tan; may wither and die back from tip															●	●	●	●			
a. Oval to irregular, dull tan spots; pink mold on crowns when wet															●						
b. Oval-to-elongated, dark-brown-to-purple spots with tan-to-gray centers																●					
c. Oval-to-elongated, blue-gray-to-brown spots with darker margin																	●				
d. Yellow-to-reddish-brown spots; leaves may turn a bleached white																		●			
C. Powdery mold on leaves; leaves may or may not turn yellow to brown, wither, and die																			●	●	
1. Orange, rust-red, or brown pustules appear during extended dry periods																			●[f]		
2. Slimy or "greasy" masses that form small crusty structures; are easily wiped off																				●	
D. St. Augustinegrass and centipede grass become more yellow, then die																					●

[a] May appear in streaks.

[b] Wet grass may be covered with white mold.

[c] Crowns and roots develop a tough, dry, dark brown to black rot.

[d] Crowns and roots are unaffected.

[e] Clusters of small, round sclerotia in thatch of dying and dead grass.

[f] Leaves turn yellow to brown, wither, and die early.

drought stress, cold injury, insect and nematode feeding damage, poor soil drainage, soil layering, use of poor topdressing mixes and/or greens construction, soil compacted by heavy traffic or rolling, and so on. The fungus commonly colonizes dead or naturally dying (senescing) leaves and stems.

Identification. The disease has two distinct phases: a foliar blight and basal-stem rot.

1. *Foliar blight.* During warm-to-hot weather (80 to 95°F; 26.7 to 35.0°C) combined with prolonged periods of leaf wetness, the lesions on individual leaves are oblong to elongated, reddish-brown-to-brown blotches often surrounded by a yellow halo. The lesions may enlarge and merge to blight entire leaves (*Color plate 36*). The leaves turn yellow and light tan to brown before dying back from the tip. Older leaves are especially prone to attack. Numerous minute, raised, dark-brown-to-black-spined fruiting bodies (acervuli) can often be seen on the dead and dying leaves with the

KEY TO COOL-WEATHER DISEASES (45° TO 60° F; 7 TO 16° C); occur in early spring and/or late fall	Downy mildew or yellow tuft	Low-temperature Pythium blight	Pink snow mold	White patch, superficial fairy rings	Yellow patch, cool-weather brown patch	Take-all patch	Bermudagrass decline and root rot	Spring dead spot	Large brown patch Zoysia patch	Fairy rings	Powdery mildew	Leaf smuts	Red thread/pink patch	Seed decay, damping off	Septoria-Stagonospora leaf spots	Ascochyta leaf spot or blight	Leptosphaerulina leaf blight	Helminthosporium leaf spots	Minor leaf spots and streaks	Bacterial wilt of Toronto creeping bent
A. Circular patches or rings in turf after green up in spring	•	•								•										
1. Yellow patches or tufts <4 in across	•																			
2. Sunken, gray to straw-colored; <3 in across; grass leaves matted or dead		•																		
3. Patches from 1 in to about 3 ft across			•	•	•	•	•	•	•											
a. Wet grass covered with white to pink mold; follows cold rainy periods			•[a]																	
b. White patches or rings with dense white mold in thatch at margins				•																
c. White to straw-colored sunken rings with green centers					•[b]															
d. Yellow to reddish brown to bronzed, finally tan, with blackened crowns and roots						•														
e. Yellow, then brown; turf is thin or bare with short, blackened roots							•[c]													
f. Bermudagrass patches sunken and straw-colored as it breaks dormancy								•												
4. Straw-colored, up to 18–20+ ft across as zoysiagrass breaks dormancy									•											
5. Rings up to 15 ft or more across with outer ring of lush, dark green grass										•										
B. Irregular patterns (usually) in turf											•	•	•	•	•	•	•	•	•	•
1. Powdery mold on leaves; leaf spots are not usually evident											•	•								
a. Milky-white to gray mold; found in shade and easily wiped off											•									
b. In leaves, long, gray-to-black streaks that split into ribbons												•								
c. Pink to reddish mycelium or threadlike growths; turf may appear scorched													•							

aid of a magnifying lens (figures 5–10, 5–12, and figure 5–13). Fruiting bodies may also be evident on young leaves when the disease is active. Infections result in stem girdling and the appearance of irregularly shaped, yellow-bronze patches varying in size from a few inches to 2 ft or more (7.5 cm to 0.6 m) in diameter with an irregular outline (figure 5–11). Diseased turf begins as reddish brown; fades to yellow, tan, and eventually brown; and thins out. Roots on diseased plants are sparse.

2. *Basal-stem rot.*   This is primarily a problem on golf greens in wet weather. The optimum temperature for this phase of the disease is 70 to 82°F (21.1 to 27.8°C),

	Downy mildew or yellow tuft	Low-temperature Pythium blight	Pink snow mold	White patch, superficial fairy rings	Yellow patch, cool-weather brown patch	Take-all patch	Bermudagrass decline and root rot	Spring dead spot	Large brown patch Zoysia patch	Fairy rings	Powdery mildew	Leaf smuts	Red thread/pink patch	Seed decay, damping off	Septoria-Stagonospora leaf spots	Ascochyta leaf spot or blight	Leptosphaerulina leaf blight	Helminthosporium leaf spots	Minor leaf spots and streaks	Bacterial wilt of Toronto creeping bent
B. 2. Seedling turf thin or bare; seedlings wilt, collapse, and die														●[d]						
3. Leaves distinctly spotted or blotched; leaves wither and usually die back from tip															●	●	●	●	●	
a. Spots yellow to gray-green or brown; dark specks (pycnidia) in older lesions															●					
b. Spots purplish to chocolate-brown then straw colored; pycnidia present in lesions																●				
c. Yellow-to-reddish-brown spots; dark specks (perithecia) in older lesions																	●			
d. White-to-tan, oval, or eye-shaped spots with a dark border																		●		
e. Other types of leaf spots or stripes; with or without a dark border																			●	
4. Toronto creeping bentgrass leaves wilt, shrivel; turf turns reddish brown and dies in irregular patches																				●

[a] Two or more patches may coalesce forming irregular bleached areas.
[b] May be evident throughout the winter.
[c] Disease is more dependent on soil moisture than temperature.
[d] Occurs at all temperatures when soil conditions are favorable for growth.

but it often occurs at temperatures much below and above this range. Infected *Poa annua* plants begin as bright yellow, turn brick red, and finally become brown and die in irregular patches up to a foot (30.5 cm) in diameter. Sometimes, in early to late spring, the patches may be a bronze or orange color. In a thin, starved turf, the discolored leaves and shoots are scattered within the patches. On closely mowed bentgrass, the diffuse patches are typically irregularly shaped and gray-green to tan at first, turning red or bronze and fading to a dull brown. The patches are at least 0.5 to 20 in (1.3 to 50.9 cm) in diameter. Root systems are generally poor. The best diagnostic feature in the later stages of infection, as in the foliar phase, is the miniature dark-brown-to-black pincushions (acervuli) on the stems (figure 5–12) and leaves. The fungus infects the bases of leaf sheaths, stem, crown, and root tissues, causing a color change to dark brown or black. Whole shoots can easily be pulled free from infected crowns. Most outbreaks occur in early spring and midsummer. On a few golf courses, infections occur off and on from spring into fall.

The anthracnose fungus is often part of a disease complex that may also involve one or more Helminthosporium fungi (including *Bipolaris sorokiniana* and *Drechslera poae*), *Magnaporthe poae* (the summer patch fungus), species of *Curvularia, Fusarium, Pythium* and *Rhizoctonia*, and nematodes. Disease complexes,

KEY TO COLD-WEATHER DISEASES (<32 TO 45° F; <0 TO 7° C); snow molds are usually visible only after a cold rain and especially at snow melt time	Winter desiccation	Water and ice damage	Spring frost	Gray snow mold	Pink snow mold	Snow scald	Coprinus snow mold	Frost scorch	Pythium snow blight or rot
A. Irregular patterns or streaks of straw-colored turf	●	●	●						
1. Mostly in dry, windswept areas	●								
2. May follow drainage patterns		●							
3. Follows freezing temperatures in spring			●						
B. Turf is killed in wettest areas or where snow is slow to melt		●		●	●	●	●	●	●
1. Circular patches of dead grass 1 in to 3 ft across				●	●	●	●	●	●
a. Wet grass covered with									
(1) White to bluish-gray mold				●		●	●		●[c]
(2) White to pink mold					●				
b. Sclerotia									
(1) present				●		●	●[a]	●[b]	
(2) absent						●	●[a]		●
c. Clamp connections									
(1) present					●		●		
(2) absent						●	●		●
d. Found only in very northern areas							●	●	

[a] Sclerotia may or may not be present.
[b] Sclerotia often in rows on leaf blades.
[c] Patches usually 1-4 in. in diameter; white mold may or may not be present.

although not common, may help explain why severe attacks of basal-stem rot are often difficult to control with fungicides aimed only at *Colletotrichum graminicola*. Another explanation could be that there is more than one biotype of *Colletotrichum graminicola* that can infect turf. The etiology of a disease that is often associated with stressed turf is difficult to prove or manage.

Disease cycle. The anthracnose fungus produces large numbers of microscopic, hyaline, one-celled, bean- or crescent-shaped spores (conidia) in the acervuli on the leaf and stem lesions (figure 5–13). The conidia are disseminated primarily by water, grass clippings, mowers, and other turf-maintenance equipment. The fungus

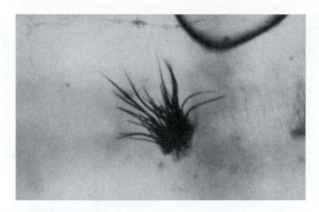

FIGURE 5–10 An acervulus of *Colletotrichum graminicola* (the anthracnose fungus) with a cluster of black, hairlike setae, as seen with the aid of a hand lens. (Courtesy J. M. Vargas, Jr.)

FIGURE 5–11 Anthracnose infecting annual bluegrass on a golf course fairway, causing irregular, yellow or bronze patches of various sizes. (Courtesy J. M. Vargas, Jr.)

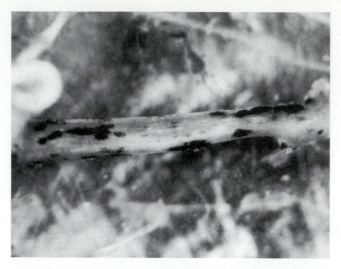

FIGURE 5–12   Close-up of black acervuli of *Colletotrichum graminicola* on a bentgrass stem. (Courtesy R. T. Kane.)

survives unfavorable periods as dormant mycelium and conidia in colonized grass debris in the thatch layer or in infected living tissues.

Importance.   Under prolonged warm-to-hot and moist weather, the fungus hyphae actively infect living grass plants under stress from temperature and moisture extremes, soil compaction, unbalanced fertility, insect or nematode feeding injuries, or other reasons. Basal rot is a serious problem on many annual bluegrass and some bentgrass golf greens in northern areas (such as in the United States, Canada, and Europe).

Management.   Follow the suggested cultural management measures outlined in table 5–1 to keep turfgrasses growing vigorously while avoiding stresses. Maintain moderate levels of nitrogen, phosphorus, and potassium; avoid late seasonal fertilization which may increase winter damage; avoid heavy topdressing mixes; keep thatch accumulation to a minimum; increase the mowing height; alleviate soil compaction and traffic by core cultivation; water the turf thoroughly (deeply) and as infrequently as possible, but avoid very dry and overly wet soil conditions. Cool golf greens one or more times during the hottest time of the day during hot, windy weather by syringing for a few minutes. A preventive fungicide program (see table 5–5 later in the chapter) aimed at controlling Helminthosporium diseases, Rhizoctonia brown patch, Pythium blight, and dollar spot; nematodes; and insects is beneficial. A number of systemic fungicides have been shown to control anthracnose effectively (table 5–5) at an early stage. Strains of creeping bentgrass, annual bluegrass, Kentucky bluegrass, perennial ryegrass, and fine-leaf fescues, tolerant to heat and drought stress and cold injury, are least affected. A mathematical model for weather-based forecasting of anthracnose on *Poa annua* during the summer months has been developed.

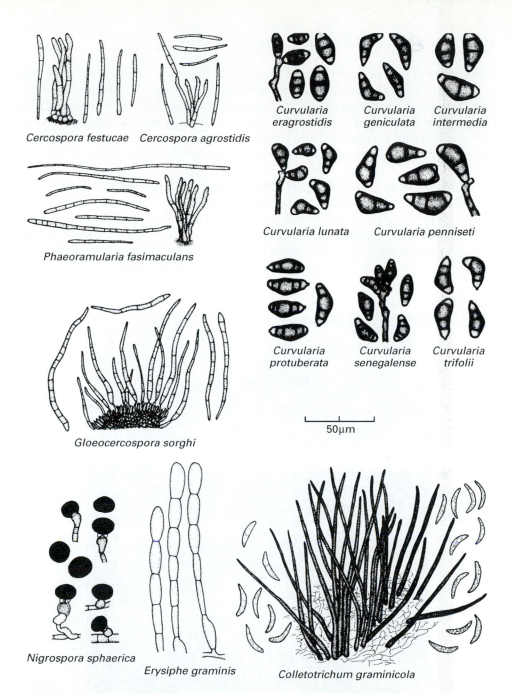

Cercospora festucae   Cercospora agrostidis

Curvularia
eragrostidis

Curvularia
geniculata

Curvularia
intermedia

Phaeoramularia fasimaculans

Curvularia lunata   Curvularia penniseti

Gloeocercospora sorghi

Curvularia
protuberata

Curvularia
senegalense

Curvularia
trifolii

50μm

Nigrospora sphaerica

Erysiphe graminis

Colletotrichum graminicola

FIGURE 5–13 Fifteen species of fungi that cause disease in turfgrasses as seen under a high-power microscope.

223

5.2.2   Ascochyta Leaf Spot or Blight

Ascochyta leaf spot or blight, caused by about 20 species of fungi in the genus *Ascochyta*, are common but minor diseases of Kentucky and annual bluegrasses worldwide; Italian and perennial ryegrasses; fescues (creeping, red, meadow, sheep, tall); bentgrasses; wheatgrasses; redtop; tall oatgrass; velvetgrass; quackgrass; and numerous other forage, wild, and weed grasses. A number of species of *Ascochyta* have wide host ranges, while several species can infect the same grass host.

Identification.   In prolonged overcast, humid, and wet weather from spring to fall, large turf areas may appear uniformly blighted, resembling damage by a dull mower; or localized pockets of infection may have a patchy appearance with healthy and infected leaves growing interspersed. Individual leaves die back from the tip, and uniform lesions may extend down to the leaf sheath (figure 5–14). Individual

FIGURE 5–14   Ascochyta leaf spot or blight. (Courtesy R. W. Smiley.)

leaf spots are small, round to elongated, and purplish to chocolate brown, often with dark borders that may girdle the leaf blades. As spots enlarge, leaves become tan and finally straw-colored. Speck-sized, brick red, or brownish-to-black fungal fruiting bodies (pycnidia) are embedded in the bleached areas of mostly dead leaves and sheaths (figures 5–15 and 5–16). The disease may closely resemble Leptosphaerulina leaf blight (sec. 5.2.15) and Septoria and Stagonospora leaf spots.

Disease cycle. *Ascochyta* fungi have the same general disease cycle as the *Septoria* and *Stagonospora* fungi. Both produce round pycnidia (figure 5–16) in older leaf lesions. The spores or conidia are hyaline to pale straw-yellow or yellow-brown, cylindrical, oblong to ellipsoid or boat shaped, and mostly two- to four-celled (figure 5–16). The spores, formed in pycnidia, ooze out in tendrils in wet weather and are carried to new leaves by splashing or running water, mowers and other turf equipment, and shoes. The fungus infects grass leaves during wet periods, usually entering grass leaves soon after mowing and later grows from the freshly cut end of the leaf blade toward the base. The pycnidia form after the

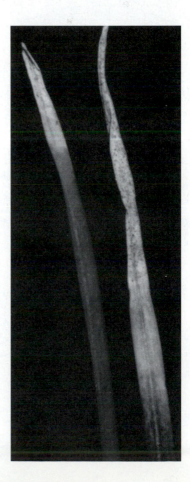

FIGURE 5–15   Two leaves of Kentucky bluegrass infected with Ascochyta leaf spot or blight. Note the dark pycnidia embedded in the dead leaf tissue. (Courtesy R. W. Smiley.)

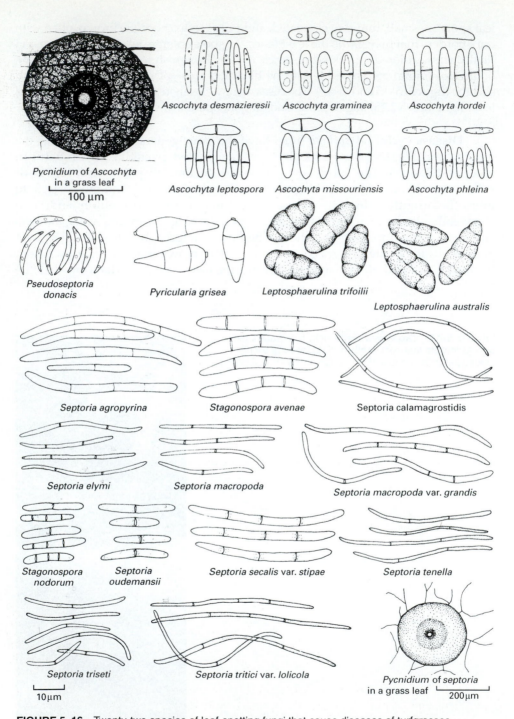

Pycnidium of *Ascochyta*
in a grass leaf
100 μm

Ascochyta desmazieresii

Ascochyta graminea

Ascochyta hordei

Ascochyta leptospora

Ascochyta missouriensis

Ascochyta phleina

Pseudoseptoria donacis

Pyricularia grisea

Leptosphaerulina trifoilii

Leptosphaerulina australis

Septoria agropyrina

Stagonospora avenae

Septoria calamagrostidis

Septoria elymi

Septoria macropoda

Septoria macropoda var. *grandis*

Stagonospora nodorum

Septoria oudemansii

Septoria secalis var. *stipae*

Septoria tenella

Septoria triseti

Septoria tritici var. *lolicola*

Pycnidium of septoria
in a grass leaf
200μm

10μm

FIGURE 5–16   Twenty-two species of leaf-spotting fungi that cause diseases of turfgrasses.

226

leaf dies. The *Ascochyta* fungi overseason in grass debris as mycelium or pycnidia. Seedling infection is likely as the fungi may be seed-transmitted.

Importance. Ascochyta leaf spot or blight seldom causes extensive damage, although infections occur throughout the growing season. Fortunately, infected turf usually recovers fairly rapidly. Periods of damp weather or frequent irrigations, especially in late afternoon and evening, favor the disease during the summer months. Frequent mowing also favors the disease by creating potential infection sites. Leaf blighting stops once the weather changes from overcast and wet to sunny and dry.

Management. Measures to control this disease are rarely necessary. If needed, follow the cultural practices outlined in table 5–1. Increasing the mowing height, mowing with a sharp mower, removal of excess thatch, correct soil compaction by turf cultivation, replacing annual bluegrass with a less susceptible turfgrass species/cultivar, and collecting clippings to remove infected leaf tips, irrigating deeply but infrequently early in the day, and applying a balanced N-P-K fertilizer in the fall can also help. Fungicides applied to control early-season Helminthosporium diseases (table 5–5) are sometimes recommended for use on golf courses. Information concerning cultivar resistance is not available.

5.2.3   Bacterial Wilt

This disease, which mainly affects closely cut Toronto or C-15 creeping bentgrass, is caused by the bacterium *Xanthomonas campestris* pv. *graminis*. The disease is evident during warm-to-hot-and-dry, sunny weather following extended periods of wet, cool-to-warm weather, mostly in spring and autumn. Young seedlings are very susceptible, and young plants are more susceptible than old ones. Other turfgrass hosts in the United States include Cohansey, Nimisilla, Old South German, Seaside, and Washington creeping bentgrasses, annual bluegrass (caused by *Xanthomonas campestris* pv. *poaannua*), and Tifgreen bermudagrass.

Other pathovars of the *Xanthomonas* organism in Europe and New Zealand include pv. *arrhenatheri*, pv. *phlei*, and pv. *poae*. These pathovars are specific for cultivars or species of colonial bentgrass, ryegrasses, bluegrasses, and fescues, as well as orchardgrass and timothy. The pathovars generally do not cause disease on other grasses.

Identification. The leaves on individual plants wilt very rapidly, turning from blue-green to reddish brown or purple within 48 hours, resulting in large, irregular patterns of twisted, withered, and dead grass (*Color plate 37*). The roots on diseased plants initially appear white and healthy but quickly decompose with the rest of the plant. Diseased turf areas often show an uneven, mottled appearance, with resistant grass cultivars or species remaining unaffected, interspersed throughout with areas of reddish brown turf. Annual bluegrass plants turn yellow, wilt, and die in small spots (about 0.5 in; 1.3 cm). The spots may merge. Affected plants turn brown, and the symptoms resemble anthracnose. In Europe, ryegrasses, tall fescue, and Kentucky bluegrass develop long, yellowish (then brown) stripes along the veins and leaf margins. In warm, dry weather, young leaves curl and wither without developing other symptoms. Affected plants die within a few days. The disease is often blamed on summer drought.

Diagnosticians in a plant clinic cut through infected plants with a razor blade. When a freshly infected grass leaf or stem is cut across and then examined in a drop of water under a compound microscope (at 100× or higher), a white cloud of bacteria can be seen oozing from the cut xylem vessels within a few seconds.

Disease cycle.   The bacterium overwinters in diseased plants, grass debris, and possibly soil. It is not seedborne. The organism is spread from plant to plant and from one area to another in water films by mowers, other turfgrass equipment, streams of irrigation water, shoes; and by planting infected sprigs, plugs, or sod. The bacterium mainly enters a grass plant through mowing or other physical wounds made by nematodes, fungi, or traffic. Infection also occurs through roots growing in infested soil or thatch. Germinating seeds may become infected through wounds caused by handling or sowing. The organism is concentrated in the water-conducting tissue (xylem) of the roots, stems, crowns, and leaves. The xylem vessels become plugged with masses of bacterial cells (0.5 × 1.2 μm) causing the plants to wilt and die (figure 5–17).

Importance.   This disease has eliminated the sale of Toronto creeping bentgrass. Certain creeping bentgrasses or other susceptible grasses may be seriously attacked, especially if other pathovars of the bacterium become widespread in North America. Large, nonuniform areas of closely cut golf and bowling greens are destroyed within several days. Fortunately, the disease causes little damage on higher-cut collars of golf greens and fairways.

Management.   Increasing the mowing height to at least ½ in (0.6 cm) dramatically reduces the severity of disease, but this remedy is not considered practical on golf greens. In late summer or early fall, infected greens should be

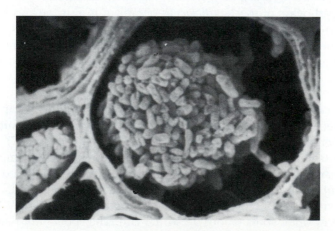

FIGURE 5–17   Scanning electron micrograph of a Toronto creeping bentgrass xylem vessel filled with bacterial cells of *Xanthomonas campestris* pv. *graminis*. (Courtesy D. L. Roberts.)

reseeded or resodded with a resistant creeping bentgrass cultivar such as Penncross or Penneagle, both of which are certified to be free of the causal bacterium. The antibiotic oxytetracycline (Mycoshield) at high rates may be used to manage the disease by suppressing symptoms during the current growing season if the disease is diagnosed early. This treatment is expensive and often does not provide an acceptable level of control. Applications are needed at 3- or 4-week intervals, and the antibiotic must be drenched in, which makes the job time-consuming.

5.2.4   Bermudagrass Decline or Take-all Root Rot

Bermudagrass in the southern part of the southeastern United States, from Florida to Texas, and Japan are commonly infected by *Gaeumannomyces graminis* var. *graminis*. The fungus causes a decline and root rot of bermudagrass cultivars and hybrids in closely mowed (<0.25 in) golf and bowling greens and, to a much lesser extent, higher-cut collars, tees, fairways, lawns, and athletic fields during prolonged wet weather. Disease symptoms appear during the rainy season, which varies from region to region. Other secondary fungi (largely saprophytes) that may be isolated from bermudagrass affected with decline or take-all root rot include *Gaeumannomyces incrustans* and species of *Phialophora*. Other hosts displaying a similar decline or root rot include centipedegrass, St. Augustinegrass, and maybe zoysiagrass, and bahiagrass. The disease is triggered by various stresses such as too short a mowing height (scalping), nematode or soil insect injury, nutrient deficiencies or imbalances, poor irrigation schedules, and low soil moisture.

   Identification.   Circular to mostly irregular, light yellow patches, 1 in (2.5 cm) to 3 ft (0.9 m) in diameter (up to 15 ft [4.59 m]) in St. Augustinegrass, coalesced patches form larger irregular shapes that appear during the rainy season. In late summer through late fall the lower leaves turn yellow, followed by the younger upper leaves. Leaf growth is reduced. The patches eventually turn dark brown, and the turf becomes thinned or bare in nonuniform areas (figure 5–18). The root systems of affected plants are short, black, and rotted along with associated rhizomes and stolons (figure 5–19). Dark brown to black strands of ectotrophic hyphae ("runner" hyphae, figure 5–106) grow over the roots, stolons, and rhizomes of susceptible plants. Dark, deeply lobed hyphopodia (appressoria) are often visible on stolons (figure 5–20). Brown to black perithecia may be observed under the leaf sheaths (figure 5–107). It is not uncommon for St. Augustinegrass to recover completely once the stress or unfavorable weather is eliminated. However, the patches may appear year after year and enlarge in size. New tillers of bermudagrass may invade the patches. They soon become colonized, decline, and die. Higher-cut bermudagrass plants are weakened and may die over a cold, winter dormant period.

   In more northern regions, bermudagrass is commonly attacked by the same fungus, often in combination with *Bipolaris cynodontis,* which causes a crown and root rot as well as leaf blotch during wet periods any time after spring green-up. Diseased plants are a brilliant yellow and form irregular areas before they die. The roots are commonly rotted and reduced to black nubs.

FIGURE 5–18   Bermudagrass decline is a serious root-rot disease in close-cut bermudagrass greens in the southernmost part of the southeastern United States. (Courtesy M. L. Elliott.)

Disease cycle.   The decline or root-rot fungus probably has a similar disease cycle as the take-all fungus, *Gaeumannomyces graminis* var. *avenae.* Optimum growth of the fungus occurs at 86°F (30°C). Survival occurs as mycelium in infected roots, rhizomes, and stolons of colonized but undecomposed grass debris in the thatch or soil and in perennial parts of living grass plants. The fungus spreads from plant to plant by ectotrophic or "runner" hyphae. It differs from the take-all fungus in having lobed hyphopodia (figure 5–20). The *graminis* variety is also disseminated in infected sod, sprigs, plugs, and infested soil, and by core cultivation and dethatching equipment. The decline and root-rot fungus may persist indefinitely in turf.

Importance.   Bermudagrass decline is a destructive disease of bermudagrass on closely cut golf course putting greens in the more southern areas of the southeastern United States during prolonged wet periods. The greatest damage occurs in the summer and fall. Bermudagrass golf greens often show symptoms of decline within 2 or 3 years after being established. Large areas of greens may become severely thinned or bare (figure 5–18). The causal fungus can readily be isolated in the laboratory from the same grasses mowed higher (such as on aprons or collars, fairways, lawns, and athletic fields), but are free of visible symptoms.

Management.   Follow the cultural management practices outlined in table 5–1, plus the cultural controls outlined for spring dead spot (sec. 5.2.31). The best treatment is to increase the mowing height, and frequency until the turf fully

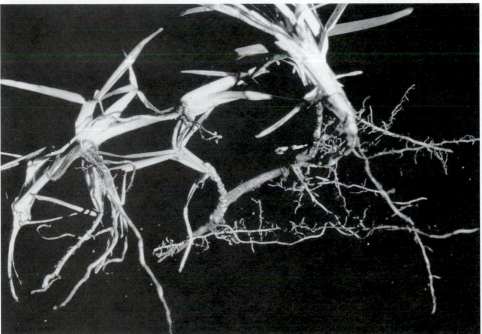

FIGURE 5–19 St. Augustinegrass affected take-all by root rot caused by *Gaeumannomyces graminis* var. *graminis*. Top, patch of dead/stunted grass; bottom, close-up of two diseased plants with black, rotted roots. (Courtesy G. W. Simone.)

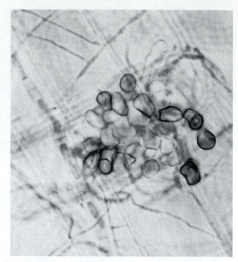

FIGURE 5–20   Hyphopodia of *Gaeumannomyces graminis* as seen under a high-power microscope. Left, lobed hyphopodia of var. *graminis;* right, simple hyphopodia of var. *avenae.*

recovers. Perform core cultivation with large tines several or more times a year, when the grass is actively growing, to alleviate compaction, promote rooting, and improve drainage. Remove the cores and then apply a pathogen-free topdressing mix containing up to 30% organic matter. Resume excessive thatch. Control nematodes and soil insects (see chapters 5 and 6). Maintain adequate and balanced fertility. Use acidifying fertilizers, such as ammonium sulfate or slow-release urea products. Apply nitrogen and potassium in a 1:1 ratio at weekly intervals using 0.5 to 1 lb (227 to 453 g) per 1000 sq ft (93 sq m). Since the root system is impaired, foliar feeding may be necessary to provide nutrients to the plant. Fungicides for controlling take-all patch (table 5–5) may be beneficial when applied preventively at the right times. However, the systemic DMI fungicides *must be used with caution* and should not be overused on bermudagrass. All bermudagrass and St. Augustinegrass cultivars and hybrids are susceptible. Use of a nomionic wetting agent improves water penetration. In addition to bermudagrass, St. Augustinegrass, centipedegrass, and zoysiagrass are also hosts for this fungus and can develop root rot and decline.

Black Leaf Spot
See the section on minor leaf spots (5.2.16).

Blister Smut
See the sections on leaf smuts (5.2.14) and minor leaf spots (5.2.16).

Brown Blight
See the section on Helminthosporium diseases (5.2.13).

Brown Patch or Rhizoctonia Blight
See the section on Rhizoctonia diseases (5.2.23).

Brown Stripe
See the section on minor leaf spots (5.2.16).

Centipedegrass Mosaic or Decline
See the section on St. Augustinegrass decline and centipedegrass mosaic and decline (5.2.25).

Cephalosporium Stripe
See the section on minor leaf spots (5.2.16).

5.2.5   Cercospora and Phaeoramularia Leaf Spots

Cercospora and Phaeoramularia leaf spots are caused by three or more species of the fungus *Cercospora* and one species of *Phaeoramularia*. The principal leaf-spotting species and the turfgrasses each infects are *C. agrostidis* (bentgrasses), *Phaeoramularia (Cercospora) fasimaculans* (mainly St. Augustinegrass), *C. festucae* (fescues), and *C. seminalis* (bermudagrass and buffalograss). The disease is most damaging to yellow-green or common selections of St. Augustinegrass in the southeastern United States and other semitropical or tropical regions when it occurs. Because of similar symptoms, Phaeoramularia leaf spot of St. Augustinegrass is often confused with the much more common gray leaf spot disease. Both diseases occur during prolonged periods of warm, very humid, and wet weather in summer.

Identification.   Small (0.5 to 1 × 1 to 6 mm), uniformly dark-brown-to-purple, oval-to-round or oblong spots form on the leaves and leaf sheaths. The enlarging, elongated-to-linear lesions develop tan to gray centers, often with conspicuous, dark purple or reddish brown margins (figure 5–21). In damp weather, the spots are covered with a whitish to gray sheen containing large numbers of spores or conidia (figure 5–13). Severely affected leaves turn yellow to brown, wither, and die back, leading to a thinned and weak turf.

Disease cycle.   The causal fungi overseason as conidia and dormant mycelium in infected leaves and grass debris. During warm, damp weather, the mostly hyaline, two- to eight-celled, long, narrow conidia are disseminated and germinate, and their germ tubes penetrate the leaves. Visible leaf spots appear a few days later. Disease activity is favored by prolonged, warm and rainy or foggy weather, and heavy dews.

Importance.   Where heavy infection takes place, particularly in warm, wet weather, leaf death and defoliation occur, causing turf thinning. Plants are seldom killed.

Management.   Follow the cultural suggestions outlined in table 5–1. Water deeply but infrequently during early-morning hours. Where turf is undernourished, apply a nitrogen fertilizer (0.5 to 1 lb of nitrogen per 1000 sq ft or 93 sq m [25 to 50 kg per hectare]). Selectively prune or remove dense trees, shrubs, and brush to improve

FIGURE 5–21 Phaeoramularia (Cercospora) leaf spot of St. Augustinegrass. (Courtesy T. E. Freeman.)

light penetration and air movement across the turf. Where cultural practices are not managing these leaf-spot diseases, apply a fungicide as suggested for gray leaf spot or Helminthosporium diseases (table 5–5). Blue-green (bitterblue) selections of St. Augustinegrass are not damaged as severely as the yellow-green (common) types.

Cercosporella Leaf Spot
See the section on minor leaf spots (5.2.16).

Char Spot
See the section on minor leaf spots (5.2.16).

Cladosporium Eyespot
See the section on minor leaf spots (5.2.16).

5.2.6   Copper Spot

Copper spot, called zonate leaf spot on bermudagrass and zoysiagrass, is a sporadic disease caused by the fungus *Gloeocercospora sorghi*. The disease is most common in turfs grown in acid soils in humid coastal regions during warm-to-hot (65 to 85°F; 18.3 to 29.4°C), moist weather. Copper spot frequently occurs together with dollar spot (sec. 5.2.8) in the same turf area. Low-cut colonial, creeping, and velvet bentgrasses; redtop; and bermudagrass are most severely attacked in infertile soils when the soil reaction (pH) is quite acid, pH 4.5 to 5.5. Velvet bentgrass is the most susceptible turfgrass to this disease.

Identification.   The disease appears as scattered, distinct, more-or-less circular patches from 1 to 3 in (2.5 to 7.6 cm) in diameter. Affected patches are coppery red to salmon pink in color (*Color plate 39*). In warm, moist weather, the grass blades are covered with mycelium and small, salmon pink-to-orange, gelatinous spore masses borne on minute sporodochia that are bright orange or red when dry. Upon close examination, infected leaves have small reddish or brown lesions that enlarge and become a darker reddish brown. Several lesions may merge, blighting an entire leaf.

Disease cycle.   The *Gloeocercospora* fungus overwinters as minute, black, lens-shaped-to-round sclerotia (0.1 to 0.2 mm) and thick-walled mycelium in infected thatch, and within older lesions on leaves and leaf sheaths. The sclerotia germinate when air temperatures reach 63 to 70°F (17.2 to 21.1°C) to penetrate leaves directly or form sporodochia. Conidia are produced, they germinate, and the resulting hyphae rapidly penetrate grass leaves in warm (68 to 85°F; 20.0 to 29.4°C), moist weather. New leaf lesions appear within a day or two, and large numbers of microscopic, elongated, needlelike spores (figure 5–13) form on the surface of the lesions 24 to 48 hours later. The spores (conidia) are disseminated to healthy leaves by splashing or flowing water, animals, and shoes, as well as by mowing and other turf equipment, when the grass is wet.

Importance.   In severe attacks, a number of spots may merge to form irregular, coppery and orange patches. Considerable turf may be lost when infection is severe. Unlike dollar spot, only the leaves are usually blighted. Diseased turf slowly recovers when the weather turns cool and dry.

Management.   Follow the same management practices as for dollar spot. Avoid mowing when the grass is wet. Apply nitrogen fertilizer and lime according to a soil test report. Avoid excessive applications of nitrogen that stimulate the copper-spot fungus. The soil pH should be between 6 and 7. Follow local cultural recommendations for the grass or grasses grown. When needed, apply a fungicide (table 5–5). Spray at about 10- to 14-day intervals in warm-to-hot, moist weather starting when the mean temperature remains in the range of 68 to 75°F (20.0 to 23.9°C) with daytime temperatures of 85 to 90°F (29.4 to 32.2°C). No information is available on turfgrass cultivars that are resistant.

Coprinus or Cottony Snow Mold
See the section on snow molds (5.2.29).

Crown and Root Rot
See the sections on Fusarium blight (5.2.11), Helminthosporium diseases (5.2.13), and Pythium diseases (5.2.21).

5.2.7   Curvularia Leaf Blight

Curvularia leaf blight or "fading-out," caused by eight or more species of *Curvularia* (figure 5–13), occurs worldwide. It is similar to diseases caused by species of *Bipolaris* (see sec. 5.2.13). *Curvularia* fungi can colonize all turfgrasses, especially bentgrasses,

FIGURE 5–22   Curvularia leaf blight or "fading-out" in a Kentucky bluegrass lawn. (Courtesy The Scotts Company.)

annual bluegrass, and zoysiagrass with isolates differing greatly in their virulence. Most attacks to cool-season grasses take place during hot weather at temperatures of 85°F (29.4°C) and above, when grass plants are growing under heat and drought stress or herbicide injury, soil compaction, or low mowing, or they are in an advanced state of senescence. On warm-season grasses, Curvularia blight attacks in early spring those turfs injured by winter temperatures. It also strikes in warm-to-hot weather under drought stress on sites with poor drainage, high populations of root-feeding nematodes, insect damage, and unbalanced fertility. For this reason, species of *Curvularia* are generally thought to be weak, secondary pathogens.

Identification.   Circular to irregularly shaped, mottled patches or streaks of thinned bentgrass may develop that often merge to form large yellow areas that turn brown (figure 5–22). Kentucky bluegrass and red fescue leaves develop indefinite, yellow and green dappled patterns that extend downward from the leaf tip. Affected leaf-tip tissue progressively turns brown and finally gray as it shrivels and dies back. A reddish brown margin may separate diseased from healthy tissue. Kentucky bluegrass and red fescue leaves may also develop tan lesions with red or brown borders. Bentgrasses are similarly affected, except that the tip dieback is first yellow, then tan, instead of brown or gray; no leaf lesions have been observed. Under very favorable conditions, a dark brown decay of leaf sheath and crown tissues may occur on all turfgrasses. On low-cut bentgrass golf greens, tan to brown patches of blighted grass 2 to 4 in (5.1 to 10.2 cm) in diameter often develop. During heat or drought stress, *Curvularia* fungi are generally associated with decline of *Poa annua*. Species of *Curvularia* sporulate profusely, causing affected and dead leaves and stems to be blackened with dark spores (conidia) and conidiophores (figure 5–13).

Curvularia lunata is a minor root-rot pathogen of creeping bentgrass, and possibly other turfgrasses, that may cause a reduction in both shoot and root growth. Infected roots are light tan and have noticeable lesions.

Disease cycle.   The disease cycle is the same as for Helminthosporium leaf, crown, and root-infecting fungi; see figure 5–42. *Curvularia* fungi are good saprophytes that colonize and decompose grass debris at or above the soil surface. Survival during unfavorable weather occurs as dormant mycelium mostly on plant debris and thatch, and as slightly or strongly curved, multicellular conidia (with one or two middle cells enlarged; see figure 5–13) on or in living and dead plants. When the weather is hot and wet, infection proceeds rapidly. Penetration of leaves occurs through cut leaf tips, crown surface or other wounds, or already existing diseased leaf tissue. Infection by *Bipolaris sorokiniana* and species of *Curvularia* (especially *C. geniculata*) results in more severe disease than either pathogen alone. Many species of *Curvularia* are seedborne on grasses.

Importance.   Turfgrass areas show a general decline, becoming thinned and ragged-looking in hot weather when the grass is growing under one or more environmental stresses. Most severely affected turf is exposed to full sun and often grows near paved surfaces or on south-facing slopes. Curvularia blight is usually considered to be a minor problem that can be controlled by cultural practices. Although it is a minor pathogen, *Curvularia* is important as a saprophyte in turf and as such is beneficial.

Management.   The management is the same as for Helminthosporium leaf, crown, and root diseases. Increase the mowing height (but not so tall as to produce a verdure that does not dry rapidly), avoid accumulations of thatch over 0.5 in (1.3 cm) on lawn-cut turf, water deeply but infrequently early in the day to avoid drought stress, aerify when soil compaction is a problem, fertilize with acidifying products such as ammonium sulfate and sulphur-coated carriers, based on a soil test and local recommendations for the grass or grasses grown, and provide for proper root-zone drainage. Where possible, cool the turf in open areas one or more times during the heat of the day in hot weather by syringing the turf for a few minutes. No information is available on cultivars resistant to *Curvularia* blight.

Damping off
See the sections on Pythium diseases (5.2.21), Rhizoctonia diseases (5.2.23), and seed decay, damping off, and seedling blights (5.2.26).

5.2.8   Dollar Spot

Dollar spot, caused by the fungus *Sclerotinia homoeocarpa*, also referred to as species of *Lanzia* and *Mollerodiscus*, attacks all turfgrasses worldwide. It may be serious on closely mowed bentgrasses (especially creeping bent), bluegrasses, fine-leaf and tall fescues, redtop, ryegrasses, bermudagrass, zoysiagrass, and bahiagrass. Dollar spot is uncommon on centipedegrass, and St. Augustinegrass. Disease attacks occur when days are warm to hot (59 to 90°F; 15.0 to 32.2°C) with cool nights and abundant dew formation, particularly on turf deficient in nitrogen. Infection periods on creeping bentgrass occur with 2 consecutive wet days with a temperature of 72° (22.2°C) or more or after 3 consecutive days if the average temperature is 59°F (15°C) or more.

Identification.   The disease on closely cut and intensely maintained bentgrass, bermudagrass, and zoysiagrass appears as circular, straw-colored, somewhat sunken spots about 1 to 2.5 in (2.5 to 6.4 cm) in diameter (*Color plate 40*). On coarse, lawn-type grasses (such as bluegrasses, fescues, and ryegrasses) maintained at a cut of 1 to 2.5 or 3 in (2.5 to 7.6 cm), the diseased areas may reach 4 to 8 in (10.2 to 20.4 cm) in diameter (figure 5–23) and are noticeably more irregular in outline. When extensive, the spots may coalesce to give large, irregular areas a drought-stricken appearance. Dollar spot is distinguished on lawn-type grasses from most other diseases by the characteristic, bleached white to light tan girdling lesions with a narrow, brown, reddish brown, or purplish border on the leaf blades of live plants at the margin of the affected area. These lesions may be up to 1 in (2.5 cm) long (figure 5–24), and they resemble those formed by the Nigro-spora blight fungus. The lesions usually extend across the blades of fine-leaved grasses. On coarser grasses such as tall fescue, the lesions tend to occur along the leaf margins. On warm-season grasses (such as bermudagrass), the tan leaf lesions may be oval shaped to oblong with a brown border. No borders occur on leaf lesions of annual bluegrass. Dieback from cut leaf tips is also common. Leaf symptoms are sometimes confused with those of red thread (sec. 5.2.22), copper spot

FIGURE 5–23   Dollar spot in Kentucky bluegrass.

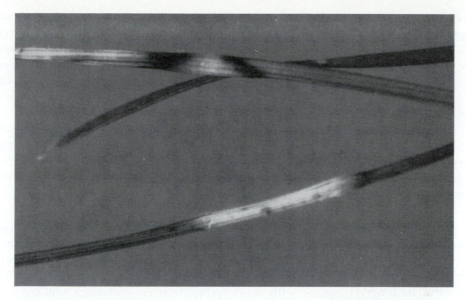

FIGURE 5–24   Dollar spot (*Sclerotinia homoeocarpa*) infecting Kentucky bluegrass leaf blades. (Courtesy D. H. Scott.)

FIGURE 5–25   Dollar spot. Mycelium of the dollar-spot fungus *Sclerotinia homoeocarpa* growing out from Kentucky bluegrass leaves.

(5.2.6), brown patch or Rhizoctonia blight (5.2.23), and Pythium blight (5.2.21). If fungicides are *not* applied to closely cut putting or bowling greens, the spots may become so numerous that they merge to produce large, irregular, sunken areas of straw-colored turf (*Color plate 40*). When early-morning dew is present and the dollar-spot fungus is active, a white to grayish, cobwebby or cottony growth of mycelium may be seen on the upper leaf blades (figure 5–25). Guttation fluid (dew), which is rich in carbohydrates and amino acids, increases the infection process. Some mycelium may persist under the leaf sheaths, but most of it disappears as it is dried by wind and sun. The mycelium of the dollar-spot fungus can be confused with that produced by species of *Pythium, Nigrospora,* and *Rhizoctonia,* and it is also mistaken for the web of a spider. Spider webs, however, are stretched across a single plane two-dimensionally and do not cause leaf lesions, whereas dollar-spot mycelium will colonize the entire volume of the verdure (three-dimensionally) and is associated with turf-displaying lesions.

The rhizomes and roots are not invaded by the fungus; however, a fungus-produced toxin affects water and nutrient uptake, causing the roots to thicken, stop growing, and turn light brown. The replacement roots are similarly affected. This is why dollar spot is most severe in dry soils. The toxin is produced at temperatures between 60 and 80°F (15.6 to 26.7°C).

Disease cycle.   The dollar-spot fungus survives unfavorable periods as black, paper-thin flakes (stromata) on foliage surfaces and in soil, and as dormant mycelium in living or dead turfgrass tissue. In spring or early summer when the temperature reaches 60°F (15.6°C), the mycelia and stromata resume mycelial growth. The fungus is spread with the clippings, mowers and other maintenance equipment, and foot traffic. It enters plants through cut leaf tips and natural openings (stomata) and by direct penetration of intact leaf surfaces when plant surfaces are wet. Maximum disease development usually occurs between 59 and 86°F (15 and 30.6°C) in dry soils where there is a buildup of thatch and where soil nitrogen and potassium levels are low. Infected tissue first appears watersoaked and dark, becoming light tan when dry. If the nights are cool and dry soon after infection has occurred, or if cultural and chemical control measures are applied promptly, damage is limited to the leaf blades, and diseased turf usually recovers quickly (especially if it is growing rapidly). If nights are warm and damp, and heavy dews persist after infection, and if fungicides are *not* applied, the dollar-spot fungus rapidly kills plant tissues, and diseased areas may require weeks or months to recover. When turfgrasses are maintained with adequate nitrogen, potash, and water, less dollar spot occurs and recovery is more rapid.

Dollar-spot infection periods consist of 2 consecutive wet days with an average temperature of 72°F (22.2°C) or more or at least 3 consecutive wet days when the average temperature is 60°F (15.6°C) or more. The fungus spreads to new areas mostly by transport of infected sod or clippings, on mowers, other turf equipment, hoses, golf carts, animals, water, wind, and shoes.

Importance.   Dollar spot is common, persistent, and very destructive to turfgrasses, especially creeping bentgrass, throughout the world. If left unchecked, affected patches frequently coalesce to produce large, irregular, sunken areas of straw-colored turf that take many weeks or months to recover.

Management.   Follow the cultural practices outlined in table 5–1. On golf and bowling greens, remove guttation water (dew) in early morning (5 to 8 A.M.) by dragging a hose or long bamboo pole over the wet grass, by mowing, or by irrigating deeply. Maintain balanced and adequate fertility during the growing season by frequent light applications. If possible, increase the mowing height.

Avoid drought stress by keeping turf near field capacity. Core cultivate to alleviate soil compaction and control thatch. Grow or overseed with cultivars of Kentucky bluegrass, fine-leaf fescues, perennial ryegrass, bermudagrass, zoysiagrass, or bahiagrass that are tolerant or resistant to dollar spot. There are no highly resistant cultivars of creeping bentgrass or annual bluegrass. Some of the newer bentgrass cultivars, such as Crenshaw 18th Green, Imperial, Century, Lopez, Southshore, and SR 1020, are highly susceptible to dollar spot. Highly susceptible cultivars should not be considered for fairways where this disease is a major problem.

Dollar spot is suppressed on bentgrass greens by 70:30 (volume to volume) mixtures of sand and organic fertilizers (such as poultry or cow manure compost and organic fertilizers composed of animal and plant meals) applied at monthly intervals. Strains of the biocontrol agents *Trichoderma harzianum* and *Pseudomonas aureofaciens* are reported to suppress the dollar-spot fungus.

Dollar spot is ubiquitous and can develop on any turf. Numerous fungicides are effective in controlling this disease, but these should only be used in combination with a sound cultural management program. Dollar spot usually develops slowly at first, giving a turf manager time to both properly identify the disease and prepare an appropriate fungicide program. In general, it is not necessary to apply fungicides preventively for the control of dollar spot.

Start fungicide applications, especially to closely cut bentgrass, when dollar-spot patches are covering about 1 to 3% of the turf and the daytime air temperature consistently reaches 65 to 70°F (18.3 to 21.1°C). Where dollar spot has been prevalent in the past and difficult to control once the symptoms have appeared, apply a protective fungicide or combination (table 5–5) at 7- to 10-day or 14- to 21-day intervals (depending on the fungicide used) during moist weather in spring, summer, and autumn when day temperatures average between 60 and 90°F (15.6 to 32.2°C). Widespread resistance to a number of fungicides (such as benzimidazoles, decarboximides, and DMI) has occurred with this disease and has become a serious problem, especially on golf courses. However, where resistance has not appeared, triadimefon, propiconazole, and other fungicides have given significant residual control of dollar spot 8 months after application to control gray snow mold. Avoid excessive use of chemically related fungicides. Note that the management of dollar spot, especially when using fungicides, will be markedly improved by concurrent application of soluble nitrogen at 13.1 to 21.8 lb/acre (14.7 to 24.4 kg/ha).

Test kits are available for the identification and prediction of dollar-spot development and severity.

5.2.9   Downy Mildew or Yellow Tuft

Downy mildew or yellow tuft is caused by the systemic water-mold fungus *Sclerophthora macrospora*. The disease occurs worldwide and is capable of attacking all turfgrasses as well as many forage and wild or weed grasses (for example, crabgrasses) during prolonged, cool (60 to 70°F; 15.6 to 21.1°C), and wet weather. The disease is generally considered to be a minor problem except on turfs maintained at close-cut mowing heights, such as golf greens, tees, and sod fields in wet, poorly drained low areas. The disease is mostly associated with seedling or immature turfs grown on poorly drained and heavily irrigated areas. Downy mildew may weaken creeping bentgrass to such an extent that additional stress will kill infected plants. Voids left by the dead plants are later invaded by weeds.

Identification.   In cool, humid weather, small, very dense clumps of slightly stunted cool-season grass appear with somewhat thickened or wider leaf blades. When severe, small yellow patches 0.25 to 4 in (0.6 to 10.2 cm) in diameter appear in the turf (yellow tuft; figure 5–26). Each spot is composed of one or two dense clusters of excessively tillered and yellowed shoots with stunted roots (*Color plate 41*). These tufted plants are easily pulled from the turf. Diseased plants commonly die from heat and drought stress, winter desiccation, or attacks of secondary fungi such as those causing leaf smuts or Helminthosporium diseases. Affected turf appears unsightly; it is yellow-spotted or mottled and has an uneven surface. During cool, wet periods, a white, downy growth may appear on the leaves. Symptoms of the disease are most prominent during late spring and autumn, especially in poorly drained soils. Large turfgrass areas may turn yellow, wither, and die during hot, dry weather. More commonly, many infected plants survive such stress periods and mask the thinning of infected turf.

FIGURE 5–26   Yellow tuft–affected plants in a Kentucky bluegrass turf.

FIGURE 5–27   Heads of tall fescue affected by the downy mildew or yellow tuft fungus *Sclerophthora macrospora*. (Courtesy J. L. Dale.)

Diseased plants of St. Augustinegrass do not appear to be seriously affected. A white downy growth appears during wet weather. White-to-pale-yellow-green, slightly raised streaks develop in the leaves (parallel to the midrib), which may become wrinkled or rippled except in damp weather in shady areas. Leaf streaks appear in the spring and remain throughout the summer, giving the leaves a yellow appearance with some tip dieback. The length and width of leaves and internodes may be reduced. Excessive tillering is absent. Growth and vigor may also be suppressed and the turf may become thinned.

If left unmowed, the heads of infected turfgrass plants assume grotesque shapes, as if injured by a phenoxy herbicide (figure 5–27).

Disease cycle.   The downy mildew fungus survives as systemic mycelium and as dormant, thick-walled oospores (50 to 75 μm in diameter and 2 to 3 times larger than those of *Pythium* species) in infected leaves, stems, and crowns (figure 5–28). In cool (41 to 75°F; 5.0 to 23.9°C), moist weather, white, stalk-like structures (sporangiophores) grow out through the stomates on leaf surfaces and leaf sheaths to bear pearly-white, lemon-shaped sporangia (figure 5–29) at the tips of their branches. The sporangia germinate in free water by releasing up to 50 or more microscopic motile spores (zoospores) that swim actively about for as long as 24 hours before settling down (encysting) on the grass and germinating to produce a hyphal strand that infects young meristematic tissue between 50 and 77°F (10.0 to

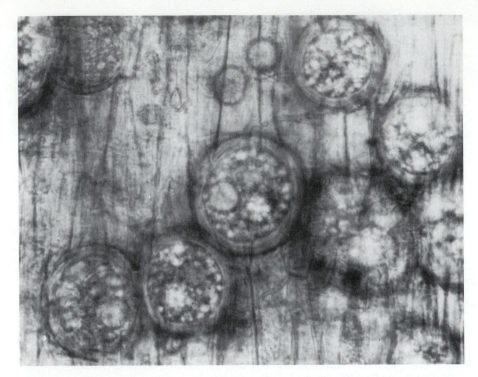

FIGURE 5–28 Oospores of the downy mildew fungus *Sclerophthora macrospora* in creeping bentgrass leaves, as seen under a high-power microscope. (Courtesy Noel Jackson.)

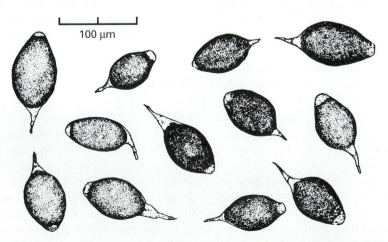

100 μm

FIGURE 5–29 Lemon-shaped sporangia of *Sclerophthora macrospora* from Kentucky bluegrass, as seen under a high-power microscope.

25.0°C). Young seedlings in water can be infected by zoospores within 6 hours. This explains why the disease is most severe in low-lying wet areas. Very young seedlings are most likely to become infected, whereas mature plants usually escape disease, because grass seedlings, during the germination process, release sugars and amino acids, which attract the swimming zoospores to where infection occurs in meristematic tissues with actively dividing cells. Numerous round oospores (larger than those of *Pythium* species; figure 5–67) are produced in infected host tissues and are evident under a microscope (figure 5–28). The oospores overseason in dead host tissues and may germinate during cool weather in the presence of moisture to produce sporangia. Spores of the downy mildew fungus spread in splashing or flowing water on turfgrass equipment and shoes, and in infected sod, sprigs, and plugs.

Importance.   Downy mildew (yellow tuft) is usually not a serious disease but may, over time, eliminate creeping bentgrass in golf greens and tees, allowing weeds to fill in the voids. The disease is greatest on seedling turfs in low-lying areas where surface drainage occurs.

Management.   Provide adequate surface and subsurface drainage when preparing a seedbed. Avoid overwatering. Follow other cultural practices (see table 5–1) that promote active but not lush growth. Mow only when the grass is dry. Preventive fungicides active against Pythium blight (see table 5–5), such as metalaxyl (Subdue) and fosetyl-Al (Aliette), effectively control the disease. Fungicides *must* be applied *before* infection occurs, for diseased plants cannot be cured. Where the disease has been serious in the past, start fungicide applications after the first mowing and continue until nighttime temperatures remain consistently above 65°F (18.3°C). Resume applications again in autumn when nighttime temperatures dip to 50°F (under 10°C). The addition of an iron compound (like Fe_2SO_4) at 10 to 20 lb acre (4.5 to 9.1 kg/ha) to the fungicide helps to mask symptoms of disease. There are no turfgrass cultivars known to be resistant to downy mildew or yellow tuft.

5.2.10   Fairy Rings

Fairy rings are found worldwide, affecting all cultivated turfgrasses by the growth of any one of about 60 species of soil- and thatch-inhabiting basidiomycete fungi [that form mushrooms (toadstools) and puffballs]. Three of the more common soil-colonizing fungi include the common field mushroom *Psalliota* (*Agaricus*) *campestris* (or *A. bisporus*), which is sold in grocery stores; the small, very common tan mushroom, fairy-ring fungus, *Marasmius oreades*; and the large, white, poisonous mushroom *Chlorophyllum molybdites* (*Lepiota morgani*); see figure 5–30. (Of course, you should never eat any mushrooms appearing in turf without first having them identified by a competent authority.) The following discussion covers soil-colonizing fairy-ring fungi. Turf with a thick thatch or under stress is more prone to damage. Sites where the soil has buried organic matter, such as stumps and roots, are most likely to develop fairy rings and related problems.

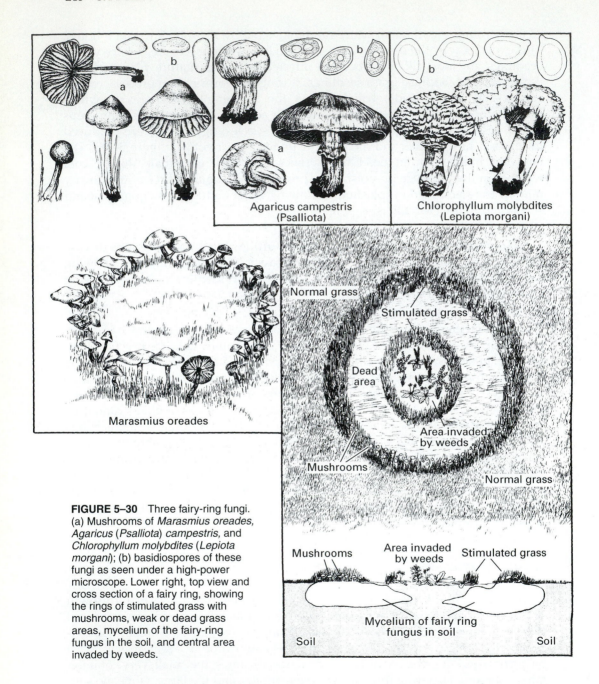

FIGURE 5–30 Three fairy-ring fungi. (a) Mushrooms of *Marasmius oreades*, *Agaricus* (*Psalliota*) *campestris*, and *Chlorophyllum molybdites* (*Lepiota morgani*); (b) basidiospores of these fungi as seen under a high-power microscope. Lower right, top view and cross section of a fairy ring, showing the rings of stimulated grass with mushrooms, weak or dead grass areas, mycelium of the fairy-ring fungus in the soil, and central area invaded by weeds.

Identification. Fairy rings usually appear as circles, arcs, or ribbons of dark green, fast-growing grass, 4 to 12 in (10.2 to 30.4 cm) wide, during the spring, early summer, and early autumn (figure 5–31). Concentric zones of thin, yellow, wilting, dead, or dormant grass may develop both inside and outside the circle (Type 1 rings). During dry weather, especially in late summer and autumn, the

FIGURE 5–31   A scalloped fairy ring showing the outer dark stimulated grass and mushroom fruit bodies.

dead ring is normally outside or between the inner and outer dark green zones. Fairy rings are highly variable in size. Most are 2 to 15 ft (0.6 to 4.6 m) in diameter but may be 50 ft (15.2 m) to 200 ft (61 m) or more across. In mild weather, after rains or heavy irrigation, large numbers of mushrooms or puffballs (the fruiting bodies of the fairy-ring fungi) may suddenly pop up in a circle at the inner edge of the fast-growing grass that outlines the fairy ring (figures 5–31 and 5–32).

Commonly, several distinct rings or arcs develop in the same general area. In the case of some fungi (such as, for instance, with *Marasmius oreades*), rings that meet partially die out and take on a scalloped outline (figure 5–32).

Generally, fairy rings are first seen as a cluster of mushrooms or as a tuft of stimulated turf. The rings enlarge radially each year from 3 in (7.6 cm) to about 2 ft (0.6 m). Some rings disappear unexpectedly for a year or more and then reappear, usually 1 ft (30.5 cm) or more larger in diameter. Fairy ring fungi survive as saprophytes in plant debris. Dissemination of the fungi occur in contaminated soil by mechanical means such as turf equipment, sod, and human activities.

FIGURE 5–32   Fairy rings in Kentucky bluegrass turf. Three rings have met, fungus activity has ceased, and the rings now have a scalloped effect. (Courtesy D. H. Scott.)

Disease cycle.   Fairy rings are initiated by the germination of a basidiospore from a fruiting body or bits of fungus-infested soil. Mycelial growth continues radially outward in soil and thatch. The fungus grows throughout the soil to a depth of 1 to 2 in (2.5 to 5.1 cm) to 12 to 20 in (30.5 to 50.8 cm) or more, depending on the soil type. In some species (such as *Marasmius oreades*), it forms an extensive, dense, white, mycelial network (spawn) that has a strong, moldy smell (figure 5–33). The fungus decomposes organic matter (for instance, thatch and plant debris) as a food supply.

The lush, dark green grass of a fairy ring is due partly to the increased amount of nitrogen (as ammonia) made available to the grass roots by the fungus as it breaks down organic matter in the thatch and soil. If this lush growth persists into autumn, the chance of pink snow mold or Microdochium (Fusarium) patch infection increases. Various bacterial species in the soil reduce the ammonia to nitrites and then to the nitrates used by the grass plants as food, resulting in the taller, darker green grass. The ring of brown, dormant grass is caused by the dense white layer of mushroom spawn, which impedes water movement into the soil and de-

FIGURE 5–33 White mycelium spawn under the outer edge of a fairy ring with the bentgrass sod peeled back. (Courtesy R. T. Kane.)

pletes nitrogen and other nutrients available for plant growth. Grass roots are invaded by the fungus, which may also produce toxic levels of hydrogen cyanide, as is suspected for *Marasmius oreades*. The grass in this area may become so weakened, yellow, and stunted that it succumbs to environmental stresses or is killed by other diseases and invaded by weeds. As the fungus grows outward radially, older mycelium in the interior of the ring dies. Bacteria degrade the aging or dead fungal spawn, releasing nitrogen and other nutrients for use by the grass plants and initiating the formation of the inner ring of lush grass. The three zones of a Type 1 fairy ring are noticeable from early spring into winter.

Fairy rings are usually most severe in sandy or gravelly, low-fertility soils that are low in moisture. Turf with a thick thatch growing in sandy soil is highly vulnerable to damage.

Importance. Fairy rings are classified into three types based on the effects they cause. All three types produce mushroom or puffball fruiting bodies.

Type 1. These are the most serious, as they commonly kill the grass in a zone at the inner edge of the outer ring of lush, dark green grass. The turf in the rings is yellow, wilted, dead, or dormant due to the infested soil being water-repellent. *Marasmius oreades* produces phytotoxic metabolites (such as agrocybin) that are believed responsible for killing of grass.

Type 2.   The grass is stimulated, forming a single ring of lush, dark green grass. No turf is killed.

Type 3.   The rings do not stimulate the grass, cause no damage, but occasionally produce fruiting bodies in the rings.

Management.   Controlling or removal of fairy rings and other mushrooms or puffballs is made very difficult because the infested soil becomes essentially hydrophobic (impervious to water filtration). Before planting a new turf area, remove tree stumps, large roots, construction lumber, and other large pieces of organic matter from which these fungi obtain nutrients. Avoid the use of root-zone mixes that contain large amounts of undecomposed organic matter. Keep the new planting well fertilized and watered. It is best to moisten the soil to a depth of 6 in (15.2 cm) or more at each irrigation. Shallow watering probably encourages the germination of many fairy-ring fungi.

In established turf, fairy rings can be effectively managed but the methods are laborious, time-consuming, and often expensive. Fairy rings can be combatted in one of three ways: suppression, eradication, and antagonism.

Suppression.   To suppress ring formation, core cultivate the entire area of the ring and 2 ft (60.9 cm) beyond it; water the turf thoroughly to a depth of 6 in (15–2 cm) or more; and fertilize well. Control excess thatch accumulation by verticutting. This is usually the most practical approach to managing fairy rings.

1. Since there are fewer rings, and these are much less conspicuous on adequately watered and fertilized turf, apply nitrogen fertilizer to the turf several times during the year. Follow local recommendations based on a soil test and the cultivar or blend of grass being grown. Avoid excessive applications of nitrogen and organic matter (manures or mulches), as they tend to stimulate the development of fairy rings and encourage other turf diseases.

2. Where feasible, the symptoms are easily masked by first core aerifying. Then pump large quantities of water (including adding a turf-wetting agent such as Aqua-gro) 10 to 24 in (25.4 to 60.9 cm) deep into the soil at 1-ft (30.5-cm) intervals for a distance of 18 to 24 in (45.7 to 60.9 cm) on either side of the outer-stimulated zone of dark green grass. Maintain the soil in a near-watersoaked condition for 4 to 6 weeks by watering every other day. Use a tree-feeding lance or root-feeder attachment on a garden hose. Repeat the treatment several months to a year or more later when the rings reappear and the turf shows drought stress and begins to wilt. Fertilization will help to mask the fairy rings.

3. The fungicide flutolanil (ProStar) is registered for the suppression of certain fairy ring fungi. It is applied as a drench and could be used to spot-treat problem rings in certain areas such as golf greens. However, suppression may be temporary. Carefully follow all label directions.

Eradication.   Fairy rings can be eradicated by soil fumigation or excavation. Both methods are time-consuming, costly, and not always successful.

1. *Soil fumigation.*   Carefully strip and dispose of the sod in an area 2 ft (61 cm) inside and outside the outer ring of grass. *Be careful not to spill any of the in-*

fested soil or sod onto the healthy turf. A better but more expensive method is to apply glyphosate (Roundup, Kleenup) to the area. This will kill the grass in about a week with little or no chemical residue remaining in the soil to affect the new planting. You still have to dispose of the dead turf after using glyphosate, since it still contains mycelium of the fairy-ring fungus. Loosen the soil underneath to a depth of 6 to 9 in (15.2 to 22.9 cm) with a spading fork or by rototilling to improve the results. Be careful not to spread the fungus with the tiller. Tools and rototiller equipment used to cultivate the soil should be sterilized with household liquid bleach (diluted 1:5 with water and then washed with plain water) or by steaming. After the soil is well loosened, fumigate with methyl bromide or a methyl bromide-chloropicrin mixture, dazomet (Basamid), or metam-sodium (Metam, Vapam Soil Fumigant; table 5–5). Dazomet and metam-sodium are the safest and easiest chemicals to handle and use, but may be difficult to obtain. Methyl bromide will probably be phased out within a few years. Soil fumigation should *only* be done by a certified golf-course superintendent or turfgrass specialist who is licensed to handle and apply restricted-use pesticides. The person doing the fumigating needs to carefully follow all the manufacturer's directions and precautions when using any fumigant.

   The soil temperature should be 60°F (15.6°C) or above for fumigation. The vapors of the fumigant are kept in the soil by covering the stripped areas with a polyethylene cover for up to 10 days, depending on the fumigant used. The plastic cover is removed after the fumigation period. Upon removing the cover, the soil should be stirred and left exposed to the air for 2 weeks or until all odor of the chemical has disappeared. Add fresh soil to the area as needed, followed by reseeding or sodding. (Use soil fumigants *only* well beyond the outer drip line of trees and shrubs.)

2. *Excavation.*   Carefully dig out and discard all infested soil in the ring at least 12 in (30.5 cm) deep and 2 ft (60.9 cm) on either side of the outer stimulated zone, including all mushroom spawn. Replace the soil in the trench with fresh soil from a source free of fairy-ring fungi. The area is then sodded or reseeded. This treatment is usually cost-prohibitive and impractical.

Antagonism.   Antagonism is most effective when a turf area is heavily infested with fairy rings. The biology of these fungi—for example, *Marasmius oreades*—ensures that two or more strains eliminate each other when occupying the same site. Figure 5–32 shows how antagonism has stopped disease development where two fairy rings meet. The method of control, while requiring the same initial steps as used for eradication, is much cheaper and easier because the costs for chemical fumigants and for a licensed applicator are eliminated. Also, there is no danger of injury from contact or inhaling chemical fumes.

1. Kill the infested fairy-ring area and 2 ft (60.9 cm) beyond with glyphosate. After about a week, strip the sod from the area and repeatedly cultivate the entire area covered by the fairy rings as deeply as possible using a heavy-duty rotary cultivator. Mix the mycelium-infested soil from several rings, blend it thoroughly, and then spread the blended mycelium evenly over the exposed

soil. Multiple cultivations with a rotary cultivator, to a depth of 6 to 8 in (15.2 to 20.3 cm) in several directions, will effectively mix the spawn of the fungi. Rake or roll the soil level.

2. Then wet the soil to a depth of 8 in (20.3 cm) or more, adding a turf-wetting agent to increase water infiltration. Fertilize based on a soil test for the grass being grown. Sod or reseed the area and keep it well watered and fertilized.

While fairy-ring fungi primarily colonize the soil underlying turfgrasses, there are a number of other basidiomycete fungi that colonize the thatch and leaf litter. Names given to these diseases include (1) superficial fairy rings, (2) localized dry spot, (3) white blight, and (4) yellow ring. Relatively little is known about these fungi and their disease cycles.

Superficial fairy rings.   Superficial fairy rings, sometimes called white patch, occur worldwide on many turfgrasses. They are most prevalent on bentgrass greens, tees, and fairways, especially sand-based greens. Superficial fairy rings are caused by a number of basidiomycete fungi, some of which are sterile and/or have clamp connections on their hyphae (figure 5–2). They include species of *Clavaria*, *Clitocybe*, *Coprinus kubickae*, *C. psychromorbidus*, *Hygrophorus*, *Psilocybe*, *Trechispora* spp. including *T. alnicola* (called yellow ring) and *T. farinases*, *Cristella unfinis* (syn. *T. cohaerens*), *Marasmius siccus*, and other fungi that do not produce spores and hence are identified with difficulty. All of these fungi primarily colonize the thatch and leaf litter of poorly nourished turf during rainy periods in cool-to-hot weather. The fungi rarely invade more than 0.5 to 1 in (1.3 to 2.5 cm) of the underlying soil. Most of these fungi are considered to be saprophytes, as they feed on dead grass debris and hence cause little damage. The thatch and underlying soil where the fungi are active commonly become water-repellent (hydrophobic). Dissemination is by fungus-infested turf and soil by mechanical means (such as turfgrass equipment, human activities, wind, and water).

Identification.   The rings or patches are white and usually range in size from 2 in (5.1 cm) to 6 ft (1.8 m) in diameter. Dense white mycelium can often be seen at the outer 1 to 2 in (2.5 to 5.1 cm) of the rings and patches, but this is sparse or diffuse within the ring. The older leaves and leaf sheaths in the border area are a bleached white and die prematurely from being covered by the mycelium. The infested thatch has a strong mushroom odor. The fungus uses the dead grass tissues for food, causing a reduction in the thatch. If severe, the turf may be somewhat thinned. Turf in the center of the patches may or may not appear healthy (figure 5–34).

Disease cycle.   These fungi probably have much the same life cycle as do the fairy-ring fungi that colonize the soil. The importance of basidiospores in the dispersal of these fungi is unknown.

Importance.   A superficial fairy-ring condition may cause little or no damage, but it may be unsightly, and it can thin the turf and disfigure or interfere with the putting surface in golf greens. Roots may be stunted and discolored. Superfi-

FIGURE 5–34   White patch in a creeping-bentgrass green. (Courtesy L. T. Lucas.)

cial fairy-ring fungi have been classified into three types based on their appearance and effects on turf.

- *Type A.*   Mycelium is sparse to abundant, and fruiting bodies may or may not form. The mycelium colonizes grass shoot bases, rhizomes, and stolons in the thatch. There is little or no effect on the growth of grass.
- *Type B.*   The fungi stimulate turf growth and/or cause discoloration of the grass. The thatch is decomposed, but the grass plants are not severely injured and later recover.
- *Type C.*   These fungi produce severe injury to the grass; earlier and adjacent turfgrass growth may or may not be stimulated.

Management.   Follow the cultural practices outlined in table 5–1. Since the rings are more common in nitrogen-starved turf, apply about 0.5 lb (2.3 kg) of a nitrogen fertilizer per 1000 sq ft or 93 sq m (25 kg N/ha) based on soil-test results and the types of grass being grown. Mechanical removal of thatch, frequent core cultivation, and top-dressing of putting greens are other suggested measures to reduce these fungi.

Localized dry spot.   Localized dry spot (LDS), caused by or associated with a number of unidentified basidiomycete (mushroom) fungi, is common on close-cut golf and bowling greens, especially those abundantly topdressed with sand, and bentgrass fairways. LDS commonly appears in new turf areas within 2 or 3 years after seeding or sprigging. Soil within patches remains very dry, even after rainy weather or frequent irrigations. Water cannot penetrate the thatch-soil interface and flows away from affected areas. Hot and dry periods predispose turf to this problem. LDS tends to decline over time. Other causes of localized dry spots may include poor mixing of the greens mix, tree roots near the soil surface, buried debris or rocks, and other physical problems.

Identification.   Solid patches of blue-green grass that soon wilt, dry out, or die are sometimes preceded by the presence of fairy rings; or numerous mushrooms can be seen during dry and hot weather. The patches are round to irregularly shaped and range from several inches (6 to 8 cm) to several feet (0.5 to 1.0 m) across. The turfgrass soil within the patches remains *very dry* despite frequent rains or irrigations because the soil becomes impervious to water filtration. (It is water-repellent or hydrophobic.)

This resistance to water is believed to be caused by older fungal mycelium breaking down and releasing organic substances that form a coating over and around the individual coarse sand particles. The sand particles become bound together, thus preventing water from entering the soil. This condition is normally restricted to the upper 1.5 to 2.5 in (3 to 6 cm) of soil.

Disease cycle.   The causal fungi are probably closely related to those causing fairy rings, white blight, and yellow ring. The fungi feed on dead turfgrass debris in the thatch.

Importance.   This is a serious problem on high-sand greens and mineral soil greens heavily topdressed with sand. Soil within the patches remains very dry, causing it to wilt and die, mostly during hot, dry periods.

Management.   Severity of symptoms can mostly be reduced by mechanically removing excess thatch, core aerifying several times a year plus water-inject cultivation, combined with a turf-wetting agent (such as Aqua-gro), at 1- or 2-week intervals during dry periods. Some turfgrass managers keep localized dry spots alive in hot, dry weather by daily syringing and water injection using a deep-root tree feeder. There is evidence that fungicides with activity against basidiomycetes (such as Prostar) can reduce the severity of localized dry spots.

White blight.   White blight is a disease of tall fescue (*Festuca arundinacea*), red fescue (*F. rubra*), and chewings fescue (var. *commutata*) in the southern United States caused by the fungus *Melanotus phillipsii*. It first appears as roughly circular blighted patches 3 to 5 in (7.6 to 12.7 cm) in diameter that may enlarge to about 12 ft (3.7 m). The patches may eventually coalesce to form large areas of blighted turf. The patches are white and commonly have a pink to salmon border. Leaf blades in the patch are often flattened to the soil surface and covered by grayish-white mycelium. Leaf blades die back from the tip. Small,

2- to 8-mm-diameter, grayish-white, stalkless mushrooms of the causal fungus are borne on the leaf blades.

Disease cycle.   Little is known about the life cycle. It is probably about the same as fairy-ring fungi. The importance of basidiospores in the dispersal of the fungus is unknown.

Importance.   This minor disease causes blighting of leaves during summer and early autumn, mostly on open, droughty sites. The disease is more severe on immature stands of tall fescue than on mature stands.

Management.   The management is the same as for superficial fairy rings and localized dry spot.

Yellow ring.   A disease of Kentucky bluegrass, yellow ring is caused by a soil-borne fungus *Trechispora alnicola*. It is evident from midspring to early autumn. It is most common on heavily thatched turf (over ¼ in; 0.64 cm) that has sufficient water and nutrients to remain lush during the entire growing season. The disease has been reported from a number of states in the midwest and northeastern United States. Disease severity is greatest at 68 to 77°F (20 to 25°C). A closely related fungus, *Trechispora confinis,* is known to attack creeping bentgrass.

Identification.   Lemon to golden yellow rings 4 in (10.2 cm) wide and up to 4.5 ft (1.4 m) in diameter occur in otherwise healthy turf (*Color plate 60*). A dense, white mycelial growth is evident, up to an inch (2.5 cm) or more thick, in the thatch beneath the yellowed leaves (figure 5–35). The white mycelium has a strong mushroom odor. Diseased turf does not die, and the rings enlarge from one year to the

FIGURE 5–35   White mycelium of *Trechispora alnicola* in thatch beneath the yellowed leaves in a yellow ring of Kentucky bluegrass.

next and may disappear after a year or more. The fungus is mildly pathogenic to the roots, which become tan to brown and stunted.

Disease cycle.   *Trechispora alnicola* is largely a saprophyte that colonizes dead grass in the thatch and is therefore beneficial in breaking this debris down into elements available to the living grass. The fungus is believed to survive as dormant mycelium in plant debris and possibly in living plants and the soil. When favorable conditions of temperature, moisture, and nutrition return, the mycelium resumes growth and heavily colonizes the thatch, causing the leaves of living grass plants to turn yellow. Spores produced by the fungus in possible spread and infection appear to be important. The causal fungus appears to be disseminated in water and by turf equipment.

Importance.   This minor disease turns living grass leaves yellow in rings but does not kill the turf.

Management.   Dethatch and core aerify affected turf in early spring and autumn when it accumulates to ½ in (1.3 cm). Use a vertical mower, power rake, or similar dethatching equipment. Kentucky bluegrass cultivars differ in resistance. Reduce the rate of thatch accumulation by reducing fertilization.

Frost Scorch
See the section on minor leaf spots (5.2.16).

5.2.11   Fusarium Blight (Fusarium Leaf Spot, Crown, and Root Rot)

Primary hosts of this disease complex are bentgrasses, annual and Kentucky bluegrasses, annual and perennial ryegrasses, fine-leaf and tall fescues, centipedegrass, zoysiagrass, and crested wheatgrass. Infection and disease development are most severe during hot and dry summer weather.

At least 12 species of *Fusarium* fungi are associated with the various phases of Fusarium blight (*F. acuminatum, F. avenaceum, F. crookwellense, F. culmorum, F. equiseti, F. graminearum, F. heterosporum, F. oxysporum, F. poae, F. sambucinum, F. semitectum,* and *F. Tricinctum*). The species primarily responsible in the United States are *F. culmorum* and *F. poae,* which are most active at 80 to 95°F (26.7 to 35°C). The same fungi may be involved with damping off of seedlings, as both fungi are commonly seedborne. Other species of *Fusarium* commonly isolated from or associated with turfgrasses in the laboratory include *F. concolor, F. merismoides, F. moniliforme, F. oxysporum, F. pallidoroseum, F. roseum, F. sambucinum, F. solani,* and *F. sporotrichoides.* In general, species of *Fusarium* grow at temperatures of 50 to 95°F (10 to 35°C).

Identification.   Oval to elongated or irregular leaf blotches that are first dark green and watersoaked and fade to dull tan, usually with a light brown or purple border, develop mostly on older leaves (*Color plate 42*). Leaf lesions are not distinct on zoysiagrasses. Typically, Fusarium leaf spot or blight occurs mostly on older, senescing leaves and on tillers weakened by drought stress or root and crown rot (see melting out and summer patch and necrotic ring spot). The disease may occur uniformly over large areas of turf in warm-to-hot, moist weather with

FIGURE 5–36   Fusarium leaf spot (caused by *Fusarium acuminatum*) in a lawn of Kentucky bluegrass. (Courtesy R. W. Smiley.)

cool nights (figure 5–36). Fusarium crown and root rot is most serious in drought-prone areas such as south-facing slopes, near paved walks, driveways, and parking lots that receive mostly all-day sun.

On higher-cut turf, the lesions, up to about ½ in (1.3 cm) long, originate at the cut tips or randomly over the leaf surface. Infected leaves may turn yellow and die back from the cut leaf tip to the base. This phase of the disease can be confused with anthracnose, dollar spot, or Helminthosporium leaf spot.

Fusarium crown and root rot (figure 5–37) produces symptoms that resemble melting out caused by *Drechslera* and *Bipolaris,* summer patch and necrotic ring spot. Affected turf may thin out in circular to irregular patterns up to 3 ft (0.9 m) in diameter. Patches of diseased plants turn tan and then straw colored as they die during hot, dry weather. The patches often have a reddish brown border, 1 to 2 in (2.5 to 5.1 cm) wide, occasionally with centers of apparently healthy grass, giving a frogeye appearance. The patches may coalesce to blight large areas. The roots and crowns develop a dry, brown-to-reddish-brown rot that later becomes firm and dark brown to black (figure 5–37).

On golf greens, the disease first appears as tan-to-light-brown, irregularly shaped areas 2 to 3 in (5.1 to 7.6 cm) in diameter. If conditions remain favorable, the blighted patches of grass enlarge into irregularly shaped areas up to 2 to 3 ft (0.6 to 0.9 m) wide. Under moist soil conditions, pinkish mycelium can be seen on the crown and root surfaces near the soil line.

FIGURE 5–37   Fusarium crown and root rot of Kentucky bluegrass. (Courtesy R. W. Smiley.)

Disease cycle.   Species of *Fusarium* overseason as dormant mycelium in or on infected plants and colonized grass debris as thick-walled resistant spores (chlamydospores; see figure 5–38) in the thatch or upper soil. When favorable conditions return (proper temperature, moisture, and nutrition), the pinkish mycelium resumes rapid growth, and the spores (large conidia and chlamydospores) germinate to produce additional mycelium that infect the grass leaves or other parts of the turfgrass plant directly. Fusarium fungi produce large numbers of spores (crescent-shaped macroconidia and smaller, oval or round microconidia) in hot weather (80 to 95°F; 26.7 to 35.0°C), with night temperatures of 70°F (21.1°C) or above, especially in dry thatch when it is rewetted (see figure 5–38).

Importance.   Fusarium blight is an important and widespread disease that may thin out and kill cool-season grasses in circular to irregular patterns or streaks during hot, dry weather. Entire tillers or plants may die when the crowns, roots, rhizomes, and stolons die.

Management.   Follow the suggested cultural measures outlined in table 5–1. Avoid drought stress and a thick thatch. Follow a balanced fertilizer program based on local recommendations for the grasses grown. Raise the mowing height, water deeply but infrequently, and core cultivate in spring and fall to relieve compaction. Increasing air movement over the turf is beneficial, as is the removal of clippings. Grow blends of well-adapted Kentucky bluegrasses in combination with blends of newer perennial ryegrasses.

Chemical controls (table 5–5) are not usually required. Fusarium crown and root rot is difficult to control with fungicides. Soil drenches to control summer patch and necrotic ring spot may be beneficial starting when night temperatures remain above 70°F (21.1°C).

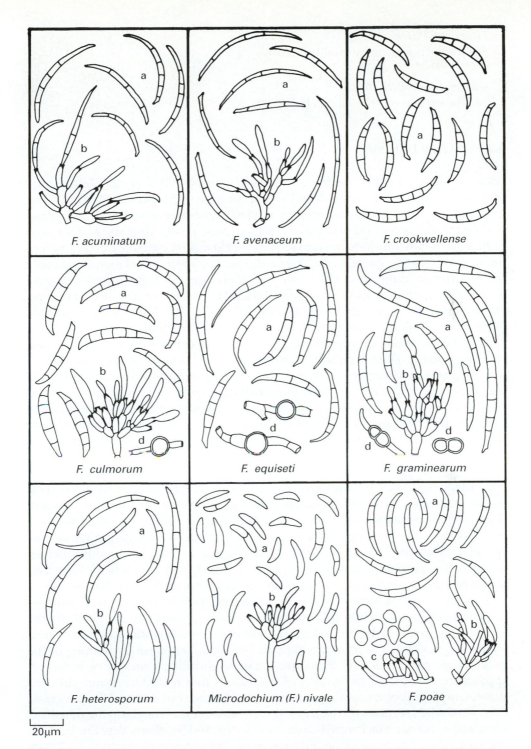

F. acuminatum

F. avenaceum

F. crookwellense

F. culmorum

F. equiseti

F. graminearum

F. heterosporum

Microdochium (F.) nivale

F. poae

20μm

FIGURE 5–38 Eight species of *Fusarium* and one of *Microdochium* that are associated with disease in turfgrasses, as seen under a high-power microscope. (a) Macroconidia; (b) conidiophores, some bearing young to mature macroconidia; (c) microconidia; (d) chlamydospores.

Fusarium Patch
See the section on snow molds (5.2.29).

Gaeumannomyces Patch
See the section on take-all patch (5.2.33).

5.2.12   Gray Leaf Spot

Gray leaf spot, caused by the fungus *Pyricularia grisea* (syn. *P. oryzae*), teleomorph *Magnaporthe grisea,* is a common and serious disease of worldwide St. Augustinegrass, annual and perennial ryegrasses, tall fescue, bermudagrass, and bahiagrass. It occasionally occurs on bentgrasses, bluegrasses, centipedegrass, fescues, and buffalograss in warm-to-temperate regions worldwide. Common weeds attacked include crabgrasses, foxtails, and barnyardgrass. Gray leaf spot is found especially in the northeast, central and southeastern United States. St. Augustinegrass is most susceptible, particularly the blue-green (bitterblue) selections. Disease attacks occur in prolonged, warm-to-hot (70 to 90°F; 21.1 to 32.2°C), humid, rainy weather. Twenty-four hours of continuous leaf wetness is needed for maximum infection. Gray leaf spot is more severe on young seedlings and in newly sprigged or seeded, rapidly growing grass than in well-established turf. However, the disease is a chronic problem in mature stands of St. Augustinegrass in subtropical regions and has become an important disease of perennial ryegrass in golf course fairways and sports fields. Shady locations with poor air movement favor disease development. Prolonged wet foliage that is high in nitrogen tends to increase the incidence, severity, and spread of gray leaf spot. The disease is worst where grass growth is slowed by growth-regulating chemicals and herbicide use.

Identification.   Symptoms of gray leaf spot vary between warm- and cool-season turfgrasses on St. Augustinegrass. Leaf lesions start as tiny, round, watersoaked, olive-green-to-brown spots that enlarge rapidly in moist weather and become oval to oblong or elongated; with blue-gray to ash gray, dirty yellow, or tan centers and definite, irregular, dark brown or purple-to-reddish-brown margins. The size of the lesions depends on the turfgrass. A yellowish halo may surround the lesions (figure 5–39). The lesions also form on the leaf sheaths, stems, and spikes. In warm, moist weather, the enlarging spots become covered with the grayish velvety mycelium of *Pyricularia* containing large numbers of spores (conidia; figure 5–16). Where lesions are numerous, severely diseased leaves turn yellow or grayish tan to brown, wither, and die in 2 days causing thinning of the grass. Large turf areas may thus have a rather unthrifty, scorched or brownish appearance from death or spotting of leaf blades that resembles severe drought stress. Gray leaf spot can cause extensive foliar damage but seldom kills the grass.

Perennial ryegrass grown on golf course fairways and roughs in the transition zone, and as far north as Pennsylvania, New Jersey, and Southern New England are blighted. The symptoms commonly resemble Rhizoctonia blight (sec. 5.2.23) or Pythium blight (sec. 5.2.21) or Melting-out (see 5.2.13), but with a lack of aerial mycelium. The leaf lesions are round to oblong shaped, grayish or reddish brown

FIGURE 5–39   Gray leaf spot of St. Augustinegrass caused by *Pyricularia grisea.* (Courtesy L. T. Lucas.)

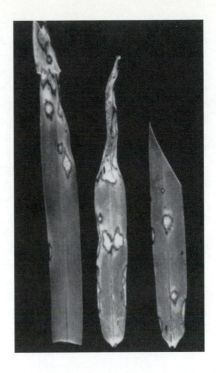

with dark brown borders, and, when moist, they produce large numbers of spores. A yellowish zone surrounds older lesions. Blighted leaves turn yellow and then brown with characteristic twisted leaf tips. Affected ryegrass seedlings may be killed by the fungus. When severe, the turf is thinned as though by drought stress.

Disease cycle.   The *Pyricularia* fungus overwinters as hyaline, pear-shaped, mostly two-celled conidia (figure 5–16); as dormant mycelium in older infected leaves, shoots, and grass debris; and possibly in or on seed. The spores are spread about by wind, water, mowers or other turf equipment, animals, and shoes. During warm-to-hot weather, when the air is moisture-saturated, the conidia germinate and invade young leaves directly or enter through stomates. Visible leaf spots appear 5 to 7 days after infection. Disease activity is favored by prolonged rainy or foggy weather, heavy dews, frequent shallow irrigations, high level of nitrogen, and a temperature between 70 and 90°F (21.1 to 32.2°C). The fungus is spread by spores via air currents, water, mowers, other turfgrass equipment, and animals.

Importance.   Gray leaf spot is a major disease of St. Augustinegrass and ryegrasses in prolonged, warm-to-hot, moist weather, especially on new plantings under stress, such as those heavily fertilized with nitrogen, inappropriate application of herbicides, drought, and soil compaction. The disease may cause thinning of the turf, which appears unsightly but seldom kills plants. Physiological races of *Pyricularia* apparently restrict the spread of the fungus from one genus of turfgrass to another.

Management. Follow the cultural practices outlined in table 5–1. Avoid stresses such as soil compaction, heavy vehicle or foot traffic, drought, a thick thatch, prolonged leaf wetness, and excessive rates of herbicides and nitrogen fertilizers. Follow local fertilizer recommendations based on a soil test for the grasses grown. Do *not* fertilize with quick-release nitrogen fertilizers in the months when gray leaf spot is active. Use very low rates of nitrogen (about 0.1 lb [45 g] of nitrogen per 1000 sq ft or 93 sq m every week or two) where gray leaf spot is a problem. If a darker green turf color is desirable, use an iron-containing fertilizer. Deep but infrequent watering, raising the mowing height, removal of excess thatch, and removal of clippings are all beneficial. Annual and perennial ryegrass cultivars differ in their resistance to gray leaf spot. Avoid overseeding ryegrass into turf affected with gray leaf spot until the severity of the disease declines. Young seedlings are very susceptible to this disease.

Where cultural practices are not checking gray leaf spot, apply a suggested, broad-spectrum fungicide (table 5–5). Repeated applications at 10- to 14-day intervals may be needed during prolonged rainy periods.

Yellow-green or common, purple-stem types of St. Augustinegrass (for instance, Roselawn, Florida Common, and Floratam) are more resistant than the blue-green, green-stem types such as Bitterblue and Floratine, which are very susceptible. Resistant types, however, lack desirable features for lawn turf. Selections of perennial ryegrass also have some resistance.

Gray Snow Mold
See the section on snow molds (5.2.29).

Hadrotrichum Leaf Spot
See the section on minor leaf spots (5.2.16).

Head Smuts
See the section on leaf smuts (5.2.14).

5.2.13   Helminthosporium Leaf, Crown, and Root Diseases

Helminthosporium leaf, crown, and root rots are probably the most common group of diseases to attack cool- and warm-season grasses. One or more of the causal fungi attack all turfgrasses (table 5–2) as well as numerous forage grasses, wild and weed grasses, and cereals throughout the world.

These diseases are favored by dry periods alternating with prolonged, cloudy, moist weather and cool-to-hot temperatures; the use of susceptible cultivars; letting clippings drop in place; close mowing; slow growth of grass; low fertility or excessive shade and nitrogen fertilization; thatch buildup; frequent light sprinklings; compaction from excessive traffic; nematode damage; and applications of certain pre- and postemergence herbicides such as 2,4-D, benefin, bensulide, dacthal (DCPA), siduron, mecoprop (MCPP), and dicamba, all of which stimulate the causal fungal pathogens by causing physiological stress and changes in the growth of the pathogens as a response to the herbicides.

TABLE 5–2   PRINCIPAL HELMINTHOSPORIUM LEAF CROWN, AND ROOT DISEASES, THE FUNGI THAT CAUSE THEM, AND THE GRASSES ATTACKED

Common name of disease	Old name(s) (*Helminthosporium*)	New name(s) (*Bipolaris, Drechslera, Exserohilum*)	Principal turfgrasses attacked
Leaf spot, melting out; leaf, crown, and rhizome rot	*H. vagans* (*H. poae*)	*D. poae* (*D. vagans*)	Bluegrasses, ryegrasses, fescues, buffalograss
Leaf spot, leaf, crown, and root rot	*H. sorokinianum* (*H. sativum*)	*B. sorokiniana* (*D. sorokiniana*)	Bluegrasses, bentgrasses, ryegrasses, fescues, buffalograss, turf timothy, bermudagrasses, wheatgrasses
Net-blotch, leaf blight; crown and root rot	*H. dictyoides*	*D. dictyoides*	Fescues, ryegrasses, Kentucky bluegrass
Zonate leaf spot or eyespot	*H. giganteum*	*D. gigantea*	Bermudagrasses, bentgrasses, bluegrasses, turf timothy, wheatgrasses, zoysiagrasses
Brown blight; leaf, crown, and root rot	*H. siccans*	*D. siccans*	Ryegrasses, tall fescue, Kentucky bluegrass
Leaf blotch; crown and root rot	*H. cynodontis*	*B. cynodontis*	Bermudagrasses zoysiagrasses, kikuyugrass, St. Augustinegrass
Red leaf spot	*H. erythrospilum*	*D. erythrospila*	Bentgrasses
Leaf spot	*H. rostratum*	*E. rostratum* (*E. halodes*)	Bermudagrasses, St. Augustinegrass, wheatgrasses
Leaf blight and crown rot	*H. catenarium*	*D. catenaria*	Bentgrasses, fescues, bluegrasses, perennial ryegrass
Stem, crown, and root necrosis	*H. spiciferum* (*H. tetramera*)	*B. spicifera* (*B. tetramera*)	Bermudagrasses, zoysiagrasses
Yellow leaf spot	*H. tritici-repentis*	*D. tritici-repentis*	St. Augustinegrass, bermudagrasses, bentgrasses, wheatgrasses
Leaf streak	*H. dictyoides*	*D. phlei*	Turf timothy
Minor diseases	*H. stenospilum*	*B. stenospila*	Bermudagrasses
	H. triseptatum	*D. triseptata*	Bermudagrasses, Kentucky bluegrass, fescues
	H. buchloës	*B. buchloës*	Buffalograss
	H. micropus	*B. micropus*	Bahiagrass
		D. nobleae	Ryegrasses

Helminthosporium fungi (now referred to species within the genera *Bipolaris, Drechslera,* and *Exserohilum;* see table 5–2) may be responsible for the gradual browning and thinning (melting out) of susceptible older Kentucky bluegrass cultivars. As the disease progresses, large, irregular turf areas may turn yellow and then brown to straw-colored before being killed (*Color plate 43*).

In some instances, the entire turf is lost. Bermudagrass may be severely attacked by a number of these fungi (table 5–2), causing leaf, crown, stolon, and root rots. The turf may be thinned, and diseased plants commonly appear in irregular patches. Fescues, ryegrasses, and bentgrasses may be similarly affected, especially under high levels of nitrogen. Once one or more of these fungi become established in a turfgrass stand, they remain an ever-present problem. It is not unusual to find two or more species of these fungi infecting a single plant or even a single leaf.

Identification.   This group of diseases produces a variety of symptoms, depending on the cultivar and kinds of grass, culture, season (primarily day length and temperature), weather conditions, and the fungus or fungi that are present.

1. *Leaf spots, blotches, or blights.*   Small, dark brown, reddish brown, brownish-green-to-black, purplish, or purplish-black spots or blotches (lesions) appear on the leaf blades from early spring to late autumn. The lesions may increase rapidly in size and become round to oblong, elongated, or irregular. Their centers often fade to an ash white to pale tan or straw color. The lesions are commonly surrounded by narrow, brown or dark-reddish-brown-to-purplish-black borders (*Color plate 44* and figure 5–40) often described as eyespots. Under moist conditions, one or more lesions may merge and girdle a leaf blade, causing it to turn yellow, tan, or reddish brown and die back from the tip. When leaf spots are numerous, leaves may be completely blighted and wither and die. Lesions may or may not occur on the leaf sheaths and stems, depending on the species of fungus. A diseased leaf sheath turns reddish to purple or brown. Leaf-sheath infection is often so severe that the entire leaf or tiller is girdled and drops prematurely. If the moisture continues, the diseases progress from leaf sheath to leaf sheath on a single plant until the plant is killed aboveground. The disease may then involve the crown, rhizomes, stolons, and roots, killing the entire plant. In severe cases, nearly all the leaves and tillers may die, resulting in a severe thinnings of turf and irregular patches up to 3 ft (1 m) across.

   In hot, humid weather, closely clipped bentgrass leaves may turn reddish brown or dark gray, giving a smoky blue cast to infected, irregularly shaped areas in golf or bowling greens. Such turf appears drought-stressed.

   Net-blotch (*Drechslera dictyoides*) gets its name because, on broadleaf species (such as tall fescues, creeping fescues, meadow fescues, and ryegrasses), irregular dark brown, purple-to-black strands develop across the reddish brown to black, irregularly shaped leaf lesions, forming a net. Infected leaves turn yellow or brown and die back from the tip. Turf thinning and irregularly shaped patches of dead shoots may occur.

2. *Crown (foot) and root rots (melting out).*   This phase of the disease usually appears in warm-to-hot, dry weather as a reddish brown, dark brown, or purplish black decay of the stems, crowns, rhizomes, stolons, and root tissues. Secondary

FIGURE 5–40   Leaf blotch of bermudagrass caused by *Bipolaris cynodontis*. (Courtesy USDA and Clemson University.)

bacteia and fungi commonly invade such tissues. The feeding roots on diseased plants are shallow, few in number, or even absent. Such plants lack vigor and often wilt during midday, even when soil moisture is abundant. Diseased turf may at first appear yellow and thin before taking on a red-brown or drought-injured appearance (figure 5–41). The damaged areas may be small and circular to large and irregular. Entire stands of bluegrasses, fescues, ryegrasses, bermudagrass, or bentgrasses may be thinned out or completely destroyed by severe crown and root rot. This phase occurs most readily when plant vigor is suppressed by one cause or another, particularly during hot weather.

FIGURE 5–41   Severe crown and root rot in a Kentucky bluegrass lawn caused by *Bipolaris sorokiniana.*

Disease cycle.   The disease cycle for all species is essentially the same (figure 5–42). The fungi survive from year to year (and periods of very hot or cold weather) as light yellow, golden brown, olivaceous to dark brown spores called conidia (figure 5–43) and as dormant mycelium in dead grass tissues or in infected leaves, crowns, roots, and rhizomes or stolons. During alternating periods of cool-moist and warm-dry weather, tremendous numbers of dark, multicelled, cigar-shaped conidia (figures 5–42 and 5–43) are produced on debris, mostly at temperatures of 38 to 82°F (3.3 to 27.8°C), with an optimum for many of 60 to 66°F (15.6 to 18.9°C). The spores are carried to healthy leaves and leaf sheaths by air currents, mowers and other turf equipment, flowing or splashing water, foot traffic, dragging of hoses, animals, and infected grass clippings. The conidia germinate in a film of moisture and infect the leaves, either directly or through stomates. Germination and infection of leaves can take place within 2 hours if the weather is favorable. These fungi are also capable of saprophytically colonizing plant debris at or above the soil surface. The fungi sporulate profusely when dry grass debris is repeatedly rewetted. The conidia, along with mycelial fragments, spread to new leaf parts and neighboring plants. Thus the cycle is repeated. New leaf and leaf-sheath infections may occur as long

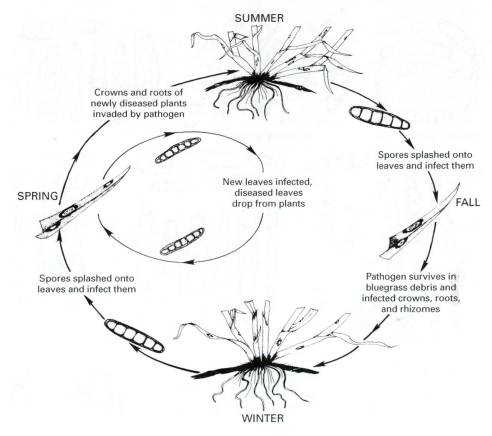

FIGURE 5–42   Disease cycle of *Drechslera poae* that causes leaf spot and leaf, crown, and rhizome rot or melting out of Kentucky bluegrass and other turfgrasses.

as the weather remains moist and temperatures are favorable. The peak of disease development varies from early or mid-spring to late summer or mid-autumn, depending on the area of the country and the species of fungus. Some species (such as *Drechslera poae, D. dictyoides, D. siccans, D. triseptata,* and *Bipolaris cynodontis*) are favored by the cool temperatures of early spring and fall, while other species such as *Bipolaris sorokiniana, D. spicifera, D. erythrospila,* and *D. gigantea* are most destructive in warm-to-hot (86 to 95°F or 30 to 35°C) and wet, overcast, summertime weather. *Bipolaris sorokiniana* is also active on old leaves during cool, wet weather in the fall and early winter.

During warm, dry weather, the leaf-lesion phase decreases (except for *D. erythrospila, D. spicifera,* and *D. gigantea*), and fungal activity may be restricted to crowns, roots, rhizomes, and stolons of diseased plants, killing round to irregular turf areas (melting out or crown and root rot). With the return of cooler, moist weather, leaf infection again becomes a problem. These fungi may also be seed-borne and cause seedling blights on new turfgrass areas.

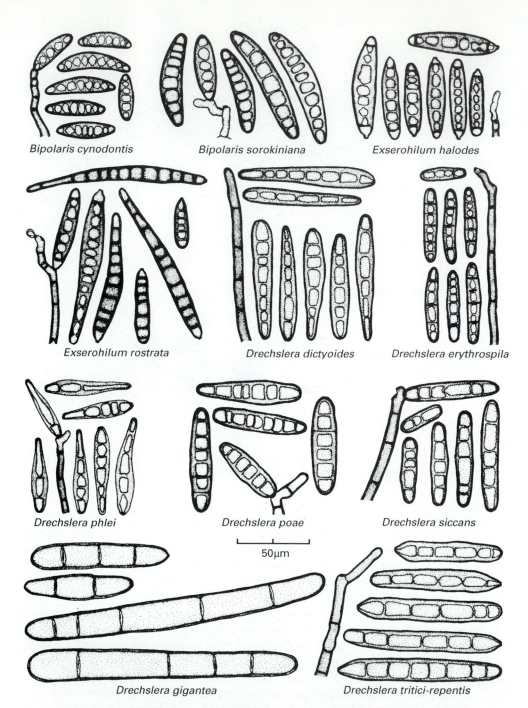

Bipolaris cynodontis

Bipolaris sorokiniana

Exserohilum halodes

Exserohilum rostrata

Drechslera dictyoides

Drechslera erythrospila

Drechslera phlei

Drechslera poae

Drechslera siccans

50µm

Drechslera gigantea

Drechslera tritici-repentis

FIGURE 5–43 Spores (conidia) and conidiophores of eleven species of *Bipolaris, Drechslera,* and *Exserohilum,* as seen under a high-power microscope, that cause Helminthosporium diseases.

Importance.   These disease reduce vigor and can be very destructive during hot and wet, humid weather, where turf is sprinkled frequently, especially in late afternoon and early evening, in poorly drained areas, and where it is shady. The turf may be severely thinned when leaves are progressively blighted with the fungus or fungi invading the stems, crowns, rhizomes, stolons, and roots, killing entire plants in small to large areas. Such turf commonly appears drought-stressed.

Management.   Follow the cultural practices outlined in table 5–1. Buy only top-quality seed, sod, sprigs, or plugs from a reputable dealer. Grow locally adapted blends or mixtures of leaf-spot-resistant grasses when possible. Most Kentucky bluegrasses released since about 1975 are resistant to one or more of these leaf-spotting fungi. There are also resistant perennial ryegrasses, fine-leaf fescues, tall fescues, and bermudagrasses (such as Ormond, Tifgreen, and Texturf 10). Check with your local county cooperative extension (or advisory) office or turfgrass specialist for suggested grass species and cultivars to grow. Renovation or overseeding with a blend or mixture of resistant cultivars is the cheapest and best approach to managing this disease complex on Kentucky bluegrass, perennial ryegrass, and bermudagrass.

Helpful management practices to reduce the severity of infection include (1) raising the mowing height to 2 to 3 in (5.1 to 7.6 cm) on lawn-type grasses and to G in (0.6 cm) on golf and bowling greens; (2) watering deeply but infrequently, preferably in the morning or early afternoon (but frequent, light watering may be needed where plants are shallow-rooted); and (3) increasing light and air flow over the turf area. Avoid applications of high rates of water-soluble nitrogen fertilizers (use slow-release nitrogen fertilizers in split applications in spring and fall), a thatch over 0.5 in (1.3 cm) on lawn-type grasses, and the use of broadleaf phenoxy herbicides and plant-growth regulators when these diseases are active. Maintain adequate levels of phosphorus (P) and potassium (K) in the soil.

If the various Helminthosporium diseases cannot be controlled adequately by cultural practices, fungicide sprays may need to be applied on a *preventive* schedule (table 5–5). For the cool, wet-weather group, begin applications in *early* spring, shortly after the first new leaves are formed, and continue at 7- to 21-day intervals until warm, dry weather develops. For the warm- to hot-weather group, begin treatment following local recommendations and continue at 7- to 21-day intervals during moist weather. Such spray programs are time-consuming, expensive, and usually impractical for the average home lawn, park, cemetery, or sports field. When only a few sprays can be applied to a turf area, it is best to apply them in the spring. Apply the first one as the grass begins to green up, the second 2 to 3 weeks later, and the third 2 to 3 weeks after that. Check label directions. If the turf area is not sprayed until late spring or summer—when leaf blight and melting out (sheath and crown rot; see *Color plate 44*) are obvious—it is practically impossible to achieve control, since the fungus or fungi are generally well established and inaccessible in the crowns, roots, rhizomes, and stolons. Granular applications or sprays of fungicides in late fall or early spring are reasonably effective in reducing pathogen inoculum on turf debris.

Best results are obtained when the fungicides are applied under pressure (25 lb per sq in or higher). The addition of a small amount of commercial spreader-sticker or surfactant to the spray solution may ensure better coverage. Ask your pesticide dealer about these materials.

The fungicides in table 5–5 for use against Helminthosporium diseases are primarily protective in action and will not control satisfactorily unless the grass plants are thoroughly coated *before* infection occurs. Avoid continuous and excessive applications of DMI and benzimidazole fungicides. Apply the fungicide in 3 to 5 gal of water, sprayed evenly over 1000 sq ft (93 sq m) of turf, carefully following the manufacturer's directions regarding dosage, timing of applications, safety, and other factors. Where feasible, spray at dusk or in the early evening, especially in hot weather. None of the fungicides listed in table 5–5 will completely control these diseases. Sprays applied at weekly intervals give better control than those sprayed at 2- or 3-week intervals. At application rates below those given on the label, certain fungicides may actually increase damage from these fungi.

Hendersonia Leaf Spot
See the section on minor leaf spots (5.2.16).

Large Brown Patch of Zoysiagrass
See the section on Rhizoctonia diseases (5.2.23).

5.2.14   Leaf Smuts, Blister or Leaf-Spot Smuts, and Head Smuts

Leaf smuts.   Stripe smut, caused by the fungus *Ustilago striiformis,* and flag smut, caused by *Urocystis agropyri,* are two kinds of leaf or foliar smuts that occur worldwide and are destructive to many cool- and warm-season turfgrasses. These smut fungi weaken the grass host, making the plant easy to kill when under severe heat and drought or other stress. Plants infected with flag smut are killed more readily than plants infected with stripe smut. Flag smut is often more prevalent in early spring, while stripe smut generally predominates in late spring and early autumn. These smut fungi grow systemically throughout a grass plant, and both smut fungi may infect the same grass plant at the same time and even the same leaf. Once infected, a plant remains infected for life. Practically all plants that arise from an infected mother plant by rhizomes or stolons are also infected. Diseased plants are weakened and often invaded by other organisms. For example, smut-infected plants are more susceptible to Helminthosporium diseases (sec. 5.2.13).

Stripe and flag smuts, together with high temperatures and drought, cause plants to exhibit stunted growth, a brown to blackish-brown appearance, a general decline, and commonly early death. Dead patches of grass often appear in heavily infected turf during midsummer; weed invasion soon follows (figure 5–44).

The stripe- and flag-smut fungi infect almost 100 species of turf and forage grasses, both cultivated and wild. A number of highly specialized varieties and pathogenic races of these fungi are restricted to certain cultivars and grasses. For example, the races of the stripe-smut fungus that attack Kentucky bluegrass do not infect creeping bentgrass.

The diseases occur most commonly on annual bluegrass and Kentucky bluegrass (especially on the cultivars Arboretum, Cougar, Delta, Fylking, Galaxy,

FIGURE 5–44   Stripe smut (*Ustilago striiformis*) infecting a Kentucky bluegrass lawn.

Geronimo, Kenblue, Merion, Newport, Park, Prato, Rugby, Troy, and Windsor), creeping bentgrass (especially Pennlu and Toronto), colonial bentgrass, and fescues. Common timothy, several wildryes and wheatgrasses, orchardgrass, perennial ryegrass, and quackgrass are also commonly infected.

Leaf smuts are favored in locations having excess thatch, frequent irrigations or rains during spring and summer, mature turf stands at least 2 to 3 years old, and a pH below 6.0; and where susceptible grass cultivars are grown.

Identification.   From a distance, infected turf appears clumpy and patchy. Smutted plants are most noticeable during overcast, cool weather (daytime temperatures under 70°F or 21.1°C) in the spring and autumn. They appear pale green to slightly yellow or brown and stunted. Single plants may be affected, or irregular patches reaching 3 ft (0.9 m) or more in diameter may occur.

Long, narrow, yellow-green streaks develop in infected leaves and leaf sheaths. These streaks soon become silvery to dull gray or black and extend the entire length of the leaf blade and sheath (figure 5–45). The grass epidermis covering the streaks soon ruptures, exposing blackish brown, dusty masses of smut spores (teliospores). After dispersal of the spores, the leaves soon split, shred into ribbons, twist and curl from the tip downward, turn light brown, wither, and die. In addition, leaves of infected plants tend to remain stiff and erect rather than lax and spreading. Affected plants do not tiller as profusely or produce as many rhizomes

FIGURE 5–45   Stripe smut (*Ustilago striiformis*) infecting leaves of Kentucky bluegrass. (Courtesy D. H. Scott.)

or stolons or inflorescences (heads) as healthy plants, nor do they develop as extensive a root system.

Smutted plants are often difficult to find during hot, dry weather because a large percentage of such plants often die during summer droughts (figure 5–44). Both smut fungi decrease leaf turgor and water potentials of infected plants under drought stress. Once infected, grass plants will rarely, if ever, recover, unless properly treated with a *systemic* fungicide.

Under close mowing, both smut fungi produce identical symptoms. Positive diagnosis can be made only through a microscopic examination of the teliospores produced by the smut fungus. The spores of *Ustilago striiformis* are single cells, round or oval in shape, dark olive brown, and covered with prominent spines (figure 5–46). The spores of *Urocystis agropyri* are smooth and roundish and composed of one to five or six dark-reddish-brown, fertile cells (teliospores) surrounded by several smaller empty or sterile cells (figure 5–47), forming a spore ball.

Disease cycle.   The stripe- and flag-smut fungi overseason as dormant mycelium in the meristematic tissue of crowns and nodes in stolons and rhizomes of infected plants and as dormant teliospores in grass debris, living plants, and soil. The teliospores are disseminated by many agents, including wind, rain, and shoes, and by mowing, watering, raking, dethatching, coring, and other turf-maintenance practices. Spores are also transported on the seed. When dry, these teliospores may lie dormant up to 3 years (4 years in stored seed) before they germinate.

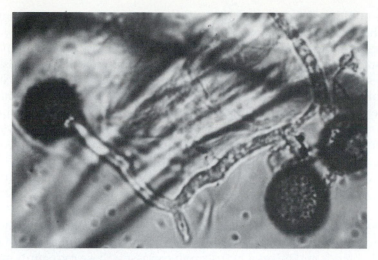

FIGURE 5–46 Teliospores of the stripe-smut fungus (*Ustilago striiformis*) germinating in a drop of water as seen under a high-power microscope. (Courtesy C. F. Hodges.)

FIGURE 5–47 Teliospore spore balls of the flag-smut fungus, *Urocystis agropyri*, as seen under a high-power microscope. (Courtesy C. F. Hodges.)

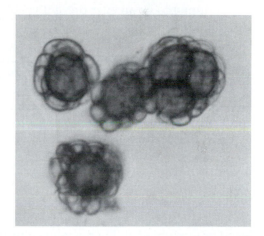

When suitable conditions occur in the spring and autumn, the teliospores germinate over a wide temperature range (39 to 95°F or 3.9 to 35.0°C) to produce a mycelialike basidium on which minute spores (sporidia) may be borne, although this is apparently rare on certain grasses such as Kentucky bluegrass and creeping bentgrass. Each sporidium then germinates and forms a germ tube. When the germ tubes of opposite mating types fuse (conjugate), an infection hypha forms that penetrates susceptible host tissue directly. Infection may occur through the coleoptile of seedling plants and actively growing (meristematic) tissues produced by the lateral or axillary buds on the crowns, stolons, and rhizomes of older plants that come into contact with germinating spores. Once inside the grass plant, smut hyphae develop systemically in the direction of plant growth, with new leaves, tillers, stolons, and rhizomes becoming infected as they form. The mycelium continues to grow within developing tissues.

Teliospore formation begins with tangled mats of mycelium within infected grass tissues. The mycelium then breaks up to form masses of blackish brown teliospores that are released when the host tissues rupture, shred, and die.

Smutted plants in new turfgrass areas are uncommon, indicating limited infection of seedling plants from soil- or seedborne teliospores. The large number of diseased grass plants in turf areas more than 3 years old is probably caused by the infection of lateral buds and the growth of smut fungi from perennially infected crowns. Once infected, watering and high fertility, which stimulate plant growth during droughts, create conditions that favor the buildup of these leaf smuts. Such practices keep the systemically infected grass plants from dying during hot, dry weather.

Importance.   Stripe and flag smuts are now less common in the United States than they formerly were due to the release and widespread use of resistant cultivars in blends and mixes. On susceptible grasses, these two smuts may cause a decline and death of small to large areas of turf during hot, dry weather. Both watering and fertilization help to keep infected plants alive during the summer, but the plants do remain infected. This increases the likelihood that leaf smut will continue to be an ever present problem unless measures are taken to renovate the turf.

Management.   Grow a blend of several improved cultivars that generally show resistance to leaf smuts and other major diseases. Contact your local county extension (or advisory) office or turfgrass specialist for suggested grass cultivars to grow. Bluegrass and bentgrass cultivars differ greatly in their resistance to stripe and flag smuts. Because numerous races of the two smut fungi exist, it is difficult to predict the relative resistance or susceptibility of a cultivar in any given location.

Start with certified, disease-free sod, sprigs, or plugs of the best adapted cultivars. During hot weather (90 to 100°F; 32.2 to 37.8°C), the smut fungi become dormant in bentgrass stolons, and the healthy-appearing stolons continue to grow. When cooler weather returns in autumn, the smut fungus resumes growth, and symptoms reappear in the stolons. Renovate and overseed infected turf with a blend or mixture of highly resistant cultivars.

Follow the suggested cultural management practices outlined in table 5–1. Where practical, remove the clippings when smutted leaves are first evident. Maintain a balanced (N-P-K-iron) fertility level based on a soil test. Keep nitrogen applications to a minimum during the hot weather months. Yearly treatment with a systemic fungicide is expensive, but will kill out the pathogens within the grass host when properly drenched into the soil in late fall just before winter dormancy or at green-up in early spring; see table 5–5.

Blister or leaf-spot smuts.   Blister or leaf-spot smuts do *not* rupture the leaf epidermis, cause shredding of infected leaves, or become dusty with teliospores, as do the stripe and flag smuts. Dark, raised blisters or spots form in the leaves. These minor diseases more closely resemble leaf spots than they do smuts and hence are discussed in the section on minor leaf spots, blights, and streaks (5.2.16).

Head smuts.   Head smuts are only a problem in seed-production fields. There are numerous fungi in the genera *Tilletia, Urocystis,* and *Ustilago* that infect inflorescences (heads or spikes) and sometimes the upper leaves and leaf sheaths. All turfgrasses, forage grasses, wild grasses, and cereals are susceptible worldwide. Head smuts with black, powdery masses of teliospores that are largely dispersed before harvest, leaving a naked rachis, are called loose smuts to separate them from covered smuts, where the dark spore masses are mostly intact until harvest.

Identification.   Smutted plants are often stunted with shortened roots. The heads are commonly abnormal in shape and color, with smut sori replacing the floral parts. The ovaries that normally contain the seeds are filled with dark-brown-to-black, powdery masses of teliospores. Some smuts, such as certain *Tilletia* species, have a rotten fish odor.

Disease cycle.   The teliospores are disseminated by wind, weathering, and harvesting and seed-cleaning equipment. The spores infest clean seed and soil following breakup of the sori, either before, during, or shortly after harvest. The teliospores usually germinate in the soil when the seed sprouts. (Some spores of *Tilletia caries, T. laevis,* and *U. bullata* are known to be viable in a dry state for 10 to 25 years.) Penetration is direct into immature seedling tissue, and the fungi then become systemic, eventually reaching the head. The fungi replace the developing seed with masses of teliospores, thereby completing the disease cycle.

Importance.   These diseases are only a problem to the seed producer.

Management.   Head smuts are largely prevented by treating the seed with the same fungicides used for cereals. Rotation with nongrass crops for a year or more is often beneficial.

Leaf Streak
See the section on minor leaf spots (5.2.16).

5.2.15 Leptosphaerulina Leaf Blight

Leptosphaerulina leaf blight, caused by the fungi *Leptosphaerulina australis* and *L. trifolii,* occurs worldwide. It is usually a minor disease in prolonged, overcast, humid or wet weather from spring to fall. The disease usually infects Kentucky and annual bluegrasses, ryegrasses, fescues, colonial and creeping bentgrasses, and zoysiagrass grown on golf courses. Leptosphaerulina leaf blight is commonly confused with Ascochyta leaf spot or blight (sec. 5.2.2), Septoria and Stagonospora leaf spots (5.2.27), dollar spot (5.2.8), Nigrospora leaf blight, and Pythium blight (5.2.21). The causal fungi grow saprophytically in thatch and leaf litter and can be seen on dying leaves or leaves damaged by other fungi or agents.

Identification.   Small to large irregular areas of turf without a well-defined border may become uniformly blighted or appear patchy with individual leaves

FIGURE 5–48   Leptosphaerulina leaf blight on Kentucky bluegrass leaves. Note the dark perithecia embedded in the dead leaf tissue. (Courtesy R. W. Smiley.)

dying back from the tip (figure 5–48). Leaf lesions and spore formation are most often seen on older leaves. Uniform yellow-to-reddish-brown lesions may extend down to the leaf sheath. In advanced stages, affected leaves turn brown, shrivel, and die. Watersoaked lesions that quickly fade to a bleached white may also occur on the leaf blades following rainy weather. Such lesions, which cause rapid death of leaves, closely resemble the nonfungal bleaching caused by high temperatures, frost, or a dull mower. Speck-sized, pale brown fungus fruit bodies (perithecia), which closely resemble the pycnidia of *Ascochyta* and *Septoria,* form in the dead tissue of older lesions. Infections occur prematurely on older, senescing leaves.

Disease cycle.   These weakly pathogenic fungi overseason as tiny, pale brown perithecia embedded in leaves and mycelium in dead grass tissue (figure 5–48). Microscopic spores, the ascospores (figure 5–16), are produced in warm, wet weather and are blown or splashed and carried on turfgrass equipment and shoes to healthy grass leaves. Germinating hyphae commonly infect through cut leaf tips and progress downward.

Importance.   Disease outbreaks are most common during warm, humid weather when the turf is stressed by close mowing, drought, other diseases, applications of preemergence grass herbicides, and excessive rates of nitrogen fertilizer, or on newly laid sod that is saturated but lacks good root contact with the underlying soil.

Management.   Follow the suggested cultural practices outlined in table 5–1. In particular, ensure that mower blades are sharp to avoid excessive leaf damage. Maintain a regular mowing schedule and ensure that the turf is no taller than 3 in. Do not lay new sod or apply herbicides before or during hot, humid weather.

Localized Dry Spot
See the section on fairy rings (5.2.10).

Mastigosporium Leaf Spot or Fleck
See the section on minor leaf spots (5.2.16).

Melting out
See the section on Helminthosporium diseases (5.2.13).

Microdochium Patch
See the section on snow molds (5.2.29).

5.2.16   Minor Leaf Spots, Blights, and Streaks

The following foliar diseases are rare on closely mown turf because the lesions, which may or may not yet be visible, are generally removed with the clippings. The diseases are sometimes important in unmowed turf or in turfgrass seed crops. In general, relatively little information is available concerning when and where infection occurs, how the fungi overseason, and other phases of their disease cycles. No special measures have been developed to manage these diseases, nor are they usually needed. Following the cultural-management practices outlined in table 5–1 should be sufficient. The fungicides applied to control Helminthosporium diseases (table 5–5) should usually do a good job where one of these diseases is important.

Blister or leaf-spot smut.   Blister or leaf-spot smut, caused by a number of species of the fungus *Entyloma,* is widespread on many grasses. *Entyloma dactylidis* is by far the most common species. It infects numerous cultivated forage and wild grasses worldwide. These fungi produce variable symptoms on individual grass blades and leaf sheaths. Kentucky bluegrass cultivars differ in resistance.

Identification.   The raised leaf blisters (or sori) are small, up to 2 mm long, round to oval or almost rectangular in shape, and olive green to brown and later almost black. They are most evident during spring and autumn on the lower leaf surface and are usually surrounded by a yellowish zone or halo. If disease is more severe, infected leaves later become yellow or whitened with raised dark spots.

Disease cycle.   These fungi are apparently *not systemic* in plants. Most infections are believed to result from variously shaped, olive-brown-to-dark-brown smut spores (teliospores) in the soil, thatch, or infected foliage. The teliospores occur in irregularly shaped groups. They germinate to produce minute, hyaline spores (basidiospores or sporidia) that are blown, splashed, tracked, or carried on maintenance equipment to grass leaves, where they germinate and produce the characteristic dark, round-to-elongated blisters (sori).

Importance.   Turf severely infected with blister or leaf-spot smut may take on an overall yellowish or grayish appearance when viewed from a distance. Tillers or even entire plants die when exposed to drought, high temperatures, and unbalanced fertility. Thick thatch and acid soils promote disease development. These smut fungi normally cause only minor damage, except in the Pacific northwest of North America during mild winters.

Brown stripe or leaf streak.   Brown stripe or leaf streak, caused by the fungus *Cercosporidium graminis* (syn. *Scolicotrichum graminis*), infects all cool-season turfgrasses, mainly bentgrasses, bluegrasses, fescues, and ryegrasses, during cool, moist weather in spring and autumn throughout much of the world. A wide range of forage and wild grasses are also infected.

Identification.   Leaf and leaf-sheath lesions are small, circular, water-soaked, and olive-gray. The lesions soon fade to a dull gray with brown or dark purple margins. On aging, the lesions may develop into gray streaks with margins of a different color. Infected leaves may die from the tip downward. Rows of dark dots (clusters of conidiophores) form in the older lesions.

Disease cycle.   The fungus overseasons as dark brown mycelial mats (stromata) in living or dead leaves and leaf sheaths. In early spring, the stromata swell when wet, and tufts of olive-brown, spore-producing structures (conidiophores) burst through the epidermis on the upper leaf surface and produce conidia (see figure 5–49). The conidia are disseminated primarily by air currents and splashing or flowing water.

Importance.   Brown stripe or leaf streak attacks a wide range of cool-season grasses during cool, moist weather, causing some dieback of the leaves. It is a very minor problem except possibly in seed-production fields.

Cephalosporium stripe.   Cephalosporium stripe is caused by the soil-borne fungus *Hymenula cerealis* (syn. *Cephalosporium gramineum*). This widespread pathogen may infect the vascular system of ryegrasses, bluegrasses, and wheatgrasses when grown as turfgrasses during cool-to-cold, wet weather from late fall into spring. The *Hymenula* fungus has also been associated with dying bentgrass leaves. The fungus, however, is much more common and serious, causing a vascular wilt of cereals and many forage and wild grasses.

Identification.   Stunted leaves with one or two narrow, yellow to yellowish brown stripes that extend from the leaf sheath to the leaf tip are uniformly and widely scattered in affected turf (*Color plate 38*). Infected leaves soon turn yellow,

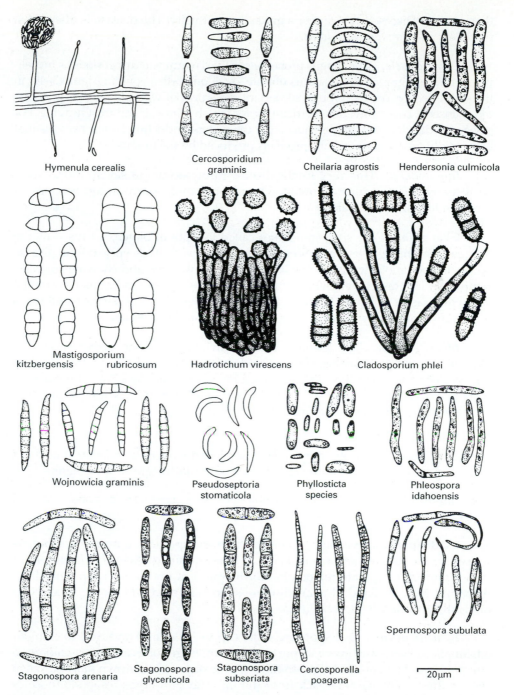

Hymenula cerealis

Cercosporidium graminis

Cheilaria agrostis

Hendersonia culmicola

Mastigosporium kitzbergensis

Mastigosporium rubricosum

Hadrotichum virescens

Cladosporium phlei

Wojnowicia graminis

Pseudoseptoria stomaticola

Phyllosticta species

Phleospora idahoensis

Stagonospora arenaria

Stagonospora glycericola

Stagonospora subseriata

Cercosporella poagena

Spermospora subulata

20μm

FIGURE 5–49 Sixteen species of fungi that cause minor leaf-spotting diseases of turfgrasses, as seen under a high-power microscope.

then brown. Diseased turf appears green but unthrifty. The disease is often most prevalent in relatively wet areas.

Disease cycle.   The fungus probably survives warm and dry periods as mycelium in infected grass tissues, and as minute, oval, one-celled conidia (figure 5–49) in the soil. Infection occurs through secondary roots or root wounds caused by nematodes, insects, core cultivation, verticutting for thatch removal, and frost heaving. For disease to develop, the fungus must enter the water-conducting (xylem) vessels that become plugged with metabolic products produced by the fungus.

Importance.   This very minor disease rarely occurs in severe form. Areas with many infected tillers grow slowly and respond poorly to applications of nitrogen fertilizer and water.

Char spot.   Char spot, sometimes called tar spot or blotch, is caused by the fungus *Cheilaria agrostis* (syn. *Septogloeum oxysporum*). This very minor disease occurs in cool, moist temperate regions on bentgrasses, red and chewings fescues, Kentucky bluegrass, perennial ryegrass, crested wheatgrass, and numerous forage and wild grasses in North America, Europe, and New Zealand. Disease outbreaks occur most commonly in late winter, early spring, and mid to late autumn.

Identification.   The lesions, which form mainly in the upper surface of older leaves and leaf sheaths, are first reddish brown and elliptical with off-white centers. The spots later become elongated and turn dull black from fungal stromata to closely resemble tar spot. Speck-sized, black fruiting bodies (pycnidia) of the *Cheilaria* fungus can sometimes be seen in the margins of older lesions.

Disease cycle.   The fungus produces large numbers of mostly hyaline and curved spores (figure 5–49) in pycnidia and probably overseason in colonized old leaves. Spread occurs mainly by splashing water and turfgrass equipment.

Importance.   This is a very minor disease on cool-season grasses in cool moist regions. It rarely occurs in severe form.

Cladosporium eyespot.   Cladosporium eyespot, or timothy eyespot, is caused by the fungus *Cladosporium phlei.* It is a common disease of *Phleum* species in cool, moist, temperate regions in North America, northern Europe, northern Asia, and New Zealand. It is much more serious in seed and forage production fields than on closely mowed turf.

Identification.   Leaf spots are small (up to 1 × 4 mm) and oval to elongated. Smaller lesions are brown or purplish brown; larger ones are straw-colored to tan or gray with a narrow, bright-to-dark-purple or purplish brown margin. When severe, the leaves may die back from the tip. The symptoms resemble those of Mastigosporium leaf spot.

Disease cycle.   The *Cladosporium* fungus overseasons on living or dead leaves. It is generally considered to be mostly a saprophyte. Conidia (figure 5–49) form on long, dark brown conidiophores, which form in groups. The spores are believed to be spread by air currents, splashing and flowing water, and probably by turf equipment.

Importance.   This is a very minor problem on mowed turf timothy, but may be a problem in seed production fields in cool, moist regions.

Frost scorch.   Frost scorch, sometimes called string of pearls disease, leaf rot, or tip blight, is caused by the fungus *Sclerotium rhizoides.* It is a minor tip or leaf blight of Kentucky bluegrass, bentgrasses, and fine-leaf fescues. The fungus also attacks numerous forage and wild grasses and winter cereals in North America, Europe, and northern Japan. This is a cold, wet-weather disease found most commonly in late winter or early in the growing season. Frost scorch is most severe where the turfgrass is undernourished and the soil pH is highly acidic.

Identification.   Infected plants may be stunted. The leaves turn light yellow and finally a bleached white starting at the tips, damage that may resemble cold injury. Later, diseased leaves wilt, wither, curl, and roll. The leaf tips may end in long tendrils, with the bases of the leaf blades remaining green. Affected leaves may also become looped. The lower portions of young diseased leaves may be covered with a white-to-light-gray mycelium, in which small, round-to-oblong sclerotia 1 to 5 mm in size commonly form. When severe, the sclerotia often develop in beadlike rows (string of pearls). The sclerotia are white at first but turn brownish to almost black when mature.

Disease cycle.   The *Sclerotium* fungus overseasons as sclerotia (viable up to 9 years when dry) in soil and as dormant mycelium in grass plants, where it infects the crowns, roots, rhizomes, and stolons. The hyphae penetrate epidermal cells, directly producing systemic infections that are perennial in the underground parts of plants. The sclerotia undergo a dormant period of 2 to 24 months before germination occurs.

Importance.   This minor, cold, wet-weather disease disappears when temperatures rise and the air is less humid. The disease tends to reappear in the same areas each year unless the bulk of the sclerotia are removed when clippings are collected.

Hadrotrichum leaf spot.   Hadrotrichum leaf spot, probably caused by the fungus *Hadrotrichum virescens,* is an unimportant disease on bentgrasses, perennial ryegrass, and, to a lesser extent, bluegrasses and wheatgrasses, mostly in Europe. The disease has been reported from spring to early winter.

Identification.   The lesions are olive green or dark brown, due mainly to dense clusters of conidiophores that arise from brown mycelial pads. These are evident with a hand lens or reading glass.

Disease cycle.   The *Hadrotrichum* fungus probably overseasons as mycelial pads in infected leaf tissue. It spreads mostly by oval-to-round, dark spores or conidia (figure 5–49).

Importance.   This disease has caused occasional injury in central Europe.

Hendersonia leaf spot.   Hendersonia leaf spot, caused by *Hendersonia culmicola* var. *minor,* occurs in North America and Europe on bluegrasses, fescues, and numerous forage and wild grasses.

Identification.   Indistinct, brown spots that turn straw-colored, sometimes with brown margins and sprinkled with black specks (pycnidia), form in the leaf lesions.

Disease cycle.   The fungus overseasons as pycnidia and probably dormant mycelium in dead leaves. In moist weather, the spores as conidia (figure 5–49) ooze out of the pycnidia and are disseminated mostly by splashing or flowing water and turfgrass equipment.

Importance.   This very minor disease rarely, if ever, requires special management measures.

Mastigosporium leaf spot or fleck.   Mastigosporium leaf spot or fleck is caused by the fungi *Mastigosporium rubricosum* (on bentgrasses) and *M. kitzbergensis* (on turf timothy) during cool-to-cold, wet weather in spring and autumn. The disease has been reported in North America, Europe, Russia, and Japan.

Identification.   Small (up to about 2 mm), oval-to-elliptical brown or black flecks that may enlarge and develop off-white centers and purplish brown or wine red borders, form on the leaf blades and leaf sheaths. The lesions may be covered with shining clusters of slimy white spores (conidia). When common, the lesions merge, causing the leaves to die back from the tips. Older leaves may die and drop prematurely. On turf timothy, the leaf lesions closely resemble those of Cladosporium eyespot.

Disease cycle.   The *Mastigosporium* fungi overseason primarily as dormant mycelium in living or dead leaves. Spread occurs through the dissemination of the shiny, hyaline conidia (figure 5–49) by splashing or flowing water and wind. The conidia germinate at 59 to 68°F (15.0 to 20.0°C) in water films, and their germ tubes penetrate the epidermal cells directly.

Importance.   A very minor problem on closely mown turf, this disease may be of some importance on unmowed grass.

Nigrospora blight.   Nigrospora blight, caused by the very common fungus *Nigrospora sphaerica,* has been observed in the northern United States during mid-summer on stressed Kentucky bluegrass, perennial ryegrass, and creeping red and chewings fescues. It is mostly a foliar blight on these grasses and only occasionally damaging. In the south it occurs on St. Augustinegrass and possibly zoysiagrass in spring and early summer. The disease is easily confused with Ascochyta leaf spot or blight (sec. 5.2.2), dollar spot (5.2.8), Leptosphaerulina leaf blight (5.2.15), and Pythium blight (5.2.21) unless you use a microscope. The *Nigrospora* fungus infects a wide range of plants besides turfgrasses. It is a common saprophyte in organic debris.

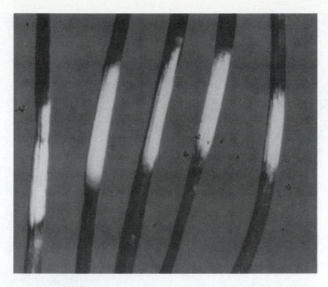

FIGURE 5–50   *Nigrospora* lesions on leaves of Kentucky bluegrass. (Courtesy R. W. Smiley.)

Identification.   Large areas of cool-season grasses may become uniformly blighted, or the disease may crop up sporadically in round to irregular patches 3 to 8 in (7.6 to 20.3 cm) in diameter (*Color plate 45*). Individual leaves die back from the tip, often with uniform lesions that extend down to the leaf sheath. Less commonly, watersoaked spots form on the leaf blade or sheath and quickly fade to tan (figure 5–50). On some cultivars, the tan lesions have a purple-to-reddish-brown border and resemble those caused by the dollar-spot fungus. The lesions may girdle the leaf blade, causing the part beyond to be tan and twisted. On other Kentucky bluegrass cultivars, the dying leaves turn a deep purple. Leaves and tillers may collapse, and the plants die.

In warm, very damp weather, early in the morning, a fluffy white mycelium forms over the blighted leaf blades (figure 5–51). This symptom is easily confused with dollar spot and Pythium blight.

Dark, often elongated lesions form on the stolons of St. Augustinegrass under close mowing. The girdling lesions cause the shoots at the end of the stolons to wilt, turn yellow, and die, resulting in a general thinning or patchiness of diseased turf.

Disease cycle.   The fungus survives unfavorable growing periods as round-to-oval, black, shiny spores (conidia) and mycelium in grass debris (figure 5–13). The spores germinate in warm-to-hot, moist weather and are splashed, blown, or carried to stressed and weakened but otherwise healthy grass plants on turfgrass equipment and shoes. New infections occur where the fluffy, white aerial mycelium contacts healthy leaf tissue and also colonizes the leaf tips (after growing in the guttated fluid or dew). The conidia are produced in or on infected leaf tissue.

On St. Augustinegrass, the disease is commonly associated with hot and dry growing conditions during the summer. Nigrospora blight also appears to be more severe on turfgrass affected by St. Augustinegrass decline.

FIGURE 5–51   Close-up of Nigrospora blight. (Courtesy R. W. Smiley.)

Importance.   This is a very minor disease where grass plants are not stressed from a combination of heat, drought, and a lack of fertility.

Management.   Follow the cultural practices suggested in table 5–1 to keep the turfgrass growing steadily in hot weather. Avoid severe moisture, nutritional, and other stresses. Apply adequate amounts of a balanced (N-P-K) fertilizer, applied mostly in the spring and fall. Irrigate thoroughly (so that soil is moist 2 to 3 in [5.1 to 7.6 cm] deep) at 3- to 4-day intervals during prolonged hot, dry periods. Water turf in the morning so that it will be dry by late afternoon. Avoid applications of herbicides or laying new sod during or just before prolonged periods of hot, damp weather. On St. Augustinegrass and other grasses, raise the cutting height in hot weather. Cultivars of Kentucky bluegrass, perennial ryegrass, and fescues differ in their resistance.

Where Nigrospora blight is severe and a recurring problem, apply a broad-spectrum fungicide such as iprodione (Chipco 26019) or chlorothalonil (Daconil) according to the manufacturer's directions.

Phloespora leaf spot or stem eyespot.   Phloeospora leaf spot or stem eyespot is caused by the fungus *Phloeospora idahoensis* (teleomorph *Didymella festucae*). It is mostly a cool, wet-weather disease that infects fescues in North America and Europe. It is an important disease in seed-production fields of fine-leaf fescues.

Identification.   Leaf lesions vary from small, round brown spots to elongated eyespots with dark brown or purple-brown borders and white or gray centers. The lesions may extend up to ½ in (1.2 cm) long. Speck-sized, round, brown fruiting bodies (pycnidia) form in the centers of older lesions. These lesions may develop yellow or buff-colored margins. When severe, the leaves die back from the tips and turn light brown.

Disease cycle.   The fungus overseasons as pycnidia in living and dead leaves. The spores or conidia (figure 5–49) are spread mainly by splashing or flowing water and turfgrass equipment.

Importance.   This disease is a very minor problem on certain closely mown fescues that may cause some dieback of the leaves in cool, moist weather. However, it is much more of a concern in seed-production fields.

Phyllachora tar spot or black leaf spot.   Phyllachora tar spot or black leaf spot, caused by about 20 species of the fungus *Phyllachora*, is widely distributed throughout the world. Most turfgrass species (especially colonial bentgrass, tall fescue, and bermudagrass) are susceptible to this minor disease, which is most common in wet, shaded areas. It is much more common on unmowed grass than on closely mowed turf.

Identification.   Heavily infected turf areas have a mottled yellow-green appearance. Infected leaves develop small, round-to-oval or elongated, shiny black spots (figure 5–52) on the upper and/or lower leaf surfaces and leaf sheaths. Young lesions may be surrounded by a yellowish zone or halo. As the leaves mature and age, the area around the black spots often remains green longer than the healthy tissue, giving a green island effect.

Disease cycle.   The *Phyllachora* fungi are obligate parasites believed to overseason as minute, black, fruiting bodies (perithecia) under crusty, black, shieldlike coverings (clypei) or spots on the leaves. Microscopic, one-celled hyaline ascospores (figure 5–52) formed in club-shaped asci within the fruiting bodies (perithecia) are released in the spring and probably infect healthy leaves during cool, wet weather.

Importance.   This is a very minor disease of closely mowed turfgrasses.

Phyllosticta leaf blight.   Phyllosticta leaf blight, caused by several or more species of the fungus *Phyllosticta*, causes a minor disease of turfgrasses worldwide during cool, moist weather. The fungal species cannot be easily separated morphologically, and some are referable to other genera including *Phoma*, *Phomopsis*, and immature spore forms of *Ascochyta*, *Septoria*, *Stagonospora*, and other genera.

Identification.   Leaf lesions are mostly circular-to-oval in shape and pale gray or white-to-straw-colored, with narrow, red-to-purple borders, sometimes surrounded by a wider yellowish halo. Rows of yellow-brown-to-black, speck-sized dots (pycnidia) form in the older lesions between the veins.

Disease cycle.   The *Phyllosticta* fungi overseason as pycnidia in infected leaves and grass debris. In cool, wet weather, the one-celled hyaline spores or conidia with rounded ends (figure 5–49) ooze from the pycnidia, spread to healthy

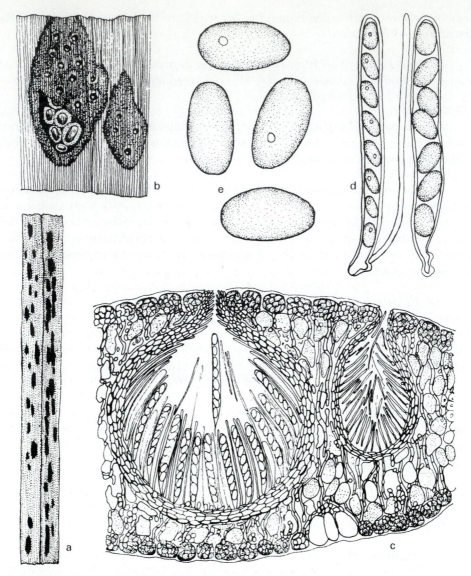

FIGURE 5–52   Tar spot (*Phyllachora graminis*). (a) Typical leaf spots on a grass blade; (b) close-up of two black lesions that are actually crusty, black, shieldlike coverings (clypei) over the embedded perithecia; (c) cross section of two perithecia within a leaf (one mature with asci and ascospores); (d) two asci, each with ascospores; (e) ascospores. Part (b) is the view through a dissecting microscope; parts (c) through (e) are views through a high-power microscope.

leaves (mostly by splashing and flowing water and turfgrass equipment), germinate, and infect the leaves.

Importance.   This is a very minor leaf blight of most turfgrasses during prolonged cool and moist weather.

Physoderma leaf spot and leaf streak.   Physoderma leaf spot and leaf streak, formerly called Physoderma brown spot, is caused by the fungi *Physoderma graminis, P. gerhardtii* (syn. *Entyloma sprageanum), P. lolii,* and *P. paspalum.* It is a relatively rare disease of bentgrasses, fescues, bluegrasses, bermudagrass, bahiagrass, turf timothy, and wheatgrasses, especially on closely mown turfgrasses.

Identification.   Small, yellowish spots or stripes form in the leaves, leaf sheaths, and stems. The lesions gradually turn purplish, brown, and finally ash gray. The lesions may be oval to somewhat elongated and may merge to form bands across the leaf blade. When severe, the crown may rot, and a wilting and shredding of the leaves can occur. Diseased plants are dwarfed with yellow, stiff, upright leaves that may be confused with yellow tuft or downy mildew (sec. 5.2.9).

Disease cycle.   The *Physoderma* fungi overseason as microscopic, round-to-elliptical, golden brown, thick-walled sporangia within infected leaves (figure 5–53) or in soil. In warm, wet weather (73 to 86°F; 22.8 to 30.0°C), swimming spores (zoospores) are released from the sporangia and move in water to infect the lower leaves and crown. The disease may be repeated at 2- or 3-week intervals, as long as prolonged warm and moist weather continues.

Importance.   This rare and minor disease is more important in seed-production fields than in closely mown turf.

Management.   No special measures are needed for its control beyond providing for good surface and subsurface drainage when preparing a new turf area. Water during the morning hours so that the grass surface will be dry before late afternoon. The fungicides suggested to control Pythium blight (table 5–5) may be effective in checking Physoderma leaf spot and leaf streak.

Pseudoseptoria leaf spot.   Pseudoseptoria leaf spot is caused by the fungi *Pseudoseptoria (Selenophoma) donacis, P. stomaticola* (syn. *Selenophoma donacis* var. *stomaticola), P. (Selenophoma) obtusa,* and *P. (Selenophoma) everhartii.* Other names for this minor, cool-and-wet-weather turf disease include red eyespot, halo spot, and frogeye. Susceptible cultivated turfgrasses include bentgrasses, bluegrasses, fescues, wheatgrasses, and weeping alkaligrass (*Puccinellia distans*). Cereals and many forage grasses and wild grasses are commonly infected in temperate regions throughout much of the world.

Identification.   Small (1 to 4 mm long), elliptical, brown-to-purple spots form in the leaves and leaf sheaths during cool, wet weather in early spring and autumn. As the often numerous spots enlarge, the centers become bleached and dotted with rows of minute, golden-brown-to-black specks (pycnidia) and surrounded by purplish margins. When severe, individual leaves and affected turf have a purplish appearance.

Disease cycle.   The *Pseudoseptoria* fungi overseason as pycnidia and possibly as dormant mycelium in infected grass leaves. During cool, damp weather (60 to 70°F; 15.6 to 21.1°C), large numbers of microscopic, hyaline curved spores or conidia

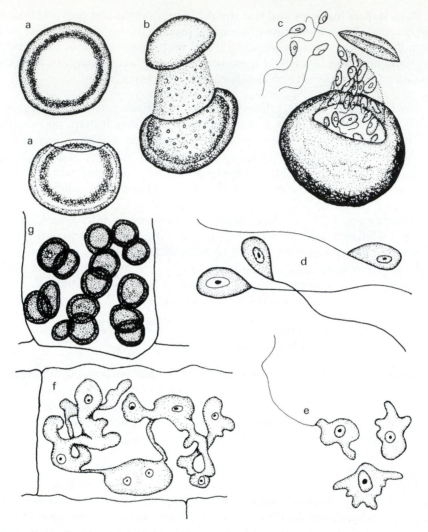

FIGURE 5–53 Physoderma leaf spot. Stages in the life cycle of *Physoderma maydis,* a pathogen of corn, as seen through a high-power microscope. It is believed to have the same basic disease cycle as *Physoderma graminis* and *P. paspali.* (a) Two thick-walled sporangia or resting spores, top and side views; (b) opening of a sporangium showing an early stage of zoospore formation with the circular cap or operculum (lid) being carried up by the enlarging sporangium; (c) mature zoospores escaping through the top of the resting spore; (d) zoospores with a single flagellum; (e) germinating zoospores, amoeboid stage; (f) *Physoderma* mycelium within a leaf cell beginning to form sporangia; (g) a cell filled with mature, resting sporangia.

(figure 5–49) are released from the pycnidia to be splashed or carried to healthy leaves by turfgrass equipment and shoes. The spores germinate in a film of water on the leaves. Infection results in the development of new leaf spots with pycnidia. The cycle may be repeated every week or two as long as the weather remains cool and moist, which promotes disease development. The *Pseudoseptoria* fungi are seedborne on grasses.

Importance.   This minor disease, when severe in prolonged cool and moist weather, may give turf a purplish appearance.

Ramularia leaf spot.   Ramularia leaf spot is caused by three species of *Ramularia: R. collo-cygni* (syn. *Ovularia hordei*), *R. (Ovularia) pusilla,* and *R. holci-lanati.* The disease, found in temperate regions over much of the world, is known to infect bentgrasses, annual and Kentucky bluegrasses, fescues, ryegrasses, wheatgrasses, and numerous other forage grasses and wild grasses in cool, humid climates. This is a widely distributed but uncommon and very minor disease, mostly affecting unmowed grasses. Infections occur on the leaves, leaf sheaths, and stems and are most noticeable in spring and autumn.

Identification.   On bluegrasses, ryegrasses, fescues, and other lawn-type grasses, the leaf spots are usually round-to-oval-shaped and tan, dark brown, or wine red, commonly found with dark centers and a yellow halo. On bentgrasses, the spots are round-to-oval and ash white, with reddish brown margins and a yellow or orange halo. The disease is commonly limited to scattered spots on the leaves. Diseased leaves may die prematurely.

Disease cycle.   The *Ramularia* fungi can infect whenever the grass is growing. They are worst during cool, humid weather in late winter, spring, and autumn. The fungi probably overseason as stromata or dormant mycelium in infected living and dead tissues and possibly as crystal white, egg-shaped spores (conidia) (figure 5–54). Little is known concerning the disease cycle and how the fungi are disseminated.

Importance.   This is a very minor disease that may occasionally cause dieback of the leaves when severe.

Rhynchosporium scald or leaf blotch.   Rhynchosporium scald or leaf blotch, caused by the fungi *Rhynchosporium orthosporum* and *R. secalis,* occurs infrequently in prolonged, cool (39 to 71°F; 3.0 to 21.7°C), damp weather on ryegrasses, bluegrasses, fescues, bentgrasses, turf timothy, and wheatgrasses. The same fungi attack a wide range of grass and cereal hosts in cool, temperate regions of the world. Ryegrasses are the most susceptible cultivated turfgrasses, especially when cut infrequently.

Identification.   Distinctly defined, oval to irregular blotches, 2 to 3 × 10 mm or more long, form on the leaves in early spring and late autumn. The blotches are grayish green to light brown or reddish brown, later turning gray, sometimes with a prominent dark brown margin. The lesions may enlarge and merge to blight large areas of a leaf. Diseased leaves may be girdled, and they may collapse and die back the tip, becoming light gray to brown. Cultivars of both annual and perennial ryegrasses differ greatly in their susceptibility to this disease.

Disease cycle.   The *Rhynchosporium* fungi overseason as dormant mycelium and flat, dark mycelial mats (stromata) in living or dead leaves and other grass

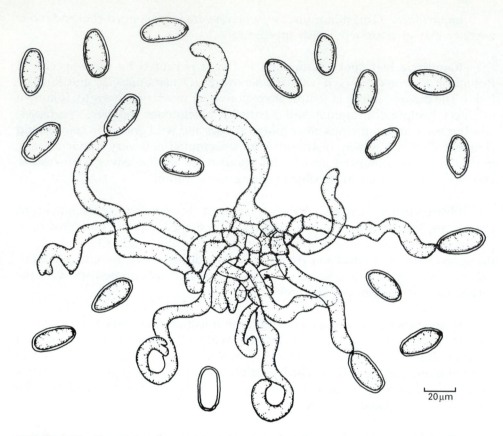

20 μm

FIGURE 5–54 *Ramularia collo-cygni*, one of the causes of Ramularia leaf spot, as seen under a high-power microscope.

debris. Large numbers of hyaline, straight, curved, or beaked spores called conidia (figure 5–55) are produced on short conidiophores that emerge through stomata on both leaf surfaces. The spores, which form in cool, wet weather in early spring and autumn, are splashed, blown, or carried to healthy leaves (on turfgrass equipment and shoes), where infection occurs directly through the epidermal cells. The cycle is repeated as long as the weather remains damp and cool. The conidia closely resemble those of *Microdochium* (*Fusarium*) *nivale* (figure 5–38), the cause of pink snow mold or Microdochium (Fusarium) patch, in size, shape, and time of appearance. *Rhynchosporium* fungi are seedborne.

Importance.   This rare and minor disease of closely mown turfgrasses may be a problem in seed-production fields. Occasionally on turfgrasses the leaves may die back from the tip.

Spermospora and Cercosporella leaf spots, leaf blight, or blast.   Spermospora and Cercosporella leaf spots, leaf blight, or blast, are caused by three species of the fungus *Spermospora* (*S. ciliata*, *S. lolii*, and *S. subulata*) and *Cercosporella* (*Spermospora*)

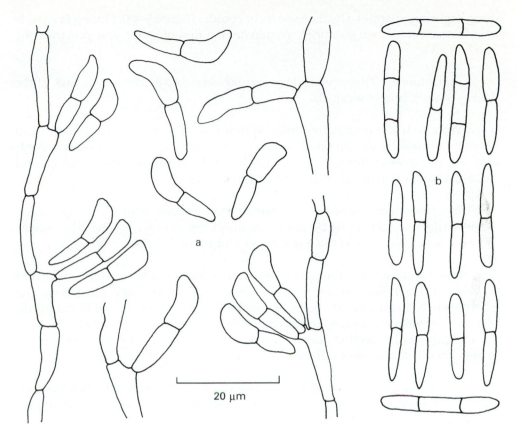

20 μm

FIGURE 5–55 Two species of *Rhynchosporium* that cause Rhynchosporium leaf blotch or scald of turfgrasses as seen under a high-power microscope: (a) *Rhynchosporium secalis;* (b) *R. orthrosporum.*

poagena. These fungi are known to infect bentgrasses, fescues, ryegrasses, rough bluegrass (*Poa trivialis*), and turf timothy in cool, moist, temperate regions of the world. Like Pseudoseptoria leaf spot, these minor diseases are found primarily in early spring and late autumn during cool, humid weather in coastal regions.

Identification. Round-to-elliptical (oval) or elongated, dull straw-colored, gray-brown, olive-gray, or reddish brown spots often appear on the leaves. Lesions on bluegrass are straw-colored with light brown borders. On ryegrasses, the lesions are off-white with gray-brown or reddish brown borders surrounded by a yellowish halo; bentgrasses have olive-gray spots with wine red borders; and on tall fescue the chocolate brown lesions turn reddish brown and have a yellowish halo. The spots often rapidly enlarge to involve much of the leaf blade, resulting in a light-to-dark-brown or straw-colored "scald" and a dieback from the leaf tips.

Disease cycle. The causal fungi probably overseason as dormant mycelium and in the dark-brown-to-dull-black stromata at the surface of infected leaves, or as long, tapering, spermlike hyaline spores or conidia (figure 5–49). During favorable conditions, conidia are produced, germinate in a film of water, and penetrate the grass blades.

Importance.   These minor diseases may occasionally cause a dieback of the leaves in cool, humid weather.

Wojnowicia leaf spot, foot rot, and root rot.   This minor but cosmopolitan disease, caused by the fungus *Wojnowicia hirta* (syn. *W. graminis*), infects bentgrasses, bluegrasses, fescues, wheatgrasses, and numerous forage grasses and wild grasses. It is generally considered to be a weak parasite.

Identification.   Occasional, indefinite brown spots form in living leaves. Brown, speck-sized fruiting bodies (pycnidia) form in older lesions. The fungus may be associated with a crown (foot) and root rot of grasses.

Disease cycle.   The fungus overseasons as speck-sized, brown pycnidia and as dormant mycelium in dead leaves. Under favorable damp conditions, the slightly curved and tapered spores or conidia (figure 5–49) ooze out of the pycnidia and are disseminated by splashing or flowing water, air currents, and turfgrass equipment. Infection occurs through stomata or directly through the cell walls. The fungi are also seedborne.

Importance.   Occasional leaves may die back from the tip, and the fungus may be isolated from decayed crowns and roots, probably as a secondary invader.

5.2.17   Minor Viral, Phytoplasmal, and Spiroplasmal Diseases

A number of viruses and the aster-yellows mycoplasma, now known as a phytoplasma, commonly infect grasses grown for nonturf purposes and in seed-production fields, but these are not generally considered to cause any significant damage to closely mowed turf. Yellow dwarf of zoysiagrasses, bentgrasses, and bermudagrasses is caused by a phytoplasma. Yellow dwarf in Japan causes grass plants to be stunted and whitish yellow with the shoots developing witches' brooms.

Another phytoplasmal disease, closely resembling yellow dwarf, is white leaf on vegetatively propagated bermudagrass. It has been reported from Israel (where it causes a major disease), Taiwan, Thailand, and Australia. Small, isolated patches of pale-yellow-to-dull-off-white turf develop up to 2 in (5.1 cm) in diameter. The leaves are bunched at the ends of the stems, giving a witches' broom effect. Little is known about the disease and its control.

Phytoplasmas are usually transmitted by leafhopper insects and then spread systemically through the plant. The susceptibility of turfgrasses to phytoplasmas and viruses has been confirmed in greenhouse inoculation studies. Perhaps phytoplasmas and viruses, by reducing root growth and photosynthesis, are making grass plants more susceptible to leaf, crown, and root-infecting pathogens and various environmental stresses. This area needs much additional research. in Japan is caused by a phytoplasma. The susceptibility of many turfgrasses to viral agents, has been confirmed in greenhouse inoculation studies. Perhaps and these viruses, by reducing root growth and restricting photosynthesis, are making grass plants more susceptible to leaf, crown, and root-infecting pathogens and various environmental stresses. This area needs additional research.

The viruses reported to infect one or more turfgrass genera in North America, Europe, east Asia, Japan, and New Zealand include agropyron mosaic, barley stripe mosaic, barley yellow dwarf, brome mosaic, cocksfoot mild mosaic, cocksfoot mottle, cocksfoot streak, cynosurus mottle, festuca leaf streak, Lolium latent, maize chlorotic dwarf, oat necrotic dwarf (necrotic mottle of Kentucky bluegrass and annual ryegrass), panicum mosaic, poa semilatent (ryegrass leaf mottle), rice dwarf, ryegrass chlorotic streak, two ryegrass mosaic viruses, soil-borne wheat mosaic, sugarcane or maize dwarf mosaic (St. Augustinegrass leaf mottle), western ryegrass mosaic, wheat spindle streak, wheat streak mosaic, and wheat striate mosaic. Wheat yellow dwarf, zoysia dwarf, and zoysia mosaic have been reported from Japan on Japanese zoysiagrass (*Zoysia japonicum*). Most of these viruses are unknown or rare on turfgrasses in the field, but the potential for infection is always present.

The viruses are transmitted from plant to plant by highly virus-specific means. The methods of transmission and animal vectors include specific species of aphids, leafhoppers, planthoppers, beetles, and mites (figure 5–6); mechanical transfer of infected sap or pollen; seed; dagger (*Xiphinema*) nematodes; and a primitive, root-inhabiting fungus, *Polymyxa graminis.* More turfgrass-infecting viruses and means of transmission will probably be discovered in the future as we learn more about double infections (for instance, a latent virus and a leaf- or root-infecting fungus or other pathogen).

5.2.18   Mushrooms and Puffballs

A large number of basidiomycete fungi that produce mushrooms (toadstools) and puffballs feed on decaying organic matter in the soil. These fungi, including those that produce fairy rings and superficial fairy rings (sec. 5.2.10), are most common around dead and buried stumps, roots, boards, or excess thatch. Small yellow-green spots 2 to 4 in (5 to 10 cm) often appear in bentgrass turfs.

Identification.   The spore-producing mushrooms and puffballs, which are 1 to 12 in (2.5 to 30.5 cm) in diameter, appear in mild weather after heavy rains or watering (figure 5–56). Some mushrooms and puffballs are foul-smelling; a few are poisonous.

Disease cycle.   These nuisance fungi overwinter as mycelial spawn in the soil and in decaying organic matter. The fruiting bodies produce large numbers of microscopic spores (basidiospores) that are spread about by air currents, water, turf equipment, and tools of all kinds.

Importance.   The mushrooms and puffballs are unsightly only.

FIGURE 5–56 Mushrooms of
Coprinus lagopus growing in a lawn.
(Courtesy B. J. Jacobsen.)

Management. The management of mushrooms and puffballs, where practical, is to carefully dig up and destroy rotting stumps, roots, or other underground sources of organic matter. If you suspect the fungi of being poisonous to children or pets, break or mow off the fruiting bodies when first seen. Mushrooms and puffballs will disappear naturally *only* when their food base in the soil is exhausted. This process may take 10 years or more for a large stump or root.

Mycoplasmal Diseases
See the section on minor viral, mycoplasmal, and spiroplasmal diseases (5.2.17).

Necrotic Ring Spot
See the section on summer patch and necrotic ring spot (5.2.32).

5.2.19   Plant-Parasitic Nematodes

The more important types of nematodes that reduce turfgrass vigor in one area or region or another and result in yellowing, stunting, and thinning of grass plants include lance (*Hoplolaimus* spp.), sting (*Belonolaimus* spp.), ring (*Criconemella* or *Mesocriconema* spp.), root knot (*Meloidogyne* spp.), lesion (*Pratylenchus* spp.), stubby root (*Paratrichodorus* and *Trichodorus* spp.), stunt or stylet (*Tylenchorhynchus* and *Merlinius* spp.), dagger (*Xiphinema* spp.), needle (*Longidorus* spp.), sheath (*Hemicycliophora* spp.), burrowing (*Radopholus similus*), spiral (*Helicotylenchus* and *Rotylenchus* spp.), awl (*Dolichodorus* spp.), pin (*Paratylenchus* spp.), false root knot (*Hypsoperine* spp.), cyst (*Heterodera* and *Punctodera* spp.), cystoid (*Meloidodera charis*), and root gall (*Subanguina radicicola* [figure 5–7]). Some of these species (for instance, sting) are much more damaging than others. Not all turfgrasses are susceptible to all of these plant-parasitic nematodes. Most species attack a wide range of different plants and can survive in weeds when turfgrasses are absent.

FIGURE 5–57   Lance (*Hoplolaimus*) nematodes feeding on a root of St. Augustinegrass.

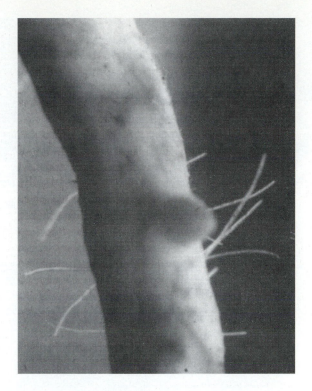

All plant-parasitic nematodes are obligate parasites and must feed on living plant cells to complete their life cycle. Almost all feed on root cells by puncturing walls with a hollow stylet (figures 5–1 and 5–8) that resembles a minute hypodermic needle, injecting enzymes into cells, and then ingesting the partially digested contents.

Plant-parasitic nematodes are divided into two major groups according to their parasitic behavior: *Ectoparasitic nematodes* spend their life cycles outside the host and feed by inserting their styles to varying depths within the root tissues (figure 5–57). *Endoparasitic nematodes* feed both externally on and internally within plant roots, often becoming only partially embedded in roots (*semiendoparasites*). Most forms remain motile throughout their life cycles and "browse" on roots. A few females (root knot, cyst, and cystoid) become immobile, assuming a sedentary parasitic habit (*Color plate* 46).

Identification.   Heavily nematode-infected turf lacks vigor and declines in growth, often appearing off-color, yellow to silvery gray, bunchy, and stunted. Grass blades turning yellow and dying back from the tips are often interspersed with apparently healthy leaves. In hot, dry weather, or at other times when the grass is under stress, such nematode-injured turf may wilt, thin out, and die, usually in scattered, round or irregularly shaped patches, or in streaks during the summer (figures 5–58 and 5–59) due to the clustered distribution of nematodes. Hybrid bermudagrass golf greens are severely damaged by sting nematodes in

FIGURE 5–58   Right, severe stunt nematode damage to centipedegrass; left, close-up of affected roots. (Courtesy W. M. Powell.)

FIGURE 5–59   Severe nematode injury; turf in the background has recently been treated with a nematicide. (Courtesy G. C. Smart.)

the south (*Color plate 47*). Such turf is also more susceptible to winterkill, insect injury, and other diseases.

The severity of symptoms varies with soil moisture, texture, and fertility, the population of plant-parasitic nematodes, and the particular species feeding on and in the roots. Symptoms are easily confused with fertilizer or pesticide injury, a thick thatch, malnutrition, poor soil aeration and compaction, drought or excessive soil moisture, heat stress, insects, diseases such as anthracnose and root rots, and other types of injury (see chapters 3 and 6).

Heavily nematode-infested grass does not respond normally to applications of water, fertilizer, fungicides, and most insecticides; nor does it respond to increased aeration (coring) and thatch removal. Feeder root systems are greatly reduced and shallow, and they may be coarse, swollen, stubby, and excessively branched and show swellings or galls. They are dark in color (figures 5–9 and 5–58). Such turf commonly is slow to green up in the spring.

Nematodes damage roots in two major ways. Root cells are injured or killed mechanically or chemically (apparently through the secretion or injection of enzymes by nematodes). Such damage renders the roots less efficient in absorbing and transporting water and nutrients. In many cases, root growth is greatly retarded, and in some cases it ceases. Nematodes also interact with various fungal and bacterial pathogens to cause disease complexes that often damage roots more severely than when either the nematode or other pathogen is present alone. Various fungi and bacteria are attracted to nematode-infested roots by the plant fluids that leak into the soil following nematode feeding. These secondary organisms can build up large populations and make it difficult to determine the real cause of a turf problem. Nematode damage is thus often difficult to separate from injury due to other pests including insects.

In more severe cases, the primary symptoms of nematode damage may include various combinations of the following:

1. *Stunting.* Suppressed plant growth. Plants are often somewhat uneven in height, usually in irregular areas, giving the turf a ragged overall appearance (figures 5–58 and 5–59).

2. *Chlorosis or other discoloration.* Loss of green color and yellowing similar to nitrogen, iron, magnesium, manganese, or zinc deficiency.

3. *Wilting.* Enhanced by wind and hot temperatures. Plants may slowly recover at night, at least for a time.

4. *Dieback of leaves.* Plants often show a slow general decline in growth and vigor. Such plants do not respond normally to water, fertilizer, or fungicides.

5. *Root lesions.* Red-to-dark-brown-or-black, often sunken, discolored areas in roots result from continued nematode feeding and invasion by other soil microorganisms. Lesion size increases from pinpoint spots to large necrotic areas (figure 5–9). Root-rotting fungi often invade tissues damaged by root-lesion nematodes, accelerating lesion enlargement.

6. *Root swellings.* Indistinct swellings, primarily of the root tips, to knotlike galls throughout the root system, often are accompanied by excessive production of branch roots (hairy root) above the swellings (figure 5–9).

7. *Injured or devitalized root tips.* Cessation of root growth caused by feeding at or near the root tips. Tissues may retain a normal white color or turn brown and die. Devitalization frequently causes excessive production of roots near the soil surface.

   • Stubby roots. Root system composed of numerous short, stubby branches often arranged in clusters (figures 5–9 and 5–58).

   • Coarse roots. Root system with few or no branch roots or feeder roots (figure 5–9).

8. *Reduced and discolored root system.*   Overall reduction in the normal size of the root system accompanied by light to dark browning or blackening (figures 5–9 and 5–58).
9. *Growth reduction.*   Growth as measured by yields of clippings or mowing frequency is greatly reduced.

All turfgrass areas throughout the world are infested with plant-parasitic, saprophytic, and predatory nematodes. It is important to know the species and numbers of plant-parasitic nematodes present and to take proper soil-root samples for diagnosis and analysis. The presence and identity of plant-parasitic nematodes can be determined *only* by taking a series of random core samples (15 to 30 cores per 500 to 1000 sq ft (45.5 to 90 sq m) from a suspect area and the same number of cores from an adjacent, apparently healthy area. The cores should be taken to a 4-in (10.2-cm) depth, excluding the thatch. These core samples should be examined by a competent nematologist in a well-equipped laboratory. Most land-grant universities provide a nematode assay service for a fee and make suggestions for their control. There are also a number of competent private laboratories that provide an assay service, again for a fee. Proper sampling and interpretation of the results are key factors in determining whether turf has a nematode problem.

Sampling should be done at near-monthly intervals during the growing season as nematode populations fluctuate greatly. Start sampling in spring after the soil temperature at the 2-in (5.1-cm) depth reaches 50°F (10.0°C). Based on the numbers of each nematode genus and species isolated, a nematologist can then make recommendations for best management approaches based on the grass, type of soil, and other factors. Within a given species of nematode, the population level needed to cause injury (the damage threshold level) varies according to the degree of stress under which the turf is growing. Damage threshold levels are difficult at best to determine when mixed populations of nematode species are present. As plant vigor decreases from close mowing (1 in [2.5 cm] or under for lawn-type grasses and ⅜ in [0.95 cm] for golf and bowling greens), high air temperatures, low soil moisture, and other stresses, the numbers of nematodes needed to cause severe damage decreases.

In nematology laboratories, there are a number of techniques used to extract nematodes from soil and roots. For extraction of endoparasites from roots, the use of mist chambers and various "shaker" techniques are very adequate. The extraction of nematodes (especially ectoparasites) from soil can be accomplished by Baermann, tray, pan, or funnel techniques. However, this procedure will not allow adequate recovery of very slow-moving nematodes such as *Criconemella* species. The centrifuge-flotation procedure is an excellent technique to use for these types of nematodes. Procedures for extracting nematodes from soil and roots are well outlined in several plant nematology laboratory manuals. Nematology laboratories are often set up with equipment to extract all types of nematodes. More details can be acquired by contacting nematologists at various land-grant universities.

To determine if nematodes are a cause of a decline in turfgrass vigor, first examine the turf carefully to eliminate other likely causes. Check for injury by insects and other animal life (chapters 3 and 6), diseases, possible soil deficiencies, fertilizer or pesticide damage, frost or high-temperature injury, too much or too little water, and other possible stress factors (chapters 2 and 3). After examination, if nematodes are still suspected, contact your local county cooperative extension (or advisory) office for further information regarding collecting soil and root samples and laboratories capable of identifying parasitic types. Soil from around the feeder roots of turfgrass plants in an *early* stage of decline is likely to contain many more nematodes than soil from around the roots of badly diseased or dead plants. Information on how to collect and mail turfgrass specimens for a nematode assay is given in chapter 1.

Disease cycle.   The life cycle of most plant-parasitic nematodes, although varying in details, is simple and direct (figure 5–9). Plant-parasitic nematodes in turf must have living cells and tissues to feed, survive, and reproduce. They overwinter in the soil, grass debris, or within living roots as eggs, cysts, adults, or juveniles (sometimes referred to as larvae). Juveniles hatch from eggs deposited by the female in the soil or within root tissue and usually resemble adults in structure and appearance. Nematodes develop to a first-stage juvenile within the egg, hatch, and undergo four molts before developing into sexually mature adults. Adult nematodes, which reproduce either sexually or parthenogenetically, lay eggs to start the next generation. Under optimum conditions, most plant-parasitic species complete their life cycle—which is largely temperature-dependent—in 20 to 60 days, although some, such as the dagger nematode (*Xiphinema americanum*), may require 9 months or longer to complete a cycle.

Water films around soil particles are needed for nematodes to move in soil, feed, and reproduce. High populations may develop, and damage is most common in high-sand golf greens, sand or sandy loam soils low in nutrients where water can be deficient or abundant but where a thick thatch is present.

The optimum soil temperature for most nematodes is 60 to 90°F (15.6 to 32.2°C). Nematode activity is stimulated by soil moisture in the range that is readily available to grass roots; their activity is restricted in dry soils and in saturated soils deficient in oxygen. Highly compacted and heavy-textured soils are less favorable than coarse-textured and sandy soils for nematode activity. Nematodes are generally most active and numerous on cool-season grasses in middle to late spring with populations often peaking in midsummer. Populations may again rebound in autumn. On warm-season grasses, activity is greatest during the summer and autumn. Symptoms of heavily infested turf in midsummer usually become apparent only when the grass is under moisture, heat, close mowing, nutritional stresses, or other conditions unfavorable for growth of the grass.

Nematodes move very slowly in soil under their own power—perhaps 1 to 2 ft (30.5 to 61 cm) per year in sand. People, however, easily spread them from one area to another in infested plant material (sod, sprigs, or plugs) and soil by core aerification cultivation, sodding or plugging, watering practices, and so on. Anything that moves soil can move nematodes!

Importance.   Nematode damage is most severe in high-sand golf greens and coarse-textured and sandy soils when the turf is under stress. Infested turf lacks vigor, declines in growth, may wilt, thin out, and be more susceptible to winter kill and serious insect or disease injury. Only by repeated thorough sampling can you be sure you have a nematode problem and the seriousness of it.

Management.   Keep the grass growing vigorously through proper watering, fertilizing, thatch removal, aerification, and especially by increasing the mowing height. Follow other recommended cultural practices (see chapter 2 and table 5–1 at the beginning of this chapter). Frequent light irrigations are needed on nematode-infested turf where the roots are stunted, devitalized, and most feeding roots are in the top 1 to 2 in (2.5 to 5.1 cm) of soil or even just in the thatch. Control diseases, insect and animal pests, weeds, and other problems.

If not starting from seed, buy *only* top-quality, nematode-free sod, sprigs, and plugs from a reputable nursery worker who realizes the necessity of producing nematode-free turf. Before planting, take soil samples and have them assayed by a competent nematologist. Follow the suggestions in the report. Grading and planting equipment should be thoroughly washed or steamed before use to prevent spread of nematodes.

If plant-parasitic nematodes are found in large numbers in areas to be established to turf, treat the soil *before* planting using a nematicide or general-purpose soil fumigant (table 5–5) that controls fungi, insects, and weed seeds as well as nematodes. These chemicals can be purchased and applied only by licensed pesticide applicators who have the proper equipment and training in handling such restricted-use chemicals. Chemical control is most effective in light, sandy soils.

On existing turf, take samples from both suspect and healthy areas and have them assayed to see if a nematode problem exists. It may be suggested that a nematicide be drenched into the root zone when the weather is cool and the temperature of the soil at the 4-in (10.2-cm) level is 60°F (15.6°C) or above. Again, this is a job for an experienced and licensed commercial applicator. On established turf only a limited number of nonfumigant nematicides are available. These should be used with extreme caution and must be watered into the root zone to be effective.

Nematicides are highly toxic chemicals that must be treated with extreme caution following the manufacturer's directions and precautions. These chemicals must be *promptly* drenched in to prevent injury to the turf, especially in hot weather. Core aerification before application is suggested to improve control. In some areas, nematicides may only be applied to commercial turfgrass sites (such as golf courses and sod farms). Turfgrass specialists also recommend *never* to apply a nematicide without first having a nematode analysis.

Proper site preparation and the use of nematode-free planting and topdressing materials is much more satisfactory than trying to eliminate high populations after the turfgrass is established. There are no turfgrass species or cultivars resistant to all nematode species or more than a very few species.

Some experienced turfgrass managers, however, who suspect nematode damage but do not wish to collect and mail soil and root samples, apply a suggested nematicide in a series of strips alternating with strips of untreated turf. If the treated turf appears obviously greener and healthier in 2 weeks or more, then nematodes could be a problem. Sometimes, however, you may get a growth response from a nematicide application that has nothing to do with nematode control. Thus, it is strongly recommended that suspected nematode problems be discussed with and diagnosed by an expert.

Net-blotch
See the section on Helminthosporium diseases (5.2.13).

Nigrospora Blight
See the section on minor leaf spots (5.2.16).

Phaeoramularia Leaf Spot
See the section on minor leaf spots (5.2.16).

Phloeospora Leaf Spot or Stem Eyespot
See the section on minor leaf spots (5.2.16).

Phyllosticta Leaf Blight
See the section on minor leaf spots (5.2.16).

Physoderma Leaf Spot or Leaf Streak
See the section on minor leaf spots (5.2.16).

Pink Patch
See the section on red thread and pink patch (5.2.22).

Pink Snow Mold
See the section on snow molds (5.2.29).

Poa Patch
See the section on summer patch, necrotic ring spot, and Poa patch (5.2.32).

5.2.20   Powdery Mildew

Powdery mildew is caused by the widespread fungus *Erysiphe* (*Blumeria*) *graminis*. It is an obligate parasite that attacks most cultivated grasses, including cereals, throughout the world. The disease occurs most commonly on Kentucky bluegrass and, to a lesser extent, bermudagrass, fine-leaf fescues, ryegrasses, and zoysiagrass growing in shaded areas. Four highly specialized *formae speciales* (f.sp.), each with numerous physiologic races of the fungus, only attack certain cultivars of one or a few turfgrass species. For example, the races of the fungus that attack Kentucky bluegrass do not attack fescues or bermudagrass.

Powdery mildew has become an increasingly important disease of Kentucky bluegrass cultivars in recent years. High-nitrogen fertilizers cause a dense growth of grass that creates an ideal environment for the mildew fungus when coupled with low light intensity, poor air circulation, and cool, damp conditions. Resistance to powdery mildew is known to exist in several cultivars of Kentucky bluegrass, bermudagrass, and in several species of bluegrass and fescues.

Powdery mildew is much more severe where air circulation is reduced and the grass is growing in shaded areas (on north and east sides of buildings, under dense trees and shrubs) than in full sun. Attacks occur chiefly in the spring, late summer, and autumn when days are mild (60 to 72°F; 15 to 22.5°C) and cloudy and nights are cool and damp.

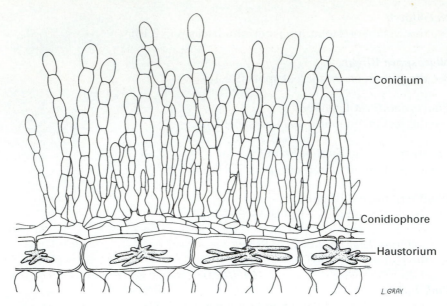

Conidium

Conidiophore

Haustorium

L. GRAY

FIGURE 5–60 Diagram of the powdery mildew fungus *Erysiphe graminis.* The fungus is on the surface of a leaf except for its feeding organs (haustoria), which invade the epidermal cells.

Powdery mildew fungi live chiefly on the outer surface of the host plant. They obtain food and water by means of small, branched, rootlike organs (haustoria) that penetrate the grass leaf or sheath and enter the surface layer of epidermal cells (figure 5–60).

Identification.   Powdery mildew appears first as small, superficial patches of white-to-light-gray, dusty growth (mycelium) found mostly on the upper surfaces of leaves and on leaf sheaths (*Color plate 48*). These patches enlarge rapidly and merge, becoming more dense. The lower leaves are often completely covered. The leaf tissue under the mildew becomes yellowed soon after infection and later turns tan or brown. Heavily infected leaves gradually dry up and die. In severe outbreaks, large areas or entire grass stands are dull white, as if dusted with flour or lime. The turf becomes severely thinned where mildew infections are heavy and persist for weeks.

Disease cycle.   The powdery appearance of mildew is due to the production of tremendous numbers of microscopic, ellipsoid spores (conidia) in chains (figures 5–13 and 5–60). The conidia are easily carried on air currents and turf equipment and may cause new infections within 2 hours after landing on a leaf during cool (55 to 72°F [12.8 to 22.2°C]; optimum about 65°F [18.3°C]), humid, cloudy weather, and where air movement is poor. With a favorable temperature and high atmospheric humidity, spores are continuously produced for 7 to 14 days, until the host tissue dies. The conidia are easily carried to other grass plants in the same or neighboring turf areas, where they produce new infections and start the cycle once again. New conidia may be produced within 4 or 5 days to a week after infection occurs. Penetration of the leaf surface is direct.

As the fungus matures, the mycelium forms dense mats. Occasionally, speck-sized, dark-brown-to-black fungus-fruiting bodies (cleistothecia) form in the woolly-to-felty mycelial mats during late autumn and early winter. These bodies are especially evident on dead grass leaves. Sexual or overwintering spores (as-cospores) are sometimes produced in 15 to 20 mostly oval asci within each cleis-tothecium. Cleistothecia, however, are not common in turfgrasses. The powdery mildew fungus survives the winter as cleistothecia on dead plant tissues and more often as mycelial mats on living grass plants. The ascospores and/or conidia are released in spring or early summer, germinate over a wide temperature range (33 to 86°F; 0.5 to 30°C), and produce initial infections that appear in about 4 days.

Importance.   Powdery mildew is an important cause of deterioration of shady Kentucky bluegrass lawns, as it significantly reduces the growth of leaves, roots, rhizomes, and stolons. A severe attack may weaken and kill the plants, especially in crowded, newly planted areas where air movement is poor, light is reduced, and atmospheric humidity is high. The surviving plants are more susceptible to winterkill, droughts, and attack by other disease-causing fungi.

Management.   Increase light penetration, air movement, and drying of the grass surface by pruning or selectively removing dense trees and shrubs that shade or border turf areas. Space landscape plants properly to allow adequate air movement and to avoid excessive shade. Certain improved, shade-tolerant cultivars of Kentucky bluegrass, fine-leaf fescues, tall fescue, perennial ryegrass, and roughstalk bluegrass (*Poa trivialis*) do relatively well in open, shady areas. Check with your county extension (or advisory) office or turfgrass specialist regarding these shade-tolerant cultivars. Where shade is dense it may be necessary to grow a shade-tolerant ground cover such as Japanese spurge (*Pachysandra*), myrtle or periwinkle (*Vinca*), or English ivy (*Hedera*).

Keep the turf vigorous and growing steadily by fertilizing on the basis of lo-cal recommendations for the grasses grown and a soil test. Recommendations will vary with the cultivar, blend, or mixture grown and its use. Avoid overfeeding, es-pecially with fertilizers that contain large amounts of nitrogen. Follow the other cultural practices outlined in table 5–1. Increasing the mowing height and avoid-ing drought stress help to promote turfgrass growth and vigor.

On high-value turf where powdery mildew consistently reoccurs, a fungi-cide program may be economically feasible. When powdery mildew is *first* evident in the spring or early autumn, two or more applications of a suggested fungicide (table 5–5) at 7- to 21-day intervals should control the disease. For ef-fective management of powdery mildew, spray 1000 sq ft uniformly with 1 to 3 gal of water containing a small amount of surfactant or commercial spreader-sticker (about ½ to 1 teaspoonful per gallon or 1 pint to 1 quart per 100 gal). Fol-low the manufacturer's directions. Thorough coverage of the grass leaves with each spray is essential for good control. A bicontrol fungus (*Ampelomyces quisqualis*), sold as AG 10 by Plant Health Care, Inc., is a hyperparasite packaged in spore form and applied as a spray when conditions for infection are favorable. The effectiveness of this biological control treatment for suppression of powdery mildew in turf is uncertain.

Pseudoseptoria Leaf Spot
See the section on minor leaf spots (5.2.16).

Puffballs
See the section on mushrooms and puffballs (5.2.18).

Pythium Blight
See the section on Pythium diseases (5.2.21).

5.2.21   Phythium Diseases

More than 30 species of the fungal genus *Pythium* have been isolated from turf-grasses. Many can thrive over a wide range of temperature, moisture, and soil conditions. *Pythium* species, which are currently thought to be more closely related to algae than fungi, are cosmopolitan in soils and waters worldwide. Most species have very broad host ranges and may cause seed rot, seedling blight, crown and root rot, and foliage blight. *Pythium* fungi are also capable of breaking down organic matter (thatch). All parts of turfgrass plants are subject to attack by one or more species of *Pythium*. Nine of these species are illustrated in figure 5–61. Other species of *Pythium* are saprophytes, weak pathogens, or secondary invaders. In general, soil-inhabiting species of *Pythium* incite the greatest damage in saturated or overly wet soils and verdure, where their germination and growth are encouraged by stimulatory seed and root exudates. All turfgrasses are susceptible to attack by species of *Pythium*.

Pythium diseases can conveniently be divided into (1) seed decay, damping off, and seedling diseases; (2) low-temperature Pythium snow blight or rot; (3) root and crown rot; (4) root dysfunction; and (5) hot-weather Pythium foliar blight.

Seed decay, pre- and postemergence damping off, and seedling diseases. There are many common soil-inhabiting and seedborne fungi that cause seed decay and seedling death before or shortly after emergence (damping off). Species of *Pythium* that may be involved include *Pythium aphanidermatum, P. aristosporum, P. arrhenomanes, P. dissotocum, P. graminicola, P. irregulare, P. multisporum, P. myriotylum, P. paroecandrum, P. splendens, P. sylvaticum, P. tardicrescens, P. ultimum, P. vanterpoolii, P. violae, P. volutum*, and possibly others. This disease complex is most common when seed germination and the growth of seedlings in the seedbed is slowed due to suboptimal conditions (of temperature, moisture, oxygen, and light); these conditions favor the growth and host invasion by pathogens.

Identification.   Seedbeds generally have a patchy emergence and thinned stands. Many seedlings develop a watery rot, turn yellow, wilt, collapse, and die in irregular patches from a rot at the soil line (postemergence damping off). Bermudagrass golf greens in the southern United States are commonly overseeded in the fall with bentgrass, ryegrass, or bluegrass. Damping off is general and may be severe on greens where preventative management practices are not taken during warm, humid conditions. The predominant species is usually *Pythium aphanidermatum*.

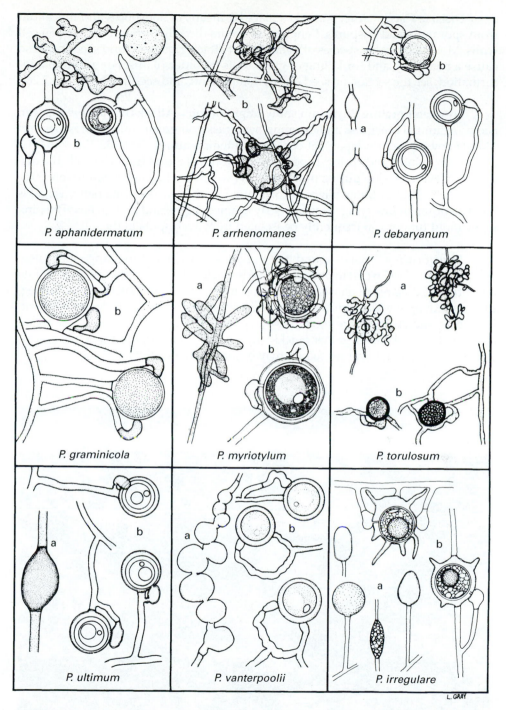

FIGURE 5–61 Nine species of *Pythium* that cause diseases of turfgrasses as seen under a high-power microscope: (a) various types of sporangia; (b) types of sex organs (oogonia with oospores and antheridia).

Disease Cycle.   Infection takes place by means of zoospores and mycelium from sporangia and oospores. Penetration occurs directly through or between cell walls. Most of the same species of *Pythium* that attack grass seed and seedlings also cause a browning and/or killing of grass roots on older plants. For additional information, see the section on seed decay, damping off, and seedling blights (5.2.26).

Low-temperature Phythium snow blight or rot.   Attacks of *Pythium* spp. can occur on annual bluegrass and creeping bentgrass golf greens fine-leaf fescues and Manila zoysiagrass (in Japan), at cool-to-cold temperatures (30 to 32°F; −1 to 0°C) in autumn, winter, and spring, especially when snow falls on unfrozen soil. The disease may even develop during prolonged wet weather when temperatures are 50 to 65°F (10.0 to 18.3°C). Snow blight or rot commonly follows drainage patterns and is most common in low-lying areas of poorly drained, high-sand golf greens. The pink snow mold fungus can frequently be isolated from such spots in the fall.

Identification.   Small, round, tan, gray, or straw-colored-to-orange spots, usually 1 to 3 in (2.5 to 7.6 cm) in diameter, appear in late winter or early spring and often resemble old dollar-spot damage. Some spots may coalesce and kill large, more irregularly sized, light to dark brown areas. The leaves are grayish tan to brown with rotted stems and crowns. The roots mostly appear normal. When crowns are rotted extensively, the plants die rather quickly (figure 5–62), and large areas of turf may be uniformly blighted. Foliar mycelium may or may not be evident.

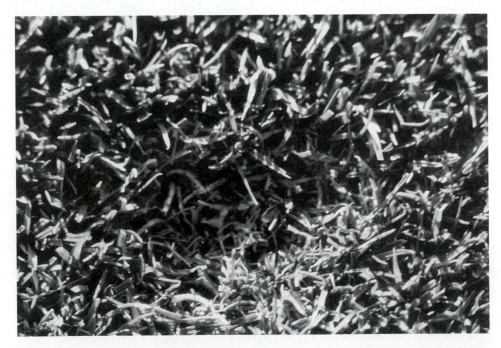

FIGURE 5–62   A small patch of rotted creeping bentgrass caused by a cool-weather *Pythium* species. (Courtesy G. W. Simone.)

Disease Cycle. See the paragraph on hot-weather Pythium blight, hereafter. Rotted tissues are filled with oospores. At least six species of *Pythium* are causal agents of snow blight. The most common are apparently *P. aristosporum, P. graminicola, P. horinouchiense, P. iwayamai, P. paddicum,* and *P. vanterpoolii.*

Importance. This is a rather uncommon and usually minor disease. in water-saturated sites with poor soil drainage and high fertility levels. In Japan, certain genotypes of Manila zoysiagrass are attacked severely.

Management. Provide for adequate surface and root-zone drainage. Maintain a moderate to low soil fertility level during autumn to the first permanent snowfall. Where disease has been damaging in the past, apply a Pythium fungicide or mixture (table 5–5) just prior to the first heavy snow in autumn or early winter. Make one or more additional applications during a midwinter thaw and again at snow melt. Combine any of these fungicides with gray- and pink-mold fungicides (table 5–5). Improving soil drainage may be beneficial.

Crown and root rot. This disease is incited or associated with *Pythium aphanidermatum, P. aristosporum, P. arrhenomanes, P. catenulatum, P. dissotocum, P. graminicola, P. intermedium, P. irregulare, P. myriotylum, P. oligandrum, P. periilum, P. periplocum, P. rostratum, P. tardicrescens, P. torulosum, P. ultimum, P. vanterpoolii, P. volutum,* and possibly other species. It is mostly a problem in highly maintained creeping bentgrass and annual bluegrass golf greens during prolonged, cool-to-hot, wet weather (>60 to 90°F or 15.6 to 32.2°C). Some species of *Pythium* are dominant at cooler temperatures (45 to 70°F; 7.2 to 21.1°C), while other species are most active at high temperatures (73 to 93°F; 22.8 to 33.9°C). It is more of a problem than is generally realized, as turfgrass stands commonly show a general decline in health, with affected turf appearing thin, off color, and slow growing.

Identification. Infected plants turn yellow, tan, brown to bronze, or reddish brown, wilt, and die in small-to-large, irregular areas that resemble anthracnose (sec. 5.2.1), pink snow mold (sec. 5.2.29), dollar spot (sec. 5.2.8), and Helminthosporium melting out (sec. 5.2.13). When the fungus or fungi are active, diseased patches, including plant crowns, may be watersoaked. Root and crown systems are commonly discolored, greatly reduced, and lacking in vigor (figure 5–63). No aerial mycelium is evident. Such turf responds poorly to applications of fertilizer. Microscopic examination of the crowns and roots is needed to determine if *Pythium* species are associated with the damage.

Disease Cycle. *Pythium* species are water-mold fungi that are common inhabitants of most soils and thatch debris. Infection probably occurs by direct penetration of root hairs, roots, and crowns by invading hyphae from zoospores and sporangia.

Importance. Losses from Pythium crown and root rot are generally underestimated and frequently misdiagnosed.

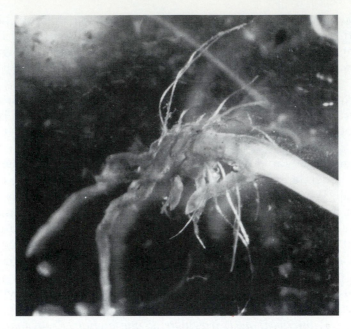

FIGURE 5–63   Pythium root and crown rot on a creeping bentgrass plant. (Courtesy R. T. Kane.)

Management.   Follow the cultural practices outlined in table 5–1. The key to management is to provide good surface and subsurface drainage *before* establishment of bentgrass putting greens. Avoid overwatering. Raise the mowing height to ¼ in (0.63 cm). Make one or more applications of a Pythium fungicide or mixture (table 5–5) when the disease is first evident. Contact your county extension (or advisory) office or turfgrass specialist for advice concerning management of this disease in your area. Chemicals alone will *not* control this phase of the Pythium disease complex without proper water and fertility management. Strains of the biochemical control agent *Trichoderma harzianum* are reported to suppress Pythium root rot and blight caused by *Pythium* spp.

Root dysfunction.   This disease is important as it can predispose turf to stresses such as heat and drought. It is common on old golf greens renovated with applications of high-sand topdressing. At least two species of *Pythium* (*P. aristosporum* and *P. arrhenomanes*) attack the secondary feeder roots. These fungi grow throughout the root tissues but do *not* produce lesions or decay, although infected roots may be a light buff color.

Identification.   Infection results in slower growth; an off color; stunted, white roots; and a thinning of grass plants. Diseased plants are killed rapidly during hot periods. Mycelium can also be found in the root hairs. Root tips are bulbous and later die. Sporangia and oospores of *Pythium* are rare in infected roots.

Management.   Follow cultural management practices (see table 5–1) that promote vigorous root growth and avoid undue plant stresses.

Hot-weather Pythium foliar blight (called Pythium Red Blight in Japan).
Other names for this disease include spot blight, cottony blight, and grease spot,
depending on the amount of aerial mycelium on the foliage. All cultivated grasses
are susceptible, especially bentgrasses, annual bluegrass, ryegrasses, and tall fescue.

There are 14 or more species of *Pythium* that have been associated with a foliar blight of mature turf. The more important species include *Pythium aphanidermatum, P. aristosporum, P. arrhenomanes, P. graminicola, P. myriotylum, P. torulosum,
P. ultimum,* and *P. vanterpoolii* (see figure 5–61). The symptoms of *Pythium aphanidermatum,* at least on creeping bentgrass, appear to be caused by a metabolite of the
pathogen called IA. Commonly, two or more species may be present and killing
turf at any one time. Since all species produce similar symptoms and signs, the
only way to determine which one or more species of *Pythium* is involved is by
growing the fungus on special laboratory media and examining the sporangia and
oospores microscopically. Pythium blight is often followed by other organisms, including blue-green and green algae, bacteria, and such fungi as *Bipolaris, Curvularia, Drechslera,* and *Exserohilum.* There is also evidence of a close association
between certain species of *Pythium* and certain plant-parasitic nematodes.

Identification. On closely mowed golf and bowling greens shaded or surrounded by trees or in low areas, small, distinct, round to irregularly shaped
sunken spots up to 0.8 to 6 in (2.0 to 15.2 cm) in diameter, sometimes as large as
12 in (30.5 cm), suddenly appear during hot, very wet, calm weather with night
temperatures above 75°F (22.8°C). The grass leaves are at first water-soaked, dark
in color, and slimy to the touch in the early morning. They quickly fade from an
orange-bronze or reddish brown to a light tan or brown as the grass blades dry out
and shrivel (figure 5–64). When the air is moisture-saturated, especially at night or
in the early morning, the watersoaked grass leaves collapse and appear matted together by a fluffy, white-to-purplish-gray, cobwebby mass of fungus mycelium, especially at the borders of the spots; hence the name "cottony blight" (*Color plate 49*).
Clusters of blighted plants may merge to form large, irregular areas, long streaks,
or serpentine patterns of the turf up to 1 ft (0.3 m) or more wide. These patterns develop because spores and mycelium of the *Pythium* fungus (or fungi) are easily
spread by surface drainage water and by mowing when the grass is wet. Pythium
blight is rare on mature Kentucky bluegrass (figures 5–64 and 5–65), fine-leaf fescue, and zoysiagrass lawns but fairly common on tall fescue (figure 5–66) and
bermudagrass in the southeastern United States.

If the growth of the *Pythium* fungus is checked by a sudden drop in temperature or humidity before entire leaf blades are blighted, distinct straw-colored-to-
reddish, girdling lesions may develop on leaves at the edges of patches. These leaf
spots may resemble those of dollar spot (sec. 5.2.8) and Nigrospora blight (sec.
5.2.16), except they lack the brown or reddish-brown borders. Pythium-infected
grass blades commonly twist, collapse, and die (figure 5–65).

Disease Cycle. Species of *Pythium* are water-mold fungi that are commonly
present in diseased turfgrass, thatch debris, and the upper soil level in the form of
delicate, nonseptate mycelium (figure 5–2) and round, thick-walled resting spores
(oospores) embedded in dead grass tissue (figure 5–67).

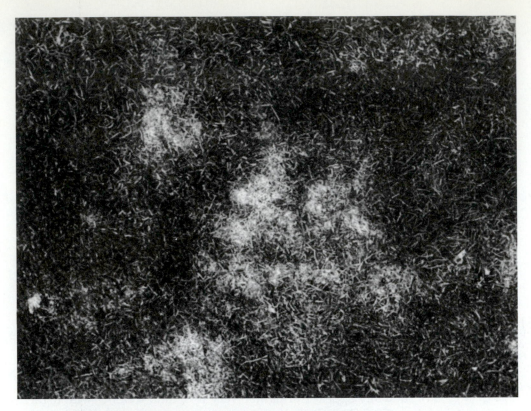

FIGURE 5–64 Pythium blight spots in Kentucky bluegrass turf.

FIGURE 5–65 Pythium blight has caused the collapse of Kentucky bluegrass leaves in this patch.

FIGURE 5–66 Shriveled, dead tall fescue leaves, mostly in the foreground, caused by Pythium blight. (Courtesy L. T. Lucas.)

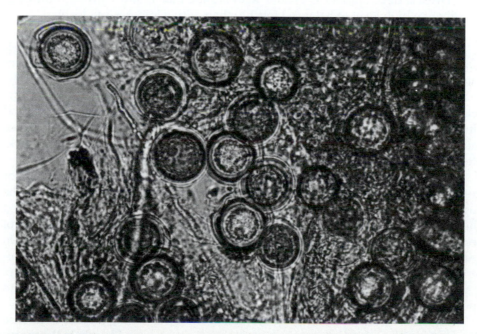

FIGURE 5–67 Oospores of *Pythium* sp. as seen under a high-power microscope. (Courtesy R. C. Avenius.)

Under favorable temperature and moisture conditions, the mycelium resumes rapid growth, while the oospores germinate to produce sporangia (bearing spores; figure 5–61) or hyphae. The sporangia in turn may germinate to produce either motile zoospores or hyphae. All spores can germinate and produce hyphae capable of infecting grass plants directly within an hour or two.

The most important and persistent survival structures are the oospores. Living and dead grass plants that were invaded earlier in the season or during the previous season commonly serve as infection centers. Disease develops rapidly from these centers by means of a cobweblike, hyphal growth of the fungus from leaf to leaf and plant to plant. Spreading of the disease occurs by very rapid growth or when oospores, sporangia, and zoospores, diseased plant parts, or Pythium-infested soil are moved by mowers or other maintenance equipment, shoes, feet of animals, or surface water.

Importance.   Pythium blight can be devastating during hot, wet, or very humid weather on heavy, poorly drained soils, low-lying areas, or near water when the grass is lush (high in sugars) and dense, there is little air movement, and daytime air temperatures are 85 to 95°F (29.4 to 35.9°C), with a minimum of 68°F (20.0°C) at night. At temperatures of 90 to 95°F (32.2 to 35.0°C), large areas of seedling or established turf may be destroyed overnight. Algae often invade diseased patches and may cause a hard, dark crust.

Bentgrass in arid regions is much more susceptible to Pythium blight where soil salinity is high. Under these conditions, *Pythium aphanidermatum* can infect at a lower air temperature and humidity than would normally occur.

Disease development is greater with a thick thatch and with high nitrogen fertility or a high level of balanced fertility than with a low-fertility level.

Management.   Follow the suggested cultural practices outlined in table 5–1. Good surface and subsurface soil drainage and air movement over the turf are very important, as is an extensive and vigorous root system. In high-cut turf, remove excess thatch over ½ in (1.3 cm). Avoid overfertilization with nitrogen, and use slow-release nitrogen sources in a balanced fertility program. Selectively prune trees and shrubs and plan the landscape design to ensure good light penetration and air movement. Golf course superintendents sometimes mount fans in trees near pocketed greens to improve air flow.

Buy top-quality seed, sod, sprigs, or plugs from a reputable dealer. During poor germination conditions, sow seed treated with a fungicide (table 5–5) and seed at the suggested rate. If possible, plant when the weather is cool and dry. The seedbed should be well prepared and fertile. Avoid overwatering, especially from the time of seeding, plugging, or sprigging to seedling emergence or plant establishment. On seedling or established turf, water early in the day so the grass can dry before late afternoon. Water infrequently and thoroughly until the soil is moist 6 in (15.2 cm) deep. If possible, do not use pond water for irrigation, as it is commonly infested with species of *Pythium;* avoid *all* nighttime watering during "Pythium weather."

Avoid mowing when the grass is wet and watering during periods of intense disease activity. On golf greens, remove dew and guttation fluid early in the morning by poling or dragging a hose over the grass.

Do not stress turf from drought, wear, or soil compaction. Reduce the soil pH, especially in the thatch and crown, to pH 6 or slightly below. If topdressing golf greens, apply organic matter to improve thatch degradation and add microorganisms (bacteria and fungi) that compete with species of *Pythium*.

There are *no* known resistant cultivars of any turfgrass species except bermudagrass. Most of the cultivars of the latter grass (such as Everglades, Florida 50, Texas 22, Texturf, Tiffine, Tiflawn, and Tifway) are highly tolerant to Pythium blight when mature.

In southern states, delay overseeding bermudagrass and other warm-season grasses with cool-season grasses until the onset of cool, dry weather. If winter grasses must be established in early fall, applications of a Pythium fungicide or mixture (table 5–5) at seeding time and again at 5- to 21-day intervals during warm, moist periods will largely eliminate disease damage. Annual bluegrass is more seriously damaged on overseeded greens than is perennial ryegrass or other grasses.

During extended periods of hot, wet weather, a fungicide program (table 5–5) may be needed when cultural practices do not check the development of Pythium blight. Although Pythium blight can develop rapidly, it is often not necessary to use a preventive fungicide program to manage this disease. By integrating cultural practices with frequent scouting of the turf during conducive conditions, a turf can be effectively managed. At the first sign of Pythium blight, the initial application of fungicide should include both a contact material and a product with systemic action. Subsequent applications may only need the systemic product. The first spray application should be made when as night temperatures are expected to remain at 65°F (18.3°C) or above; daytime air temperatures are 80°F (26.7°C) or higher; the forecast is for continued wet or very humid (>85%) weather; and the disease is visible. Depending on the fungicide or mixture used, repeat applications are needed at 5- to 21-day intervals as long as the weather stays hot and humid. For the most effective control of Pythium blight on established turf, uniformly spray 1000 sq ft (93 sq m) of turf with the proper fungicide mixture of unrelated chemicals in 5 or more gal of water to ensure wetting the thatch and surface layer of soil.

Disease predictor models (e.g., Envirocaster, sold by the Neogen Corporation) have been developed. Disease is likely to appear at maximum daily temperatures greater than 82°F (27.8°C) and a minimum of 68°F (20.0°C) at night followed by 14 hours or more of relative humidity greater than 90%.

To control Pythium blight on new seedings, apply a fungicide mixture (table 5–5) at the suggested rate immediately after seeding. (Do not use chloroneb [Terraneb-SP, Teremec SP], which may affect seed germination.) Repeat within 5 to 7 days if the soil is wet and environmental conditions remain favorable for disease development. When mixing or applying any fungicide, carefully follow the manufacturer's directions and precautions.

Ramularia Leaf Spot
See the section on minor leaf spots (5.2.16).

Red Leaf Spot
See the section on Helminthosporium diseases (5.2.13).

5.2.22   Red Thread and Pink Patch

Red thread and pink patch are widespread diseases of most cool-season turf-grasses in North America, northern Europe, Japan, Australia, and New Zealand. The fungi that cause these two diseases are commonly found together infecting the same plant, even the same leaf, mostly during cool (65 to 75°F; 18.3 to 23.9°C), damp, and overcast weather coupled with slow growth of the grass. Both diseases, however, may occur during warm, very humid or drizzly, cold summer weather. If abundant surface moisture is present, and even at snow melt in winter or very early spring. Both diseases may be widespread on turfgrasses during mild and wet winter weather. The diseases are most common on low-fertility turf and where applications of growth regulators retard growth.

Red thread is caused by the fungus *Laetisaria fuciformis,* while pink patch is incited by the fungus *Limonomyces roseipellis.* Red thread and pink patch infect colonial, creeping, and velvet bentgrasses, annual and Kentucky bluegrasses, fine-leaf fescues (red, chewings, sheep), hard and tall fescues, annual and perennial ryegrasses, bermudagrass, velvetgrass, quackgrass, redtop, and turf timothy. They are most damaging to perennial ryegrass, the older common types of Kentucky bluegrass, and fine-leaf fescues. Less damage occurs to improved cultivars of Kentucky bluegrass as well as bentgrasses, tall fescue, and bermudagrass. Of the bentgrasses, velvet bentgrass cultivars are more susceptible than colonial and creeping bentgrasses.

Although red thread and pink patch rarely kill plants outright, they do weaken turfgrasses and contribute to the decline and death of grass plants from subsequent stresses or diseases. Red thread is the more important disease, especially of slow-growing, nitrogen-deficient turf. It has even been found on unfrozen turf under the snow. Red thread is also becoming more prevalent on lawns serviced by lawn-care companies and sometimes on golf course fairways.

Pink patch for many years was believed to be a form of red thread. Now it is known to be a distinct but generally minor disease sharing the same turfgrass hosts and most of the same symptoms of red thread. Pink patch rarely develops where red thread is absent. It is much more severe on unmowed or infrequently mowed grasses grown under low-nitrogen fertilization than on highly maintained turfgrasses.

Identification.   Small patches of infected leaves and sheaths appear watersoaked at first when infected by the red-thread and pink-patch fungi. They shrivel and die rapidly, fading to a bleached tan to a light straw color when dry. Death usually progresses from the leaf tip downward. Where infection is severe, diseased turf is yellowed or "scorched" in circular to irregular patches, which are usually 1 to 6 in (2.5 to 15.3 cm) in diameter but may be up to 3 ft (0.9 m) in diameter. Dead leaves are generally interspersed with apparently healthy leaves, giving diseased turf a scorched or ragged appearance (figure 5–68) due to numerous apparently healthy leaves interspersed among the dead leaves. The patches may be scattered within a turf area, or a number may merge to form large, irregular areas of blighted turfgrass with a reddish brown, tan, or pinkish cast. Only the leaves and sheaths are infected.

On the infected parts, the red-thread fungus forms conspicuous, gelatinous mycelial masses colored coral pink, orange, or red. During prolonged periods of

FIGURE 5–68   Red thread in red fescue turf. (Courtesy D. H. Scott.)

overcast, moisture-saturated air, the gelatinous growth may completely cover the leaves. When moisture is abundant, the faintly pink-to-reddish web of mycelium of the red-thread and pink-patch fungi may also mat the leaves and leaf sheaths together (figures 5–69 and 5–70).

The gelatinous red threads of *Laetisaria*, usually ¹⁄₁₆ to over ⅖ in (0.1 to 1 cm) long, are formed by strands of branched hyphae. They often protrude from the cut tips of leaves or connecting leaf blades as needlelike or branched, antlerlike appendages (*Color plate 50*). The bright-coral-pink-to-blood-red mycelial mats harden, become threadlike (red threads) and brittle when they dry, and function as sclerotia.

The gelatinous, pink-to-red mycelial film of *Limonomyces* forms on the leaves when the air is very humid after a prolonged rain. The mycelium tends to form first along the leaf margins bordered by light-green-to-yellow bands of discolored leaf tissue; later, the entire width of the leaf blade develops a pinkish cast. Such affected leaves die back from the tip. The disease is slow developing and rarely causes severe turf thinning. On frequently mowed turf where the clippings are removed, the light green patches that later turn brown rarely reach 2 ft (60.9 cm) in diameter. Where the turf is mowed infrequently or left unmowed, the fungus may kill the plants to the soil line and turn them pinkish. Usually the pink-patch fungus causes only mild, largely superficial symptoms. The fungus survives as mycelium in plant residues; it is also seedborne.

FIGURE 5–69 Close-up of red thread. (Courtesy R. W. Smiley.)

FIGURE 5–70 Pink patch infecting creeping bentgrass cv. 'Penncross.' Note the mycelium of the fungus growing over the grass blades. (Courtesy P. H. Dernoeden.)

The following diagnostic characteristics are to separate pink patch and red thread:

The *red-thread fungus* produces distinctive "threads" that grow past the tips of the leaf blades and also may form cottony pink flocks of arthroconidia. Seen with a microscope, the mycelium is multinucleate and lacks clamp connections.

The *pink-patch fungus* does *not* produce red threads or cottony flocks of spores. Microscopic examination of the mycelium reveals clamp connections (figure 5–2) and binucleate cells.

Disease cycle.   The red-thread fungus overseasons as dried, dormant mycelium (red threads or sclerotia) on the leaves and in the thatch of previously infected plants. If kept dry, the sclerotia remain viable for up to 2 years. When humidity is high, pink, cottony aggregates of hyphae may form in the patches, which, on drying, fragment to form hyaline, irregularly shaped arthroconidia. The fungus is disseminated to healthy turf areas by bits of red threads, arthroconidia, and minute basidiospores, and as dormant mycelium in infected leaf tissue and plant debris. Spreading may occur mechanically on mowers and other turf equipment, or by shoes, wind, and splashing or flowing water. The fungus can infect healthy leaves through the stomates but commonly invades the cut leaf tips. Leaf death can occur within 2 days after infection occurs.

The pink-patch fungus grows saprophytically on turfgrass debris. It overseasons as dormant mycelium in the thatch and as small, reddish, waxy pads of mycelium on the leaves, sheaths, and stems. The *Limonomyces* fungus grows more slowly than the red-thread fungus and is thus more damaging to grass left long or unmowed. The hyphae penetrate the leaves directly. Spread to healthy turf, mostly by turf equipment, occurs largely by diseased leaf tissues and colonized turf debris.

Infection and disease development by both fungi is favored by air temperatures of 65 to 75°F (18.3 to 23.9°C), coupled with prolonged periods of overcast weather, light rains or drizzle, fog, heavy dews, and moisture-saturated air. Slow-growing and poorly nourished turf are generally the most severely attacked. Growth of the fungi and disease development ceases below 33°F (0.6°C) and above 86°F (30.0°C). (Optimum growth of the red-thread fungus in culture is between 57 and 70°F [14.1 to 21.1°C]; for the pink-patch fungus, it is somewhat higher [68 to 74°F; 20.0 to 23.3°C]. However, the pathogens remain alive and continue their activity when conditions again become suitable for disease development.)

Importance.   Red thread is considered a much more damaging disease than pink patch. Infected patches generally are a bleached tan to straw brown or reddish brown. When disease is severe, diseased turf is yellowed or scorched in small-to-large, ragged, irregular patches of all sizes.

Management.   Follow the cultural-control suggestions outlined in table 5–1. Maintain adequate and balanced fertility based on soil tests and the recommended turfgrass fertilization program for your area and the turfgrasses grown. Red thread is *most* severe where potassium, phosphorus, calcium, and especially nitrogen are deficient. Avoid overstimulation with fertilizer, particularly with a quickly available, high-nitrogen product. Water infrequently but deeply early in the day to prevent drought stress. Avoid frequent light sprinklings in late afternoon and early evening.

Increase light penetration and air movement over the turf by selectively pruning nearby trees and shrubs or better arranging of the landscape design.

Where red thread has been a problem in the past, maintain a soil reaction (pH) between 6.5 and 7.0. Test the soil pH, and treat the soil accordingly, if practical.

Mow frequently at the recommended height of cut for the grass or grasses grown. Collecting the grass clippings during periods when the grass is growing slowly and the disease is active may reduce the number of red threads that eventually fall back into the turf.

Where red thread and pink patch have been particularly devastating, apply one of the suggested fungicides (table 5–5) at 10- to 14-day intervals during prolonged, moist weather in the spring and autumn when daytime temperatures average between 65 and 75°F (18.3 to 23.9°C). Start applications when disease is *first* evident. Apply the fungicide in 5 or more gal of water per 1000 sq ft (93 sq m) of turfgrass to ensure proper wetting.

Red-thread-resistant perennial ryegrasses are available, as are fine-leaf fescues, tall fescues, and Kentucky bluegrasses. Check with your county extension (or advisory) office or turfgrass specialist regarding the availability of resistant cultivars.

Rhizoctonia Blight, Rhizoctonia Brown Patch, Rhizoctonia Leaf and Sheath Blight, Rhizoctonia Large Patch

See the section on Rhizoctonia diseases (5.2.23).

5.2.23   Rhizoctonia Diseases

Rhizoctonia diseases of turfgrasses are caused by the common soil-borne fungi *Rhizoctonia solani, R. cerealis, R. oryzae,* and *R. zeae.* One or more of these cosmopolitan, soilborne fungi are present in practically all soils throughout the world. The fungi feed equally well on living plant tissues or on organic matter present in the thatch and soil. As soil inhabitants, they compete well with other saprophytic microorganisms. The fungi are composed of a large number of strains or races that attack more than 1800 different species of plants, including most ornamentals, vegetables, and field crops.

All species of *Rhizoctonia* can live in the soil as saprophytes, but not all can attack a living turfgrass plant. Even within a species like *R. solani,* some strains cause severe brown patch while others cause very little disease. In general, *Rhizoctonia* fungi, because of their saprophytic ability, can be isolated easily and maintained on laboratory culture media.

Symptoms of turfgrasses infected by species of *Rhizoctonia* vary widely and are easily confused with the symptoms of diseases produced by other pathogens. The symptoms vary with the specific combinations of turfgrass cultivar or species, height of cut, soil and air environmental conditions, and the specific species and strains (or races) of *Rhizoctonia.*

One or more species of *Rhizoctonia* infect all known turfgrasses, causing foliar blights as well as seed rot and seedling blights. The most susceptible turfgrass species appear to be annual bluegrass, bentgrasses, bermudagrass, tall fescue, ryegrasses, St. Augustinegrass, and zoysiagrass when grown at its northern limits of adaptation.

Species of *Rhizoctonia* produce several forms of hyphae that vary with the age of the hyphae. Diagnosticians with access to a good compound light microscope distinguish species of *Rhizoctonia* by the mature hyphae, which usually branch at right angles for *R. solani*, *R. zeae*, and *R. cerealis*. Young hyphae of *R. oryzae* branch at an acute angle. The hyphal branches of all species are somewhat constricted where they originate, and a septum separates the hyphal branch from the parent hypha close to its point of origin (figures 5–2 and 5–71).

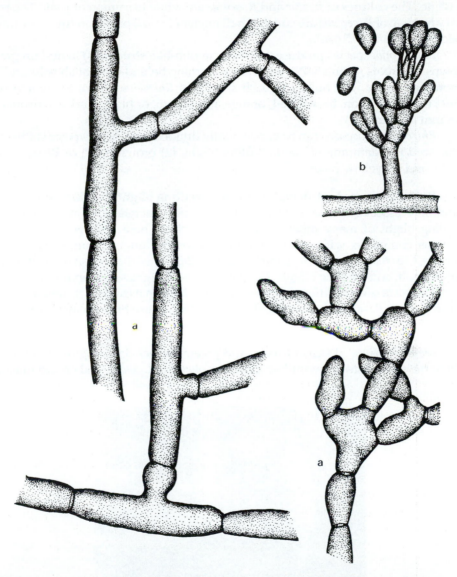

FIGURE 5–71 *Rhizoctonia solani*, teleomorph *Thanatephorus cucumeris*, is a cosmopolitan soil inhabitant that causes a variety of diseases on over 1800 kinds of plants. (a) Young mycelium (left) of *Rhizoctonia* and older mycelium (right). Note the long and wide hyphal cells, near right-angled branching, slight constriction at the branch junctions, and the septa near the main hypha. Older mycelial cells tend to be barrel shaped. (b) Mycelium with several immature and one mature basidia bearing four basidiospores (*Thanatephorus* state). Two basidiospores have been discharged. (Sketches presented as viewed through a high-power microscope.)

Trying to distinguish between the species of *Rhizoctonia* is difficult and requires special staining procedures to assess the nuclear condition of hyphal cells. The mycelial cells of *R. solani* and *R. zeae* mostly contain an indefinite number of nuclei (multinucleate); mature cells of *R. oryzae* usually contain four nuclei per cell; and mycelial cells of *R. cerealis* have two nuclei. When grown on laboratory media, mature cultures of *R. solani* are usually some shade of brown, and *R. cerealis* are buff to white. The cultures of *R. zeae* and *R. oryzae* are white to salmon or pink. The primary hyphae of *R. cerealis* are smaller in diameter (2 to 6.2 µm) than the other three species (5 to about 15 µm).

Rhizoctonia species produce bulbils (sclerotia-like structures) in and on grass crowns and roots (figure 5–72), sometimes forming behind the leaf sheaths and in laboratory media. The bulbils of each species of *Rhizoctonia* vary in color (from white to buff, salmon, brown, red, orange, red-brown, or black) and size (from 0.1 to 5 mm in diameter).

Rhizoctonia diseases can be conveniently divided into three types: (1) Rhizoctonia seed rot, damping off, and seedling blight; (2) brown patch or Rhizoctonia blight; and (3) yellow patch.

Rhizoctonia seed rot, damping off, and seedling blight.   *Rhizoctonia* spp. are commonly associated with seed rot, pre- and postemergence damping off, and seedling blight of many different turfgrass species. These pathogens commonly occur as complexes with other seed and soil-borne fungi, including species of *Bipolaris, Curvularia, Drechslera, Fusarium,* and *Pythium* (see the section on seed decay, damping off, and seedling blights ([5.2.26]) to cause turfgrass seed and root rots).

There are several different species of *Rhizoctonia* that cause turfgrass diseases. These include *R. solani; R. oryzae* and *R. zeae,* which cause brown patch or Rhizoctonia blight: and *R. cerealis,* the cause of yellow patch.

Identification.   Decay of seeds and preemergence damping off is common where the soil is infested with *Rhizoctonia,* especially *R. solani,* and environmental

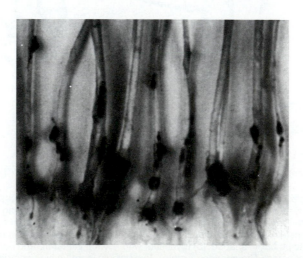

FIGURE 5–72   Bulbils of *Rhizoctonia solani* on creeping-bentgrass seedlings.

FIGURE 5–73   Brown patch or Rhizoctonia blight infecting seedlings of tall fescue. (Courtesy L. T. Lucas.)

conditions favor growth of the pathogen over the turfgrass plants; that is, the turf matures slowly. Seedlings that manage to emerge have a brown rot at the soil surface followed by a withering and "pinching" of the shoot (sometimes called wirestem) that causes seedling plants to collapse and turn light brown (postemergence damping off; see (figures 5–73 and 5–86). Less severely diseased seedlings may develop eyespot lesions at or close to the soil line.

Disease Cycle.   Bulbils (sclerotia) of *Rhizoctonia solani* can be mixed in with bentgrass and ryegrass seed. When these bulbils germinate, they may serve as a source of inoculum, but invasion of seed and seedling tissues is much more likely to result from thick-walled mycelium, bulbils, and monolioid cells that survive in the soil. The *Rhizoctonia* fungi overseason as mycelium and bulbils on organic residues and in soil, optimum temperature for seedling infection is 60 to 75°F (15.6 to 23.9°C). Infected seeds provide nutrients that allow the mycelium to grow and attack adjacent seeds and seedlings. *Rhizoctonia* spp. generally invade turfgrass tissues by direct penetration of epidermal cells and root hairs.

Rhizoctonia solani is believed to be the dominant species of *Rhizoctonia* where turf is grown. It attacks all turfgrasses. *R. zeae*, which grows best at a higher temperature (optimum 91°F [32.8°C]) than *R. solani*, is often associated with brown patch or Rhizoctonia blight in the southeastern United States. *R. oryzae* is common in subtropical regions, but may attack turfgrasses in temperate and transitional climatic zones. *R. solani* will grow at any soil pH, temperature, and moisture level that will support the growth of turfgrasses.

Importance and Management.   See the section on seed decay, damping off, and seedling blights (5.2.26).

Brown patch or Rhizoctonia blight, Rhizoctonia leaf and sheath blight, large brown patch of zoysiagrass.   Brown patch, caused by anastomosis group (AG) AG-1 of *R. solani,* is common in dense, highly fertilized, cool-season turfgrasses during extended periods of hot (above 85°F; 29.4°C), moist, overcast weather when the temperature at night is above 68°F (20.0°C) and the leaf surfaces are covered with water for 8 hours or more. Under these conditions, large areas of turf may be involved in less than 24 to 48 hours. Bentgrasses, annual bluegrass, fine-leaf and tall fescues, and perennial ryegrass can be severely damaged. Colonial and velvet bentgrasses are more susceptible than creeping bentgrass. Bentgrasses cut at golf-green height (⅛ to 3⁄16 in; (0.3 to 0.8 cm) and grown under a high level of maintenance are particularly susceptible to brown patch. Tall fescue and perennial ryegrass are the most seriously damaged amenity turfs. Perennial ryegrass recovers much more quickly than tall fescue.

Warm-season grasses are infected by AG-2-2 of *Rhizoctonia solani.* Disease is most severe from spring through fall usually at temperatures between 80 and 85°F (26.7 and 29.4°C) and inactive above 90°F (32.2°C). St. Augustinegrass, centipede-grass, zoysiagrass, carpetgrass, and hybrid bermudagrass are damaged more than other grasses. High nitrogen levels and thatch buildup favor disease.

Identification.   On *closely mowed turf* such as golf and bowling greens, the disease appears as roughly circular brown patches that vary in size from 1 in (2.5 cm) to 2 or 3 ft (0.6 to 0.9 m) in diameter (*Color plate 51*). Infected grass blades first appear watersoaked and dark purplish green to black; they rapidly dry, wither, and turn light brown. The dead leaves remain upright. A dark-purplish-to-grayish-brown-or-black "smoke ring," ⅜ to 2 in (1.0 to 5.1 cm) wide and composed of wilted, recently infected grass blades, may mark the advancing margin of a patch (figure 5–74) under warm-to-hot summer conditions. This ring, composed of watersoaked grass, sometimes webbed with mycelium, is most noticeable when the air is very humid or during the early morning; it disappears as the grass dries. No smoke ring is evident on warm-season grasses infected by AG-2 of *R. solani.* Usually, only the leaf blades are killed. The fungus is incapable of killing the crowns and roots. After several mowings, new but thinned-out grass appears in the affected areas. Algae often invade diseased patches and may cause a hard, dark crust. Brown patch also may occur together with Pythium blight.

On *amenity* (higher-cut, cool-season) *turf,* such as home and industrial lawns, parks, athletic fields, and golf course fairways, diseased patches may be roughly circular to irregular, dull tan to light brown, and up to about 12 in (30.5 cm) in diameter (figure 5–73). The patches may enlarge up to 3 ft (0.9 m). The grass with these patches may be matted or slimy. Grass in the center may appear less affected, leading to the formation of a ring- or doughnut-shaped patch. The patches may coalesce to form irregular areas of blighted turf up to 50 ft (15 m) in diameter that often appear sunken. Dieback from the leaf tips is common.

FIGURE 5–74   Brown patch on creeping bentgrass. Note the "smoke ring" at the margins of the patches.

On *warm-season grasses*, circular to irregular, straw-colored brown patches up to 20 ft (6.1 m) or more in diameter may develop (figures 5–75, 5–76, and 5–77) at any time during the growing season. Disease severity can be extreme during hot, humid weather. At such times the patches may develop dark purplish borders (bright orange on zoysiagrass) 2 to 6 in (5.1 to 15.2 cm) wide. In some cases a ring pattern is not seen. Diseased areas may appear doughnutlike as grass recovers in the centers of the circles. Patches may appear yellow when the turf begins growth after winter dormancy. On zoysiagrass, bermudagrass, centipedegrass, carpetgrass, and St. Augustinegrass, the lower parts of the leaf sheaths, leaves, and stems may develop a soft, dark-brown-to-purplish rot that turns reddish brown with the return of drier weather. The rot

FIGURE 5–75 Brown patch (*Rhizoctonia solani*) seriously affecting St. Augustinegrass turf. During prolonged humid weather, the patches may develop dark purplish borders up to 6 in (15.2 cm) wide. (Courtesy The Scotts Company.)

FIGURE 5–76 Brown patch, caused by *Rhizoctonia zeae,* infecting bermudagrass on a croquet court. (Courtesy Monica L. Elliott.)

FIGURE 5–77   Large brown patch of zoysiagrass in a golf course fairway as it appears in the spring (above) and in the fall (below).

causes the upper leaf blades to wilt, turn yellow to orange or reddish, die, are easily pulled up, and sometimes detach . Large areas of such affected turf may be thinned and/or killed (figures 5–75, 5–76, and 5–77). Extensive death is common on St. Augustinegrass growing in shady, wet areas or under conditions of prolonged high humidity. Irrigated warm-season turfs, or cool-season turfs grown in warm, humid climatic regions, are especially prone to severe brown patch.

Symptoms of brown patch in St. Augustinegrass may be confused with chinch-bug damage. Chinch-bug-damaged areas, however, are not as well defined and occur mostly in sunny turf areas. In addition, chinch bugs do not cause expanding leaf lesions with a red-brown margin.

A foliar blight is common on centipedegrass, and St. Augustinegrass, causing light tan lesions on the leaves, especially on leaf blade tips. Optimum nighttime temperatures of 78°F (25.6°C) or more with daytime air temperatures from the low 80s to low 90s (26.7 to 33.0°C) favor disease development on cool-season grasses.

Mildly severe brown patch on cool-season grasses generally recovers in 2 or 3 weeks. When disease is severe, however, the crowns, rhizomes, and stolons may turn brown and rot. Such turf is thinned or killed out in large areas.

On higher-cut Kentucky bluegrass, perennial ryegrass, tall fescue, centipede-grass, zoysiagrass, and other grasses, leaf and sheath lesions (called Rhizoctonia leaf and sheath blight) may also result from infection by *Rhizoctonia solani, R. zeae,* and *R. oryzae*. The symptoms vary with the type of turfgrass, height of cut, and environmental conditions. Elongated to irregular watersoaked spots appear first. The center turns a straw or ash gray to light chocolate brown and is surrounded by a narrow, dark reddish brown or dark brown border (*Color plate 52* and figure 5–78). The size of the lesion varies with the turfgrass species, ranging from a large spot

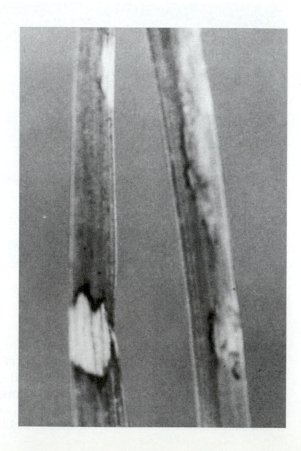

FIGURE 5–78   *Rhizoctonia solani* lesions on two leaf blades of tall fescue. The irregular dark borders may be wine red. (Courtesy L. T. Lucas.)

(up to 10 mm long) on tall fescue to small lesions (about 1.5 mm) on bentgrass and bermudagrass. Turfgrass species on which such leaf lesions have been observed include bentgrasses, bermudagrass, bluegrasses, perennial ryegrass, tall fescue, St. Augustinegrass, and zoysiagrass.

Large brown patch of zoysiagrass.   Figure 5–77 shows an unusual type of brown patch that develops on zoysiagrass in transitional climatic regions of the United States, Japan, and Taiwan. In addition, the *Rhizoctonia solani* most often associated with the disease is type AG-2-2, which is different than the mating types commonly isolated from brown patch of cool-season grasses. Large brown patch also develops during the fall and spring of the year in the United States as the turfgrass is entering or breaking winter dormancy and air temperatures are between 50 and 77°F (10.0 to 25.0°C) and the soil is moist. Patches may also develop in shaded areas during unusually cool, moist, midsummer weather. In the spring, the disease appears as large patches 1 to more than 8 m in diameter with few healthy shoots that grow slowly. As the temperature rises much above 75 to 80°F (23.9 to 26.7°C), disease development is suppressed, and zoysiagrass in the patch produces new shoots and eventually repairs the diseased turf, but this may take several months. If cool temperatures and wet soil persist for several weeks or more, the large, roughly circular brown patches expand rapidly (more than 10 cm a week) and develop an orange border several inches thick. The patches may merge to blight large areas of turfgrass.

In autumn, when cool, wet conditions return but before the zoysiagrass becomes dormant, large brown patch may again become active. Patches observed in the fall develop bright orange margins and blighted sheaths. Large brown patch may or may not have been evident in such turf the previous fall. The disease symptoms now consist primarily of roughly circular rings of matted, bright orange turfgrass 3 to 8 in (7.6 to 20.3 cm) thick. The center of each patch fades to tan or dull brown and appears unaffected or slightly thinner (figure 5–77). Small, watersoaked, reddish-brown-to-black lesions form on the leaf sheaths but not on the leaf blades, stolons, or roots. Optimal infection of leaf sheaths occurs at 68 to 77°F (20 to 25°C).

In general, this disease does not result in the death of the turf but instead kills the leaves and some crowns, resulting in a progressive thinning that makes the turfgrass unsightly. Zoysia patch, caused by *Gaeumannomyces incrustans*, is often confused with large brown patch. The foliar symptoms of zoysia patch are very similar to those of large brown patch, but in the case of zoysia patch there will develop a concurrent root rot. There is also evidence to suggest that zoysia patch and large brown patch can develop in the same turf. *Ophiosphaerella herpotricha* can also be isolated infrequently from diseased zoysiagrass stolons and roots.

Disease Cycle.   *Rhizoctonia* fungi overseason in plant debris and soil in the form of hard, often rounded, dark-brown-to-black resting bodies (bulbils) up to 5 mm in diameter (figure 5–72). The bulbils occur largely in the thatch, diseased grass tissues and the top ½ in (1.3 cm) of soil associated with organic debris. Bulbils (sclerotia) of *R. zeae* differ from other species of *Rhizoctonia* in being a distinctive orange in color. They are also uniquely round and ½₂ to ⅟₁₆ in (0.1 to 0.3 cm) in diameter. They can be found at the bases of blighted plants. Bulbils, like sclerotia, are extremely resistant to heat, cold, droughts, and fungicides. Each bulbil may germinate, cause infection up to 30 times or more, and survive in soil for a number of years.

The teleomorphic state of *Rhizoctonia solani* (*Thanatephorus cucumeris;* figure 5–71) is uncommon and usually found on the lower leaf surfaces of Kentucky bluegrass, perennial ryegrass, tall fescue, crabgrasses, common dandelion, plantains, and white clover. It appears as easily removable white patches.

During moist periods when the soil-thatch temperature is 46 to 104°F (7.8 to 40°C) with an optimum of 82°F (27.8°C), the bulbils may germinate by sending out hyphae radially through the upper soil and thatch until turf plants are intercepted. The result is a roughly circular spot of diseased grass. The optimal temperature for infection and disease development varies with the different Rhizoctonia species and strains from 70 to 90°F (21.1 to 32.2°C). The hyphae penetrate and infect the grass plants through leaf pores (stomates), directly through leaf and stem tissue, or via mowing wounds. The lower, older leaves that touch the soil or turf mat are the first ones attacked. The fungus grows up and over the grass leaves and sheaths and throughout the leaf tissues. The spread from one grass blade to another on closely mowed turf commonly occurs through droplets of dew or guttated water exuded at the cut leaf tips. Nutrients and organic compounds (primarily amino acids and sugars) in the guttated water stimulate rapid fungus growth.

As with bulbils, the mycelium within grass clippings or thatch may resume growth and initiate infections as long as temperature and moisture conditions are favorable. Infection and disease development are comparatively slow for *Rhizoctonia solani* at air temperatures below 70 to 75°F (21.1 to 23.9°C). When the air temperature is 80 to 85°F (26.7 to 29.4°C) or somewhat above and the air is moisture-saturated, the fungus grows rapidly. Large areas of turf can become completely blighted. The pathogenic activity of *R. solani* ceases when the air temperature reaches about 90°F (32.2°C). Temperatures from about 85 to 100°F (29.4 to 37.8°C), with an optimum about 90°F (32.2°C), are ideal for growth of *R. zeae* and *R. oryzae*. The optimal temperatures for infection and disease development vary, however, for the different strains or races of these species.

Although brown patch may occur at low humidity levels, warm-to-hot nighttime temperatures in the 70s (21.1 to 26.1°C) and daytime temperatures in the mid-80s (28.9 to 30.6°C) or above; rainy weather; or a saturated atmosphere greatly accelerate disease development.

The severity of the disease is greatest in lush, succulent turf that has been highly fertilized with nitrogen and watered especially in late afternoon and night. Turfgrasses are more susceptible when grown at a moderate to high fertility level than at a low level of nitrogen fertilization. Resistance increases when the levels of phosphorus, and especially potassium, are increased.

Four conditions are necessary for brown patch to develop on cool-season grasses: the presence of the fungus in an actively growing state; a dense and well-watered stand of grass; prolonged periods of dew or the presence of a film of moisture on the lower foliage for 8 hours or longer with a relative humidity of at least 95%; and a temperature of 70 to 95°F (21.1 to 35.0°C) for at least several hours.

If any of these conditions are lacking, the attack of brown patch or Rhizoctonia blight will not be severe.

Importance.   Species of *Rhizoctonia* may cause turf to be thinned or killed out in large areas from a crown, stem, rhizome, and stolon rot. More commonly, there is a destruction of the leaves with affected turf recovering in 2 to 3 weeks.

Management.   Brown patch can usually be controlled by following the cultural practices outlined in table 5–1. Good surface and subsurface drainage are essential. Avoid overseeding. Keep the turf mowed to the desired height, and mow frequently enough to prevent the lower leaves from remaining wet during much of the day. Avoid applying nitrogen fertilizer when the disease is active. High levels of nitrogen and low levels of phosphorus or potassium contribute to disease severity. Irrigate early in the day, and increase air movement to the turf by selectively pruning or removing dense trees and shrubs. Reduce thatch when it is more than ½ in (1.3 cm) thick. Golf course superintendents remove the dew and guttated water on golf greens in the early morning by brushing the grass with long, limber bamboo poles or dragging a mat, hose, or rope over the turf. Avoid fall nitrogen applications on St. Augustinegrass and centipedegrass to reduce the risk of winter outbreaks of brown patch or large brown patch.

No species of turfgrasses are known to be highly resistant to brown patch. Perennial ryegrass and tall fescue cultivars and selections of St. Augustinegrass are reported to have some resistance. Check with your county extension (or advisory) office or turfgrass specialist regarding the desirability of growing these cultivars in your area.

When cultural practices do not check the development of brown patch, a preventive fungicide-spray program (table 5–5) may be needed. This is especially true of grass cut at golf-green height where a history of disease exists.

The first fungicide application on cool-season grasses should be made when the temperature at night is expected to remain at 66 to 70°F (18.9 to 21.1°C) or above, the daytime temperature 82°F (27.8°C) or above, and the air is near the saturation point for 8 hours or longer. Repeat applications are needed at 5-day intervals during hot, humid weather. When the turf receives over 1.5 in (3.9 cm) of water in a week as rain or irrigation, the interval between applications should be shortened to 3 days. If night temperatures remain below 64°F (17.8°C), extend the spray interval to 10 days. Where feasible, the fungicide should be applied in late afternoon or early evening, when the temperature is 80°F (26.7°C) or lower.

For warm-season grasses, apply the first spray when the disease first appears in autumn and continue until the grass goes dormant. Commence again as the plants start to green up in the spring.

For the most effective fungicidal control of brown patch, apply the fungicide using 5 gal of water per 1000 sq ft (93 sq m). Use the lower fungicide rates listed on container labels in a routine *preventive* program; use the higher rates for a *curative* program, when disease is evident. It is strongly suggested that one or more systemic fungicides be alternated or combined with one of the other protective-contact chemicals listed in table 5–5 to aid in avoiding future fungicide-resistance problems. The use of any one of these products at the recommended rate per 1000 sq ft (93 sq m) usually gives effective control for at least 2 weeks.

When brown patch and Pythium blight are active coincidentally, alternate or combine two or more of their respective fungicidal controls (table 5–5).

Since fungicide use and restrictions are subject to change without notice, always carefully read and follow all the manufacturer's directions and precautions.

Environment-based forecasting systems and the use of Rhizoctonia immunoassay (ELISA) kits can assist in the diagnosis of the disease, but accurate prediction is complicated by variability of the pathogen. A disease prediction system (Envirocaster, sold by Neogen Corporation) has been developed as an aid in the timing of fungicide applications. The biological control agent *Trichoderma harzianum* is reported to suppress *Rhizoctonia solani.*

Yellow patch.   Yellow patch or Rhizoctonia yellow patch, sometimes called winter brown patch, is caused by *Rhizoctonia cerealis.* Like *R. solani,* it is a common soil-borne fungus that infects mainly bluegrasses and bentgrasses (figures 5–79 and 5–80). The fungus also attacks ryegrasses, tall fescue, bermudagrass, zoysiagrasses, and probably other turfgrasses. In Japan, where it attacks zoysiagrasses, it is called Rhizoctonia spring dead spot and Rhizoctonia patch where brown patches 12 to 20 in (30 to 50 cm) in diameter occur in golf greens and tees and more irregularly shaped patches in higher-cut zoysiagrass. Disease attacks occur from autumn to early spring during prolonged, wet conditions when air temperatures are to 68°F (5.0 to 20.0°C). Fungal infections are usually superficial in that the grass crowns, stolons, and rhizomes are infected and stressed but usually not killed.

Identification.   Yellow patch is commonly seen on cool-season grasses as light yellow, tan, straw- or bronze-colored, concentric rings, crescent-shaped patches, or

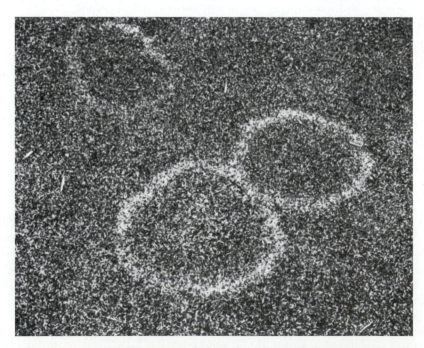

FIGURE 5–79   Yellow patch (*Rhizoctonia cerealis*) on a creeping-bentgrass golf green. (Courtesy D. H. Scott.)

FIGURE 5–80   Yellow patch infecting a Kentucky bluegrass lawn. (Courtesy R. E. Partyka.)

streaks ranging from 1 in (2.5 cm) to about 3 ft (0.9 m) in diameter. The larger patches often have apparently healthy green grass in the center of the circles (figures 5–79 and 5–80). Where numerous, the patches may coalesce to involve large turfgrass areas. Smaller patches usually result from infections that occur under cold, wet conditions. The patches often have a distinctly sunken appearance on cool-season grasses due to rapid decomposition of the thatch. On warm-season grasses, the disease is usually only seen as leaf yellowing. Individual leaves of both cool- and warm-season grasses turn yellow to light tan, starting at the tips. Mottled, grayish tan lesions with a darker brown border may develop on the lower parts of leaves. The leaves of infected Kentucky bluegrass plants near the margins of diseased patches frequently have a characteristic reddish or reddish purple appearance (*Color plate 53*) that begins at the leaf tip. Damage is usually superficial, but thinning of turf may occur during prolonged cool, wet weather. In cool, humid weather, white-to-tan mycelium commonly appears on the crowns, outer leaf sheaths, and lower leaves. The crowns and roots of cool-season grasses may later turn brown or black and eventually kill plants. No smoke rings form in diseased turf, and leaf lesions are rare. Turf affected by yellow patch often takes several months to recover.

The symptoms of yellow patch may resemble pink snow mold, summer patch, and necrotic ring spot. Attacks of summer patch and necrotic ring spot, however, occur in hot weather, while pink snow mold is usually evident from late autumn into midspring.

Disease Cycle.   *Rhizoctonia cerealis* survives from year to year, much as does *R. solani* and other species of *Rhizoctonia*, primarily in the form of minute, dark-brown-to-black bulbils (0.25 to 2 mm in diameter) or as thick-walled mycelium or monilioid cells in the thatch, diseased grass tissues, and soil near the surface. The *Rhizoctonia* fungus is spread mostly by mechanical means.

Importance.   The rings, often with green grass in the centers, commonly appear sunken in cool-season grasses. They are unsightly, but the damage is usually superficial. Thinning of the turf may result when stem and crown infections kill plants. Affected areas commonly remain light yellow for several weeks until the weather turns warm and dry. All plants in these areas may completely recover.

Management.   If yellow patch is serious, applications of nitrogen fertilizer should be reduced in the fall and use a slow-release form of nitrogen. Balanced fertility (N-P-K) speeds recovery of the grass when climatic conditions are drier and either cooler or warmer. Once the unaffected turf is growing, cultural practices, including fertilization, should be used to encourage vigorous regrowth into diseased patches. Do not let the root zone dry out during the winter.

Fungicide applications have *not* been completely successful in controlling yellow patch. Two or more preventive sprays of a DMI or sterol-inhibiting fungicide or ProStar (tables 5–4 and 5–5) have given good results when applied in the fall. Applications need to be implemented *before* infections occur and symptoms appear in fall. No fungicide or combination of cultural practices are known that prevent the formation of yellow rings and patches.

Rhynchosporium Scald or Leaf Blotch
See the section on minor leaf spots (5.2.16).

Root Dysfunction
See the section on Pythium diseases (5.2.21).

5.2.24   Rusts

All turfgrasses are attacked worldwide by one or more rust fungi in the genera *Puccinia, Uromyces, Physopella* (table 5–3), *Phakopsora,* and a single species of *Uredo* in China. Bentgrasses are usually not affected. Some of these rust fungi may cause severe damage on very susceptible species and cultivars, especially when grown for seed production in the Pacific northwest of North America and elsewhere.

Rust fungi are obligate parasites and infect only living grass plants. Two or more rusts may attack the same grass plant at the same time. Grass plants are commonly infected under stressful growing conditions, when water, fertility, and soil compaction are less than adequate for good growth. Most rust problems occur in mid to late summer and autumn on Kentucky bluegrass, perennial ryegrass, tall fescue, and zoysiagrass when grass growth has stopped. These diseases are found wherever susceptible grasses are grown.

Most rusts do not usually become a growth-limiting problem except during extended, warm-to-hot, humid, but droughty periods when grass is growing slowly or not at all, and nights are cool with heavy dews.

TABLE 5–3   COMMON RUST FUNGI THAT INFECT CULTIVATED TURFGRASSES

Rust fungus	Turfgrasses infected
Puccinia	
brachypodii var. *arrhenatheri*	Bluegrasses, fescues
brachypodii var. *poae-nemoralis*	Bentgrasses, bluegrasses, fescues, turf timothy
chaetochloae	Bahiagrass, saltwater couch (*Paspalum notatum*)
cockerelliana	Fescues
coronata (crown rust)	Bentgrasses, bluegrasses, fescues, ryegrasses, turf timothy
crandallii	Bluegrasses, fescues
cynodontis	Bermudagrasses
dolosa var. *circumdata*	Bahiagrass, saltwater couch
festucae	Fescues
graminis (stem rust)	Bentgrasses, bermudagrasses, bluegrasses, fescues, ryegrasses, weeping alkaligrass, wheatgrasses
graminis var. *graminicola* (stem rust)	Bentgrasses, fescues, ryegrasses, turf timothy, wheatgrasses
hordei	Perennial ryegrasses
kansensis	Buffalograss (*Buchloë dactyloides*)
levis	Bahiagrass, saltwater couch
montanensis	Annual ryegrass, crested wheatgrass
pattersoniana	Annual ryegrass, wheatgrasses
piperi	Meadow and tall fescues
poarum	Bentgrasses, bluegrasses, fescues, turf timothy
pygmaea	Bentgrasses, fescues
recondita (leaf rust)	Bentgrasses, bluegrasses, fescues, perennial ryograss, crested wheatgrass, weeping alkaligrass
sessilis	Fescues
stenotaphricola	St. Augustinegrass, kikuyugrass (*Pennisetum clandestinum*), weeping alkaligrass
striiformis (stripe rust)	Bentgrasses, bluegrasses, fescues, turf timothy, perennial ryegrass, weeping alkaligrass, crested wheatgrass
substriata	Bahiagrass, kikuyugrass, saltwater couch
zoysiae	Zoysiagrasses
Phakopsora apoda	Kikuyugrass
Physopella compressa	Bahiagrass, tropical carpetgrass (*Axonopus compressus*)
Uromyces	
dactylidis	Bluegrasses, fescues
dactylidis var. *poae*	Bentgrasses, bluegrasses, weeping alkaligrass
fragilipes	Bentgrasses, wheatgrasses
setariae-italicae	St. Augustinegrass, kikuyugrass
tenuicutis	St. Augustinegrass

Sources: Adapted from Cummins (1971); Smiley, Dernoeden, and Clark (1992); and *Fungi on Plants and Plant Products in the United States* (1989), Farr et al. For illustrations of the urediniospores and teliospores of many of these fungi, see figures 5–83 and 5–84.

The most serious rust diseases are stem rust and leaf rust on Kentucky bluegrass, crown rust on perennial ryegrass, rust on zoysiagrass, and bermudagrass
rust. Stripe or yellow rust (*Color plate 54*) can be a serious problem during the cool,
wet months of winter, early spring, and fall in northern humid areas of Europe, the
northwestern United States, and other northern regions with a similar climate. The
pustules of stripe rust are a bright golden yellow and occur in long yellowish
streaks (*Color plate 54*) on the leaf blades, sheaths, and stems.

Identification.   A continuous, heavy rust infection causes many grass
blades to turn yellow to brown, wither, and die. Such turf may be yellowish,
thinned, weakened, and also made more susceptible to winterkill, drought, weed
invasion, and other diseases.

During early infection, a close examination of the grass blades, leaf sheaths,
and stems will show small, light yellow flecks. These soon enlarge and turn yellow.
In several days, the leaf epidermis ruptures and tears away to expose the round,
oval, or elongated powdery, spore-filled pustules (uredinia). These pustules may
be reddish to chestnut or cinnamon brown, brownish yellow, bright orange, orange-
yellow, or lemon yellow (figures 5–81 and 5–82), depending on the species of rust.
The powdery material rubs off easily on equipment, hands, shoes, clothing, and
animals. Severe, rust-affected leaves or even entire plants may turn yellow (orange
on zoysiagrasses), wither, and die. Affected turf becomes weakened, stunted, thin,
and unsightly.

FIGURE 5–81   Stem rust infecting a leaf
blade of Merion Kentucky bluegrass.
(Courtesy The Scotts Company.)

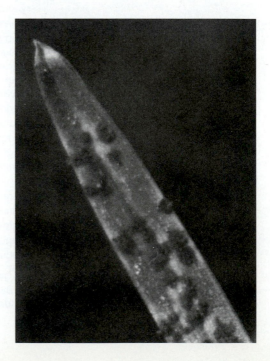

FIGURE 5–82   Leaf rust infecting three tall fescue leaves. (Courtesy L. T. Lucas.)

Severe rust development for several rust fungi is favored by poorly nourished and droughty turf, close mowing, shade, 4 to 8 hours of low light intensity with temperatures of 66 to 86°F (17.3 to 30°C), high humidity, heavy dews or light rains, followed by 8 to 16 hours of high light intensity and temperatures of 68 to 86°F (20.0 to 30.0°C), with slow drying of leaf surfaces. The cool-season crown-rust fungus (*Puccinia coronata*) is favored by temperatures of 50 to 68°F (10.0 to 20.0°C).

Disease cycle.   The cycle of development for these rust fungi is very complex because of the many species involved (about 40; see table 5–3 and figures 5–83 and 5–84) and the numerous alternate hosts, most of which are woody shrubs and herbaceous ornamental plants.

The powdery material that rubs off is composed of tremendous numbers of yellow, orange, brownish yellow, dark brown or brick red microscopic spores (urediniospores, urediospores or uredospores; (figures 5–83 and 5–84), the reproductive structures of the rust fungi. A single pustule (figures 5–81 and 5–83) may contain 50,000 or more spores, each capable of producing a new pustule. These spores are readily transported by air currents, water, shoes, turf equipment, and infected sod, plugs, or sprigs. Some spores land on susceptible leaf tissue, where, in the presence of moisture, they germinate and infect by developing germ tubes that penetrate the leaves and sheaths through open stomates within a few hours. The spores of *Phakopsora apoda* infect kikuyugrass leaves directly through the cuticle.

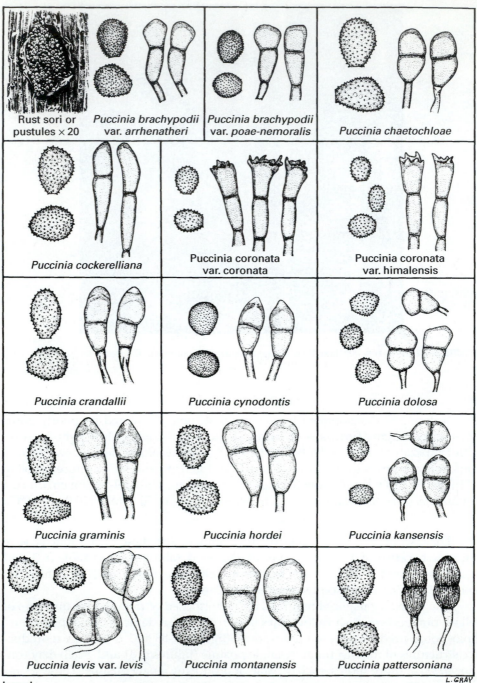

FIGURE 5–83 Fifteen species of *Puccinia* that cause rust diseases of turfgrasses; for each species the urediniospores are on the left and the teliospores are on the right. (After Cummins [1971].)

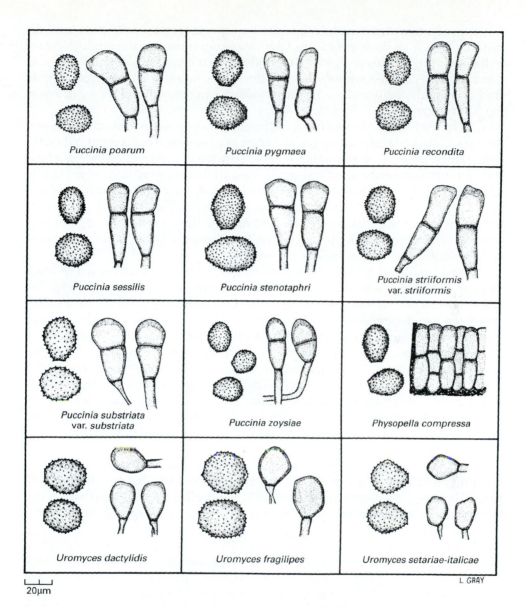

Puccinia poarum

Puccinia pygmaea

Puccinia recondita

Puccinia sessilis

Puccinia stenotaphri

Puccinia striiformis
var. striiformis

Puccinia substriata
var. substriata

Puccinia zoysiae

Physopella compressa

Uromyces dactylidis

Uromyces fragilipes

Uromyces setariae-italicae

20μm

L GRAY

FIGURE 5–84 Twelve species of fungi that cause rust diseases of turfgrasses; for each species the urediniospores are on the left and the teliospores are on the right. (After Cummins [1971].)

A new generation of rust pustules and urediniospores appear 7 to 15 days later, depending largely on the temperature. Urediniospores constitute the repeating stage of the rust fungus. This cycle of spore production, release, penetration, and infection may be repeated a number of times during the summer and autumn or until environmental conditions become unfavorable for the growth and reproduction of the rust fungus.

In mild climates, the rust fungi overwinter as dormant mycelium and as urediniospores in or on infected turfgrass foliage or crowns and equipment. When the temperature and moisture conditions are conducive to regrowth of the mycelium and germination of the urediniospores (usually between 60 and 90°F [15.6 to 32.2°C]), the leaves and leaf sheaths become infected, and a new generation of uredinial pustules and their urediniospores are formed. These spores are readily transported over long distances by air currents, and those from warm regions may serve as sources of wind-blown inoculum for northern regions, where mycelium and urediniospores cannot survive the winter.

Most rust fungi also produce another spore type, the dark brown or black teliospores (figures 5–83 and 5–84), when the leaves senesce or dry slowly. The brown-to-black telial stage, however, is minor on mowed turfgrasses growing under a good cultural management program. The teliospores, if produced, may serve as overwintering structures in the north, germinating in the spring to produce a third spore type, basidiospores. Basidiospores are transported by air currents to the leaves of nearby, alternate hosts (mostly woody shrubs and herbaceous ornamentals), where they may germinate and produce new infections that result in two more spore types, sporidia, or pycniospores, and later, the aeciospores. Cluster cups or aecia form on the alternate hosts and release aeciospores, which are then capable of infecting only grass plants, giving rise to urediniospores, thus completing the disease or life cycle. Only the urediniospores (figures 5–83 and 5–84) are important in producing rust infections on turfgrasses. The fungus *Pithomyces chartarum* can also be found in turfs that are affected with rusts. This fungus will also produce a black, multicelled conidiospore, which can be confused with rust teliospores. *Pithomyces chartarum* is a saprophyte that colonizes the thatch.

Importance.   Rusts rarely kill mown turfgrasses, especially if they are kept growing steadily by proper watering and fertilizing. Rusts may cause severe leaf damage, especially when the leaves are also attacked by other fungi (such as by species of *Drechslera* and *Bipolaris*) under droughty conditions. Seed-production fields of turfgrasses may be severely damaged (losses of 90% or more) unless protected by fungicide sprays.

Management.   Grow locally adapted, rust-resistant grasses, preferably as blends or mixtures. Check with your county extension (or advisory) office or your state turfgrass specialist for suggested grass species and cultivars to grow. Turfgrasses with several or more resistant cultivars to one or more rusts include Kentucky bluegrass, fine-leaf fescues, tall fescue, perennial and annual ryegrasses, and bermudagrass. Resistance to rusts is limited, however, by the presence of numerous physiological races of the rust fungi. A cultivar in one location may be resistant, whereas it appears quite susceptible in another turfgrass area. This is why planting blends or mixtures of resistant grasses is usually the best solution. Fertilize and irrigate to keep grass growing at a steady rate (about 1 in or 2.5 cm a week for most turfgrasses) during summer and fall drought periods and where the spring and fall weather is cold and wet for long periods (like the Pacific coast of North America). The growth of the grass blades pushes the rust-infected leaves outward, where they can be mowed off and removed. To increase vigor, maintain a proper balance of nitrogen, phosphorus, and potassium (N-P-K), according to local recommendations based on a soil-test report. These recommendations will vary with the grasses grown and their use. If you fertilize, do it lightly, especially with a readily available high-nitrogen source when disease is active. Keep the phosphorus and potassium levels high. Light fertilization makes the grass somewhat more succulent and susceptible but better able to grow out of the damage, provided that other measures, such as frequent mowing, are used.

During summer or early autumn droughts, water established turf thoroughly early in the day so that the grass can dry before late afternoon. Water infrequently and deeply, moistening the soil at each watering to a depth of 6 in (15.2 cm) or more. Avoid frequent light sprinklings, especially in the late afternoon or evening. Free water on the leaf surfaces as dew or guttated water for several hours or more, enhances the development of rusts and many other diseases. Other important management practices include raising the cutting height, increasing mowing frequency, removing the clippings, aerating to alleviate compaction, removing excessive thatch, and increasing the amount of light exposure and air movement over the turf by opening up the tree and shrub canopy and possibly redesigning landscape plantings.

Follow other cultural practices outlined in table 5–1. These cultural practices and those outlined above should provide for a steady, vigorous growth of grass during extended warm-to-hot, dry periods, when rust attacks are most common.

If rusts are serious year after year, especially in seed- and sod-production fields, these practices may need to be supplemented by a preventive fungicide spray program (table 5–5). The initial application should be made when rust is first evident on the grass blades. Repeat applications are needed at 7- to 14-day intervals as long as rust is prevalent. Sterol-inhibiting fungicides such as triadimefon (Bayleton) and propiconazole (Banner) will provide several weeks of protection with a single application. For best results, apply the fungicide soon after mowing and removal of the clippings. Good coverage of the leaf surfaces is necessary for control. The addition of about ½ teaspoonful of a wetting agent or surfactant, such as liquid household soap or detergent, to each gallon of spray mixture will help spread the spray droplets over the grass surface. For the most effective control of rusts, uniformly spray 1000 sq ft (93 sq m) with 1 to 3 gal of water.

When mixing or applying any fungicide, always carefully follow the manufacturer's directions and precautions.

Superficial Fairy Rings
See the section on fairy rings (5.2.10).

5.2.25   St. Augustinegrass Decline and Centipedegrass Mosaic or Decline

St. Augustinegrass decline (SAD) and centipedegrass mosaic or decline are caused by the panicum mosaic virus strain group (CMMV) subdivided into serological types. This disease causes serious damage to St. Augustinegrass. SAD and a different strain or strains of panicum mosaic virus affecting centipedegrass are the only important viral diseases of cultivated turfgrasses. For minor viral diseases of turfgrasses, see sec. 5.2.17.

SAD is a potential threat wherever St. Augustinegrass is grown in the Gulf coast and southeastern United States. Symptoms are more pronounced where the turf is under stress from lack of nitrogen, cold injury, drought, nematode and insect injury, and shade. St. Augustinegrass and centipedegrass are the only turfgrasses presently known to be affected by the virus, which is spread through diseased sod or sprigs and from infected to healthy grass plants by contaminated mowers.

Identification.   The panicum mosaic virus causes a decline of both St. Augustinegrass and centipedegrass. The first symptoms appear as pale green or yellowish spots, blotches, and a speckling or stippling of the leaves (figure 5–85). SAD and centipedegrass mosaic in their early stages may be confused with iron or zinc deficiencies, mite-feeding damage, and yellow tuft or downy mildew. Infected grass blades have a mottled or mosaic pattern with bright yellow stippling or streaking. Deficiency symptoms, on the other hand, appear as *continuous* yellow or pale stripes in the leaves parallel to the veins.

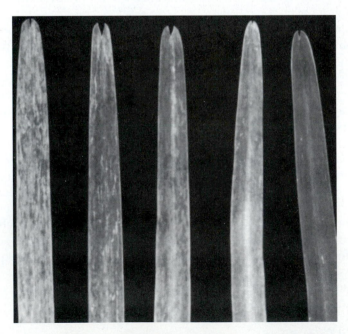

FIGURE 5–85   St. Augustinegrass decline (SAD) caused by the panicum mosaic virus; the right leaf is healthy. (Courtesy R. W. Toler.)

With SAD, the mild mottling or yellow-green mosaic gradually becomes more yellow as the spots and blotches merge, and, by the second year, infected turf has a uniform bright yellow appearance. Such turf is stunted, lacks vigor, and becomes thin as stolon growth is retarded. During the third year, infected plants are severely chlorotic. Eventually they wither and die, producing large, irregular, thin or dead areas of turf up to 30 ft (9.1 m) across. Such areas are soon invaded by native weed grasses and broadleaf weeds. *Entire* turf areas may be killed by the end of the third year. Diseased St. Augustinegrass is much more susceptible to winterkill, drought, and pesticides.

Disease cycle.   The panicum mosaic virus is introduced into new turfgrass stands by means of contaminated mowers, turf cultivation, vertical cutting, and diseased sprigs and sod. There are no insect or other vectors of the virus.

Importance.   The virus causes a decline, thinning, and eventual death of St. Augustinegrass turf, which is then invaded by weeds.

Management.   Resprig, plug, or resod infected St. Augustinegrass turf areas with a SAD resistant or tolerant cultivar such as Bitterblue, Floralawn, Floratam, Raleigh, or Seville if adapted to the region. Floratam is also resistant to chinch bugs but lacks winter hardiness in the cooler areas where St. Augustinegrass is grown. It also lacks good leaf texture and shade adaptation. All existing cultivars of centipedegrass are susceptible.

When establishing a new turf area, purchase only disease-free sod that has been certified by the state department of agriculture. When SAD or centipedegrass mosaic has been reported in your area, *use only your own mowing and other turfgrass equipment.* Other *temporary* beneficial practices include (1) raising the mowing height, (2) avoiding mowing when the foliage is wet, (3) applying a balanced N-P-K fertilizer plus iron based on a soil-test report, and (4) applying adequate irrigation to prevent drought stress.

Sclerotial LTB (SLTB) Snow Mold
See the section on snow molds (5.2.29).

Sclerotinia Snow Mold
See the section on snow molds (5.2.29).

Sclerotium Blight
See the section on southern blight (5.2.30).

5.2.26   Seed Decay, Damping Off, and Seedling Blights

Seeds of different turfgrasses require various periods of time to germinate. Under ideal conditions, cultivars of Kentucky bluegrass germinate in 2 to 3 weeks or more; centipedegrass, in about 2 weeks; bermudagrass and buffalograss take 3 to 4 weeks; bentgrasses and perennial ryegrass, about a week; fine-leaf fescues, a week to 10 days; and tall fescue, 10 to 12 days or more. To ensure a good stand and seedling vigor, the properly prepared seedbed should be moist (not too wet or dry) during the entire germination period. Other requirements for the establishment of a healthy turf include adequate sunlight and oxygen. Optimum day-night temperatures for the germination of cool-season grasses is 60 to 85°F (15.6 to 29.4°C) and for warm-season species 68 to 95°F (20.0 to 35.0°C). As environmental factors (primarily moisture, light, oxygen, and temperature) become less favorable for seedling growth, the risk of seed decay, damping off, and seedling blights increases.

This disease complex is most common when prolonged, hot and wet periods or overwatering follow seeding, where the stand is too dense and seeds are buried too deep or shallow; and when there is poor contact of seeds with soil due largely to a poorly prepared seedbed, excessive nitrogen fertilization, poor soil-water drainage, and poor air circulation; and when seeding is done at other than suggested times (for instance, in late summer or spring for most of the United States, and with old and slow-to-germinate seed).

Seed decay, damping off, and seedling blights are caused primarily by numerous species of *Pythium, Rhizoctonia, Fusarium, Bipolaris, Drechslera, Exserohilum, Curvularia, Colletotrichum, Microdochium, Cladosporium, Nigrospora, Septoria, Stagonospora, Ascochyta, Botrytis, Acremoniella, Cladochytium, Cylindrocarpon, Idrella, Trichodoma,* and *Alternaria* (figures 5–2, 5–3, 5–13, 5–16, 5–38, 5–43, 5–61, and 5–71).

Each of these fungal pathogens has its own requirements for optimal growth. For example, *Pythium irregulare, P. ultimum* (and *P. debaryanum*), *P. vanterpoolii, Microdochium nivale, Fusarium culmorum,* and *Rhizoctonia cerealis* grow best at a low temperature, 50 to 72°F (10.0 to 22.2°C), while other species of *Pythium* (*P. aphanidermatum, P. arrhenomanes, P. graminicola,* and *P. myriotylum*), *Rhizoctonia solani,* and *Bipolaris sorokiniana* are most pathogenic at a higher temperature (80 to 95°F; 26.7 to 35.0°C). Other seed- and seedling-blight fungi are capable of causing disease at low or high temperatures. The pathogenic capabilities of the various fungi mean that the prudent turfgrass manager should plant at a time and under conditions where seed germination and seedling vigor will be strongly favored.

For problems associated with overseeding of warm-season turfs during the winter in the southern United States, see the section on Pythium diseases (5.2.21). Seeds sown into an existing turf face additional problems such as competition for light, water, space, and nutrients with larger, established plants; exposure to high levels of certain pathogens (especially crown and root rot organisms) in the thatch; and toxins produced by the decomposition of organic matter in the thatch.

Identification.   Stands of newly seeded or overseeded turfgrass are thin, stunted, patchy, off-color, and slow to fill in when seed germination and seedling growth are slowed by environmental stresses (too cold, hot, dry, or wet) that favor the growth of pathogenic fungi. Seeds and seedlings of all cultivated turfgrasses rot in the soil (seed decay and preemergence damping off) or young seedlings that do emerge become watersoaked at the soil line and below, turn yellow or bronzed to brown, collapse, wither, and die (postemergence damping off; see figure (5–86). Surviving plants are stunted, weakened, and more susceptible to other diseases. The grass stand is thin, slow growing, and weak.

Disease cycle.   Although most of the fungi that cause seed decay, damping off, and seedling blights are common soil inhabitants and are present in all

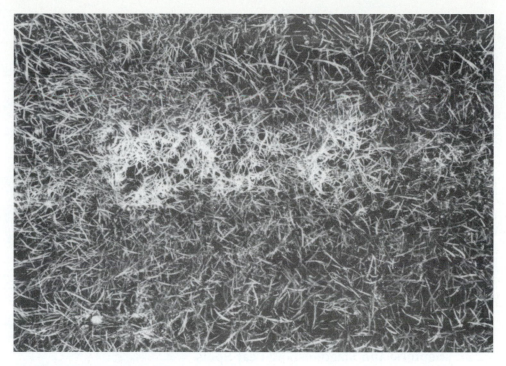

FIGURE 5–86   Close-up of seedling blight or postemergence damping off of creeping-bentgrass seedlings caused by *Rhizoctonia solani*.

unfumigated seedbeds, many others are transported on or within seed. Infection occurs before emergence, especially when the seed lacks vigor and is slow to sprout due to unfavorable temperatures, an excess or lack of moisture, or other suboptimal conditions for seedling growth, either above or below the soil surface. For additional information, see the sections on Pythium diseases (5.2.21) and Rhizoctonia diseases (5.2.23).

Importance.   Stands of seedling grass are thin and patchy, off color, slow growing, and weak. Losses may be severe under cold or hot wet conditions.

Management.   Provide for good surface and subsurface drainage and avoid compaction when establishing a new turf area. Fill in depressions where water may stand. Test the soil reaction (pH) and follow the soil-test report to obtain a pH between 6 and 7. Provide for good air movement over the turf.

Buy and sow only top-quality (vigorous, high-germinating), fresh, plump seed of locally adapted, disease-resistant grasses or combinations (blends and mixtures) of grasses. Except for southern grasses, which germinate best at somewhat warmer temperatures, plant when the weather is cool and dry (night-day temperatures of about 60 to 85°F or 15.6 to 29.4°C) and therefore ideal for rapid seed germination and growth. The seedbed should be fertile and well prepared and the seed sown at the recommended rate and depth. Ensure that the soil-seed contact is good to speed germination. If you anticipate a problem, treat the seed with a seed-protectant fungicide such as captan or thiram plus metalaxyl (Apron) or etridiazole (Koban). See table 5–5. Avoid overwatering and inadequate soil moisture, especially from seeding to seedling emergence or plant establishment; overfertilization with nitrogen; and deep planting.

Apply a suggested fungicide (table 5–5) uniformly over the seedbed, using 5 or more gal of water to cover 1000 sq ft (93 sq m). Repeat immediately after seeding, when seedlings are beginning to emerge, and at the *first* evidence of a patchy, uneven stand. One or two additional applications at 14- to 21-day intervals may be beneficial. Fungicide applications can be important in obtaining satisfactory stands of turfgrasses!

Mow the new turf area when the seedlings reach a height one-third greater than the anticipated cutting height. A light application of nitrogen fertilizer when grass seedlings reach 1.5 to 2 in (3.8 to 5.1 cm) tall encourages vigorous growth. Apply about ½ lb of actual nitrogen per 1000 sq ft (93 sq m), and water the fertilizer into the top 1 to 2 in (2.5 to 5.1 cm) of soil.

Seedling Blight
See the section on seed decay, damping off, and seedling blights (sec. 5.2.25).

5.2.27 Septoria and Stagonospora Leaf Spots

Septoria and Stagonospora leaf spots, sometimes called tip blight, are caused by more than 15 species of the fungus *Septoria* and several species of the fungus *Stagonospora*. This is usually a minor disease complex that attacks most cultivated turfgrasses, including bentgrasses, bluegrasses, fescues, ryegrasses, bermudagrasses, and crested wheatgrass. Septoria and Stagonospora leaf spots commonly occur during cool, wet weather in the late winter or early spring and autumn. The diseases are rare in warm to hot weather. Some of the species are seedborne.

Identification. The overall appearance of a diseased turfgrass area may appear scorched, resembling damage from a dull mower except that the leaf tips are not frayed. The grass leaves are often a pale yellow and mottled from the tip downward. The spots and blotches, up to 3 mm or more long, may be scattered near the leaf tips. They are a yellowish green to dark gray to gray-green, brown, or dark purple. On aging, the lesions fade to a light straw color. On perennial ryegrass, the lesions are yellowish green, later turning a chocolate brown. All *Septoria* and *Stagonospora* fungi produce yellowish-brown-to-dark-brown-or-black, speck-sized, fruiting bodies (pycnidia) embedded in the older lesions (figures 5–16 and 5–87). They can usually be seen with a magnifying lens or reading glass and help to distinguish these leaf spots from a dull mower, dollar-spot infection, drought, or winter injury.

Disease cycle. The *Septoria* and *Stagonospora* fungi overseason as mycelium and pycnidia in grass debris. During cool rains or watering in spring and autumn, microscopic, needlelike or elongated, oblong, one- to six-celled spores formed

FIGURE 5–87 Septoria leaf spot (*S. macropoda*) lesions on Kentucky bluegrass leaves. Note the dark pycnidia embedded in the dead leaf tissue. (Courtesy R. W. Smiley.)

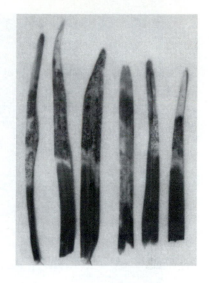

within the pycnidia (figures 5–16 and 5–87) ooze out and are splashed and washed or carried on turfgrass equipment to healthy leaves. Here, the spores germinate in a film of water, and infection occurs, often in the freshly cut tips of the grass blades. The cycle can be repeated every week or two as long as the weather remains cool (60 to 75°F; 15.6 to 23.9°C) and moist.

Importance. Damage is most severe on unmown or infrequently mown grass in shaded, moist sites during cool, wet weather in early spring. Some fungal species occur only on dying and dead leaves and hence are generally considered as saprophytes. The disease usually disappears after several mowings early in the season.

Management. No special measures are usually needed to keep Septoria and Stagonospora leaf spots in check. Where feasible, follow suggested cultural practices as outlined in table 5–1. Fungicides applied to control Helminthosporium diseases and dollar spot should provide adequate control (table 5–5). Kentucky bluegrass cultivars differ in resistance.

5.2.28 Slime Molds

Slime molds (figures 5–88 and 5–89) cause concern to many turfgrass managers when they occur suddenly in warm, overcast weather after heavy rains or watering. Numerous species of myxomycetes may be involved, including the common *Physarum cinereum,* and occasionally *Mucilago spongiosa, M. crustacea,* and *Didymium crustaceum.* Other species of *Didymium, Fuligo, Mucilago, Physarum,* and *Stemonitis* have also been reported on all species of turfgrasses (figure 5–89).

Slime molds are primitive organisms that lack cell walls and creep or flow slowly like amoebae over low-lying objects and vegetation, such as turfgrasses, plantains,

FIGURE 5–88 Slime mold. Plasmodium of *Physarum cinereum* forming immature sporangia. (Courtesy Noel Jackson.)

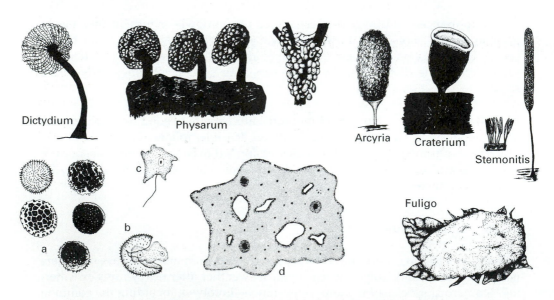

FIGURE 5–89 Common slime molds are cosmopolitan Myxomycetes that fruit on lawngrasses, ground covers, other low-growing vegetation, and numerous substrates in warm, wet weather (*Arcyria, Craterium, Dictydium, Fuligo, Physarum,* and *Stemonitis*): (a) various types of spores with different spore-wall ornamentations; (b) spore germination and release of a myxamoeba (in other cases, motile flagellate swarm cells are released); (c) flagellate swarm cell; (d) young plasmodium that has engulfed various minute bacteria and protozoa. (Sketches presented as seen under a high-power microscope.)

dandelions, clovers, strawberries, bedded flowers, ground covers, mulches, bases of trees, and even sidewalks and driveways. These organisms are not parasitic on turf but feed on decaying organic matter, fungi, bacteria, and other organisms in the thatch layer and soil. Generally, slime molds cause little damage to living turfgrass, other than being unsightly, although they may cause some yellowing by shading the affected leaves.

Identification.   In its slimy, amoebalike stage, the slime mold is a watery white, creamy-to-light-yellow, orange, gray, red, violet, blue, green, or greasy purple-brown mass that occurs in round to irregularly shaped patches from 0.5 in to 2 ft (1.5 to 60.9 cm) in diameter. This stage consists of a naked mass of protoplasm called a plasmodium that simply engulfs its food. The plasmodium soon "heaps up," and the crusty, ball-like fruiting bodies (sporangia), about 1 to 2 mm in diameter, envelop individual grass blades. The sporangia are variously colored and range from white, gray to creamy white, yellow, purplish brown, bluish gray, tan to orange, brown or black (*Color plate 55* and figure 5–89). The sporangia are filled with numerous microscopic purplish, blue-gray, gray, black, dirty yellow, or white powdery spores (figure 5–89). Moist, warm weather and high soil moisture favor the fruiting of slime molds on turfgrasses. Slime molds commonly reappear in the same areas each year and last for several days to a week or longer. They are easily rubbed off the leaf blades.

Disease cycle.   Slime molds survive unfavorable conditions primarily as microscopic spores in the soil and turfgrass thatch. The spores are spread about primarily by air currents, water, mowers and other turf equipment, people, and animals. During or after cool, wet weather or deep watering from spring to autumn, the spores absorb water, crack open, and a motile swarm spore emerges from each (figure 5–89). The amoebalike swarm spores feed on fungi, bacteria, other microorganisms, and decaying organic matter in the soil and thatch while they undergo various changes and numerous fissions. Finally, they unite in pairs to form zygotes and become shapeless, slimy plasmodia that increase in size. The plasmodium works its way to the soil surface and creeps over vegetation or other objects in round to irregularly shaped patches. Here, the crustlike fruiting or reproductive stage occurs, which is the only stage that most of us ever see. The masses of powdery spores (figure 5–89) are easily rubbed off the leaves and stems. Slime molds are nonparasitic and technically do not cause disease.

Importance.   Slime molds are much more unsightly than harmful and merely use grass leaves and stems as a means of support for their reproductive structures. However, slight damage may occur when leaves are smothered or shaded for several days to a week or longer. The weakened and somewhat yellow grass leaves are more susceptible to infection and killing by secondary fungi and bacteria. An abundance of thatch favors slime molds by providing a ready source of organic matter plus high populations of microorganisms.

Management.   No measures are usually necessary. Remove excess thatch and provide for adequate surface and subsurface drainage. You can break up the unsightly spore masses by vigorous raking, brushing, or hosing down with a stream of water. Washing is suggested only after the onset of dry weather. Mowing the grass usually removes the spore masses. Spraying the turf with any suggested fungicide listed in table 5–5 should also be beneficial in checking their growth but is rarely, if ever, necessary.

Snow Blight or Rot
See the section on Pythium diseases (5.2.21).

5.2.29   Snow Molds

Snow molds are cold-tolerant fungi that grow at freezing or near-freezing temperatures. Snow molds may damage turfgrasses from late autumn to mid or late spring, especially in shaded or wet areas where the turf remains wet or is covered by snow on unfrozen ground and is slow to melt. Leaves, stems, crowns, and roots may be rotted over a wide range of temperatures (about 30 to 60°F [−1.1 to 15.6°C]). Some snow molds may injure turf at snow melt or during cold, drizzly periods when snow is absent. When the grass surface dries out and the weather warms, snow-mold fungi cease to attack; however, infection tends to reappear in the same areas year after year.

Snow molds tend to occur in regions with more than 40 days of snow cover. They are favored by excessive, early fall applications of quick-release, nitrogenous fertilizers; excessive shade; a thick thatch; or mulches of straw, leaves, or other moisture-holding debris on the turf. Disease is most serious when air movement and soil drainage are poor and the grass stays wet for long periods or where the snow is deposited in drifts or piles. Snow molds also may develop on turf covered with a geotextile or plastic blanket. When these are being used, it may be necessary to treat the turf that will be covered with a preventive fungicide.

Most turfgrasses are susceptible to one or more snow-mold fungi. They include bentgrasses, Kentucky and annual bluegrasses, fescues, ryegrasses, bermudagrass, and zoysiagrass. Bentgrasses, annual bluegrass, and perennial ryegrass are often more severely damaged than are other turfgrasses.

There are two principal snow molds over most of the colder areas where turfgrasses are grown: pink snow mold or Microdochium (Fusarium) patch (caused by the fungus *Microdochium* [*Fusarium*] *nivale,* teleomorph state *Monographella nivalis*), and gray or speckled snow mold, also known as Typhula blight (caused by several species or subspecies of the fungus *Typhula*). The two types are found in roughly the same geographical areas of temperate and subboreal regions of North America, northern Europe. The more widely distributed pink snow-mold fungus also occurs in Iceland, Japan, southern Australia, and New Zealand, and in somewhat warmer regions than gray snow mold.

There are other low-temperature-tolerant fungi that attack turfgrasses. Coprinus or cottony snow mold is caused by *Coprinus psychromorbidus.* Snow scald, or Sclerotinia snow mold, is incited by *Myriosclerotinia (Sclerotinia) borealis.* Other cold-tolerant fungi are *Sclerotium rhizoides,* the cause of frost scorch, a *Rhizoctonia* species, snow rot or snow blight caused by several species of *Pythium,* and a brown root rot incited by *Phoma sclerotiodes.* The weakly pathogenic or saprophytic *Neophodium (Acremonium) boreale* (teleomorph *Nectria tuberculariformis*) is often an antagonist of snow-mold fungi.

Different snow-mold fungi also occur as complexes, even attacking the same plant. They are commonly found together with non-snow-mold fungi, including species of *Rhizoctonia, Cladosporium, Fusarium,* miscellaneous leaf and crown pathogens, and saprophytes. All of these fungi compete for the same available nutrients, susceptible grass tissue, space, and moisture. Certain snow-mold fungi are commonly antagonistic to each other, which affects their development.

Gray snow mold or Typhula blight. Gray or speckled snow mold is caused by the fungus *Typhula incarnata,* three varieties of *Typhula ishikariensis:* var. *canadensis,* var. *idahoensis,* var. *ishikariensis,* and *T. phacorrhiza.* A deep and prolonged snow cover on unfrozen soil, tall grass matted down, leaves, straw mulch, or plastic desiccation covers, and unbalanced nitrogen fertilization all produce favorable conditions for disease development. The fungus (or fungi) are less active while the turf and soil are frozen. In winter or early spring when the snow melts and the turf thaws, the fungus may again become active, and diseased patches may enlarge. As the weather warms and the turf dries, *Typhula* becomes dormant until middle to late autumn. The optimum temperature for infection is between 30 and 45°F (−1.1 to 7.2°C). All cool-season turfgrasses are susceptible, with bentgrasses, annual bluegrass and ryegrasses being most susceptible. Kentucky bluegrass cultivars differ in resistance to the *Typhula* fungi. The fine-leaf fescues are commonly more resistant than are Kentucky bluegrasses. Gray snow mold may occur with or without snow cover; damage, however, is usually minor when snow is absent. Turf injury is aggravated when the snow is compacted by walking, skiing, snowmobiling, sledding, and vehicle traffic.

Identification. During and after snow melt, gray snow mold appears in roughly circular light yellow to white-to-grayish-white patches, 2 to 4 in (5.1 to 10.2 cm) in diameter, with regular margins that coalesce under snow cover to form areas up to 2 to 3 ft (0.6 to 0.9 m) in diameter. The disease is most active where the snow is melting. A number of spots may merge, forming large, irregular, straw-colored, or grayish brown dead areas (figure 5–90). Wet grass may be matted together and covered at first with a fluffy, white-to-grayish-white mold (mycelium) speckled with numerous pale-to-dark-brown, orange-brown, or black sclerotia (figure 5–91) 1 to 5 mm in diameter. The mold soon turns bluish gray to almost black. At other times, a silvery white, brittle, membranous crust develops over the injured turf. When conditions favor disease development, large turf areas may be killed when the fungus (or fungi) attack the crowns and even the roots. More commonly, only the leaves are killed, and new leaves emerge from the overwintered plant crowns. Old gray mold scars may be evident until mid or late spring.

A mild form of gray snow mold may also develop in winter with little snow cover or the absence of a snow cover at temperatures of 36 to 40°F (2 to 5°C), first appearing as small (about 1 to 2 in; 2.5 to 5.1 cm) circular patches of light yellow-brown grass up to 6 in (15 cm) across.

FIGURE 5–90 Gray snow mold or Typhula blight infecting Kentucky bluegrass at the edge of melting snow.

FIGURE 5–91 Gray snow mold. Close-up of the mycelium and sclerotia of *Typhula incarnata*. (Courtesy Noel Jackson.)

Disease Cycle.   After the period of active mycelial growth when the snow melts, *Typhula incarnata* produce small (up to 5 mm), roundish or irregular, flattened, wrinkled, faintly pink, orange-yellow or brown-to-tan, or reddish, reddish brown, chocolate brown, or black sclerotia.

The sclerotia of *T. ishikariensis* are smaller (0.2 to 2 mm), generally circular, and dark brown to black, so that the turf appears to be sprinkled with pepper. This species occurs most commonly in the western half of North America and other areas where winters are longer, more severe, and the snow cover remains for 4 months or longer. The sclerotia of *T. phacorrhiza* are pear shaped to irregular, often up to [0.25 in (7mm)] in diameter, reddish to dark brown, and stalked. Sporocarps are threadlike, dark when mature, and up to [4 in (10 cm)] tall. The sporocarps are seldom found in the field. They are mostly found growing on dead plant tissue and on patches of snow molds caused by *T. incarnata* and *T. ishikariensis* or a combination of each.

The hard sclerotia of *Typhula* spp. are embedded in or attached to the leaves and crowns of diseased plants (figure 5–91) and lie dormant during the following summer and early autumn. The cortical cells of the sclerotia under a microscope look like interlocking pieces of a jigsaw puzzle and are an aid in separating the *Typhula* species from other sclerotium-forming fungi that are associated with grasses.

The sclerotia germinate in cold (optimum 50 to 65°F; 10.0 to 18.3°C), wet weather in the fall or after autumn snow melt to produce delicate, pink-to-grayish-white, clublike spore-bearing sporocarps 1 to 2 cm tall (figure 5–92) or hyphae with clamp connections (figure 5–2) that infect all tissues of the grass plant and start the disease cycle once again. The fungi spread by reduce growth of mycelium, movement of sclerotia, or by windblown basidiospores produced by the sporocarps, splashing or flowing water, turf equipment, and shoes. *Typhula* fungi that infect turfgrasses are not seedborne.

Gray snow-mold fungi start the disease cycle as saprophytes feeding on dead thatch debris. Under snow, however, the mycelia spread, producing infections earlier and penetrating leaves and sheaths and may later invade the crowns and roots.

Importance.   When environmental conditions are favorable (a deep and prolonged snow cover on unfrozen soil), gray snow mold may kill large turf areas. In warmer regions, only the leaves are usually killed, and gray-mold scars may be evident until late spring, when new leaves emerge from the overwintering plant crowns.

Management.   An integrated pest management (IPM) program is needed. Follow the cultural practices outlined in table 5–1. Keep the turf as dry as possible by good soil drainage and by selectively pruning trees and shrubs to increase air movement over the turf canopy. Rake off fallen leaves. Hold the thatch to ½ in (1.3 cm) or less on higher grass. Follow a suggested fertilizer program for your area and the grass or grasses being grown. The risk of snow-mold damage can be reduced by using moderate, *balanced* fertilization. Encourage new growth in spring by lightly fertilizing the damaged turfgrass using 4 oz (114 g) of readily available nitrogen per 1000 sq ft (93 sq m). *Avoid* straw or other mulches, heavy topdressings, and fertilization with quick-release nitrogen fertilizers in late summer or autumn within about 6 weeks of a killing frost or when the first heavy snow is expected.

FIGURE 5–92 Gray snow mold. Close-up of sporocarps of *Typhula incarnata*. (Courtesy Noel Jackson.)

Slow-release forms of nitrogen fertilizer are recommended, although soluble forms of nitrogen (such as ammonium sulfate) should work just as well in most situations. Turfgrasses should *not* go into the winter in a succulent condition. Use lime *only* when the need is indicated by a soil test; avoid excessive use of lime. Continue to mow late into the fall. Remove the clippings where feasible.

Prevent large snow drifts and excessive accumulation of snow on high-value turf areas by proper placement of snow fences, living evergreen windbreaks or similar barriers, and the use of snow blowers. Divert skiers and snowmobiles around golf greens, tees, and other highly maintained turf to avoid snow compaction. Turfgrasses somewhat resistant to one or more *Typhula* species include well-adapted, winter-hardy cultivars of Kentucky bluegrass and perennial ryegrass. Red fescues are, in general, more resistant to gray snow mold than are annual bluegrass and bentgrasses.

Just before the first heavy snow or cold, drizzly weather are isolates of forecast in autumn, apply one or more of the suggested turf fungicides or mixtures (see table 5–5) to areas with a history of snow-mold infection. *Carefully follow the manufacturer's directions and precautions.* Reapply one or more times during late autumn, early winter, or midwinter during a thaw.

A promising biocontrol agent for gray snow mold is *Typhula phacorrhiza* applied in pellets made from sodium alginate. Treatment is made in late fall for several consecutive seasons. The effect lasts for 3 years or longer. Aside from the possible pathogenicity of *T. phacorrhiza*, another concern is that treatment may not control other snow mold diseases that often coexist with Typhula blight.

Repair snow-mold damage in spring by raking the matted grass and fertilizing. Reseed or resod as necessary. Fungicide sprays may be needed. See the section on seed decay, damping off, and seedling blights (5.2.26).

Pink snow mold, Microdochium (Fusarium) patch.   Pink snow mold and Microdochium patch, caused primarily by *Microdochium (Fusarium) nivale*, damages nearly all grass species. Other fungi often associated with *M. nivale* include *Fusarium avenaceum, F. culmorum, F. equiseti, F. semitectum,* and *Trichoderma viride*. This disease is common and troublesome worldwide, wherever prolonged periods of wet, cool-to-cold, overcast weather occur from autumn to middle or late spring and early summer; or any time of the year during cool-to-cold, wet weather. Microdochium (Fusarium) patch often occurs in the absence of snow when grass growth is retarded. Patches of the disease, which persist until a snow cover develops, may increase in size, especially if the snow falls on unfrozen ground. At snow melt, on exposure to light, the fungus on diseased turf turns pink, hence the name "pink snow mold" (*Color plate 56*). Infection, spread, and disease development occur most rapidly when the turf moisture and air humidity are high and temperatures are 32 to 46°F (0.0 to −7.8°C) with a maximum about 65°F (18.3°C). Some isolates of the fungus will even grow at −6°C. Most isolates will grow at temperatures, at least in culture, up to 86 to 90°F (30.0 to 32.2°C). Nearly all cool-season turfgrasses, bermudagrass, zoysiagrass, forage grasses, and cereals are susceptible. Fine-leaf fescues and tall fescues are usually not damaged as severely as annual bluegrass, bentgrasses, Kentucky bluegrasses, and ryegrasses.

Identification.   Pink snow mold or Microdochium (Fusarium) patch disease first appears as roughly circular, watersoaked spots, 1 to 3 in (2.5 to 7.6 cm) in diameter, that soon become yellow, orange-brown, or reddish-brown-to-dark-brown matted patches. Varying amounts of white or pale pink, cobwebby mycelial growth *without* clamp connections (figure 5–93) are evident under wet conditions. Later, the patches may enlarge and become light gray or light tan rings up to about 1 to 2 ft (30.6 to 60.9 cm) across, with orange-brown or reddish brown borders. The patches are usually more circular and smaller than those of gray snow mold. Where there is snow cover, the spots often enlarge up to 8 to 12 in (20.3 to 30.5 cm) across and commonly merge to blight large areas. Shortly after snow melt, the white, often feltlike mycelium turns pink and can be seen at the advancing margins of the spots (*Color plate 56*). Diseased plants later collapse and die. The matted leaves have a tan color. After snow recedes, the patches are a bleached white and may or may not have a pinkish margin. Most plants in such patches under snow are killed, and recovery of apparently healthy plants may be slow if the weather in spring is dry.

Disease Cycle.   The *Microdochium* fungus is inactive when the grass is dry and the weather is warm. It overseasons as dark brown aggregates of dark

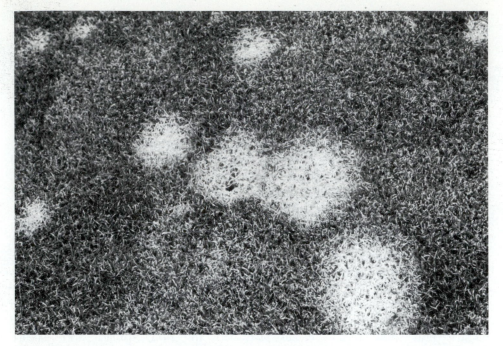

FIGURE 5–93 Pink snow mold or Microdochium (Fusarium) patch.

mycelium or spores in infected grass plants, grass debris, and soil. There is also evidence that the fungus is systemic within the grass plant. When temperature and moisture conditions are favorable, the fungus produces large numbers of somewhat curved, two- to six-celled microscopic spores (macroconidia) in sticky masses on fruiting bodies called sporodochia (figure 5–38). The white-to-salmon-pink sporodochia appear as flecks on moist, dead leaf tissue. The spores are carried to grass leaves primarily by splashing or flowing water, mowers and other equipment, and foot traffic. Infection occurs through stomates, cut leaf tips, and directly into injured epidermal cells. The fungus can exist and attack grasses in all soils from sands to heavy clays. It is favored by alkaline turf surfaces.

Importance. Disease attacks of pink snow mold weaken and may kill turfgrass plants in large areas, especially under a prolonged snow cover. Such areas are then commonly invaded by weeds.

Management. The management is the same as for gray snow mold. Follow a suggested fertilizer program for your area and the grasses grown. Use adequate amounts of a balanced N-P-K fertilizer based on local recommendations for the grasses being grown. Where the soil is alkaline, excellent control of Microdochium patch has been obtained with a sulfur program similar to that outlined for take-all patch (sec. 5.2.33), which maintains the pH of the soil in the acid range (below 6.5). Avoid sudden changes in surface pH, especially on *Poa annua* turf. Maintain high potassium and phosphorus soil-test values, and avoid overstimulation with *any* source of nitrogen. Continue to mow until the grass stops growing in autumn. Somewhat resistant Kentucky bluegrass, ryegrasses, fine-leaf fescues, and bentgrasses are available.

Fungicide treatments are required at 2- to 8-week intervals during cold, wet weather from mid to late autumn through the winter and into spring. See table 5–5. Triadimefon does *not* control pink snow mold.

Coprinus or cottony snow mold (low-temperature basidiomycete [LTB] and sclerotial LTB [SLTB] snow molds). Coprinus or cottony snow mold, caused by different strains of the fungus *Coprinus psychromorbidus*, is most common in Canada and Alaska on lawns and athletic turf, where snow covers turfgrasses for prolonged periods from the middle of autumn well into spring. Two vegetative phases of the *Coprinus* fungus are known: nonsclerotial (LTB) and sclerotial (SLTB). The former is recognized only as a snow mold with a fine, sparse-to-abundant cottony white mycelium with clamp connections (figure 5–2) at the septa (such as *Typhula*). The individual patches often do *not* coalesce with wavy ribbons of healthy grass between them. The sclerotial strain produces a similar mycelium, but it has irregularly shaped sclerotia that are white at first and brownish black when mature, and 1 to 3 mm in diameter (figure 5–94).

Most common turfgrass species grown in Canada and Alaska (for instance, bentgrasses, annual and Kentucky bluegrasses, fescues [red, chewings, sheep, and tall]), as well as forage grasses, wheat, rye, forage legumes, and weeds are susceptible to one or both phases of *Coprinus.*

Identification. Round to irregular snow-mold patches, 6 to 12 in (15.2 to 30.5 cm) in diameter, are evident when the snow recedes in spring. The patches are pale

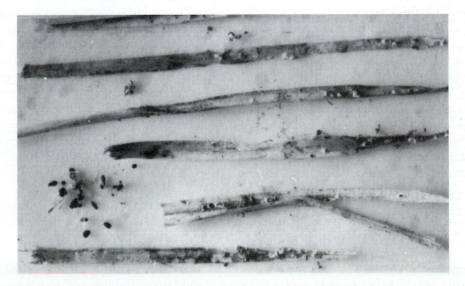

FIGURE 5–94 Sclerotia of the SLTB phase of *Coprinus psychromorbidus*, the Coprinus snow-mold fungus. (Courtesy J. D. Smith.)

yellow at first but soon become bleached. In the case of LTB, the grass blades are covered with a white-to-gray, cottony or sparse mycelium on leaves and plant debris, especially at the margins of the patch (*Color plate 57*). Under humid conditions, saprophytic fungi (such as *Alternaria* and *Cladosporium*) may colonize the leaf tissues killed by the LTB fungus, giving the patches a gray appearance.

In the SLTB, much of the cobwebby growth on the leaves and plant debris appears as white, cream, or tan mycelial knots that develop into small, dark-brown-to-black, irregularly shaped sclerotia (figure 5–94). Turfgrasses may be completely killed by the LTB phase, especially bentgrasses and annual bluegrass, or they may recover very slowly (as with Kentucky bluegrass). The SLTB phase is less pathogenic.

Disease Cycle.   The *Coprinus* fungus grows at temperatures a few degrees below freezing to about 68°F (20.0°C). Sclerotia in grass debris and soil may carry over the sclerotial (SLTB) phase of the pathogen until the following autumn or winter. The method of oversummering of the nonsclerotial or LTB phase probably occurs as dormant mycelium in infected living and dead plants. Small white-to-grayish mushroomlike structures (cap 7 to 12 mm wide; stalk 4 to 7 cm long) of *Coprinus* have been found in wet weather during late summer and early fall, but the function of the spores (basidiospores) they produce in causing disease is unknown. The fungus is spread by the movement of infested grass debris and soil. The disease commonly recurs in the same sites for years.

Importance.   Patches of bentgrasses, annual bluegrass (*Poa annua*), red fescues, and very susceptible cultivars of Kentucky bluegrass, attacked by the LTB fungus, usually fail to recover, and the space is invaded by grassy and broadleaf weeds. Large turf areas may be killed under a prolonged snow cover.

Management.   Follow the cultural control practices outlined in table 5–1 and those given for gray snow mold. Early snow melt should be encouraged on high-value turf areas, as for other snow molds, by spreading snow drifts, by selective placing of snow fences or other barriers, by snow removal with snow blowers, and by applying a light coating of a dark-colored topdressing or charcoal over the snow. Dormant applications of a complete (N-P-K), balanced fertilizer aids in spring recovery.

Two or three applications of fungicide mixtures as for gray snow mold (that is, quintozene [PCNB], chlorothalonil, chloroneb, fenarimol, and other DMI fungicides; table 5–5), are needed during autumn and into early winter, with the last application being made just before the first permanent snow cover of winter. Resistant cultivars of Kentucky bluegrass and red fescue have been identified (figure 5–95). Crested wheatgrass (*Agropyron cristatum*) is very resistant to LTB.

Snow scald or Sclerotinia snow mold.   This disease, caused by the fungus *Myriosclerotinia* (*Sclerotinia*) *borealis*, affects all cool-season turfgrasses (primarily Kentucky bluegrass, rough bluegrass, red and tall fescues, perennial ryegrass, and bentgrasses) as well as forage and weed grasses in very northern areas of the continental United States. It may be severe in areas of Canada, Alaska, Scandinavia, other areas of northern Europe, and Japan. The disease, also known as Sclerotinia patch, is favored by a damp autumn and the early accumulation of

FIGURE 5–95 Coprinus snow mold, LTB phase, showing the differential effect on cultivars of Kentucky bluegrass. (Courtesy J. D. Smith.)

a deep, long-lasting snow cover on frozen or unfrozen soil. The fungus is capable of causing damage at lower temperatures (−5°C) than other snow molds. Red fescues, bentgrasses, and perennial ryegrasses are very susceptible.

Identification.   Patches of dead grass up to about 6 in (15.2 cm) across appear as the snow melts in the spring. The patches often enlarge and coalesce to kill very large, irregular areas of turf (figure 5–96). Infected leaves are watersoaked at first and covered with a sparse, dark grayish mycelium without clamp connections. Later, the leaves turn bleached to almost white, wrinkle, and mat together. When severe, crown tissues may rot. Wrinkled, variously shaped or almost flakelike tan sclerotia 0.5 to 8 mm long and dull black when mature, eventually form on and in leaves, leaf sheaths, leaf axils, and crowns. Small (up to 5.5 mm in diameter), saucer-shaped, spore-producing structures (apothecia) with stipes (stalks) up to 6 mm long arise from the black sclerotia (figure 5–97).

Disease Cycle.   The *Myriosclerotinia* fungus grows best when there are prolonged and frequent rains, and air temperatures range between 49 and 59°F (9.4 to 15.0°C) followed by a deep and persistent snow cover on unfrozen soil. Sclerotia are the means by which the fungus oversummers. The sclerotia germinate during cool-to-cold, humid weather in autumn to produce the stalked, cup-shaped apothecia. Airborne ascospores produced in the apothecia and dispersed in the fall

FIGURE 5–96 Snow scald or Sclerotinia snow mold. (Courtesy J. D. Smith.)

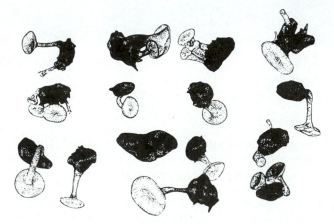

FIGURE 5–97 Apothecia on stalks from germinated sclerotia of the snow-scald fungus *Myriosclerotinia borealis.* The apothecia are rare in nature. (Adapted from a photograph by J. D. Smith.)

serve as the principal means of infection. Hyphal penetration occurs through wounds, stomata, and by direct penetration between the epidermal cells.

Importance. Large, irregular areas of turf may be killed in northern regions where the snow cover is deep and prolonged, especially if the snow forms on unfrozen or lightly frozen soil. Water-saturated soils reduce snow scald development.

Management. Same as for Coprinus or cottony snow mold and gray snow mold. Maintain the soil pH between 5.6 and 6.0. Quintozene (PCNB) is very effective, especially when combined with chloroneb (table 5–5). Some Kentucky bluegrass cultivars from northern latitudes have fair resistance. Bentgrasses, red fescues, and perennial ryegrass are very susceptible, with Kentucky bluegrass the most resistant species. Where bentgrass turf is severely attacked, brush the turf to dislodge the sclerotia and then suck most of them up with a vacuum sweeper.

Snow Scald
See the section on snow molds (5.2.29).

5.2.30   Southern or Sclerotium Blight

Southern or Sclerotium blight, caused by the soil-borne fungus *Sclerotium rolfsii*, teleomorph *Athelia rolfsii*, attacks bentgrasses, annual and Kentucky blue-grasses, fescues, bermudagrass, annual and perennial ryegrasses, and broadleaf turfs (lawnleaf [*Dichondra carolinensis*] and *Cotula*), mostly in the southeastern and western United States, the tropics, and in warm, temperate regions through-out the world during warm-to-hot (daytime temperatures of 86 to 95°F [30.0 to 35.0°C] and nighttime temperatures of 75°F [23.9°C] or above) and very humid weather with frequent rain showers. The *Sclerotium* fungus attacks the stems, crowns, and roots of more than 500 species of plants, including field crops, fruits, ornamentals, vegetables, and weeds. It also survives in soil as a sapro-phyte. The disease is most severe on heavily thatched turf that has previously been under drought stress. Dry conditions, followed by a prolonged rainy or very humid period, enhances disease development. Sclerotium blight is mostly a disease of annual bluegrass, creeping bentgrass, and bermudagrass on golf courses as well as Kentucky bluegrass and perennial ryegrass turfs (such as lawns and athletic turf).

Identification.   Yellow-to-white, thin, circular-to-crescent-shaped patches up to about 12 in (30.5 cm) across appear in mid to late spring or early summer. As the disease progresses, the patches increase rapidly in size. The grass at the border of the ring dies and turns reddish brown. The rings of dead grass grow outward up to 1 or 2 in (2.5 to 5.1 cm) per day during hot, moist, overcast weather. Mature rings may be 3 ft (0.9 m) or more in diameter (*Color plate 58* and figure 5–98). Patches on golf greens are yellow or reddish brown. Some patches commonly have tufts of apparently healthy grass in the centers. During very humid weather, an abundant, fluffy white-to-grayish mycelium is usually evident in wet thatch near the soil surface, growing on the dying or dead grass and soil surface at the edge of the patch. Clusters of small (1 to 3 mm in diameter), round, white or yellow (turning tan to dark brown) sclerotia form on the mycelium, the dead and diseased grass, and the thatch in the ring. Sclerotia are formed in largest numbers when the soil is hot (86°F; 30.0°C), quite acidic (pH 4), well aerated, and where organic matter is abundant.

Disease cycle.   The numerous small, round, mature, dark brown sclerotia in the thatch and upper soil serve as overwintering structures. Infection by

FIGURE 5–98   Southern or Sclerotium blight in a Kentucky bluegrass fairway. (Courtesy L. T. Lucas.)

basidiospores of the *Athelia* state is possible (figure 5–99). The sclerotia germinate in warm-to-hot (>75°F or 23.9°C), moist weather. The hyphae (figures 5–2 and 5–99) grow rapidly through the thatch or soil and penetrate grass leaves, stems, and crowns directly or through wounds. The foliage turns yellow and then reddish brown as it dies. Sclerotia formed on the dead grass and thatch may persist in the soil for several years. Cool temperatures and poorly aerated soils that are neutral to alkaline, tend to restrict the growth of *Sclerotium rolfsii*. Aerifying and dethatching equipment spread the sclerotia.

Importance.   Southern or Sclerotium blight is becoming a more important disease in warm, temperate-to-subtropical regions, especially on heavily thatched turf following prolonged wet weather alternating with short summer dry periods.

Management.   Follow recommended cultural practices to reduce thatch and drought stress and to maintain vigorous growth. Aerify to avoid compaction. Add lime, if needed, to raise the pH of the soil to 7.0 or 7.5. Renovate with less susceptible turfgrass species. The fungicides triadimefon, flutolanil, and cyproconazole have given the best control of this disease (see table 5–5) when applied at 2-week intervals during the spring and summer. Check with local authorities as to when to start treatments (in the eastern United States, about mid to late June; in California, the second or third week in May).

Speckled Snow Mold
See the section on snow molds (5.2.29).

Spermospora Leaf Spot, Leaf Blight, or Blast
See the section on minor leaf spots (5.2.16).

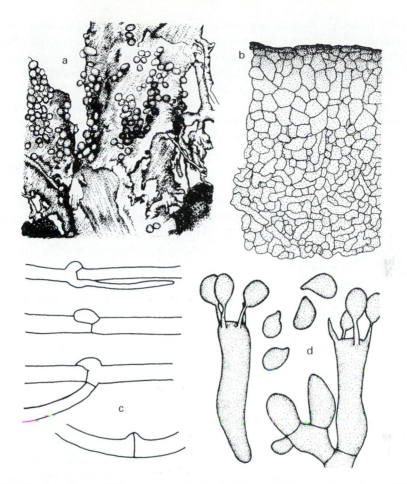

FIGURE 5–99 *Sclerotium rolfsii,* teleomorph *Athelia* (*Corticium*) *rolfsii,* is a cosmopolitan soil inhabitant in warm regions. It causes southern blight on a wide variety of hosts. (a) Base of plant covered by a fan of white mycelium and clusters of small, spherical, white (young) to dark brown (old) sclerotia the size of mustard seed (*Sclerotium* state); (b) cross section of a sclerotium; (c) hyphae showing various types of clamp connections; (d) *Athelia* state with two basidia bearing basidiospores and four discharged basidiospores. Sketches (b), (c), and (d) presented as viewed through a high-power microscope.

Spiroplasmal Diseases
See the section on minor viral, mycoplasmal, and spiroplasmal diseases (5.2.17).

5.2.31   Spring Dead Spot or Take-all Root Rot of Bermudagrass, Buffalograss, and St. Augustinegrass

Spring dead spot (SDS) is an important disease or disease complex of dormant bermudagrasses. Usually minor hosts include carpetgrass, kikuyugrass, and zoysiagrass. It is caused by the fungi *Gaeumannomyces graminis* var. *graminis, Ophiosphaerella* (*Leptosphaeria*) *herpotricha, Ophiosphaerella* (*L.*) *korrae,* and *O.* (*L.*) *narmari,* depending on geographic location. On some occasions, two of these fungi can be found in the same patch. All four fungi attack the root system and produce "runner" or ectotrophic fungi, which attack bermudagrass turf during the latter part of the growing season, weakening the plants that are then killed by cold-weather stress of winter. The grass appears healthy in the fall, goes into dormancy after a frost, and, upon the return of warm spring temperatures, the grass is dead. SDS is a widespread and serious disease of bermudagrass throughout its northern limits of adaptation, where winter temperatures are cold enough to induce the grass to go dormant. In the United States, SDS occurs mostly in the transition zone, where the average daily temperature during November is between 45°F (7.2°C) and about 57°F (13.9°C). The colder and longer that low temperatures are present, the greater the disease damage. The disease also occurs in Japan and Australia, where similar temperatures (54 to 57°F [12.1 to 13.9°C]) occur during winter (June to August). *Ophiosphaerella herpotricha* colonizes bermudagrass at soil temperatures between 59 and 77°F (15 to 25°C).

Spring dead spot is most common on compacted soils where the turf has been closely mowed, heavily fertilized with nitrogen in late summer or autumn, and watered. The best maintained turf is often very susceptible. Golf course fairways are usually the most severely infected. The disease can occur on greens that have been overseeded, frosted, or covered with straw or plastic. Spring dead spot usually appears in turf 3 to 5 (or more) years old that has developed a thick thatch. The disease may be confused with snow molds, winter injury, or damage by insects (chapter 6). SDS is usually more severe following very cold winters.

Identification. Sunken, circular, well-defined, bleached and dead patches appear with regrowth in the spring, varying in size from 1 in (2.5 cm) to more than 3 ft (0.9 m) in diameter (figure 5–100). The centers may remain alive after several years, resulting in doughnutlike rings. The patches tend to enlarge and reappear in the same spots for several years before disappearing. The sunken, straw-colored patches usually remain dead for several months to a year or more, becoming invaded by weeds, such as crabgrasses and annual bluegrass. The grass in the center of patches may start to grow outward resulting in a ring symptom (figure 5–100). The patches may coalesce, resulting in serpentine arcs of diseased grass. On closer examination, the crowns, roots, rhizomes, and stolons of SDS-affected bermudagrass plants develop a severe, dry, brown-to-black rot. Small, flattened, disc-shaped-to-irregular dark brown sclerotia (of *Ophiosphaerella* [*Leptosphaeria*]) and pseudothecia may form on or under basal leaf sheaths and stem bases and on the stolons and roots. The sclerotia, with a commonly associated network of dark brown septate, ectotrophic "runner" hyphae, are easily separated from plant surfaces. Infected grass plants can be readily lifted from the turf. The dark brown runner hyphae grow over the roots (see figure 5–106) and stolons. In the warmer areas of the bermudagrass belt where good broadleaf weed management is practiced, the patches are later covered by September with bermudagrass stolons growing in from the margins of the spots.

On greens, the recovered patches appear as sunken or thin areas with very little thatch. A number of other pathogenic fungi (for instance, *Bipolaris, Curvularia, Drechslera, Fusarium,* and *Pythium*) are frequently found in association with diseased bermudagrass plants.

FIGURE 5–100   Spring dead spot in a bermudagrass fairway. (Courtesy L. T. Lucas.)

Disease cycle.   The spring dead-spot fungi are believed to infect and colonize the roots and stolons of bermudagrass in late summer or early autumn when the soil is moist and daytime temperatures are in the low to mid 70s (21.1 to 23.9°C) by influencing acclimatization to low temperatures. This probably does not occur, however, until the plants are fully dormant with air temperatures of 50 to 60°F (10.0 to 15.6°C) or less. Primary infections come from the dark brown mycelium, sclerotia, and pseudothecia on and in the surfaces of the roots and stolons. Hyphal penetration of the epidermal cells is direct. The vascular tissues are invaded later and filled with a brown substance and dark, spindle-shaped sclerotia.

Importance.   Spring dead spot is a widespread, serious, and unsightly disease of mature, usually highly maintained turf of bermudagrass. This disease occurs mostly in northern areas, where winters are cold enough to induce dormancy. The disease is worst at low cutting heights and where the thatch is over ½ in (1.3 cm) thick.

Management.   Remove excess thatch at least once a year during the summer when it accumulates to ½ in (1.3 cm) or more. Aerify to relieve compaction and promote deep root development. Follow the suggested fertilizer program for bermudagrass in your area based on a soil test. Maintain *balanced* fertility and a soil pH of about 6.5. Avoid overfertilizing with nitrogen and potassium after late August. Maintain optimum soil levels of potassium (K) and phosphorus (P).

The soil should be well drained. Good surface and subsurface drainage should be provided when establishing a new turf area. Low spots where water may stand should be filled in or drained. Eliminate weeds from infected turfs, as this slows recovery of the bermudagrass. Mow at the maximum height.

U-3 bermudagrass, Tifton hybrids (such as Tifdwarf, Tiffine, Tifgreen, Tiflawn, Tifway), Sunturf, Texturf 1F, Tufcote, and other improved cultivars are usually more severely attacked than common bermudagrass, Midfield, Midiron, Midlawn, and Vamont. Cultivars with a high level of winter hardiness are less affected by spring dead spot.

One or more systemic fungicide sprays and drenches during the autumn (table 5–5) are effective in reducing but not completely suppressing stolon and root rot. Three or four monthly treatments may be needed, starting in September or October. Carefully follow the manufacturer's recommendations regarding rates and timing of applications.

Stagonospora Leaf Spot
See the section on Septoria and Stagonospora leaf spots (5.2.27).

Stem Eyespot or Phloeospora Leaf Spot
See the section on minor leaf spots (5.2.16).

5.2.32   Summer Patch and Necrotic Ring Spot

Summer patch (Poa patch on *Poa annua*) and necrotic ring spot (NRS) are distinct diseases that can develop in the same turf area (sward). They are serious and widespread diseases in the United States of established Kentucky bluegrass, annual bluegrass, and fine-leaf fescue turfs that are managed as amenity or sports turfs. Other grasses mildly susceptible to one or both of these diseases include rough bluegrass; colonial and creeping bentgrasses; annual and perennial ryegrass; tall fescue; bermudagrass; zoysiagrass; centipedegrass; many forage, wild, and weed grasses; and small grains.

Summer patch is a disease of the roots, crowns, stolons, and rhizomes that can kill entire grass plants. In addition, once a turfgrass plant is infected, it is nearly impossible to eradicate the fungus; that is, the disease is perennial.

Summer patch is serious on Kentucky bluegrass and annual bluegrass where nighttime temperatures stay above 70°F (21.1°C). Disease symptoms are most severe in sunny, exposed slopes or other heat-stressed areas (for instance, next to sidewalks, driveways, and parking lots) on turf with a thick thatch that is mowed low.

The normally weakly pathogenic, soil-borne fungus *Magnaporthe poae* (anamorph is a species of *Phialophora*) causes summer patch disease, and symptoms are generally seen when hot (82 to 95°F; 27.8 to 35.0°C), sunny days follow warm-to-hot, very-wet-then-dry periods. However, initial infections occur in the spring when soil temperatures stabilize at 65 to 70°F (18.3 to 20.0°C) at the 2-in (5.1-cm) depth. Only the outer cortical tissue of the roots are affected initially. *Magnaporthe poae* remains in the cortical tissue until the soil temperature reaches 70°F (21.1°C) and water pushes oxygen out of the soil, weakening the grass plants. The fungus then enters the vascular tissue of the roots and interrupts the flow of water and nutrients to the foliage, thus stressing the plant and predisposing it to drought and infection by other weak pathogens.

Necrotic ring spot is caused by the soil-borne fungus *Ophiosphaerella* (*Leptosphaeria*) *korrae*. The fungus causes damage primarily to Kentucky bluegrass turf in cooler areas of the United States (northeast, upper midwest, Pacific northwest, and Rocky Mountain states). Other hosts include annual and rough bluegrasses and fine-leaf fescues. Necrotic ring spot attacks the roots and crowns during cool, wet weather in spring. However, symptoms appear most frequently in the spring and autumn, when wet weather is followed by hot, dry periods. The disease is most severe in 2- and 4-year-old turfs.

Ophiosphaerella korrae differs from the causal agent of summer patch (*Magnaporthe poae*) in a number of ways. Most notably it is more aggressive in how it attacks turf. Because of this aggressiveness, symptoms of necrotic ring spot are observed in the spring and fall and are not greatly dependent on conditions that stress the turf.

Necrotic ring spot can occur over a pH range of 5.0 to 8.0 and is more severe in compacted soils. The disease can be confused with yellow patch and pink snow mold or Microdochium (Fusarium) patch, since all three diseases appear in wet spring (or early summer in more northern areas) and autumn weather and have similar foliar symptoms.

Summer patch and necrotic ring spot may occur in the same turf and produce confusingly similar symptoms. Both diseases are also difficult to control.

Magnaporthe and *Ophiosphaerella* (*Leptosphaeria*) have similarities that also resemble *Gaeumannomyces graminis* var. *avenae* (the take-all patch fungus). These fungi produce dark-brown-to-black "runner" or ectotrophic hyphae (figure 5–106) that grow on the roots, crowns, and rhizomes of grass plants.

Also associated with these diseases are other minor pathogens and species of *Fusarium* that are widely distributed but generally considered as secondary pathogens (see figure 5–38). These fungi may increase disease severity by stressing the infected turf, but they are not the primary cause of the patch diseases.

Summer patch and necrotic ring spot can become severe when turfgrass, especially Kentucky bluegrass (figure 5–101), is under one or more stresses and entering summer dormancy, although infection occurred several months earlier during cool weather. Moisture and heat stress, excessive watering, close mowing, a thatch exceeding ½ in (1.3 cm) with a soil reaction (pH) of above 7 or below 5, unbalanced applications of fertilizer (an excess of nitrogen and low levels of potassium and/or phosphorus), high populations of parasitic nematodes, such as the stunt nematode (*Tylenchorhynchus dubius*), soil compaction, and other factors that predispose turf to stress conditions can increase the severity of these diseases. The abiotic factors that most influence disease development appear to be related to the soil and thatch conditions of temperature and moisture. In particular, the factors

FIGURE 5–101   Severe summer patch in a Kentucky bluegrass lawn.

responsible for the rate of thatch decomposition, including good aeration under warm, moist conditions and an acidic soil pH (5.5 to 5.8), are strongly implicated.

Identification.   Summer patch and necrotic ring spot produce symptoms that are largely indistinguishable from each other except that necrotic ring spot is seen sooner and later in the growing season than summer patch. The first symptoms on Kentucky bluegrass are scattered, yellow, light green or darker gray-green, circular-to-irregular patches of slow-growing, thinned, or wilted grass typically 2 to 6 in (5.1 to 15.2 cm) in diameter. In prolonged hot weather, the plants soon wither and rapidly fade to a dull reddish brown or bronze, then a tan, and finally to a light straw color (figures 5–101 and 5–102). The younger, inner leaves of plants infected by the necrotic ring-spot fungus may turn purple to wine red before fading to brown and then straw-colored. The patches may become sunken rings or craters, elongated streaks or crescents, 1 to 3 ft (30.5 cm to 0.9 m) or more in diameter, that may coalesce to blight large irregular areas (figure 5–101).

The most characteristic symptom on Kentucky bluegrass in many areas of the United States is a roughly circular, doughnut-shaped ring of dead or stunted grass up to 2 or 3 ft (0.6 to 0.9 m) across, with tufts of apparently healthy grass in the center, giving a frogeye pattern (*Color plate 59* and figures 5–102 and 5–103). Plants die as the basal stem, crown, root, and rhizome tissues are destroyed by a hard and tough, black or dark brown, dry rot. The vascular tissues of these organs become filled with a brown, gumlike material. Affected plants can be easily lifted from the soil. With summer patch, serious turf damage occurs in a 7- to 10-day hot and dry period followed by a very wet period. Under these conditions, the numerous

FIGURE 5–102 Summer patch or necrotic ring spot in Kentucky bluegrass turf. (Courtesy R. E. Partyka.)

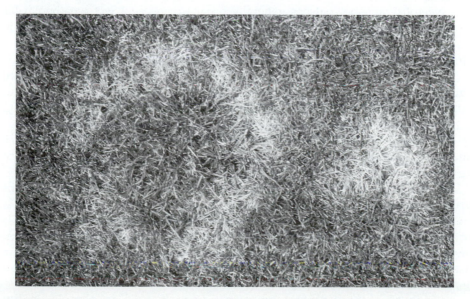

FIGURE 5–103 Summer patch in Kentucky bluegrass. A young patch is on the right, and an older, doughnut-shaped patch is on the left.

FIGURE 5–104   Summer (Poa) patch affecting annual bluegrass in a bentgrass green.

blighted areas (dead ring spots) overlap and coalesce (figure 5–101). The circular patches tend to increase in size (about 3 to 8 in; 76 to 20.3 cm) in diameter each year for several years and then subside or disappear. The dead turf is commonly invaded by weeds.

Necrotic ring-spot patches generally occur during cool wet weather in late spring or early summer, but heat and/or drought stress in July and August greatly intensify the damage, which becomes very obvious. The patches commonly persist over winter and are slow to green up in the spring, with affected plants being stunted.

Annual bluegrass (*Poa annua*) alone, or in creeping bentgrass turf, is killed, generally leaving creeping bentgrass unaffected. Affected *Poa annua* plants turn reddish brown, brown, or yellow and die in patches up to 12 in (30.5 cm) in diameter (figure 5–104). The patches may merge to blight large, irregular turf areas.

Diseased plants of creeping red fescue turn bronze or reddish brown in patches 3 to 6 in (7.6 to 15.2 cm) in diameter. The patches die with the leaves turning straw-colored.

Frequently, more subtle symptoms develop. Irregular patches that resemble drought, anthracnose, nematode damage, Pythium root rot, sod webworm, or chinch-bug injury contain living plants scattered among dead or weak plants. The turf in these areas is stunted and often a pale green to various shades of red, yellow or tan, eventually turning dull tan or brown, but not usually developing into distinct rings or patches. Such turf does not readily recover from mowing or adverse weather conditions.

Diagnosis of the summer patch–necrotic ring-spot disease complex is difficult and laboratory identification is needed which may take up to several months.

Samples should be sent to a plant disease clinic for accurate diagnosis (see Appendix C for a listing of land-grant institutions).

Disease cycle.   The causal fungi are believed to have the same disease or life cycle as for the take-all fungus (sec. 5.2.33). The necrotic ring-spot organism overseasons as dark brown, flattened sclerotia and black, flask-shaped pseudothecia or mycelium in colonized plant debris; the summer patch fungus, as mycelium in infected tissues. The two fungi spread from plant to plant by growing along roots and rhizomes up to 1 in (2.5 cm) or more per week. Dark hyphal strands progressively invade the crowns, roots, and rhizomes, which blacken as colonization proceeds. Both the necrotic ring spot and summer patch fungi are spread over long distances by infected sod, dethatching, and aerification equipment. Any one patch may die out or disappear after reaching a certain size.

Importance.   Summer patch (Poa patch on annual bluegrass) and necrotic ring spot are very important patch diseases of Kentucky and annual bluegrasses in the United States.

Management.   Follow the suggested control measures outlined in table 5–1 to avoid as many environmental stresses as possible. Promote root development and keep the grass growing steadily. Heat stress can be reduced on hot and humid days by briefly sprinkling (syringing) the turf one or more times from midday to 4 or 5 P.M., keeping the grass mowed during the summer months to the recommended height (2.5 to 3 in [6.4 to 7.6 cm] for lawns, parks, and the like, and 0.25 in [0.6 cm] for golf and bowling greens). Remove no more than one-third of the leaf surface at one cutting. Keep the soil and thatch pH between 5.5 and 5.8 by applying small amounts of sulfur or lime at frequent intervals based on a soil test. Avoid excessive applications of lime and high pH sand topdressings. Maintain adequate moisture on south- and west-facing sunny slopes and other sites (near parking lots, driveways, and sidewalks, for example) that dry faster and have higher thatch and soil temperatures than slopes facing east or north. Use soaker hoses on slopes where water tends to run off instead of slowly infiltrating the soil. Water in the morning or afternoon so the grass will dry before dusk. Adequate levels of nitrogen are required of a slow-release form (for example, sulfur-coated urea) every 3 weeks or use sewage sludges or animal waste products. Apply most nitrogen fertilizer in a slow-release form in the fall. Avoid close mowing heights, soil compaction and heavy foot traffic. Restrict the use of herbicides during hot weather.

Avoid pure stands of very susceptible cultivars of Kentucky bluegrass to summer patch and/or necrotic ring spot. Planting a blend of several improved Kentucky bluegrass cultivars that have good tolerance or resistance is recommended. The blending of 15 to 20% seed (by weight) of a turf-type perennial ryegrass mix with 80 to 85% seed of resistant Kentucky bluegrass cultivars will reduce the symptoms of summer patch and necrotic ring spot. Overseeding diseased turf with a blend or mixture of perennial ryegrasses, turf-type tall fescues, or resistant cultivars of Kentucky bluegrass also reduces symptoms. Cultivars of Kentucky bluegrass, perennial ryegrass, fine-leaf fescues, bentgrasses, and tall fescue differ in their resistance to *Magnaporthe poae* and *Ophiosphaerella korrae*. Check with your county extension (or advisory) office or turfgrass specialist regarding the availability of resistant turf-grasses. Heavily damaged turf may require complete renovation.

Where the cultural management practices just outlined and those in table 5–1 do not adequately control this disease complex, a number of systemic fungicides are available (tables 5–4 and 5–5). The fungicides should be applied and drenched into the root zone *before* disease symptoms become active again that season. Applications should start in mid to late spring when the soil temperature at the 2-in depth reaches 65 to 68°F (18.3 to 20.0°C) and when night temperatures first exceed 70°F (21.1°C). Repeat at 21- to 30-day intervals during the summer. Carefully follow the manufacturer's directions. Certain DMI fungicides and thiophanate materials are the only fungicides that control *both* summer patch and necrotic ring spot (table 5–5). Triadimefon (Bayleton) is effective only against summer patch. To control the disease on annual bluegrass on golf greens, apply a DMI fungicide at 21- to 28-day intervals, starting when the soil temperature at the 2-in level reaches 68°F (20.0°C), about 2 to 3 weeks after crabgrass seed germinates. Continue applications through August. Coring the turf and applying about ¼ to ½ in (1.3 to 0.6 cm) of water the day prior to treatment aids in movement of the fungicide into the root zone.

Superficial Fairy Rings
See the section on fairy rings (5.2.10).

5.2.33   Take-All Patch

Take-all patch, sometimes called Gaeumannomyces patch, is caused by the fungus *Gaeumannomyces graminis* var. *avenae.* It is a major disease of creeping bentgrass in cool, moist climates wherever bentgrasses are grown. Bentgrasses are commonly and seriously attacked, especially newly established turfs, during cool (50 to 65°F [10.0 to 18.3°C]) wet weather with a high sand content and/or a pH above 6.5 weather in fall and spring. In many cases take-all does not occur until 2 to 4 years after planting. Perennial ryegrass, annual bluegrass, and rough bluegrass are also susceptible. Other cultivated turfgrasses that may become mildly infected include Kentucky bluegrass, annual ryegrass, fine-leaf fescues, tall fescue, and crested wheatgrass.

The disease is most noticeable in the spring and autumn, on closely mowed bentgrasses that are excessively limed, improperly fertilized, and overly wet. Take-all patch is more severe and the causal fungus more aggressive on turf growing in light, sandy soils and on newly cleared land that was formerly in a forest or peat bog. It can be serious on turf seeded into soil recently fumigated with methyl bromide.

Identification.   Take-all patch usually first appears in early spring or autumn as scattered, more or less circular, sunken, light yellow, reddish brown, orange or bronze patches of dead grass that continue to enlarge, fade to a dull brown, and become gray in winter. Affected patches are 2 to 4 in (5.1 to 10.2 cm) in diameter at first and expand to 2 to 3 ft (0.6 to 0.9 m) or more across. Patches may coalesce to form large, irregular areas of dead turf. Typically, the dead sunken centers are invaded by annual bluegrass and other resistant grasses, or such broadleaf weeds as dandelions, clovers, and chickweeds, giving a frogeye or doughnut appearance (figure 5–105). Around the margin of a patch, where the fungus is active, a bronzed-

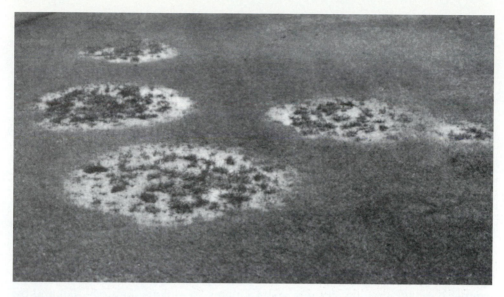

FIGURE 5–105   Take-all patch in a golf green. (Courtesy L. T. Lucas.)

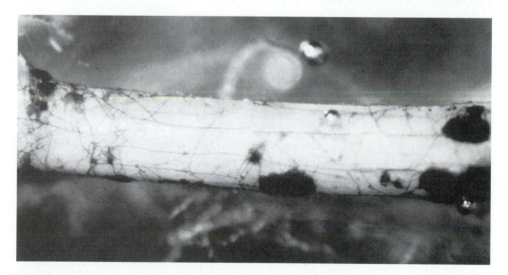

FIGURE 5–106   Dark ectotrophic or "runner" hyphae growing on a bentgrass root, together with black mycelial plates or mats of *Gaeumannomyces graminis* var. *avenae*. (Courtesy R. T. Kane.)

to-reddish-brown or yellow ring of infected grass plants have dark-brown-to-blackened crowns, stolons, and rhizomes; and dark brown, shallow, brittle, and rotted roots. Dark-brown-to-black runner or ectrotrophic hyphae and mycelial plates or mats of the fungus can be seen under the base of the leaf sheaths and crowns and on the rhizomes, stolons, and roots (figure 5–106). Dead plants toward the center of the patch can be easily peeled off. Several patches may merge to kill fairly large, irregularly shaped areas. The patches continue to enlarge up to 6 to 10 in

(15.2 to 25.4 cm) per year for several years before gradually declining, probably due to a buildup of antagonistic or competing microorganisms in the soil. In mixed stands, bentgrasses become thinned and yellow-brown to brown and dominated by weeds and nonsusceptible grasses.

The base of shoots, crowns, roots, and stolons in affected areas becomes dark brown to black before dying; thus, *Gaeumannomyces*-killed turf is very slow to recover. The results of disease may still be evident a year or more after the attack is first visible.

Disease cycle. The *Gaeumannomyces* fungus (anamorph a species of *Phialophora*) overseasons as mycelium in colonized but undecomposed grass debris in the thatch or soil and on perennial parts of living grass plants. The fungus may also be seedborne. Healthy crowns, stems, and underground parts of turfgrass plants are invaded by brown or colorless hyphae (under simple hyphopodia; see figure 5–20) during cool, moist weather (50 to 65°F; 10.0 to 18.3°C) in early spring and late fall. As colonization proceeds, the vascular tissues become filled with fungal hyphae. The fungus spreads from plant to plant by strands of dark brown mycelium (runner hyphae) growing *over* new roots, crowns, or stolons (figure 5–106) and by infested soil and colonized plant debris adhering to mowers and cultivation turf equipment.

Associated with the mycelial mats, often embedded in leaf sheath or crown tissue and on roots, are dark-brown-to-black, flask-shaped and beaked fruiting bodies (perithecia) that, when mature, contain large numbers of hyaline, septate, elongated ascospores (figure 5–107). Germinating ascospores can penetrate root epidermal cells and root hairs. Infected but symptom-free sod, sprigs, and plugs also serve to spread the fungus, which may persist indefinitely in turf as a saprophyte where the soil is alkaline and where the pH of irrigation water has a pH above 7.5.

Importance. Take-all is a major and widespread disease in cool, moist climates on bentgrasses and sometimes on perennial ryegrass that are overly wet, improperly fertilized, with an alkaline soil pH or high pH sand topdressings. Diseased patches commonly increase in size for several years before gradually declining.

During spring, in mixed stands of creeping bentgrass and annual bluegrass, the bentgrass can be weakened and killed by take-all patch, allowing the generally unaffected annual bluegrass to invade the bentgrass. During the summer months, take-all patch is inactive, but the annual bluegrass is attacked by one or more *Helminthosporium* fungi and *Magnaporthe poae* (the summer patch fungus), allowing the bentgrass to invade the weakened annual bluegrass. This back-and-forth process can go on for years. In most cases, however, annual bluegrass prevails, due to its prolific seed production and aggressive growth habits.

Management. Provide good surface and subsurface drainage when establishing a new turf area. Fill in depressions where water may stand. Avoid overwatering. Plant a blend or mixture of grasses. Topdressing mixtures for greens should have a pH below 5.5 to 6.0. In areas where take-all patch is a problem, incorporate 3 to 4 lb of *actual* elemental sulfur per 1000 sq ft (93 sq m) into the top several inches of soil prior to planting.

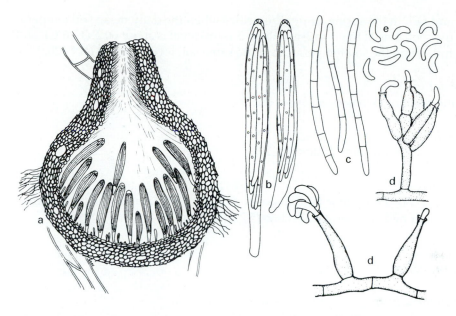

FIGURE 5–107   Take-all fungus, *Gaeumannomyces graminis*, as seen under a high-power microscope: (a) erumpent, dark-brown-to-black perithecium; (b) two asci with eight ascospores each; (c) three ascospores; (d) anamorph or *Phialophora* state with conidiophores and conidia; (e) nine conidia. The perithecia may be evident in or under the leaf sheaths and crowns.

Practice *balanced* soil fertility based on a soil test. Follow local recommendations for the grass or grasses grown. Avoid overliming! Apply lime *only* when absolutely needed on bentgrasses (pH below about 5.0), using a slowly dissolving type (about 20 mesh). For fertilizer, use ammonium sulfate, ammonium chloride, monoammonium phosphate, muriate of potash (KCl), or other sulfur-bearing material containing nitrogen to maintain the pH at 5.5 or slightly below. Fertilizer applications are usually called for in late March, mid-May, late June, and early September, using an ammonium-based nitrogen (N) fertilizer applied with ones containing adequate phosphorus (P) and potassium (K) in a 3:1:2 ratio. Three to 5 lb/sq per 1000 sq ft (93 sq m) of *actual* elemental sulfur (S) in several split applications is often recommended for each growing season where take-all patch is a serious problem. The disease frequently survives the first season of sulfur or ammonium-based fertilizer applications, which must be continued for 2 or 3 years to be effective. If serious, acid injection of irrigation water may be needed to keep the pH in the acid range (below 6.5).

Take-all patch is often more severe where thatch has built up and the soil has been fumigated. This is probably due to the killing of organisms antagonistic to or parasitic on the take-all fungus.

Once take-all patch has invaded a turf, it is nearly impossible to eliminate. You can reduce severity of the disease by using a combination of cultural methods and systemic fungicides, such as fenarimol and propiconazole (table 5–5). Application of fungicides should coincide with the activity of the pathogen, namely, spring and fall. In the spring, apply fungicides when the daily mean soil temperature of the root zone at the 1- to 2-in (2.5- to 5.1-cm) level is about 60°F (15.6°C), and in autumn, when the soil temperature at the same soil level is about 68°F (20.0°C).

Tar Spot
See the section on minor leaf spots (5.2.16).

Typhula Blight
See the section on snow molds (5.2.29).

Virus Diseases
See the sections on minor viral diseases (5.2.17)and St. Augustinegrass decline
    and centipede mosaic and decline (5.2.25).

White Blight
See the section on fairy rings (5.2.10).

Wojnowicia Leaf Spot, Foot Rot, and Root Rot
See the section on minor leaf spots (5.2.16).

Yellow Leaf Spot
See the section on Helminthosporium diseases (5.2.13).

Yellow Patch
See the section on Rhizoctonia diseases (5.2.23).

Yellow Ring
See the section on fairy rings (5.2.10).

Yellow Tuft
See the section on downy mildew (5.2.9).

Zonate Leaf Spot
See the section on Helminthosporium diseases (5.2.13).

5.2.34   Zoysia Patch

Zoysia patch, a disease caused by the soil-borne fungus *Gaeumannomyces incrustans*, has been observed on sod farms and golf course fairways in the Mississippi Valley area of Missouri as well as in Arkansas, Illinois, Indiana, Kansas, Oklahoma, and Tennessee in the transition zone. This disease can be grouped with those collectively known as take-all diseases. A similar fungus, *Gaeumannomyces graminis* var. *graminis,* also will infect and kill the roots of zoysiagrass, but it is distinct from zoysia patch in that no patch-type symptoms have been reported to be caused by it in zoysiagrass turf.

Identification.   Roughly circular patches of yellowed and thin turf appear in the spring as the zoysiagrass begins to break dormancy. The affected turf is stunted and slightly sunken with yellow leaves and weakened or rotted roots. A diagnostic key to differentiate zoysia patch from large brown patch (caused by *Rhizoctonia solani*) is the presence of rotted roots resulting from infection by *G. incrustans*. The turf is thinned to where only 10 to 15% of the shoots remain alive. The disease is again active in the fall as cooler temperatures start to slow growth of the zoysiagrass. The patches grow out radially each season at the rate of up to 3 ft (0.9 m) or more. Diseased patches, which persist for several years, can become irregular as individual patches merge and may later be invaded by grassy or broadleaf weeds. The disease is often more severe in poorly drained sites.

Disease cycle.   The *Gaeumannomyces* fungus is like the take-all fungus in producing black "runner" hyphae over infected roots (see figure 5–106). It also has a *Phialophora* state. A temperature range of 60 to 70°F (15.6 to 21.1°C) and moist soil appear optimal for root-rot development.

Importance.   Zoysia patch can seriously thin and kill roughly circular patches. It can be serious on sod farms and golf course fairways in the more northern limits of zoysiagrass adaptation in the United States. This disease is active under conditions similar to those associated with large brown patch of zoysiagrass (*R. solani*). Both fungi can simultaneously attack the same turf.

Management.   Keep the turf vigorous when zoysiagrass is actively growing. Fertilizing properly, increasing the mowing height to 1½ in (3.8 cm) if at all possible, watering, dethatching, and core aerifying are suggested management practices. Aerify and verticut in late spring or early summer to rejuvenate the diseased turf. Improve drainage by filling in low areas and installing slit trenches to drain excess water away. Do not apply more than 2 lb of nitrogen per 1000 sq ft (93 sq m) per year (100 kg N/ha/yr). Avoid overwatering in spring and fall.

Fungicides listed for take-all patch and spring dead spot of bermudagrass are available and may help in suppressing disease development (table 5–5). Preventive applications need to be made in early fall, *before* zoysia-patch symptoms become evident and the thatch temperature falls below 70°F (21.1°C). Fall applications of fungicide may suppress or delay development of the disease in spring.

5.3   *TURFGRASS DISEASE MANAGEMENT*

Turfgrass disease management should include frequent inspections combined with a rapid and accurate diagnosis of the observed problem. Diagnosis is based on knowledge of the turfgrasses, growth habits and requirements, and possible problems, such as noninfectious or abiotic disorders (chapter 3), weeds (chapter 4), diseases (the current chapter), and insects (chapter 6).

Turfgrass diseases can vary in severity from year to year and from one neighborhood, area, or region to another, depending on environmental conditions (principally the amount and distribution of rainfall or irrigation, temperature, humidity, grass nutrition, and other cultural practices followed), the relative resistance or susceptibility of the turfgrass host, and the causal fungus or other pathogen. For disease to develop, these three factors must be conducive. For example, if the environment is favorable for a disease, and the disease-producing fungus, bacterium, or virus is present, but the host plant is highly resistant, little or no disease will develop. Similarly, if the causal agent is present, and the host is susceptible, but the environment is unfavorable, disease will be mild or nonexistant. A disease results from the right combination of a susceptible grass plant, a virulent pathogen, and an environment that favors infection, colonization of the host, reproduction, and spread of the pathogen, plus the necessary time for the disease to develop.

We can put this relationship in the form of a simple equation:

Susceptible grass + Disease-causing pathogen + Proper environment
+ Spread of the pathogen + Time to develop = Disease

If any one of the foregoing key ingredients is lacking or not conducive, then it is highly probable that disease will either not develop at all or be mild if it does. Effective disease-management measures are aimed at breaking this equation in one of three basic ways: (1) The susceptible turfgrass plant is made more resistant; (2) the air and soil environment is made less favorable for the causal pathogen and more favorable for growth of the grass plant; and (3) the pathogen is killed or prevented from reaching the plant, penetrating it, and producing disease.

5.3.1   Three Basic Methods of Disease Management

1. *The grass plant is made more resistant.*   This is the ideal method of management. Turfgrass breeders cooperating with plant pathologists are working to develop more resistant grasses. Much progress has been made. We now have regionally adapted grass cultivars that have at least limited resistance to dollar spot, red thread, snow molds, one or more Helminthosporium diseases, one or more rusts, powdery mildew, summer patch and necrotic ring spot, gray leaf spot, St. Augustinegrass decline (SAD), spring dead spot of bermudagrass, bacterial wilt, and other diseases. This important management measure is progressing but is still in its infancy. For some diseases such as brown patch, Pythium blight, and rusts, where the causal fungi are composed of a number of species and many biotypes, strains, or races, the development of highly resistant or immune grass cultivars is many years into the future. Before such grasses can be developed and released, sources of resistance in wild or cultivated grasses must be found or genetically engineered. Then comes further years of breeding to get this resistance into otherwise desirable grass cultivars. The resistant cultivars may become diseased in a few years, however, if new races of the pathogen arise that overcome the resistance.

   Grass cut at the ideal height of 2½ to 3 in (6.3 to 7.6 cm) for bluegrasses, ryegrasses, and fescues, and ½ to 1 in (1.3 to 2.5 cm) or somewhat less for bentgrasses, bermudagrasses, and zoysiagrasses, usually has less severe disease and recovers more rapidly from a disease attack than does turf cut too short (less than ¼ in [0.6 cm] for golf or bowling greens and less than 1 to 1½ in

[2.5 to 3.8 cm] for lawn-type grasses). Without sufficient green-leaf area to manufacture food to produce new leaves, roots, rhizomes, and stolons, the grass is definitely weakened or stressed. Close mowing makes most diseases worse, particularly anthracnose, dollar spot, Helminthosporium diseases, brown patch or Rhizoctonia blight, rusts, summer patch, and injury caused by plant-parasitic nematodes. Grasses maintained as a mowed turf, as in a lawn, park, athletic field, cemetery, fairway, or golf green, are more subject to attack by disease-producing pathogens than they are in their natural wild environment. Exceptions are the rust diseases and pink patch, which are worse on unmowed grasses. Healthy, vigorously growing, well adapted turfgrasses that are properly managed can best reduce the severity of disease as well as insect, weed, and other problems.

Another means of increasing plant vigor and tolerance to attack by pathogens is through properly balanced plant nutrition. Dollar spot, red thread, pink patch, pink snow mold or Microdochium (Fusarium) patch, gray snow mold or Typhula blight, Pythium blight, powdery mildew, rusts, brown patch, summer patch, necrotic ring spot, anthracnose, take-all, spring dead spot, gray leaf spot, Helminthosporium, and other diseases are less serious when a balanced level of soil nutrients based on a soil-test report is maintained in the root zone. This may mean more frequent and lighter applications of fertilizer and keeping the three major nutrients—nitrogen (N), phosphorus (P), and potash or potassium (K)—in balance. When N is high in relation to P and K, there may be disease trouble ahead, especially in hot weather. At least 50% of all the nitrogen applied should be from a slow-release source such as isobutylidene diurea (IBDU), methylene urea, sulfur-coated urea, and composted sludges or animal waste products. A high level of potassium and, to a lesser degree, phosphorus, helps reduce injury from brown patch, dollar spot, pink snow mold, gray snow mold, red thread, pink patch, take-all patch, Pythium blight, and Helminthosporium-caused diseases. Application of acidifying nitrogen fertilizers (such as ammonium sulfate or chloride) has been shown to reduce the severity of pink snow mold, spring dead spot, and take-all patch, and keep the soil pH between 5.5 and 6.0.

2. *The environment is made less favorable for the pathogen.* Most fungi that cause turf diseases thrive in the turf-thatch-soil environment given an adequate supply of food, moisture, oxygen, and a favorable temperature. The basic concept of management here is to grow grass in an environment that will be unfavorable to the growth, multiplication, spread, and infection by disease-producing pathogens. Try the following techniques.

a. *Keep the grass blades as dry as possible for as long as possible.* Fungi, with the exception of powdery mildew, require free moisture on the grass plant for 1½ to 12 hours or longer to infect a plant. Poling, brushing, and hosing the grass early in the morning will remove the dew and guttated water in which these organisms thrive.

The timing, duration, and frequency of irrigation may greatly affect the intensity of disease. Frequent and light irrigations discourage deep root development. Such turf is much more susceptible to damage during extended droughts. Frequent and light irrigations greatly increase summer patch of

Kentucky bluegrass and creeping red fescue during periods of high temperature. However, such irrigation is sometimes recommended where insects, nematodes, or necrotic ring spot have caused extensive damage to roots.

Water established turf thoroughly during a drought. Moisten the soil to a depth of 6 in (15.2 cm) or more with each irrigation. Water as infrequently as possible. Avoid daily light sprinklings, especially in mid to late afternoon or evening, that result in grass blades remaining wet overnight. Another way to speed drying of the grass is to increase light penetration and air movement by selectively pruning or removing dense trees, shrubs, or other barriers that border the turf area. Golf course superintendents sometimes install fans around pocketed, low-lying greens to increase air flow over the grass surface.

Poor surface and subsoil drainage lead to compaction and soil-aeration problems. Roots suffocate from lack of oxygen, setting up an anaerobic relationship that can lead to other problems such as black layer, algae and mosses, and the germination of weed seed. The indirect result, too frequently, is disease (such as Pythium diseases, snow molds, Rhizoctonia diseases, downy mildew, or yellow tuft). Stagnant, humid air over a pocketed turf area causes disease problems due to a lack of wind to dry off the grass blades. If we could keep grass dry, we would have few foliar disease problems. Root rots that cause wilt of golf greens in the summer and early fall are commonly due to overwatering of the root zone. Keeping the soil at or near the saturation point prevents normal root growth and favors the growth of organisms that are common water molds (such as *Pythium*). Proper water management is the single biggest environmental factor in keeping diseases in check on frequently watered and highly maintained turf areas. For a few turf problems, where summer root growth is short or severely diseased (for instance, in summer patch, necrotic ring spot, and bermudagrass decline), or where soil nematode populations are high, it may be necessary to water affected areas lightly one or more times on a daily basis when disease is active.

b. *Remove excessive thatch.* Most disease-causing organisms grow and thrive in thatch. Removing it takes away the food supply for the fungi that cause dollar spot, Fusarium blight, Pythium blight, brown patch, Helminthosporium diseases, fairy rings, red thread, pink patch, summer patch, yellow ring, white patch, and many other diseases. This forces the causal fungi to compete unfavorably with the multitude of bacteria and fungi in the soil (4 to 8 billion microorganisms per pound of topsoil). Many of these soil-inhabiting organisms are antagonistic (suppressive) to, or may even parasitize, the disease-producing fungi that attack grass.

Periodic removal of thatch can greatly reduce the use of fungicides and other chemical-management measures. Remove thatch when the grass is growing vigorously in early spring or early autumn and it accumulates to ½ in (1.3 cm) or more in higher-cut, lawn-type grasses and ⅛ in (0.3 cm) or less on golf and bowling greens. You can buy or rent a vertical mower, power rake, or similar dethatching machine at equipment-rental or large garden-supply stores. (*Note:* Excessive thatch accumulation may reflect excessive fertilizer and pesticide applications.)

Although clippings do not add significantly to a buildup of thatch, their removal may help in the control of several important diseases, including Pythium blight, rusts, leaf smuts, brown patch, dollar spot, various leaf spots, red thread, and Helminthosporium diseases. The removal of clippings from infected plants can also reduce the number of pathogens, such as those causing rusts and leaf spots, and hasten recovery.

c. *Avoid competition.*   Keep large trees as far away from golf greens, tees, and other heavily shaded turf areas as possible; or install root barriers to keep the roots from "robbing" the grass of nutrients and moisture during drought periods.

d. *Avoid injuring the grass.*   Careless use of pesticides and fertilizers, use of a mower that is dull or out of adjustment, compaction—walking or riding on turf that is soggy, removal of more than one-third of the foliage at one mowing, and the like, can cause serious injury. Avoid dumping piles of mowed clippings, especially on wet turf, and remove tree leaves promptly. Remember that anything that weakens or stresses grass may lower its natural resistance, allowing a disease-producing fungus to take over. Compacted turf areas should be aerified using a power machine that can be purchased or rented at many of the stores that carry dethatching equipment.

e. *Eliminate excess traffic and soil compaction.*   Foot and vehicle traffic produce wounds, allowing fungal pathogens easy infection sites. Compaction from heavy traffic prevents air and water movement into the soil, restricts root function, and results in a decline of both plant vigor and disease resistance. Anthracnose, Helminthosporium diseases, and summer patch are just several of the many diseases that are more damaging in highly trafficked or compacted turf sites. Core aerification alleviates compaction and reduces the severity of disease.

3. *The pathogen is killed or prevented from reaching the plant or developing and producing disease.*   This can be done by removing moisture from the grass blades, thus preventing a fungus from penetrating and establishing a disease relationship; using sand or other amendments to improve surface and subsurface drainage and aeration; buying only top-quality seed, sod, sprigs, or plugs from a reputable dealer who has followed a sound disease-management program; and by following suggested weed and insect control practices (see chapters 4 and chapter 6). The principal means of management here is chemical.

5.4   CHEMICAL MANAGEMENT OF TURFGRASS DISEASES

The pesticides used to protect against pathogens of turfgrasses include soil fumigants, nematicides, and fungicides (or fungistats). For meaningful disease-management strategies, it is necessary to have an understanding of the nature of these three types of pesticides.

5.4.1   Soil Fumigants

A soil fumigant (or biocide) can be applied to the turf area *before* planting to kill simultaneously soil-borne fungi, bacteria, nematodes, insects, other soil fauna, and weed seeds. Chemicals such as methyl bromide, methyl bromide-chloropicrin mixtures, metam (Vapam), and dazomet (Basamid) perform this function. The expense is high, but fumigation is being done before seeding, sodding, sprigging, or plugging golf and bowling greens, turfgrass nurseries, and athletic fields.

Before application of a fumigant, the soil should be (1) at a moisture level for good seed germination (not too dry or wet); (2) thoroughly prepared as for a fine seedbed and free of large clods, roots, other coarse organic debris, rocks, and stones; and (3) at a temperature between 60°F (15.6°C) and 80°F (26.7°C) at the 2-in (5.1-cm) depth in the soil. A gas-proof plastic cover is generally placed over the treated area for several days or longer to retain the toxic fumes of the fumigant. *Only experienced and certified and/or licensed pesticide applicators are permitted to buy and apply these toxic chemicals.* Because it takes 1 to 3 weeks for the fumes to dissipate in the soil, seeding, sodding, sprigging, or plugging must be delayed for 10 days to several weeks, depending on the fumigant, temperature, and type of soil. (This is why fumigation is often done in early fall.)

Following fumigation, disease and nematode problems may become more severe due to the lack of competitive and antagonistic fungi, bacteria, nematodes, and other beneficial soil organisms in the treated area. Once a pathogen is introduced (blown, washed, tracked, or hauled) into a treated area, there is no natural biological check and balance. As far as disease control is concerned, soil fumigation is most useful for controlling the fungi that cause brown patch or Rhizoctonia blight, summer patch, necrotic ring spot, take-all, bermudagrass decline, leaf smuts, gray snow mold, southern blight, Pythium blight, red thread, and dollar spot.

5.4.2   Nematicides

Nematode problems in soil are greatly reduced before planting by application of a soil fumigant. A fumigant applied just to control nematodes is justified only where the turf has had a problem or history of high populations of plant-parasitic nematodes. When turfgrass is established, a granular form of fenamiphos (Nemacur; see table 5–5) properly applied uniformly to the turf will sharply reduce nematode numbers for as long as 12 weeks. Fenamiphos rapidly causes nematodes to stop feeding by blocking neuroenzymes that interfere with its movements, resulting in paralysis and then death. Most nematicides also control such insects as chinch bugs, cutworms, grubs, sod webworms, and other pests (chapter 6). The nematicide-insecticide is drenched *immediately* into the thatch and soil using ½ to 1 in (1.3 to 2.5 cm) of water (300 to 600+ gal) per 1000 sq ft (93 sq m) of turf. Treatment is best done in the autumn or spring. (Treat for both seasons if nematodes are a serious problem, as they can be on high-sand golf greens and other turf areas when the soil temperature at the 2-in [5.1-cm] depth is 60°F [15.6°C] to 80°F [26.7°C]. Do *not* apply a nematicide at temperatures above 80°F [26.7°C] or 85°F [29.4°C] due to the high potential for turf injury.) Core aerification of the turf before application improves the results. *This is not a treatment for newly established turf areas.* Since nematicides are highly toxic chemicals, it is important that all the manufacturer's directions and precautions be carefully followed. *These chemicals are for use only by certified and/or licensed pesticide applicators.* Most turfgrass specialists say, "Never apply a nematicide without first having a nematode analysis."

5.4.3   Fungicides

Effective management of turf diseases depends on a rapid and accurate diagnosis. Experienced golf course superintendents, sod growers, lawn-care-company personnel, and other turfgrass managers can usually recognize or predict the occurrence of common diseases and hence can promptly initiate proper management practices, including fungicides, where appropriate. Homeowners, however, are usually unable to diagnose turf diseases until substantial damage has occurred, when the application of fungicides is a waste of money and inappropriate.

There are a variety of chemicals available to manage turfgrass diseases caused by fungi. Fungicides are formulated and sold to kill certain groups of fungi (true fungicides) or to inhibit infection and colonization or reproduction of fungi (fungistats) for a period of days, weeks, or months. Table 5–4 lists the available generic (common or coined) and trade (or brand) names of chemicals used to manage fungi that cause turf diseases. Fungicides that act as fungistats include the benzimidazoles (benomyl and thiophanate-methyl compounds), and the demethylation inhibitor fungicides (DMI), also called sterol-inhibitor fungicides, which include cyproconazole, fenarimol, mycobutanil, propiconazole, terbuconazole, and triadimefon.

The fungicides used to control diseases of turfgrasses are of two general types: protectant-contact and systemic or eradicant fungicides. *Protectant-contact* fungicides are applied to the foliage as a barrier against germinating, disease-causing fungal spores and to check the growth of fungal hyphae before they can enter susceptible grass tissue. This type of fungicide prevents disease and usually cannot kill fungi after invasion of grass tissue has occurred. These fungicides must be applied fairly frequently to turf (usually at 5- to 10-day intervals) since mowing and weathering (washing away, breakdown by sunlight) remove much of the surface protective shield barrier soon after application. Relatively high spray volumes (2 to 5 or 10 gal of water per 1000 sq ft [93 sq m]) are required to supply uniform and continuous coverage of the foliage and thatch with the fungicide. Adding a spreader-sticker (or surfactant) to the spray mixture often facilitates better foliar coverage. Many available fungicides for turf use are the protectant-contact type; some also have very limited systemic activity. These include anilazine, chlorothalonil, chloroneb, ethazole or etridiazole, iprodione, maneb, mancozeb, propamocarb, quintozene (PCNB), thiram, vinclozolin, and mercury chlorides where these materials can still be used.

Systemic fungicides (some researchers prefer to call them penetrants) give a protective exterior shield against infection. After penetration, they are absorbed and translocated within the grass plant in toxic quantities, destroying or suppressing growth to established infections of the target fungus.

TABLE 5–4   GENERIC (OR COINED) NAMES, TRADE (OR BRAND) NAMES, AND MANUFACTURERS OR FORMULATORS OF TURFGRASS FUNGICIDES

Generic, coined, or common name	Some trade names (manufacturers/formulators)
Anilazine	Dyrene (Bayer)
Azoxystrobin	Heritage (Zeneca)
Benomyl	Benomyl (Bonide)
	Benlate (Agrochemicals)
Captan	Captan (Bonide, Drexel, Gustafson,* Micro Flo, SureCo, Zeneca)
Chloroneb	Proturf Fungicide V (The Scotts Co.), Tee Time 75% Chloroneb (Andersons)
	Teremec SP (PBI/Gordon)
	Terraneb SP (Kincaid)
Chloroneb + thiophanate-methyl	Fungicide IX (The Scotts Co.)
Chlorothalonil	Daconil 2787, Daconil Ultrex (ISK Biosciences)
	Echo 500 (Sostram), Daconil 2787 (Fermenta)
	Flo-130, Ornatek, Omathal, Thalonil (Terra), and Thal-o-nil (NCH)
	Lawn Fungicide 2787 (Rockland)
	Manicure (LESCO), Tee Time 5% Daconil Fungicide (Andersons)
	Thalonil (Riverside/Terra International)
	Turf Fungicide (Lebanon Seaboard)
Chlorothalonil + fenarimol	TwoSome Flowable Fungicide (LESCO)
Chlorothalonil + thiophanate-methyl	ConSyst (Regal)
Cyproconazole	Sentinel (Sandoz)
Etridiazole (ethazole)	Koban (The Scotts Co.)
	Terrazole (Uniroyal)
Fenarimol	Rubigan (Dow/Elanco)
Flutolanil	ProStar (AgrEvo)
Flutolanil + triadimefon	ProStar Plus (AgrEvo)
Fosetyl-Al	Chipco Aliette (Rhône Poulenc), Aliette T&O (Terra)
	Prodigy (LESCO)
Fosetyl-Al + mancozeb	Chipco Aliette (Rhône Poulenc)
Iprodione	Chipco 26019 (266 T Rhône Poulenc)
	Fungicide X (The Scotts Co.)
Mancozeb	Fore (Rohm and Haas)
	Formec 80 (PBI/Gordon)
	Mancozeb (Bonide, LESCO)
	Penncozeb (Elf Atrochem)
	Protect T/O (Cleary)
Metalaxyl	Apron* (Gustafson, Ciba)
	Pythium Control (The Scotts Co.)
	Subdue (Ciba)
Mefenoxam	Subdue (Novaris), Quell (Uniroyal)
Metalaxyl + mancozeb	Pace (Ciba)
Mycobutanil	Eagle (Rohm and Haas)
Oxadixyl	Anchor* (Gustafson)
Propamocarb hydrochloride	Banol (AgrEvo)
Propiconazole	Banner (Ciba)

TABLE 5–4   GENERIC (OR COINED) NAMES, TRADE (OR BRAND) NAMES, AND MANUFACTURERS OR FORMULATORS OF TURFGRASS FUNGICIDES (continued)

Generic, coined, or common name	Some trade names (manufacturers/formulators)
Quintozene (or PCNB)	Defend, PCNB (Cleary)
	Engage (Turfgo/United Horticultural Supply)
	Penstar, Penstar Flo (The Scotts Co.)
	Revere (LESCO)
	Terraclor, Turfcide, Turfcide 400 (Uniroyal)
Trifloxystrobin	Compass (Novaris)
Terbuconazole	Lynx (Bayer)
Thiophanate-methyl	Cleary's 3336 (Cleary)
	Fungo 50 Proturf Systemic Fungicide (The Scotts Co.)
	SysTec 1998 (Regal)
Thiophanate-methyl + iprodione	Fluid Fungicide (The Scotts Co.)
Thiophanate-methyl + mancozeb	Duosan (The Scotts Co.)
Thiram	Lawn Disease Control (Bonide)
	Spotrete (Cleary)
	Thiram (Gustafson,* LESCO, Micro Flo, UCB, Rhône Poulenc)
	Thiramad (Sierra)
Thiram + triadimefon	Fluid Fungicide III (The Scotts Co.)
Triadimefon	1% Turf Fungicide with Bayleton (Bonide, Rockland)
	Accost (Turfgo/UHS)
	Bayleton (Bayer), Tee Time Bayleton Fungicide (Andersons)
	Fungicide VII (The Scotts Co.), Bayleton 1% Granular (Lebanon)
	Granular Turf Fungicide (LESCO)
	Turf Fungicide 1% Bayleton (Lebanon Seaboard)
Triadimefon + metalaxyl	Fluid Fungicide II (The Scotts Co.), Drench Pak (Cleary)
Vinclozolin	Curalan (BASF)
	Touché (LESCO)
	Vorlan (The Scotts Co.)

*Seed treatment

Note: Mention of a trade name or proprietary product does not constitute warranty or endorsement, a guarantee of its effectiveness, or approval of the material to the exclusion of comparable products that may be equally suitable. No discrimination is intended against products marketed under trade names not mentioned. Always, consult a current label for information concerning rate, application frequency, tank mixing, compatibility, and other critical facts. Some states or countries may not have approved all the fungicides or the specific labeling for all diseases noted. Some products may also have been withdrawn from the market in certain states or countries. In addition, labels are constantly being rewritten for a number of reasons, including recent changes in fungal taxonomy, adding or removing diseases, and changes in the percentage of the active ingredients.

Except where indicated on the label, all materials should be applied in 2 to 5 gal of water per 1000 sq ft (93 sq m), 90 to 228 gal/acre, or 840 to 2100 liters/hectare.

There are three types of systemic, disease-managing chemicals or penetrants: (1) *localized penetrants* (such as iprodione, propamocarb, and vinclozolin), which are absorbed into the immediate area of entry; (2) *acropetal penetrants,* are only enter plant tissue, move across the leaf tissue, and translocated upward in plants via xylem (water-conducting) vessels—these systemic materials include azoxystrobin (Heritage), benomyl, thiophanate-methyl, and -ethyl cyproconazole, fenarimol, flutolanil, metalaxyl, mycobutanil, propamocarb, propiconazole, terbuconazole, and triadimefon; and (3) *systemic penetrants,* which are translocated in both xylem and phloem (food-conducting tissue) and distributed uniformly throughout the grass plant in quantities toxic to the target fungus. The only true systemic fungicide is fosetyl-Al, which is marketed mainly to manage Pythium diseases. Trifloxystrobin (Compass) is a mesosystemic fungicide that acts on the plant surface being absorbed by waxy layers on the plant surface being absorbed by waxy layers on the plant and has "translocational movement" by water and vapor action.

Systemics have the advantage of long residual action, protection of plant crowns and roots, movement within plants to protect newly formed tissues, eradication of fungi already inside plants, and protection from washoff and weathering. Certain products control specific diseases for several weeks before being degraded, leached from the soil, or bound up within the plant or soil. Then the activity of the systemic fungicide is lowered to where it is no longer effective and a new application is needed. The period of time a systemic fungicide gives control depends on the soil type, the disease-producing fungus, growth of the grass plant, and other factors. Systemic fungicides are applied to the foliage or roots and are absorbed by the plant and distributed internally. For certain diseases (such as leaf smuts, summer patch, necrotic ring spot, Pythium root rot, and take-all patch), the fungicide needs to be drenched or watered in immediately for best results, preferably following core cultivation.

Both protective-contact and systemic fungicides are available in an expanding number of formulations for application as sprays or soil drenches. These include wettable powders (WP), liquids (L), flowables (F, Fl, Flo, DF, WDG), soluble (S) formulations, aqueous suspensions (AS), emulsifiable concentrates (E, EC), and water-soluble plastic pouches (WSP), all of which contain measured amounts of fungicides and are dropped into the spray tank. Some fungicides are also available as granular (G) formulations. Granular fungicides are mostly used on dormant grass to control snow molds and other cold-weather diseases or on closely cut golf and bowling greens. Granular products are *not* very efficient in covering the leaf surfaces of lawn-type turfs. Presumably they act by suppressing fungal growth and spore production in the thatch, although several are systemic and taken up by the grass leaves and roots.

Some fungicides are sold as combination products (table 5–4) containing two or more active ingredients to give broad-spectrum control.

Most chemical management of turfgrass diseases is through the use of turf fungicides that are applied preventively *before* disease strikes. Applying fungicides after a disease has significantly damaged turf is bound to be an expensive failure. Curative applications when early symptoms of disease are *first* evident save money but must be made *before* severe damage is widespread. *The manufacturer's directions on the container label should be carefully followed regarding rates to use, intervals between applications, compatibility with other chemicals, grasses on which the fungi-*

cide is to be used, and safe use and handling. The key to whether to use a fungicide is vigilant scouting on a daily to weekly basis, depending on the grass, weather conditions, and maintenance level of the turf (low, moderate, or high).

The method of application is also very important. If spraying, use 2 or 3 to 5 or more gal of spray per 1000 sq ft (93 sq m) or 90 to 225 gal per acre (840 to 2100 liters per hectare) to adequately wet the grass blades, thatch, and top ¼ in (0.6 cm) or so of soil. Use 1 to 3 gal of spray against such foliar diseases as powdery mildew, dollar spot, Rhizoctonia blight, red thread and pink patch, leaf spots, and rusts. Others such as Pythium diseases, Helminthosporium melting out (crown, root, and rhizome rot), and snow molds attack the crown and root area before growing over the grass surface and within the plant. Here, 3 to 5 gal per 1000 sq ft (93 sq m) is usually adequate. For control of summer patch and necrotic ring spot, Fusarium blight, and leaf smuts, a soil drench using ½ to 1 in (1.3 to 2.5 cm) of water (300 to 600+ gal) per 1000 sq ft is needed to move the fungicide down into the root zone. This is best done by first spraying the turf with the recommended rate of fungicide, followed *immediately* by irrigation to deliver the desired amount of water.

5.4.4   Some Negative Effects of Fungicides

The widespread and indiscriminate application of fungicides as frequent preventive sprays for many turf diseases, especially minor ones, should be discouraged. The reasons include (1) fungicides may reduce populations of beneficial (antagonistic or parasitic) bacteria and fungi in the soil, which could result in an excessive buildup of thatch; (2) fungicides may upset the delicate balance among microorganisms in the thatch and upper level of soil that compete with and antagonize fungi that cause disease. This could well explain why certain diseases reoccur more commonly or rapidly and cause increasing amounts of damage in turf previously treated with fungicides; (3) continuous usage of a single fungicide or chemically related fungicides may lead to the development of fungal strains or biotypes resistant to a particular fungicide or class of fungicides; (4) a fungicide may control a particular disease while increasing the development of other diseases; and (5) phytotoxic or undesirable hormonal effects are possible—these include less dense shoot growth, reduced leaf and/or root growth, lowered chlorophyll content, delayed senescence, and altered carbohydrate content.

The repeated use of certain fungicides (e.g., benomyl and cadmium products, Bromosan, Duosan, mancozeb [Fore, Dithane T/O], thiram, and iprodione) commonly leads to a buildup of thatch. This is believed to be caused by (1) acidifying the soil, which inhibits decomposition of the thatch by beneficial microorganisms; (2) enhancing a buildup of thatch by directly inhibiting or killing the microorganisms that break down thatch into humus; or (3) increasing root and rhizome production.

Benzimidazole fungicides (benomyl and thiophanate-methyl) are toxic to earthworms, probably due to their anticholinesterase activity. Earthworms are generally considered to be beneficial, as they mix soil with organic matter and feed on the thatch.

The chief disadvantage of systemic fungicides has been the problem of increasing resistance—perhaps more correctly called tolerance—to these fungicides by many important turf pathogens. This is of increasing concern to turfgrass managers. Strains, races, or biotypes of fungi that incite anthracnose, dollar spot, Pythium blight, pink snow mold, powdery mildew, red thread, summer patch, necrotic ring spot, and other diseases have become tolerant or resistant to one or more types of fungicides. Resistance to systemic fungicides occurs because these chemicals generally poison fungi at only a single location (site) in their growth and development cycle. Therefore, it is relatively likely that some tolerant individuals will be present in naturally occurring populations of disease-causing fungi that are able to circumvent the poisoned site. Such individuals are able to grow and increase in numbers while the fungicide is present. With repeated applications of the same or a closely related systemic fungicide, over one or more growing seasons, the naturally resistant individuals in a fungal population multiply until the population is composed primarily of those that are fungicide-resistant. Disease control then fails! This has happened worldwide, where systemic fungicides have been used. In the United States, many disease-control failures from resistance to systemic fungicides have occurred on turfgrasses. There are published reports of resistance control failures of benzimidazoles (benomyl and thiophanate-methyl) to the fungus or fungi that cause anthracnose, dollar spot, powdery mildew, red thread, summer patch, and necrotic ring spot; metalaxyl (Subdue) to Pythium blight; and iprodione (Cleary's 26019) for dollar spot, pink snow mold, and Curvularia leaf blight. Dollar spot resistant strains of iprodione are also resistant to vinclozolin and PCNB.

The continued development of tolerant strains of turfgrass-infecting fungi is probably due to mutations and a natural selection process. Resistance to some single-site action fungicides (such as benzimidazoles, dicarboximides, and metalaxyl) is controlled by a single gene, while resistance to DMI fungicides is controlled by several genes. Essentially no resistance has occurred where the older protective-contact fungicides have been used continuously for a period of years. Why? Because these fungicides affect many vital fungal systems. A number of natural mutations must then occur for the numerous genes that control these multi-action sites to become resistant—a most unlikely phenomenon.

The single-site fungicides eliminate only sensitive strains or biotypes of a turfgrass fungus, while the older, multisite protective-contact fungicides keep down the entire, naturally occurring population of a fungus. The latter chemicals generally prevent germination of spores, while single-site fungicides may (1) inhibit cell division (benzimidazoles), (2) cause fungal hyphae to burst (carboximides and dicarboximides), and (3) prevent the development of cell membranes (DMI fungicides). The single-site systemic fungicides all act on sensitive fungi after the spores have germinated.

Wild-type strains of the dollar-spot fungus, resistant to the benzimidazoles or to both benzimidazole and DMI fungicides, have also developed tolerance to the carboximide and dicarboximide fungicides. Persistence of resistant or tolerant strains in fungal populations differs among the various fungicide groups. For example, the degree of resistance among fungi to benzimidazoles, carboximides, and dicarboximides is high, while resistance to DMI fungicides is low. The persistence of tolerant fungal strains lasts for several or more seasons with the benzimidazoles and DMI fungicides, while the persistence among the carboximide and dicarboximide fungicides apparently disappears after a single growing season. Resistance is also more common in fungal strains or biotypes of the dollar-spot fungus (*Sclerotinia homoeocarpa*) and possibly other fungi that have developed resistance in another fungicide group, such as the benzimidazoles and DMI fungicides or the benzimidazole-DMI-carboximide and dicarboximide groups.

Discovering, developing, testing, and marketing of new fungicides is costly and time-consuming. We *must* learn to use systemics intelligently to prolong their useful lives. To prevent or delay fungicide resistance in populations of disease-causing fungi, it has been suggested that systemic fungicides groups be alternated or used in mixtures. There are an increasing number of published reports on which recommendations for preventing or delaying fungicide resistance can be based. Alternating fungicides from different fungicide groups is effective in cases when resistant individuals in fungal populations are not as competitive as sensitive individuals. The population thus fluctuates, with the resistant component increasing when a resistance-prone fungicide is applied and the more vigorous sensitive component increasing when fungicide selection pressure is not present and an alternate fungicide is used. Unfortunately, the resistant individuals in fungal populations are commonly just as competitive and vigorous—or more so—than the sensitive ones. In such cases, an alternating program results in an explosive increase in the proportion of resistant individuals until the population is predominantly or wholly resistant, a condition we are trying desperately to avoid. In populations of equally fit, resistant and sensitive components, fungicide mixtures have been effective in keeping resistant populations stable.

Assuming that fungicide mixtures are able to keep resistance levels stable in fungal populations, they still must be effective in managing disease. We cannot use full rates of each fungicide in mixtures due to increased financial and possibly environmental costs. We must be sure that reduced rates of fungicides in mixtures will give satisfactory control in the field. Field and greenhouse studies have repeatedly shown that reduced-rate mixtures also give disease control equal to or sometimes greater than the additive control of the individual fungicides in the mixture alone at the reduced rate. Reduced-rate mixtures also give acceptable field disease control of Pythium blight and dollar spot, and they delay resistance problems. Consult local turfgrass specialists for recommended reduced rates to control these or other diseases.

When selecting fungicides for use in alternating or reduced-rate mixtures, consider using *only* fungicides that have different methods of poisoning the target fungus. This way, you can delay or prevent control failures resulting from fungicide resistance in fungal populations.

For example, three systemic fungicides registered for Pythium blight control (propamocarb [Banol], fosetyl-Al [Chipco Aliette], and mefenoxan [Subdue]) all have different modes of action and can therefore be used alternately or as two-component, half-rate mixtures to prevent resistance and give disease control. Three-component, one-third-rate mixtures of Banol, Chipco Aliette, and Subdue have proven effective in controlling Pythium blight and preventing resistance. The same is true in controlling dollar spot in creeping bentgrass turf where various reduced-rate, three-component mixtures (containing chlorothalonil; benomyl or thiophanate-methyl, propiconazole, fenarimol, or triadimefon; and iprodione) provided season-long control equal to that of standard single fungicides (see table 5–5). These balanced, multicomponent mixtures should be useful in reducing fungicide selection

TABLE 5–5   FUNGICIDES

Diseases controlled

Disease	Anilazine	Benomyl	Captan	Chloroneb	Chloroneb + thiophanate-methyl	Chlorothalonil	Chlorothalonil + fenarimol	Chlorothalonil + thiophanate-methyl	Cyproconazole	Etridiazole (ethazole)	Fenarimol	Flutolanil	Flutolanil + triadimefon	Fosetyl-AL	Iprodione	Mancozeb	Mercury chlorides[c]	Metalaxyl	Metalaxyl + mancozeb	Myclobutanil	Oxadixyl
Zoysia patch																					
Yellow patch												●									
Take-all patch						●			●		●										
Summer patch						●			●		●									●	
Spring dead spot		●				●			●		●									●	
Southern or Sclerotium blight				●					●				●	●						●	
Leaf smuts		●							●		●										
Slime molds																●					
Septoria leaf spot																				●	
Rusts	●					●	●	●	●		●					●				●	
Red thread/Pink patch	●					●	●	●	●		●	●	●	●	●					●	
Pythium blight				●	●					●				●				●	●		●
Powdery mildew						●			●		●		●							●	
Pink snow mold (Fusarium patch)		●			●	●			●		●				●	●	●				
Necrotic ring spot						●			●		●				●						
Helminthosporium diseases	●					●	●	●							●	●			●	●	
Gray snow mold (Typhula blight)	●			●	●	●			●		●	●			●	●	●				
Gray leaf spot		●		●		●	●	●	●												
Fusarium leaf spot + blight		●				●			●						●	●					
Fairy rings														●							
Downy mildew (Yellow tuft)													●						●	●	
Dollar spot	●	●				●	●	●	●		●		●		●	●			●	●	
Damping off + Seed decay			●							●								●	●		
Curvularia leaf blight						●	●									●					
Coprinus snow mold + Snow scald																	●				
Copper spot	●					●	●	●	●		●		●		●	●				●	
Brown patch (Rhizoctonia blight)	●	●			●	●	●	●			●	●	●						●	●	
Anthracnose	●					●	●	●	●											●	
Mode of action — Systemic		●		●	●		●	●	●		●	●	●	●	●			●	●	●	●
Mode of action — Contact	●		●	●	●	●	●	●		●					●	●	●		●	●	
Fungicide group	Triazine	Benzimidazole	Dicarboximide	Substituted aromatic	Substituted aromatic + benzimidazole	Substituted aromatic	Substituted aromatic + pyrimidine (SI)[b]	Substituted aromatic + benzimidazole	Triazole (SI)	Thiadiazole	Pyrimidine (SI)	Benzanilide	Benzanilide + triazole (SI)	Organophosphate	Dicarboximide	Dithiocarbamate	Inorganic mercury	Phenylamide	Phenylamide + dithiocarbamate	Triazole (SI)	Phenylamide
Fungicide[a] (Common or generic name)	Anilazine	Benomyl	Captan	Chloroneb	Chloroneb + thiophanate-methyl	Chlorothalonil	Chlorothalonil + fenarimol	Chlorothalonil + thiophanate-methyl	Cyproconazole	Etridiazole (ethazole)	Fenarimol	Flutolanil	Flutolanil + triadimefon	Fosetyl-AL	Iprodione	Mancozeb	Mercury chlorides[c]	Metalaxyl	Metalaxyl + mancozeb	Myclobutanil	Oxadixyl

Table: Fungicides and diseases controlled. Dots (•) indicate the disease is controlled by the fungicide. Columns are the diseases controlled (plus mode of action: Systemic / Contact).

Fungicide[a] (Common or generic name)	Fungicide group	Zoysia patch	Yellow patch	Take-all patch	Summer patch	Spring dead spot	Southern or Sclerotium blight	Leaf smuts	Slime molds	Septoria leaf spot	Rusts	Red thread/Pink patch	Pythium blight	Powdery mildew	Pink snow mold (Fusarium patch)	Necrotic ring spot	Helminthosporium diseases	Gray snow mold (Typhula blight)	Gray leaf spot	Fusarium leaf spot + blight	Fairy rings	Downy mildew (Yellow tuft)	Dollar spot	Damping off + Seed decay	Curvularia blight	Coprinus snow mold + Snow scald	Copper spot	Brown patch (Rhizoctonia blight)	Anthracnose	Systemic	Contact
Propamocarb	Dithiocarbamate												•											•						•	•
Propiconazole	Triazole (SI)[b]		•	•	•	•		•	•		•	•		•	•		•	•	•	•	•		•					•	•	•	•
Quintozene (PCNB)	Substituted aromatic							•							•		•	•					•					•			•
Thiophanate-methyl	Benzimidazole				•	•		•	•			•			•	•	•			•			•				•	•	•	•	
Thiophanate-methyl + iprodione	Benzimidazole + dicarboximide														•		•						•					•	•	•	•
Thiophanate-methyl + mancozeb	Benzimidazole + dithiocarbamate										•	•			•		•		•				•				•	•	•	•	•
Thiram	Dithiocarbamate										•				•			•	•				•					•			•
Thiram + triadimefon	Dithiocarbamate + triazole (SI)										•	•			•		•	•		•			•				•	•	•	•	•
Triadimefon	Triazole (SI)			•	•		•	•			•			•					•				•					•	•	•	•
Triadimefon + metalaxyl	Triazole (SI) + phenylamide												•										•					•		•	•
Vinclozolin	Dicarboximide											•			•		•	•					•					•			•

[a] Fertilizer-fungicide combinations are not included.

[b] Sterol-inhibiting fungicide.

[c] Restricted-use fungicide. Use for specific diseases approved only in some states or countries.

Some of the fungicides listed may not be registered for a particular disease as listed or may be labeled for use only in selected states or countries or on certain turfgrasses. Mention of a trade name or proprietary product does not constitute warranty of the product by the authors or the publisher and does not imply approval of this material to the exclusion of comparable products that may be equally suitable. Except where indicated, all materials should be applied in 1 to 5 gal of water per 1000 sq ft. Use lower fungicide rates on container labels in *preventive* programs and higher rates for *curative* programs. Fungicide use and restrictions are subject to change without notice. Always read and follow all the instructions and precautions on the *current* label. Check state extension or advisory recommendations each year as new turfgrass fungicides are introduced and older ones are removed from the market.

pressure on populations of the dollar-spot fungus and, therefore, effectively reduce the risk of fungicide resistance in field populations of the fungus.

The broad-spectrum systemic fungicides that control turf diseases fall into three groups according to their mode of action: (1) the benzimidazoles (benomyl and thiophanate-methyl), (2) the carboximides and dicarboximides, and (3) the sterol or demethylation inhibitor (DMI) fungicides. Any fungus that is resistant to one of the benzimidazole fungicides will be resistant to them all. The same is true within the carboximide, dicarboximide, and DMI or sterol-inhibitor groups of fungicides. Therefore, to prevent resistance, broad-spectrum systemic fungicides must be mixed or alternated *between* but not *among* groups. Systemic fungicides may also be mixed or alternated with any protective-contact fungicide that provides the desired disease control.

In addition to mode-of-action differences, the length of disease control provided by components in mixtures must be matched to avoid resistance selection. If a short-residual fungicide is included in a mixture for delaying resistance, an interspray of the short-residual chemical will probably be necessary.

If available, it is much better to use systemic fungicides in mixtures for resistance management. The reason is that the turfgrass plant itself can unmix mixtures of contact and systemic fungicides. If you apply a protective-contact/systemic mixture, both fungicides in the mixture will be present on plant surfaces, but only the systemic fungicide will be present inside the plant. For example, in the case of a Subdue-Fore mixture, Subdue alone will be active against *Pythium* that already has invaded the plant. For this reason, mixtures of systemics are safer to delay resistance than protective-contact/systemic mixtures.

Managing fungicide resistance in populations of disease-causing fungi is an area of research still in its infancy. Additive, synergistic, or antagonistic effects may be possible with certain fungicide mixtures. It is important, therefore, to test alternating chemicals and mixtures of various fungicides for disease management and for resistance delay in as many different field situations and turfgrass/pathogen systems as possible.

There is much more we need to know about how to best use systemic fungicides to avoid resistance failures. One thing is clear, however: We cannot safely use *any* systemic fungicide repeatedly and exclusively for disease control. Prudent use of systemic fungicides dictates that we use chemicals having different modes of action. Be very skeptical of any recommendations suggesting that a systemic fungicide can be used alone and continually without risk of resistance problems.

Fungicides for lawn-type grasses.   The use of fungicides is commonly discouraged in most home-lawn situations (unlike chemical control of weeds and insects) for several reasons: (1) Proper diagnosis and selection of the right fungicide to apply is difficult; (2) it is usually too late for recovery after extensive damage has occurred; (3) homeowners often lack the proper application equipment or cannot purchase the suggested turfgrass fungicides locally; (4) it is probably less expensive and more satisfactory in the long run to overseed or resod a badly diseased turf area with a mixture or blend of disease-resistant cultivars or species; and (5) the typical homeowner lacks application experience and/or proper certification to handle certain pesticides.

Many lawn-care companies do not offer a disease-management program. The employees may not be sufficiently trained in diagnosis or are called in too late to effectively manage disease problems, particularly where the roots and crowns have been extensively invaded. Can a lawn-care company afford to show up every 7 to 10 or 21 days when repeat applications of fungicides need to be made? Only a small percentage of sales by lawn-care companies goes into applications of fungicides and nematicides.

Lawn-maintenance managers are at a decided disadvantage. They see each lawn only a few times each year, not every day, as do golf course superintendents or grounds-maintenance managers. Homeowners must also realize *their* responsibilities: proper watering to minimize disease, mowing lawn-type grasses at 2½ to 3 in (6.4 to 7.6 cm), and decreasing shade and increasing air movement over the turf. Responsibilities of the homeowner *and* lawn-care company may include applying fertilizer at the proper rates and times, keeping the thatch to ½ in (1.3 cm) or less, seeding or overseeding at suggested rates using the most resistant cultivars as blends or mixtures, watering properly after seeding, and maintaining the soil pH between 6 and 7.

One or more broad-spectrum turf fungicides such as anilazine, chlorothalonil, Duosan, iprodione, mancozeb, propiconazole, or vinclozolin may be used to maintain a lawn-type grass such as a home or industrial lawn, municipal or state park, athletic turf, school or church grounds, cemetery, sod farm, fairways, tees, and the like. All of these fungicides control *Helminthosporium diseases and dollar spot* as well as other less important ones (table 5–5).

If *powdery mildew* is a problem in the shade, you may wish to add cyproconazole, fenarimol, mycobutanil, or triadimefon to the list above.

If *red thread* is serious in cool, wet weather, a number of fungicides may be used, including anilazine, chlorothalonil, Duosan, iprodione, mancozeb, or a DMI fungicide (see table 5–4).

If *rusts* are serious in warm-to-hot and dry weather, anilazine, chlorothalonil, cyproconazole, Duosan, fenarimol, mancozeb, mycobutanil, propiconazole, and triadimefon provide good to excellent control.

Snow molds (primarily pink snow mold or Microdochium [Fusarium] patch and gray snow mold or Typhula blight) damage turf in shady areas and wherever else snow is slow to melt. Products that give control of both diseases include combinations of PCNB, cyproconazole, fenarimol, flutolanil, mycobutanil, propiconazole, or triadimefon, iprodione, chlorothalonil, chloroneb, and vinclozolin. Combinations of two or three fungicides may be necessary for snow-mold management where the snow cover is prolonged.

Summer patch and/or necrotic ring spot are becoming more severe each year in the United States, especially in sunny, droughty sites where turf has a thick thatch. Cyproconazole, fenarimol, mycobutanil, propiconazole, or thiophanate-methyl applied preventatively, all give good to excellent control of both diseases as well as Fusarium blight and other crown-root diseases. For maximum effectiveness, the thatch should be reduced, nitrogen should be applied in adequate but not excessive amounts, and the turf should be thoroughly watered during dry periods. Triadimefon is a good preventive fungicide, but it should be combined with one of the materials mentioned above (see table 5–5).

Seed decay (rot) and seedling blights are usually problems only with poorly germinating seed or seed planted in an excessively wet seedbed and/or buried too deeply, perhaps at unseasonable temperatures. Treat seed with a captan or thiram seed protectant plus Apron or Koban (to control *Pythium* spp.) before planting. Spray just after seeding, at early seedling emergence, and 7 to 14 days later using a mixture of Koban or Banol or Subdue or Terrazole plus either captan or iprodione.

Leaf smuts (largely stripe smut and flag smut) are of concern mostly in Kentucky bluegrass and creeping bentgrass turf areas over 2 or 3 years of age. The only materials to provide lasting (or eradicative) control are the fungicides cyproconazole, fenarimol, mycobutanil, propiconazole, thiophanate-methyl, or triadimefon. These materials must be applied in late autumn just before the *grass* goes into dormancy, or in spring at early green-up. These products need to be drenched into the soil (table 5–5) using 1 in (2.5 cm) of water (600+ gal) per 1000 sq ft (93 sq m) *immediately* after application. Make two applications 14 to 21 days apart. Carefully follow the label directions.

If you stock only *one* broad-spectrum fungicide, choose between chlorothalonil, Duosan, fenarimol, iprodione, propiconazole, or vinclozolin. When other disease problems are expected, supplement the broad-spectrum product with one or more disease-specific chemicals given in table 5–5.

Fungicides for bentgrasses.   If you are growing bentgrasses on golf greens, low-cut tees, fairways, or other areas, choose a broad-spectrum fungicide or mixture to control *Helminthosporium diseases, dollar spot, and Rhizoctonia blight or brown patch* (such as chlorothalonil, Duosan, iprodione, propiconazole, or vinclozolin).

In hot, humid weather, when *Pythium* is active, ethazole, fosetyl-Al, metalaxyl, propamocarb, and Heritage do an excellent job. (But read the section on resistance problems, 5.4.4.). In Pythium weather, one of these fungicides is often mixed with one or more of the following fungicides: anilazine, chlorothalonil, cyproconazole, fenarimol, iprodione, mycobutanil, propiconazole, thiophanate, triadimefon, or vinclozolin, to give more broad-spectrum disease control.

Snow molds damage turf in shady areas where snow is slow to melt and where no protective fungicide has been applied. Use a combination of the products previously mentioned for snow-mold control.

Where *dollar spot* is a serious problem on bentgrasses, even where the turf has been adequately fertilized, the systemics (cyproconazole, mycobutanil, propiconazole, thiophanate materials, triadimefon) and vinclozolin provide the longest and best protection. Be alert, however, for evidence of resistant or tolerant strains of the dollar-spot fungus or fungi. Using a systemic fungicide in combination with anilazine, chlorothalonil, iprodione, or vinclozolin, and perhaps in conjunction with alternating fungicides that are chemically unrelated, should solve the resistance problem. It is often suggested that maneb and mancozeb *not* be added in the same spray tank, since these products tend to stimulate dollar spot when used at normal rates (2 to 6 oz per 1000 sq ft), probably because they kill or suppress fungi that are antagonistic to the dollar-spot fungus or fungi.

Leaf smuts are an uncommon problem. For control, see the paragraph on leaf smuts under "Fungicides for lawn-type grasses" earlier in this section.

5.4.5 Getting Ready to Spray[2]

Wettable powder (WP) products must be kept in suspension if they are to be applied accurately without equipment containing an agitator. The new liquid (L), flowable (F, Fl, Flo), dry flowable (DF), soluble powder (SP), aqueous suspensions (AS), water-dispersible granules (WDG), and water-soluble packets (WSP) that you simply drop into the spray tank stay in suspension or solution indefinitely, no matter what ground or air application equipment is used. Emulsifiable concentrate (EC) formulations and other water-soluble materials are easy to apply but may cause injury, especially in hot weather. Unlike insecticides, few fungicides are formulated as emulsifiable concentrates. In general, wettable powders, flowables, and solubles should *not* be mixed with emulsifiable concentrates unless the product label states otherwise.

The time interval between spray applications will vary with the temperature, target diseases, cultural management practices, grass cultivar and species, chemical used, and the amount of rainfall or irrigation. The spray interval may be as short as 5 days in hot, wet weather or 2 weeks and longer if the weather is cool and dry. Some systemic fungicides give protection for 3 weeks or longer, even when 4 to 6 in (10.2 to 15.2 cm) of water has fallen as rain or has been applied by sprinkler. A protective-contact fungicide may last only a few days under similar conditions. The problem is complex and one that has to be felt out, based on knowledge of the chemical and its past performance, the problem turf area involved, its history with fungicides and other control measures, and knowledge of the factors that cause a particular disease to flourish. It is only through adequate record keeping that you can hope to determine why a certain chemical failed or did a good job. Remember that even the best fungicide or fungicide mixture is only one tool in managing turfgrass diseases. *All the fungicides in the world cannot compensate for or replace a poor turf management program.*

5.5 STUDY QUESTIONS

1. How would you go about demonstrating that nematodes are causing a problem in your turf?
2. It has been determined that you have a complex of dollar spot, Rhizoctonia blight or brown patch, and Pythium blight on your golf course. What cultural and chemical management practices should you be following?
3. Resistance to certain fungicides has suddenly appeared, and you cannot seem to manage Pythium blight and dollar spot. What can you do to solve the problem?
4. How would you go about selecting a fungicide (or fungicides) to manage snow molds? When should you apply fungicides to manage this disease complex?
5. You cannot seem to keep Kentucky bluegrass vigorous in moderate shade. What management practices should you follow?
6. How does having a thick thatch (over ½ in on lawns) affect turfgrass diseases?

[2] For a complete discussion of safe pesticide application and equipment, read chapter 7.

7. Keeping the soil in golf greens on the wet side could lead to what kinds of disease problems?

8. Can you ordinarily cure a crown-root rot complex with fungicides after symptoms appear in the foliage? Why?

9. You cannot seem to decide which of two or three diseases are present in a lawn. What can you do to get an answer to the problem?

10. Why do we not have chemical controls for minor turfgrass diseases?

11. You feel the need for more education (by attending classes, workshops, and conferences) to become better acquainted with issues concerning diseases. Where would you go for help?

12. What do you feel is the most serious disease problem on Kentucky bluegrass? Bermudagrass? Zoysiagrass? Why?

13. What would you suggest to someone who has a serious rust problem in a lawn but does not want to apply a chemical?

14. Why is a microscope essential in identifying the cause of some diseases?

15. As a lawn-care representative, you have been asked by a homeowner to recommend cultural practices that will help to prevent diseases. What would you tell that person?

16. In what ways are disease-causing fungi disseminated from plant to plant and from one turfgrass area to another? What can you do to help prevent this?

5.6 SELECTED REFERENCES

BEARD, J.B. *Turfgrass Science and Culture.* Englewood Cliffs, NJ: Prentice-Hall, Inc., 1973.

BEARD, J.B. *How to Have a Beautiful Lawn,* 2nd ed. College Station, TX: Beard Books, 1979.

BRUNNEAU, A.H. (ed.). *Turfgrass Pest Management: A Guide to Major Turfgrass Pests and Turfgrasses.* Raleigh, NC: North Carolina State University Agricultural Extension Service Publication AG-348, 1985.

CLARKE, B.B., and A.B. GOULD (eds.). *Turfgrass Patch Diseases Caused by Ectotrophic Root-Infecting Fungi.* St. Paul, MN: APS Press, 1993.

COUCH, H.B. *Diseases of Turfgrasses,* 3rd ed. Malabar, FL: Krieger Publishing Co., 1995.

CUMMINS, G.B. *The Rust Fungi of Cereals, Grasses, and Bamboos.* New York: Springer-Verlag, Inc., 1971.

FARR, DAVID F., G.F. BILLS, G.P. CHAMURIS, and A.Y. ROSSMAN. *Fungi on Plants and Plant Products in the United States.* St. Paul, MN: APS Press, 1989.

FREEMAN, T.E. *Diseases of Southern Turfgrasses.* Gainesville, FL: Florida Agricultural Experiment Station Technical Bulletin 713A, 1969.

HANSON, A.A., and F.V. JUSKA (eds.). *Turfgrass Science.* Agronomy Monograph 14. Madison, WI: American Society of Agronomy, Inc., 1969.

LARSEN, P.O., and B.G. JOYNER (eds.). *Advances in Turfgrass Pathology.* New York: Harcourt Brace Jovanovich, 1980.

SANDERS, P.L. *The Microscope in Turfgrass Disease Diagnosis,* 2nd ed. University Park, PA: The Pennsylvania State University, 1993.

SHURTLEFF, M.C., and R. RANDELL. *How to Control Lawn Diseases and Pests.* Kansas City, MO: Intertec Publishing Corp., 1974.

SMILEY, R.W., P.H. DERNOEDEN, and B.B. CLARKE. *Compendium of Turfgrass Diseases,* 2nd ed. St. Paul, MN: APS Press, 1992.

SMITH, J.D., N. JACKSON, and A.R. WOOLHOUSE. *Fungal Diseases of Amenity Turf Grasses.* London: E. & F. Spon., 1989.

SNEH, B., L. BURPEE, and A. OGOSHI. *Identification of Rhizoctonia Species.* St. Paul, MN: APS Press, 1991.

TONI, T., and J. B. BEARD. *Color Atlas of Turfgrass Diseases.* Ann Arbor, MI: Ann Arbor Press, 1997.

TURGEON, A.J. *Turfgrass Management,* 4th ed. Upper Saddle River, NJ: Prentice-Hall, Inc. 1996.

VARGAS, J.M. JR. *Management of Turfgrass Diseases,* 2nd ed. Boca Raton, FL: Lewis Publishers, 1994.

WATSCHKE, T.L., P.H. DERNOEDEN, and D.J. SHETLAR. *Managing Turfgrass Pests.* Boca Raton, FL: Lewis Publishers, 1995.

WEIHING, J.L., M.C. SHURTLEFF, R.E. PARTYKA, J.M. VARGAS, JR., and J.E. WATKINS. *Lawn Diseases in the Midwest,* 4th ed. Lincoln, NE: North Central Regional Extension Publication 12, 1978.

Biology and Control of Insects and Related Pests of Turfgrass

6.1 INTRODUCTION

Insects are animals. Many people do not think of insects as belonging to the animal kingdom, but they do belong, and in staggering numbers. Of the over 800,000 species of insects in the world, less than 1% are considered pests. In the United States, there are about 100,000 insect species, with less than 1% as pests of crops or people. In the area of turfgrass management, the number of pest-insect species is fewer than 100.

Not all insectlike animals are classified as insects, but they may be close relatives. Insects have three pairs of legs and three distinct body parts—head, thorax, and abdomen—as adults. Many adult insects have two pairs of wings; others have one pair, and some have none (figure 6–1).

Close relatives of insects include centipedes, millipedes, spiders, mites, ticks, sowbugs, and pillbugs. Centipedes have long antennae and a pair of legs on each of many body segments. Millipedes are elongated, like centipedes, with many body segments but have short antennae and two pairs of legs on each body segment. Spiders, mites, and ticks have two body regions and four pairs of legs as adults (figure 6–2). As larvae, mites and ticks have only three pairs of legs. More distant insect relatives include earthworms, land snails, and slugs.

Insects are shaped differently than most other groups of animals. The body form is somewhat cylindrical and divided into segments. The outer covering or body wall of the insect acts as the skeleton, entirely supporting the internal muscles and organs.

Insects use a respiratory system to breathe. Air is aspirated through openings along the sides of the body wall. The nervous system consists of a brain located in the head, a nerve cord running the length of the body, and a system of connecting nerve cells. Sensory nerves are found in the antennae, eyes, mouth, and sometimes on the feet or tarsi of some insect species.

The method by which insects feed varies from group to group. Certain species devour grass roots by chewing on them, whereas others suck plant sap from the leaves. There are a variety of feeding methods and mouthparts among the groups of insects. Some have chewing mouthparts (such as beetles and caterpillars) fitted with mandibles, which are toothlike structures that aid in tearing off pieces of plant material. Another group has sucking mouthparts with an elongated beak or syringe to insert into plant or animal material to extract food. Aphids, leafhoppers, and mosquitoes are examples. Many variations of these two basic types exist. One is siphoning, found in most moths and butterflies. These insects do not pierce plant tissue, but

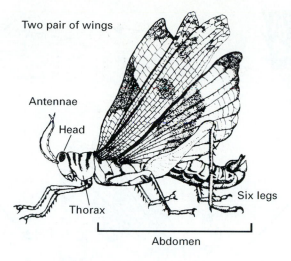

FIGURE 6–1 Parts of an adult insect. (Courtesy University of Illinois.)

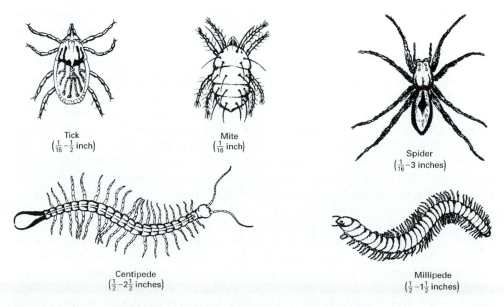

Tick
$(\frac{1}{16}-\frac{1}{2}$ inch$)$

Mite
$(\frac{1}{16}$ inch$)$

Spider
$(\frac{1}{16}-3$ inches$)$

Centipede
$(\frac{1}{2}-2\frac{1}{2}$ inches$)$

Millipede
$(\frac{1}{2}-1\frac{1}{2}$ inches$)$

FIGURE 6–2 Insect relatives. (Courtesy University of Illinois.)

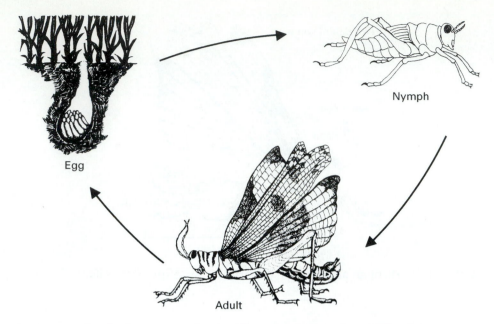

FIGURE 6–3   Gradual metamorphosis. (Courtesy University of Illinois.)

suck or siphon liquids such as nectar from flowers and dew from foliage. Other suck-ing mouthpart variations occur in insects that are not turfgrass pests.

Insect growth and development varies from one species to another but again is quite different from that of other animals. Many insect species emerge from eggs laid by adults. The emerging young, or immature insect, may bear a similar appearance to the adult form except for being smaller and without wings. Examples of this kind of development are found in aphids, leafhoppers, and chinch bugs. This growth or de-velopment is often called gradual, simple, or incomplete metamorphosis (figure 6–3). Immature insects with gradual metamorphosis are commonly referred to as nymphs (figure 6–4).

Some newly hatched young may be completely different from adults of the species. The young or immature stages are called larvae, and the development or metamorphosis is said to be complete. The life stages of these insects pass from egg to larva to pupa and then transform into an adult (figure 6–5). Butterflies, moths, beetles, and flies grow from egg to adult through complete metamorphosis. Cater-pillars, grubs, and maggots are examples of immature or larval stages in this com-plete development (figure 6–4).

Another unique characteristic of insect growth is its accomplishment not by a gradual increase in size but by a shedding of the external skin. Throughout its immature life, there are commonly four or five stages, or instars, of increasing size before transformation into a pupa, or resting stage, prior to emergence as an adult. Adult insects usually do not change in appearance, size, or form. Some adults change color. For example, black turfgrass ataenius beetles change from a reddish color to shiny black after emergence from the soil.

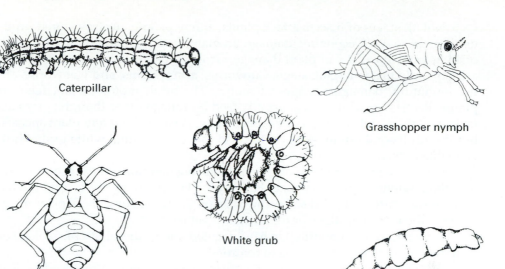

FIGURE 6–4   Immature insects. (Courtesy University of Illinois.)

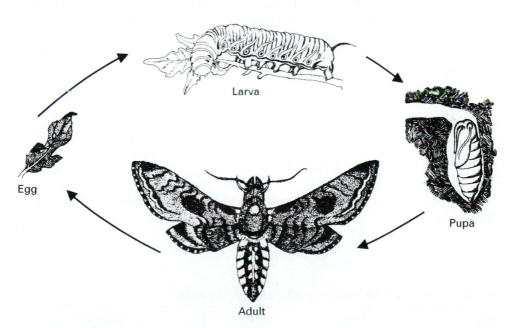

FIGURE 6–5   Complete metamorphosis. (Courtesy University of Illinois.)

Reproduction is quite variable among insect species. Length of development from egg to adult can vary greatly. For example, certain species of wireworms require 3 or more years to complete immature growth. At the other extreme, certain aphids complete more than 10 generations per year. Fruit or pomace flies may pass through 20 or more generations annually.

Not all stages of insects attack plants, annoy people, destroy stored products, or cause other economic damage. In many instances, the immature stage of an insect is the cause of plant damage. Grubs, the immature stage of beetles, are generally the damaging stage. Cutworms, armyworms, and webworms are the immature, damaging stages of moths. The adult moths do not feed on plants. Beetles, or adult grubs, usually feed on foliage other than the grass attacked by their grubs. The Japanese beetle feeds heavily on many plant species, but the adult southern masked chafer, also called the annual white grub adult, is unable to feed.

Larval stages vary in the degree of food consumption, which can translate into the amount of plant damage. Full-size or last-instar larvae of caterpillars, armyworms, cutworms, and sod webworms probably consume more plant material than their previous three or four instars combined.

In summary, it is important to have detailed knowledge about the biology of the pest insect you are attempting to control.

Animals discussed in this chapter are grouped as follows: root feeders; leaf and stem chewers; leaf and stem suckers; mound- and hole-making insects; turf animals that do no damage; turf animals that annoy, bite, or sting people; and mammals and birds that damage turf. At the beginning of each section is a table that helps distinguish these animals from each other for easy field identification. Within each section appear specific details concerning the biology, description, life history, habits, and potential for damage to turfgrasses or the people occupying these areas.

The final portion of this chapter deals with control. View this discussion and the enclosed table as alternative methods to reducing a pest-insect population. Insect populations are reduced or suppressed, not eradicated. In fact, no insect pest has ever been eradicated from the world.

6.2 ROOT-FEEDING INSECTS

Many insects feed on turfgrass roots as larvae in both immature and adult stages (table 6–1).

TABLE 6–1   INSECTS THAT FEED ON ROOTS OF TURFGRASSES*

Pest	Legs	Color	Other traits
White grub larvae	Present	White	C shaped
Billbug larvae	Absent	White	Straight, chunky
Wireworm larvae	Present	Yellowish, rust colored	Hard shelled, slender
Mole crickets	Jumping back legs	Brownish	Spadelike front legs
Ground pearls	Absent	Yellow-purplish	Rounded, pearl-like
Cranberry girdler larva	Present	Whitish	Slender caterpillar

*Turf damaged by these insects dies in irregular areas.

6.2.2   White Grubs

Several species of beetles have an immature or larval stage known as a grub, white grub, or grubworm. White grubs as a group include the immature stages of the Japanese beetle *(Color plate 17)*, European chafer, northern masked chafer, southern masked chafer (also called annual white grub) *(Color plate 18)*, Oriental beetle, Asiatic garden beetle, green June beetle (figure 6–7), black turfgrass ataenius, and May beetles. They feed on the roots of turfgrass, causing the grass blades to turn brown and die. Attacked turf will pull up easily due to the eaten roots. Continued feeding will kill the turf.

Identification.   White grub larvae, when fully grown, are robust and white or cream colored, with three pairs of legs, a brown head, and a dark area at the posterior end of the body *(Color plate 16)*. Most white grubs are identified specifically by the raster pattern of coarse hairs or spines on the underside of the last abdominal segment (figure 6–6). These larvae are usually in a C-shaped position when discovered damaging turfgrass. The size of larvae can vary between species and according to age (figure 6–7), but most grubs range in length from ¼ to 1 in (6 to 25 mm) (table 6–2).

These species usually complete their life cycle in 1 year, except for black turfgrass ataenius and most May beetles. Adults of all species emerge from the soil, eggs are laid on or beneath the soil surface and hatch, and the larvae feed on the roots of turfgrass during its growing season *(Color plate 19)*, causing extensive damage. The adult beetles are all robust, heavily built scarab beetles

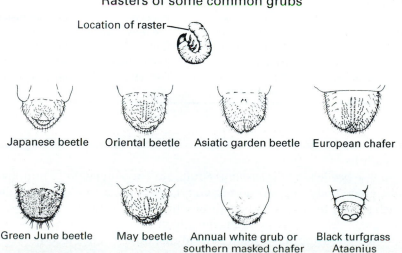

FIGURE 6–6   Rasters of some common grubs.

FIGURE 6–7　White grubs under rolled-back sod.

TABLE 6–2　WHITE GRUB LARVAL IDENTIFICATION

Pest	Raster pattern	Anal slit	Other traits
Japanese beetle	V shaped	Transverse	—
European chafer	2 parallel rows	Y shaped	Raster rows diverge posteriorly
Masked chafer	None	Transverse	—
Oriental beetle	2 parallel rows	Transverse	Slender raster spines
Asiatic garden beetle	Crescent shaped	Y shaped	—
Green June beetle	2 parallel rows	Transverse	Short legs
May beetles	2 parallel rows	V or Y shaped	—
Black turfgrass ataenius	None, 2 padlike structures	Transverse	Less than ¼ in long

(table 6–3). Many of them cause extensive feeding damage to woody landscape plants as adults.

Japanese beetle *(Popillia japonica)* adults are ⅜ to ½ in (10 to 13 mm) long with a metallic green body, coppery-to-brown wing covers, a row of five white spots along each side of the body, and a pair of white spots at the tip of the rear end of the body *(Color plate 17)*. These adults fly only during the daytime. Eggs are laid in the soil of turfgrass and other plants. As adults, these beetles feed heavily on the leaves, flowers, and fruit of more than 200 different plants and are easily noticed. Field crops, flowers, and small fruits are favorite host plants for adults.

TABLE 6–3   WHITE GRUB ADULT IDENTIFICATION

Pest	Length	Color
Japanese beetle	⅓–½ in (10–13 mm)	Metallic green; coppery wing covers
European chafer	½ in (13 mm)	Light brown to tan
Masked chafer	½ in (13 mm)	Tan
Oriental beetle	⅓–½ in (8–13 mm)	Straw to brown; black markings
Asiatic garden beetle	⅜ in (8–11 mm)	Brown
Green June beetle	1 in (25 mm)	Metallic green; brownish wing covers
May beetles	½–1 in (13–25 mm)	Reddish brown to blackish brown
Black turfgrass ataenius	¼ in (6 mm)	Black

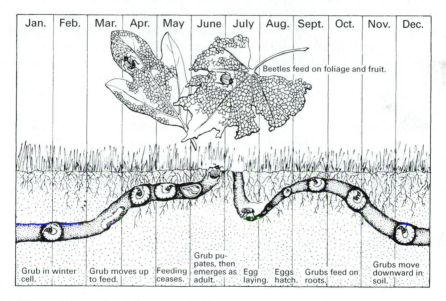

FIGURE 6–8   Life cycle of the Japanese beetle.

Japanese beetles complete a life cycle similar to other annual grubs. Eggs are laid in midsummer; the eggs hatch, the larvae feed on the roots of turfgrass and other plants, hibernate down to 11 in (28 cm) deep in the soil during the winter, and return to feed on grass roots in the spring. Larvae are distinguished from other species by a raster pattern in the shape of an inverted V (figure 6–6). The adult beetles emerge in time to feed, mate, and begin the cycle over again (figure 6–8). Damage is most extensive in late summer to early fall.

Infestations of Japanese beetles are common in the northeastern United States, with scattered infestations in the midwest and as far south as Georgia. In many areas, the adults are noticeable, but little or no damage to turf is observed.

The European chafer *(Rhizotrogus majalis)* was first observed as a pest problem in New York in 1940. It is believed that this insect entered the United States in the 1920s, but is now present in many of the northeastern states and westward into Michigan. The eggs are laid in midsummer, 2 to 6 in (5.1–15.2 cm) deep in the soil, just before the adults die. The eggs hatch and larvae feed until winter, when they move below the frost line. The grubs are identified by a Y-shaped anal slit and two parallel rows of raster spines that diverge posteriorly (figure 6–6). The grubs feed on the fibrous roots of many plants other than turfgrasses and may thus cause the decline of ornamental trees and shrubs as well as turf. Turfgrass root feeding is most extensive in late summer and early fall, but if moisture levels from rain and/or irrigation are high during this time, browning turf may not appear until spring.

After a short period of feeding by the grubs in spring, the adults emerge in large numbers and swarm with a buzzing sound around trees and shrubs at sunset. Adults are most numerous when catalpa (*Catalpa bignonioides*) is in full bloom. Although adults may be very numerous on landscape trees and shrubs, they cause very little damage. The adults are oval, about ½ in (13 mm) long, and light brown or tan. Shallow grooves run lengthwise on their shell-like wing covers. Every European chafer grub does not change to and emerge as an adult beetle at the end of its first year. Some larvae spend a second winter as a grub and mature in early summer of the second year.

The northern masked chafer (*Cyclocephala borealis*) extends as far south as Florida and westward to California, but is most prevalent and damaging to lawns and other turfgrass areas in the northeastern United States. Eggs of the northern masked chafer are laid 4 to 6 in (10 to 15 cm) below the soil surface during midsummer. The resulting grubs have no pattern of rastral spines. Damage by the grubs increases throughout the summer and fall months. The grubs hibernate below the frost line in the soil and return to feed the following spring. Chestnut brown, ½-in-1 long (13-mm) adults emerge in early summer. They remain in the soil during the day and become active at about dusk. The adults are attracted to lights and are unable to feed on foliage.

The southern masked chafer (*Cyclocephala lurida*) is common in the southeastern and midwestern United States. In many areas, this grub species is called the annual white grub. The adult beetles cannot feed or "chafe" on the foliage; they emerge from the soil, mate, lay eggs, and starve.

The life cycle of the southern masked chafer is similar to that of the Japanese beetle and northern masked chafer. Adult, ½-in-1 long (13-mm) tan beetles (*Color plate 18*) emerge from the soil in midsummer and mate. The eggs are laid in the soil after dark. The eggs hatch, and grubs begin to feed on grass roots. The grubs have no pattern of rastral spines (figure 6–6). They overwinter deep in the soil and return to feed a short time in the spring months (figure 6–9).

The Oriental beetle (*Anomala orientalis*) is a pest grub in several northeastern states and in the Carolinas. The adults are about ⅓ to ½ in (8 to 13 mm) long, straw-colored to black, with black markings on the wing covers and just behind the head. The adult beetles feed on flowers. The immature or grub stage emerges from eggs laid 6 in (15 cm) deep in the soil. Grubs have two parallel rows of spines on the raster similar to that of May beetle grubs. However, the spines are not as thick and distinct as that of May beetles, and the anal slit is transverse (figure 6–6). Grub damage occurs in late summer and fall, and feeding occurs on the roots of many plants as well as turfgrass. The grubs overwinter in the soil and return to the turf root zone in the spring.

The Asiatic garden beetle (*Maladera castanea*) is found in scattered areas along the Atlantic coast, from Massachusetts to South Carolina. The adult is a chestnut brown beetle about ⅜ in (8 to 11 mm) long. Its ventral or underside is covered with short yellow hairs. The grub stage hatches from eggs in early summer and feeds

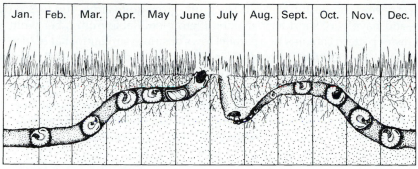

| Jan. | Feb. | Mar. | Apr. | May | June | July | Aug. | Sept. | Oct. | Nov. | Dec. |

Jan.–Feb.: Grub in winter cell.
Mar.–Apr.: Grub comes up near surface to feed.
    May: Grub forms cell and begins to pupate.
   June: Grub changes to pupa and then to adult which emerges
         from ground.
   July: Beetle lays eggs in ground, preferably in grass sod.
   Aug: Eggs hatch. Young feed on living roots of plants.
Sept.–Oct.: Grubs continue to feed and grow rapidly. Injury to roots of
         grass most common at this time.
Nov.–Dec.: Grubs are mostly full grown and go down below frostline.

FIGURE 6–9   Life cycle of the annual white grub.

FIGURE 6–10   Green June beetle adult.

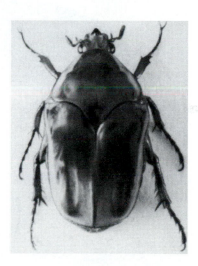

only on the small roots, causing little serious damage to turfgrass. The raster pattern of the grub has one crescent-shaped row of spines (figure 6–6). The grubs migrate deeper into the soil to avoid freezing temperatures. They migrate upward in spring to feed on roots before pupating. Adults emerging in early summer are active only after dark and feed on various kinds of plants, causing damage to the leaves of many trees, shrubs, flowers, fruit, and vegetables.

The green June beetle (*Cotinis nitida*) as an adult is sometimes called the fig eater due to its habit of feeding on soft, fleshy fruit (figure 6–10). This insect is present in the southeastern states northward to central Illinois and Long Island. The larvae emerge from eggs laid in July. The eggs are often laid in high-organic-matter soils, tobacco beds, or rotting compost in the soil. The grubs grow quite large—up to 2 in (5.1 cm) long—and have very short legs, making them identifiable from other white grubs. Their raster pattern contains two parallel rows of spines similar to that of May beetles (figure 6–6). They burrow through the soil humus, drying it out, and in periods of heavy rains will come to the surface. Holes, ½ in diameter, with a small amount of fine soil next to it, are evidence of grub emergence to the turf surface during the night. They go back into the soil by morning. Grubs have a habit of always crawling on their backs. The adults emerge in June and July. They are almost 1 in (2.5 cm) long and are metallic green with brownish-yellow wing covers. The ventral side of the body is a shiny greenish yellow-orange.

May beetles (*Phyllophaga* spp.) are quite numerous, but only a few species in each geographic region are serious pests of turfgrass. Many species attack the roots of flowers, field crops, vegetables, and other plants. The raster pattern of the grubs contains two parallel rows of spines. The anal slit is V or Y shaped (figure 6–6). Many species have adults that are similar in size and appearance, being about 1 in long (2.5 cm) and red-brown to black-brown (figure 6–11). One species common in the midwestern United States is ½ in (13 mm) and tan. It is commonly confused with adult masked chafers, but emerges in late spring rather than early summer.

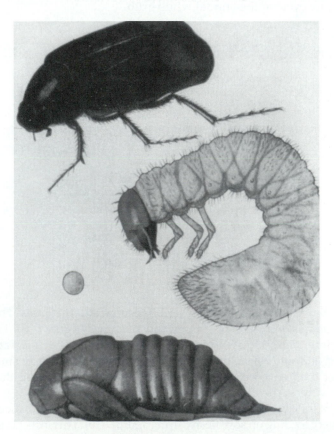

FIGURE 6–11   Life stages of true white grub: top to bottom—adult, larva or grub, egg, and adult.

Some species feed heavily on tree foliage, whereas other species feed very little. Oak, willow, apple, poplar, birch, hickory, and other trees have been observed with May beetles feeding on the foliage, primarily at night.

Most species of May beetles that attack turfgrass require 2, 3, or even 4 years to complete their life cycle from egg to adult. Species with a 3-year life cycle emerge from the soil as adults in April and May, feed on tree foliage, mate, and lay eggs in the soil beneath the sod. Young grubs hatch, feed, and grow during the summer and fall months. Winter is spent below the frost zone. The grubs return to the root zone to feed during the spring, summer, and fall months, then return to below the frost-free area. In the third spring, the full-grown grub feeds and then pupates about 1 ft (30 cm) or more below the soil surface. It emerges as an adult, but stays below ground through the winter and then leaves the soil the following spring to mate, lay eggs, and initiate the next generation. Damage is observed only in the second and third years of the grub's life cycle, but there can be overlapping generations (figure 6–12).

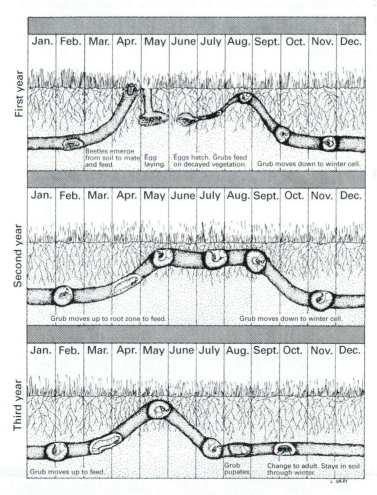

FIGURE 6–12 Life cycle of the May beetle.

The black turfgrass ataenius *(Ataenius spretulus),* although occasionally observed and written about for the past 60 years, was first observed damaging golf course turf near Minneapolis in 1932. In 1973, the grubs were found near Cincinnati, feeding on and pruning the roots of annual bluegrass and bentgrass in fairways. Since 1973, this species has been found damaging golf course fairways in at least 25 of the northern states. It is found throughout the continental United States.

Black turfgrass ataenius grubs are similar to all other grubs, being C shaped and white with a brown head. When full grown, however, they are only one fourth or less the size of the common May beetle or chafer grubs. They are distinguished from other young grubs by having two padlike structures just in front of the anal slit (figure 6–6).

Adult ataenius beetles hibernate beneath the soil surface in the woods, usually bordering or near a damaged golf course. In the spring these ¼-in-long (6-mm), shiny black beetles migrate to the golf course and can be observed flying over greens and fairways, often sitting on the leaves of closely mowed bentgrass greens *(Color plate 20).* They are most active during the warmest period of the day, from noon until 3 P.M. The eggs are laid beneath the soil surface in late May and June. Newly hatched grubs soon begin to feed on the roots. Annual bluegrass seems to be their favorite host plant. Although the ataenius grubs are small compared with other soil-infesting grubs, their numbers can range from a harmless few to 500 or more per square foot (30 cm sq). Adults mate in July and lay eggs for a second generation. During the early fall period, the adults can be observed flying to hibernation sites on the border of wooded areas (figure 6–13). There is usually only one generation per year in northern regions of the United States.

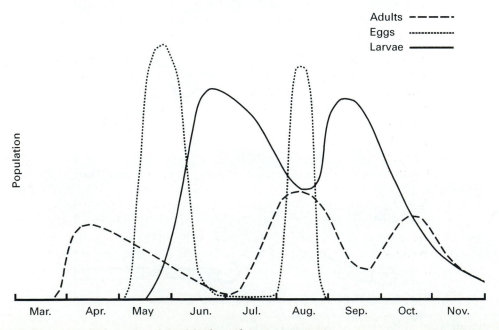

FIGURE 6–13 Life cycle of the black turfgrass ataenius.

Importance.   White grubs attack most species of turfgrass, as they are quite general in their feeding habits. Their feeding on the roots of turfgrass greatly reduces the movement of water and nutrients into the rest of the plant. This results in wilting of the grass blades with their associated decline and browning. Turf areas with high populations of white grubs will initially be brownish with small patches of green that soon disappear as grub feeding continues (*Color plate 19*). Although most species of white grub are not very active as larvae and thus do not move easily into other areas, areas that are infested tend to have enough grubs present to turn the entire area brown.

By the time large areas have turned brown, enough roots have been eaten that the sod can be easily pulled back from the soil like a carpet from a floor. If the moisture level is adequate, the white grubs will be easily seen in the root-soil interface. In dry soil conditions, the grubs will migrate lower into areas containing more moisture.

Even heavily damaged turf that has turned completely brown and can easily be lifted up will frequently regenerate within a few weeks provided that the grubs are killed and sufficient irrigation is supplied. New roots and blades will be produced from the existing crowns and rhizomes of the damaged turf. In many cases, recovery is complete enough that overseeding or plug introduction is not needed to obtain a high-quality turf.

Numbers of grubs necessary to cause damage vary with the soil type, moisture availability, turfgrass species, maintenance schedule, heaviness of use, and other factors. Even so, 10 to 12 grubs per square foot (30 cm sq) of turf are usually sufficient to cause at least some noticeable damage. With the small grub species black turfgrass ataenius, 50 grubs per square foot (30 cm sq) is an effective measure. An additional complicating factor is that grub-feeding birds such as starlings and robins, as well as mammals such as skunks, raccoons, and armadillos, will commonly be attracted to numbers as low as 3 to 5 grubs per square foot (30 cm sq). The associated turf damage caused by these predators in feeding on the grubs is usually greater than the damage done by the grubs themselves.

Management.   Some species of white grubs are strongly attracted to moist soil areas in which to lay their eggs. Particularly in low-maintenance areas, reduced irrigation during egg-laying periods will make it unlikely that the resulting numbers of white grubs will be high enough to cause damage.

Irrigation and sufficient rainfall will also help reduce the need for other control measures where numbers are marginally high enough to cause damage. The few uneaten roots may be able to adequately supply the needed moisture and nutrients to the turf if the soil is kept moist. In most parts of the country, grub damage rarely occurs in the spring because spring rainfall is sufficient to offset the short period of feeding that many grub species do in the spring before pupation.

There are very few natural enemies of white grubs. They are attacked by a few species of parasitic wasps and flies. Green June beetle grubs are attacked by scoliid wasp larvae heavily enough that the 1-in (2.5 cm) long, black and reddish wasps sometimes become a nuisance flying around turf areas. Typhiid wasps and tachinid flies are important parasitoids of Japanese beetle in areas of the northeast United States. Research is being conducted to determine their usefulness in other areas.

Sometimes diseases reduce the number of grubs. Bacterial agents, species of *Bacillus,* are sometimes found infecting grubs in the soil, which is commonly called milky disease. It is specific to each grub species in which it is found and has been observed on *Phyllophaga, Cyclocephala,* and *Ataenius* spp., plus Japanese beetle grubs.

The milky disease that attacks the Japanese beetle, *Bacillus popilliae,* has been packaged and is available to home and commercial users for controlling primarily Japanese beetle larvae. Once introduced into the turf's soil, this bacteria will maintain itself for several years without reapplication as long as sufficiently high numbers of Japanese beetle grubs are present yearly to maintain it. This commercially prepared material is not effective against most other white grub species.

Another preparation from *Bacillus thuringiensis japonica* or *Bt. bui bui* is being developed for commercial release. It is effective against a wider range of white grubs, including northern masked chafer, southern masked chafer, and Japanese beetle. In this case, the toxin preparation will be applied when needed and will not maintain itself from year to year.

Warm-blooded animals such as moles, raccoons, and skunks, and sometimes birds, damage high-quality turfgrass that is grub-infested. The tunnels or runs of moles, the plowing and turning over of sod by raccoons and skunks, and the pecking and scratching by birds cause much more physical damage to the turf than the infestation of grubs. These animals, acting as grub-control agents, often cause more damage to turfgrass than the grubs themselves.

Trapping has been attempted to reduce or suppress populations of adult beetles or grubs. A common trap contains a combination of a sex or pheromone lure plus a food attractant. These traps are marketed for reducing numbers of Japanese beetles and may be effective in areas where the numbers of Japanese beetles are very low. In most situations, these traps only serve to attract more beetles into the area than find their way into the traps. These numbers may be higher than would have occurred naturally, making it possible for increased damage to ornamental plants from the adults as well as increased turf damage from the grubs.

Grub control primarily consists of using chemical insecticides by (1) applying a preventative treatment to the turf area prior to the appearance of damage, or when grubs are first detected; or (2) treating the grub-damaged turfgrass after damage is visible. The chemical insecticides are applied annually if there has been a history of grub infestation. Some are applied near to the period of egg hatch and the chemical may need to be irrigated or watered into the soil surface at the site of the damaging grubs. Specific chemical control suggestions are given in table 6–12 at the end of the chapter.

6.2.2   Billbugs

Billbugs are a group of weevils or beetles with bills or snouts. The snouts have a pair of jaws at the end and are used for chewing tunnels into plant material in which to lay eggs. Various species are pests of turfgrass. The Phoenix billbug *(Sphenophorus phoeniciensis)* attacks bermudagrass in the southwestern United States. The hunting billbug *(Sphenophorus venatus vestitus)* is a pest insect of zoysiagrass. The bluegrass billbug *(Sphenophorus parvulus)* is primarily a pest of Kentucky bluegrass in regions where this grass is grown in home lawns or other turf areas, but it can also infest other species of turfgrass.

FIGURE 6–14   Bluegrass billbug adult.

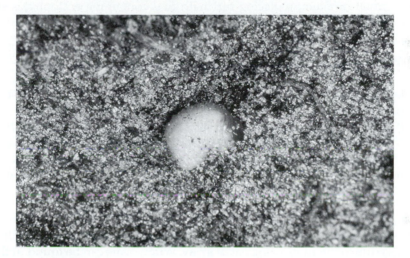

FIGURE 6–15   Hunting billbug larva.

Identification.   The adult beetles (figure 6–14) range from ¼ to ¾ in (6 to 19 mm) long, and the color varies between species from black or brown to a dull yellow. Adult bluegrass billbugs spend the winter in or near turfs infested the previous year. In the spring, they can be seen crawling on sidewalks and driveways. Eggs are laid in stems above the crown of bluegrass plants. White, legless larvae with brown heads feed and tunnel into the stems, eventually migrating into the root zone. Full-grown larvae are about ⅓ in (8 mm) long (figure 6–15). Peak feeding activity occurs in July as they continue to feed on the roots. There is one generation per year. New brown or black adults emerge in early fall and are visible again on walks and driveways before hibernating for the winter.

Importance.   Billbug damage can be distinguished from white grub feeding. Billbug-damaged plants pull loose from the crown. They chew the stem (*Color plate 21*), leaving sawdustlike material (frass) present in the chewed area. Late in the season, dead areas appear in the grass. The sod cannot be easily rolled back as with grub feeding, but the legless grubs will be present if the sod is removed. Detection of bluegrass billbug adults can be made in the spring or fall of the year. A pitfall trap is particularly effective in detecting billbug adults. A simple pitfall trap consists of a plastic drink cup buried to the rim in soil. Insects walking along fall into the cup and cannot crawl up the slippery sides. Adults crawling about on driveways, patios, sidewalks, and other paved areas are an indicator of the potential for future larval damage during the summer.

Management. Chemical control of billbugs is suggested where there has been a past history of damage. Control of adult beetles is achieved by placing the insecticide on grass foliage and thatch in the spring when adults are moving about. Larval or grub control is similar to white grub control. The insecticide is drenched or irrigated into the soil site, where the billbug larvae live during midsummer. For specific control suggestions, refer to table 6–12. Where billbugs are a perennial problem, replacing turf with that containing endophytes should provide control.

6.2.3 Wireworms

Identification. Wireworms (Elateridae) are the larval or immature stages of insects commonly called click beetles, snapping beetles, or skipjacks. These ½-in long (13-mm), narrow-bodied, hard-shelled beetles have the ability to flip themselves into the air when placed on their backs. The beetles commonly live for almost a year, laying eggs on the roots of grasses. Wireworm larvae (figure 6–16) usually spend the next 2 to 6 years in the soil, depending on the species. The brownish or rust-colored, slender, hard-shelled worms are about 1½ in (38 mm) long when full grown.

Importance. Wireworms occasionally damage turf by eating or boring into the roots of grasses. Wireworm-damaged turf has irregular areas of wilting or dead grass due to the roots being eaten away. Wireworms can be found in the soil near the surface in the spring but farther down as the soil temperature rises.

FIGURE 6–16 Wireworm.

When working the soil, it is common to find wireworm larvae, but they are rarely present in high enough numbers to cause important damage.

Management.   Many of the insecticides that are effective in controlling white grubs are also effective against wireworms. For current insecticide suggestions, refer to table 6–12.

6.2.4   Ground Pearls

Identification.   Ground pearls (*Margarodes* spp.) are tiny scale insects that attack the roots of bermudagrass and centipedegrass primarily in the southeastern and southwestern states. The eggs are laid by pinhead-size, pinkish adult females in a group enclosed in a white, waxy sac. The tiny scale nymphs hatch, move out, and attach themselves to the grass roots. At the same time, they secrete a pinhead-size, pearl-like, yellowish-purple shell or covering about themselves while they feed on sap (figure 6–17). Ground pearls may require more than a year to complete a life cycle. Adult males are tiny, gnatlike insects that are rarely seen.

Importance.   Feeding damage to the roots causes the plants to turn yellow and then brown. Infested turf turns brown and eventually dies in late summer, leaving irregular brown patches (figure 6–18). Damage is common in the southeastern United States.

FIGURE 6–17   Close-up of ground pearls. (Courtesy Milwaukee Sewerage Commission.)

FIGURE 6–18   Ground-pearls injury to a bermudagrass lawn. (Courtesy Milwaukee Sewerage Commission.)

Management.   Irrigation and fertilization to maintain good grass vigor is a practical method of control. Some of the newer soil insecticides (table 6–12) have given routine results when drenched into infested soil.

6.2.5   Mole Crickets

Mole crickets *(Scapteriscus* and *Neocurtilla* spp.) may be found burrowing in the soil of turfgrass. They often burrow 6 to 8 in (15.2–20 cm) or more below the soil surface in burrows ½ in (13 mm) in diameter. They are most common in the south Atlantic and Gulf coast states, but are observed in areas farther northward.

Identification.   There are at least four species of mole crickets: the tawny mole cricket *(S. vicinus,* figure 6–19), the southern mole cricket *(S. borellii),* the short-winged mole cricket *(S. abbreviatus),* and the northern mole cricket *(N. hexadactyla).* Mole crickets are light brown with a front pair of spadelike feet developed for digging that also act as a pair of scissors in cutting small roots as they move through the soil. Their hind legs are enlarged for jumping. Adult crickets are about 1¼ in (32 mm) long. All stages of immature and adult crickets feed and move through the soil.

Adult mole crickets emerge to the soil surface in the spring to mate. During this time, they will open their burrow to the surface, and males call to attract mates. These open burrows can be identified by the mound of loose soil around them. Eggs are deposited in newly constructed burrows in the soil. Nymphs remain in the soil and increase in size throughout the summer and fall. Overwintering adult crickets may also produce surface burrows. There is only one generation per year.

FIGURE 6–19   Adult tawny mole cricket.

Importance.   Mole crickets damage turf areas in two ways. They feed on the roots, which reduces the ability of the turfgrass to obtain water and nutrients. They also physically lift up the soil when burrowing near the soil surface. The uplifted turf feels soft or fluffy when walking across it. The burrows reduce the aesthetic appearance of the turf and may cause golf balls to change course. The burrowing and root pruning will dry out grass areas, causing wilting and death *(Color plate 22)*.

Damage is heaviest in the spring when adults are present and in late summer when nymphs are older, larger, and hungrier than ever. Mole crickets can be very destructive to new turfgrass seedings. The tawny and southern mole crickets are serious pests of bahiagrass, bermudagrass, and other turfgrasses of the southeastern United States. They are particularly a problem in sandy soils.

The short-winged mole cricket is not considered to be a major problem because it is primarily predaceous, feeding on other insects. The northern mole cricket occurs outside the southeastern United States, but is rarely a pest. It is occasionally found feeding in very damp areas such as along golf course water hazards and near ponds, where it is unclear whether the turf damage is due primarily to mole cricket feeding or overly wet growing conditions.

Scouting for mole crickets can be achieved by drenching damaged turf areas with an irritant. Mix 1 tablespoon of dishwashing detergent in 1 gal of water and spread it over 1 sq ft of turf. The mole crickets will become irritated by the solution and come onto the soil surface. This is particularly effective in late spring in determining the presence and number of nymphs. Treatments at other times are less effective, particularly when checking for adults.

The presence of adults in order to identify the pests or their relative numbers can be achieved by listening for the male's trill-like call for approximately 1 hour after sunset. The adults are strong fliers that are immediately attracted to lights at night, which also helps identify their presence. It is important to realize that adults may fly as far as 6 miles from their source.

Management.   Mole crickets spend much of their time in burrows below the soil surface, thus complicating control. June applications, drenched in after the young mole crickets hatch, seem to be the best control. For specific insecticide suggestions, see table 6–12.

Mole crickets are attacked by several species of fungi, including *Metarhizium anisopliae* and *Beauveria bassiana*, which are being considered commercially and may become available as biological insecticides. The insecticidal nematode, *Steinernema scapterisci*, has shown promise as another biological control agent against tawny mole cricket and southern mole cricket.

Parasitoids of mole crickets include a tachinid fly and a sphecid wasp. The larvae of a bombardier beetle have been found to specialize on tunneling through soil to feed on mole cricket eggs.

6.3   LEAF- AND STEM-CHEWING INSECTS

A variety of caterpillars and other chewing insects attack the leaf blades and stems of turfgrass (table 6–4).

6.3.1   Sod Webworms

Sod webworms are the larvae of several species of moths. Important species include the vagabond webworm *(Agriphila vulgivagella)*; silver-striped webworm *(Crambus praefectellus)*; bluegrass webworm *(Parapediasia teterrella)*; tropical sod webworm *(Herpetogramma phaeopteralis)*; and the lawn or larger sod webworm

TABLE 6–4   INSECTS THAT CHEW ON GRASS LEAVES AND STEMS*

Pest	Markings	Color	Other traits
Sod webworm larvae	Brown spots or none	Grayish to greenish	Slender
Armyworm larvae	Striped	Brown to black	Crawling in grass
Cutworm larvae	Striped or thin lined	Brown to black	Thick bodied, coiled on soil surface
Grasshoppers	Stripes or none	Green or brown	Jumping back legs
Lucerne moth larvae	Spotted	Grayish to greenish	Slender, active when disturbed
Striped grass looper larvae	Black and white spots and stripes	Various	These insects loop measure as they move
Fruit fly larvae	None	Yellow-white	Legless maggots

*These insects cause irregular brown or dead turf areas.

TABLE 6–5   CATERPILLAR IDENTIFICATION

Pest	Markings	Length	Other Traits
Larger sod webworm	Brown spotted	1 in (25 mm)	Lives in silk tube covered with grass
Tropical sod webworm	No spots or stripes	¾ in (19 mm)	Spends day curled on soil surface
Bluegrass webworm	Brown spotted	½ in (13 mm)	Lives in silk tube covered with pieces of turf
Burrowing webworm	No spots or stripes	1 in (25 mm)	Lives in silken tube without debris on it
Cranberry girdler	No spots or stripes	¾ in (19 mm)	Root feeder
Lucerne moth	Spotted	1 in (25 mm)	Very active when disturbed
Black cutworm	Thin light stripe down the back	1½ in (38 mm)	Heavy bodied; spends day curled on soil surface
Bronzed cutworm	Broad yellow stripe on back and each side	2 in (51 mm)	Bronze sheen on body
Armyworm	Orange and brown stripes	1½ in (38 mm)	—
Fall armyworm	3 thin, yellow-to-white stripes on back and 1 on each side	1½ in (38 mm)	Light-colored inverted Y on front of head
Striped grass loopers	Brown and yellow stripes	2½ in (64 mm)	These insects move by "looping"
Variegated cutworm	Several elongated spots on back; orangish-brown stripe on each side	1½ in (38 mm)	—

(*Pediasia trisecta*). Sod webworms do extensive damage to turf areas. The burrowing webworm (*Acrolophus popeanellus*) is rarely a pest, but it causes concern for turf owners. The cranberry girdler (*Chrysoteuchia topiaria*) is similar in appearance to sod webworms, but causes the type of damage usually associated with white grubs.

Identification.   The majority of adult webworm moths listed in table 6–5 wrap their wings around the body when at rest (figure 6–20). The buff-colored adults of the larger sod webworm, vagabond, and silver-striped webworms all have a wingspan of about 1 in (25 mm). The moths are easily flushed from their hiding places when grass is being mowed or shrubbery is disturbed. They fly in a jerky fashion for a few feet, then dive, and finally rest on a grass blade. The moths are readily attracted to lights at night. The oval eggs, dropped among the grass blades by the female moths as they fly above the turf, are tiny, dry, and nearly impossible to find.

Larger sod webworm larvae are about 1 in (25 mm) long when fully grown and gray to dusky green with a brown head and brown-spotted body (figure 6–21). These caterpillars often hide during the day in silk-lined burrows in the thatch layer or on the soil surface. Larval excrement appears as a cluster of small, pale-to-dark-green pellets, some the size of a pinhead. The resting stage between the larva and adult is a dark brown, torpedo-shaped pupa about ½ in (13 mm) long.

FIGURE 6–20 Sod webworm adult.

FIGURE 6–21 Sod webworm in burrow.

The larger sod webworm and some of the other species pass the winter as larvae, tightly coiled in a closely woven silk case covered with particles of soil. In the spring, the larvae resume feeding, grow rapidly, and pupate in cells in the soil. The adult moths emerge in about 2 weeks. A day after emerging, the female moths begin to lay eggs as they fly in a zigzag pattern a few inches above the turf, usually in the early evening. Each female lays about 500 eggs. In hot weather, the eggs hatch in about 6 days. The larvae or webworms require 4 to 5 weeks to complete their full growth. The entire life cycle from adult to adult usually requires 6 to 8 weeks. Thus, under normal conditions, there are two or three generations a year, with additional ones in the south.

The tropical sod webworm feeds on most species of southern turfgrasses and is found in the southeastern states from Louisiana to Florida. Generations are continuous throughout the year, with the highest numbers in late summer and fall. The larvae are cream to greenish in color with yellowish-brown heads. Younger larvae eat the edges of grass blades, making them ragged looking. When larvae approach the full-grown size of ¾ in (19 mm) long, they consume entire grass blades, with sizeable populations eating large areas of turf overnight. The larvae pupate in shapeless silken bags coated with bits of debris. The resulting moth has a ¾-in (19-mm) wingspan and brownish wings with dark markings. Unlike other sod webworms, it does not fold its wings against the body but holds them out in a triangular shape like most other moths (figure 6–22).

The bluegrass webworm is a pest of Kentucky bluegrass in the eastern United States. It is especially numerous in the bluegrass regions of Kentucky and Tennessee. On golf courses, the larvae have damaged bentgrass roots in greens, similar to grub feeding. The larvae pupate vertically in the thatch layer, with the small moths emerging from the upper end of the pupal case. They can be found resting on the closely mowed turf *(Color plate 24)*. Damaged bentgrass sod can be rolled back to expose the larvae.

FIGURE 6–22   Tropical sod webworm adult.

FIGURE 6–23   Two burrow linings of the burrowing webworm.

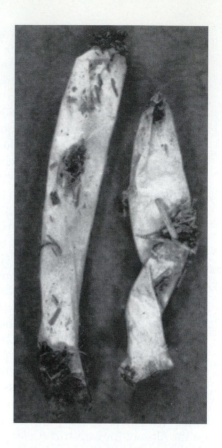

The burrowing webworm occasionally infests lawns and golf course fairways in the northeastern and north-central states. This insect lives in a 2-in (5-cm) vertical burrow in the soil. The larvae feed at night on the grass blades surrounding their burrow. When the larvae become fully grown, they line their burrow with a silk, paperlike, white sac in which to pupate (figure 6–23). Birds pull out the burrow lining, eat the webworm, and leave the paperlike sac laying on the grass. This insect is not a severe pest, but the scene of many white paper tubes scattered across the lawn can be dramatic to the homeowner. This feeding by birds commonly accomplishes all of the control that is needed.

The cranberry girdler is similar in appearance to webworms, and it also feeds on the roots of Kentucky bluegrass. It has been reported in grass-production fields in the west and midwest. Feeding by this insect severs the roots, allowing the infested, wilted sod to be rolled back. The larvae are found in the remaining roots at the soil surface. The cranberry cirdler overwinters in the soil similar to sod webworms but produces injury to turf like grubs. Control measures applied for grubs also control this insect.

Importance.   Sod webworm larvae feed on more than one species of grass. They clip off the blades of grass just above the crown. Irregular brown spots appear in the turf area where the larvae are numerous. Close examination of the turf reveals that the grass blades are missing with exposed thatch being responsible for the brown spots. Heavy infestations result in large brown areas *(Color plate 23)*. Because the roots and crowns or rhizomes are still present, control of the webworm population and irrigation will usually result in regrowth, although it may be too spotty to be acceptable.

A well-kept turf, fertilized and watered, supports more webworm feeding than one with poor management. Four or more webworms per square foot will do severe damage to any turf area.

Early detection of webworms in turf areas is important in preventing serious feeding damage. Large numbers of moths flying over the lawn at dusk, collecting on doors and windows and around outside lights, mean that the caution sign is out. Egglaying time in midsummer, for the second generation, usually results in the most moths. Egg hatch, which is the ideal time to apply chemical controls, occurs about 2 weeks after peak moth flight. Additional moth flights continue into September, with overlapping generations in some areas.

The presence of unusual numbers of birds feeding on the lawn—especially robins, starlings and other blackbirds, flickers, and other insect-feeding species—may indicate a webworm infestation. You can detect webworms simply by carefully digging through the sod, searching for silken burrows, frass, grass clippings, or the webworms themselves.

A simple, quick survey method is to mix 1 tablespoon (15 ml) of 2% pyrethrin in 1 gal (3.8 l) of water. Sprinkle this mixture over 1 sq ft (30 cm sq) and count the number of webworms leaving the crown area to climb up onto the grass blades (figure 6–24). The pyrethrin is very irritating to the worms and other insects hiding in the turf. Armyworms and cutworms also crawl out when the solution is applied. The mixture can be

FIGURE 6–24   Sod webworm flushed to the surface by pyrethrin solution. Note the bird feeding holes.

made of the same dilution with other concentrations of pyrethrin (2 tablespoons or 30 ml of 1%, or 1 teaspoon or 5 ml of 6%). This chemical is sometimes sold in agricultural pesticide outlets as a fly spray or vegetable garden spray. One tablespoon (15 ml) of dishwashing detergent per gallon of water and applied to 1 sq ft (30 cm sq) of turf will also bring sod webworms to the surface.

Management. Microsporidia diseases help to reduce sod webworm populations. After the second year of peak webworm populations, as many as 90% of the adult moths have been found to be infected. They continue to lay eggs, and the eggs hatch, but many of the diseased larvae do not survive the winter. These microsporidia tend to be more of a factor when damp, cool weather conditions prevail at some time during the growing season. Sod webworms are an infrequent pest in irrigated turf. In many areas of the northeastern and midwestern United States, sod webworms are usually pests only during hot, dry years in unwatered turf areas.

Predatory insects such as ground beetles, ants, wasps, and rove beetles consume large numbers of sod webworm larvae. Insect-eating birds such as starlings and other blackbirds, robins, flickers, and cowbirds eat large numbers of larvae as well, with large flocks being common for several days straight on infested turf areas.

Insecticides for sod webworm control are listed in table 6–12 at the end of the chapter. The insecticide can be applied as a granular formulation or as a spray. If granulars are used, ¼ in (6 mm) of irrigation is applied to dissolve and activate the insecticide off of the granule without flushing it into the soil. Turfgrasses containing endophytes can be grown to provide protection without applying insecticides.

6.3.2   Cutworms

Cutworms are a group of thick-bodied caterpillars. They hatch from eggs laid by robust, dark, night-flying moths. Cutworm damage is similar to that of other leaf-eating caterpillars—chewed-off grass blades above the crown.

Identification. The black cutworm (*Agrotis ipsilon*) overwinters in the southern states and migrates northward each spring. The eggs are laid by the sooty black moths, especially on winter annual weeds such as chickweed and yellow rocket, and bentgrass sod in golf course greens and tees. The eggs hatch into gray-to-black caterpillars. These caterpillars feed on the grass blades at night and hide in burrows in the soil during the day. Full-grown cutworms are 2 in (50 mm) long (*Color plate 25*). The life cycle of black cutworms is completed about every 30 to 40 days. Several life cycles are completed per year starting in early spring and continuing into fall.

The bronzed cutworm (*Nephelodes minians*) is dark brown or almost black across the top and lighter on the underside (figure 6–25). Both the larvae and adult moths have a distinct copperlike or bronze appearance. This cutworm infests bluegrass turfs in numbers, leaving large, straw-colored or dead areas in early summer.

Other species, such as the variegated cutworm (*Peridroma saucia*), may simply hide in the grass during the day and feed only at night on adjoining plants in flower beds. Bedding plants that are cut off indicate the presence of this cutworm.

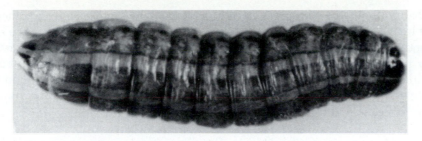

FIGURE 6–25   Bronzed cutworm.

Importance.   Cutworm larvae have a wide range of hosts including many turfgrasses. Black cutworm larval survival is high on tall fescue and perennial ryegrass, regardless if endophytes are present. Survival is also very high on bentgrass, making their presence common on golf course greens. On Kentucky bluegrass, less than 10% of black cutworm larvae are able to survive.

The larvae feed on grass blades at night; and large cutworms, fifth and sixth instar, can eat all of the grass blades in a small, roughly circular area from their burrow. The exposed thatch results in a brown area in that location. Younger black cutworm larvae, third and fourth instar, frequently live in taller grass near a golf course green and migrate 30 ft (9 m) or more onto the green at night to feed. In many areas of the country, this type of damage is small compared with the damage caused by birds feeding on the cutworms.

Many insect-feeding birds such as robins, cowbirds, starlings, and other blackbirds feed extensively on cutworm larvae. Most of the time this feeding is not a problem and helps reduce the likelihood of the feeding damage by cutworms as described. In closely mowed, high-quality turf areas such as golf greens and some fairways, the feeding and scratching of the turf by birds in their search for cutworms can cause as much or more damage than the cutworms alone. Holes poked in turf by birds can make golf balls change course, and flocks of birds can tear out large areas of turf in search of larvae.

The presence of birds on greens and other closely mowed areas is commonly used to justify a treatment for black cutworm. Before treatment, these observations should be backed up by scouting for the larvae. An irritant solution like the one described for sod webworms should be applied to determine the number of cutworm larvae present. For most golf courses, four or more cutworms per square foot is high enough to justify control. Courses demanding higher quality turf or those with special events such as tournaments may wish to treat lower populations.

Management.   Control is rarely needed in lawns, roughs, and other less closely mowed turf areas. In closely mowed turf areas, control treatments are usually applied at the first symptom of bird damage and as needed thereafter throughout the growing season. Cutworms can be controlled by sprays containing one of the insecticides listed in table 6–12. Repeat applications are often necessary to control black cutworms, as there are multiple generations.

Black cutworm eggs are laid singly at the top of the grass blade and hatch in 3 to 10 days. Daily mowing and clipping collection removes most of these eggs before they hatch. Avoid dumping grass clippings near bentgrass golf greens and other susceptible turf areas as most of these clipped eggs will hatch. Fewer larvae are found on sand-topdressed greens, but topdressing will not reduce the number of larvae already present. Core aerifying does not reduce the number of cutworms, but the larger larvae will live in the core holes. Growing Kentucky bluegrass around bentgrass greens should reduce the number of black cutworm larvae migrating onto the greens to feed at night.

6.3.3   Armyworms

Large areas of turf in which all of the grass blades disappear overnight are likely to have been attacked by armyworms, particularly if those areas are adjacent to small grain fields. Both true armyworm and fall armyworm larvae feed primarily in late evening and at night.

Identification.   Two species of armyworm sometimes infest turf areas. The true armyworm *(Pseudaletia unipuncta)* adult is a night-flying moth about 1 in (25 mm) long, tan to grayish brown, with a tiny white dot in the center of each forewing. The female deposits her eggs, which are small white globules, in rows or groups on the leaves of grasses and then rolls the blade around the egg mass. Full-grown larvae are about 1½ in (38 mm) long with two orange stripes on each side of the mostly brown or black body (figure 6–26). Another pale orange stripe passes down the back of the insect. Within this stripe is a fine, light-colored, broken line. Small armyworms, less than half grown, appear dark or almost black with small stripes.

Armyworms overwinter in the southern United States. The moths migrate northward during the spring to most states east of the Rocky Mountains. Three generations are found in the south and commonly two farther north.

The fall armyworm *(Spodoptera frugiperda)* attacks a more diversified group of plants than the true armyworm, feeding on corn, alfalfa, clover, tobacco, and various grasses. The adults resemble cutworm moths, with dark forewings mottled with light and dark spots. They are also active only at night. The eggs are deposited on grasses in groups of 100 or more that are covered with fuzz from the adult moth's body. The eggs hatch into young worms in 2 to 10 days. Newly hatched larvae are white with black heads. Their bodies become darker as they feed. During growth, they may curl up in leaf sheaths, suspend themselves from plants by threads, or crawl about on the ground. Fall armyworms become fully grown in 2 to 3 weeks.

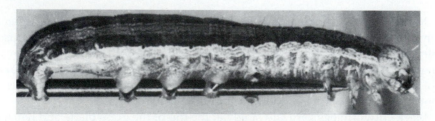

FIGURE 6–26   Armyworm.

The full-grown larvae attain a size similar to that of the true armyworm, which is 1½ in (38 mm) long. The larval body is dull black with several stripes along the body. There is a distinct, inverted light-colored Y on the front of the head. Fall armyworms overwinter in the southern United States. They migrate northward in late summer to northern states from Montana to the northeast. There are as many as six generations in the south but only one in its northern limits.

Importance.   The true armyworm feeds on agricultural crops such as small grains and corn as well as many species of turfgrasses. Armyworms commonly build up to large numbers on wheat, rye, or oats and then invade adjoining turf areas. Sometimes a large number of larvae may suddenly appear moving across a lawn area like an army. When this occurs, large turf areas will simply disappear overnight, leaving only the thatch, crowns, and root system behind. Most of these high armyworm populations and their resulting damage occur after a cold, wet spring. Night feeding by lower numbers of this insect causes the grass to appear ragged, uneven, and often bare.

In the south, the fall armyworm can be one of the most destructive turf insects, particularly to bermudagrass. Young fall armyworm larvae feed on the lower surfaces of leaves. Later, as they mature, entire leaves are eaten, leaving only the stem. High populations of full-grown fall armyworms can eat grass down to the crown.

Management.   In humid regions, armyworm populations are sometimes greatly reduced by a fungus disease that prevents further generations. A parasitic fly may also attack armyworms by attaching its egg by the armyworm's head. These white eggs are easily seen against the darker color of the caterpillar. The hatching maggot feeds inside the armyworm and kills it. Specific chemical controls are listed in table 6–12.

6.3.4   Grasshoppers

Identification.   Adult grasshoppers (*Melanoplus* spp.) may be yellow with brown or black markings. They are winged and can fly as well as hop to turf areas (figure 6–27). Depending on the species, they can range in size as adults from 1 to 3 in (25 to 76 mm) long. The eggs are laid in uncultivated areas in the fall, overwinter in the soil, and hatch in early summer. The immature hoppers, called nymphs, are wingless, but they otherwise resemble the larger adult. There is one generation a year.

Importance.   Grasshoppers are not a common pest of turfgrass. They may appear abundant in turf as young nymphs, but survival is poor on turfgrasses. Very large numbers of adult grasshoppers may migrate to well-kept turf located in a new subdivision or near rural areas where pasture or hay crops are growing rather than remain in a nearby mature, dried-up, or dormant crop. In these cases, feeding on the grass blades can be considerable.

Management.   Applied or chemical control of grasshoppers is necessary only if adult grasshoppers have migrated to a turf to feed on grass leaves and stems.

FIGURE 6–27   Grasshopper.

If moisture levels are high, many of the grasshoppers will succumb to disease, although they may cause considerable damage before they die. For insecticides that effectively control grasshoppers, see table 6–12.

6.3.5   Lucerne Moth

Identification.   Larvae of the lucerne moth *(Nomophila noctuella)* are larger than sod webworm larvae, but their bodies are similarly spotted. They wriggle actively when disturbed. The webwormlike larvae construct small horizontal silken tubes near the base of clover plants.

Adult moths are a mottled gray-brown with two pairs of dark spots on the frontwings, which span about ½ in (13 mm). The eggs are laid on clovers and sometimes on grasses. Newly hatched caterpillars feed on the clover or grass on which the eggs were laid. There are two to four generations through the growing season.

Importance.   The caterpillars commonly feed on clovers and similar legumes and occasionally on turfgrasses. If damage appears, it is usually in late summer.

Management.   Control is usually not necessary (table 6–12), but the larvae may be mistaken for sod webworms.

6.3.6   Striped Grass Loopers

Several species of striped grass loopers *(Mocis* spp.) feed on grass areas. When numerous, they may invade home lawns.

Identification.   The loopers (measuring worms, inch worms) are about 2½ in (64 mm) long when fully grown. They range in color from cream to blue-gray, brown, black, or orange, depending on the species. The loopers have large, black-and-white spots and stripes that extend the length of the body.

Importance and Management.   Control measures are rarely needed, and then only if the caterpillars are causing visible damage. See table 6–12 for specific insecticides.

6.3.7   Frit Fly

The frit fly *(Oscinella frit)* is probably more of a turfgrass nuisance than a damaging insect on golf courses. The larvae or maggots infest the stems of wheat, rye, and other grains, causing the seedheads to be white and empty. Other species of *Oscinella* infest the seedheads of Kentucky bluegrass, causing silver top, a general term used to describe a condition in which the seedhead is pale or white, with few or no viable seeds. In a Kentucky study, nine species were found in sweeping bluegrass fields. The frit fly has long been a pest of grains and grasses in Europe.

Identification.   The adult frit fly is a minute black fly with yellow on its legs, about ⅙ in (4 mm) long. The flies often hover over golf tees or an area of the fairway. They may be found in collected clippings of greens mowers along with ataenius beetle adults. Both insects frequent high-quality bentgrass greens— ataenius beetles in the early afternoon and frit flies in the midmorning. Frit flies are also attracted to white objects such as golf balls, towels, and golf carts. The adult flies do no damage to grass but are a nuisance, laying eggs on the leaves and leaf sheaths. Yellow-white, legless maggots hatch and burrow into grass stems to feed. There are about four generations per year during the summer months.

Importance.   Damage may appear as scattered yellow or dead leaves throughout the turf. Close examination reveals a central dead leaf, or possibly the entire shoot will be dead. Frit fly maggots may be found feeding at the base of the dead leaf on the dead shoot. The yellow-white, legless worms move about when disturbed.

Management.   Control of frit flies is sometimes justified to keep the adult flies from bothering people on the golf course or elsewhere. As with most insects, large numbers of adults may indicate a damaging infestation of larvae in a few days to a week or two. Chemical controls applied to golf greens and collars will reduce the population of this fly (table 6–12).

6.4 Leaf- and Stem-Sucking Pests

Sucking insects and mites feed in both the immature and adult stages on the leaf blades and stems on turfgrasses (table 6–6).

6.4.1 Chinch Bugs

Chinch bugs, like many turf insects, are often common pests of field crops. The common chinch bug *(Blissus leucopterus leucopterus)* has long been a serious pest of corn, wheat, rye, oats, and barley; it also feeds on some forage grasses but is rarely a serious problem. It is occasionally a pest of bermudagrass in the southern United States.

The southern chinch bug *(Blissus insularis)* is a common turfgrass pest in Florida and Louisiana on St. Augustinegrass. Southern chinch bugs are often found damaging zoysiagrass turf as far north as Indiana, Illinois, and Missouri.

The hairy chinch bug *(Blissus leucopterus hirtus)* has long been a pest of turf in the northeastern states. It feeds on bentgrasses, fescues, and Kentucky bluegrass and has damaged lawns in the upper midwest as well as the northeast.

The buffalograss chinch bug *(Blissus occiduus)* is a major pest of buffalograss in Nebraska. It appears to feed primarily on buffalograss, and is likely to become a pest wherever this grass is grown in the midwest.

Identification.　The life cycles and appearances of all species of chinch bugs are similar. Adult bugs are ⅕ in (5 mm) long and black with white wings overlapped over the back (figure 6–28). The wings usually cover the tip of the abdomen, but some adults have shorter wings that are half the length of normal wings.

The adults lay eggs over a period of time with the egg hatch varying from 10 to 25 days later depending on the temperature. Hatching bugs, called nymphs, are tiny, red with a white band across the body, and wingless. There are usually five immature stages before the winged adults emerge (figure 6–29). Chinch bug nymphs change from red and white to orange, brown, black, and white as they become adults. There

TABLE 6–6　INSECTS AND MITES THAT SUCK JUICES FROM GRASS LEAVES AND STEMS

Pest	Size	Color	Other traits
Chinch bugs	⅕ in (5 mm) or less	Black and white (red nymphs)	Live in thatch
Big-eyed bugs	⅛ in (3 mm)	Brownish	Bulging eyes
Aphids	1/16 in (1.6 mm)	Green	Orangish-brown turf
Scale insects	1/16–⅛ in (1.6–3 mm)	White, cottony tufts	—
Leafhoppers	⅛–¼ in (3–6 mm)	Greenish to brownish	Jump and fly when disturbed
Spittlebugs	⅛–¼ in (3–6 mm)	Greenish	Covered in spittle mass
Mites	1/64 in (0.4 mm) or less	Whitish to red-brown	Eight legged

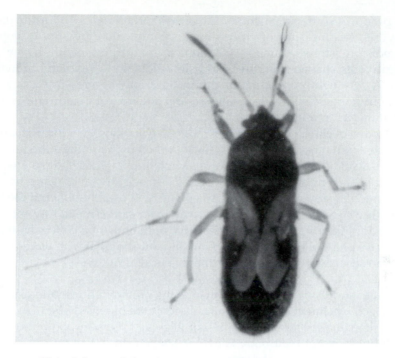

FIGURE 6–28   Chinch bug adult.

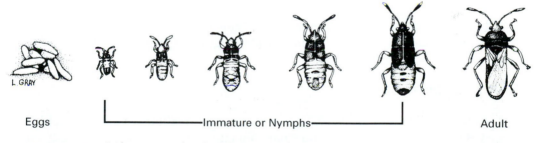

Eggs   ├──────────Immature or Nymphs──────────┤   Adult

FIGURE 6–29   Life stages of a chinch bug.

are usually two generations per year in the midwest and one in more northern states. Adults hibernate during the winter in clumps of grass and under debris.

Southern chinch bugs can complete five or six generations annually in Florida. Overlapping generations are common after early summer. Farther north, the adults hibernate and pass through two to three generations. Adults pass the winter in dead grass, leaves, and other debris.

Importance.   Chinch bugs, both the adults and all sizes of nymphs, damage grass by injecting salivary fluids into the plants as they suck sap from them. Many bugs mass on one plant and then leave to attack adjoining plants. This type of feeding produces scattered patches in a turf area rather than a uniformly damaged one. Favorite feeding sites of chinch bugs are turfs that are sunny, sloping, or under moisture stress. Damage is common in turfgrass with heavy thatch. Damaged plants change from green to pale yellow and eventually brown *(Color plate 26)*. A very light population of chinch bugs may be present with no visible damage, but more than 25 bugs of all sizes per square feet in a droughty area can severely damage or kill the turf.

Detection of chinch bugs involves more than walking over the turf. Bugs may be crawling about on a driveway or sidewalk when high numbers are present and the grass is dead or almost dead. Vigorous scratching in the damaged area will often cause the bugs present to scatter. Forcing a strong stream of water from a garden hose into the crown area of the grass plants will usually float out the bugs.

Another effective method is to drive an open can or metal ring into the soil surface and fill the ring with water. (Remove both ends of a coffee or shortening can and drive one end through the turf into the soil and fill it with water.) If chinch bugs are present, they will float to the top in 5 to 10 minutes.

Management.   Predators of chinch bugs quite often run about in the turf area damaged by chinch bugs. One such insect is the big-eyed bug *(Geocoris* spp.*)*, and more than one species may be present. Big-eyed bugs are about the same size as chinch bugs, except their body is oval in shape rather than narrow. Their eyes are also very large; hence the common name. This predator, along with lady beetles, earwigs, and ants, can reduce chinch bug populations.

A fungus disease *(Beauveria bassiana)* is sometimes an effective natural control of chinch bugs. High humidity plus a high population of the bugs are necessary before the fungus can increase, spread among the population, and reduce the numbers of bugs. Planting endophyte-containing turfgrasses will also provide protection from chinch bug damage. Specific insecticide suggestions are given in table 6–12.

6.4.2 Aphids

Aphids, or plant lice, attack many agronomic and horticultural crops. Rose aphids suck the sap from rose foliage and stems. Tomato aphids can increase rapidly on tomatoes, and corn-leaf aphids may dry up the tops of corn plants. Certain aphids infest stands of grain; these include the English grain aphid, apple grain aphid (also a pest of apples), oat-bird cherry aphid, and the greenbug. The bird cherry-oat aphid *(Rhopalosiphum padi)* has been observed feeding on stands of bluegrass. The only other aphid recognized as a turf pest is the greenbug.

Identification.   The greenbug *(Schizaphis graminum)*, sometimes called the spring grain aphid, is a destructive aphid, attacking turfgrasses and grain crops. Greenbugs damage barley, oats, and wheat from Texas to Nebraska. They are yellowish to greenish with a dark green streak running up the back. These pear-shaped, soft-bodied insects are the size of a pinhead when fully grown (figure 6–30).

Aphids such as the greenbug have a rather unusual life cycle. Adult females give birth to young during the grain-growing season or the summer months. Con-

FIGURE 6–30   Greenbug adult and young.

tinuous generations are produced in the south. Farther north, both winged males and females are produced in the fall; and black, shiny overwintering eggs are laid.

The greenbug is one of many species of grain aphids that migrates into the midwest on southerly or southwesterly prevailing winds. In a midwestern study of migrating grain aphids during the 1960s, the greenbug was the most common species trapped in water pans and wind socks.

Greenbugs have been reported feeding on turfgrass throughout the midwest and northeastern United States. Infestations reappearing in the same turf early in successive years are typically due to overwintering eggs and occasionally overwintering adults. In some communities, the number of infested lawns increases from year to year, up to 50% of all lawns, before disappearing completely. It is likely that infestations of greenbugs in midwestern and northeastern turf are due to both overwintering eggs and migration from southern states and other midwestern locales.

Importance.   Greenbug damage commonly first appears in early July, but may occur earlier from overwintering eggs. Turf beneath trees and next to buildings becomes rust-colored from their feeding. The rust-colored areas expand as the numbers of the aphids increase (*Color plate 27*). The females give birth to three to five young per day, which reach maturity in 1 to 2½ weeks and are then capable of producing young. No males or eggs are present during this increase in numbers during the summer. The adult females and young aphids mass around the border or rim of the damaged area (*Color plate 28*).

Greenbugs feed on the sap of grass plants; as such, they are commonly found on the grass blades as well as on the stem. While feeding, these aphids inject a toxic saliva that causes the area around the feeding site to die and turn rusty orange. The toxin travels throughout and weakens the entire plant, including the roots.

The relationship of greenbug damage and trees is not clearly understood. One can speculate that migrating winged adults strike objects and end up in the shade, giving birth to living young. Greenbugs have been found under trees, next to buildings, and even around the base of utility poles along highways.

Damage commonly is skewed to the northeast of trees and is commonly found east of buildings and posts. This leads one to believe that the greenbugs avoid the hot afternoon sun from the southwest during the summer. This is supported by reports that greenbugs on southern crops tend to be found lower on the plants, being uncommon on top of the plants where it is sunniest and hottest. However, heavy infestations will move out from the shade and damage large areas of turf in full sun.

Management.   It is common to find predators and parasites in the greenbug population. Lady beetles, syrphid fly larvae, and parasitic wasps attack the pest insects. With a rapid population growth, greenbugs often increase in numbers in spite of the beneficial insects. In humid periods, fungal diseases can reduce the populations of greenbugs.

Greenbugs are often resistant to the insecticides commonly used for aphid control. This is not surprising, since much of the grain acreage infested with greenbugs is treated on a regular basis, and resistance to more than one insecticide has been verified. New outbreaks appearing in an area may be resistant to the insecticide used for other foliage-feeding insects. If treatments are ineffective, other labeled insecticides listed in table 6–12 will control greenbugs.

6.4.3   Leafhoppers

Identification.   Adult leafhoppers are tiny and wedge shaped, pale green, yellow-gray, or mottled in color. The young or nymphs resemble the adults except they are wingless. Both adults and young can hop and crawl about on foliage. Adult leafhoppers usually fly a short distance when disturbed. Leafhoppers feed on plant juices with their sucking mouthparts.

The most common species associated with turfgrass damage is the potato leafhopper *(Empoasca fabae,* figure 6–31). It is a serious pest of alfalfa in the eastern United States but also feeds on over 100 kinds of plants, including potatoes, beans, young shade trees in nurseries, and turf areas. Adult potato leafhoppers are about ⅛ in (3 mm) long and pale green. This species has a tendency to swarm in large numbers, moving about from area to area. They are strongly attracted to lights at night.

FIGURE 6–31   Potato leafhopper.

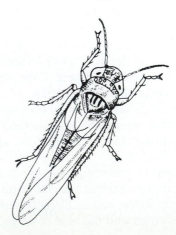

Importance.   Large numbers of leafhoppers can cause the turf to be mottled or whitened in areas. Sucking of the plant sap causes drying out of the grass leaves and stems, which can be mistaken for drought or disease. A close examination of hopper-damaged leaves reveals tiny, white feeding punctures that cause the entire leaf to appear mottled. Turfgrasses are not affected as severely, with only mottled, pale foliage.

Detection of leafhopper injury to grass may have to be made without the presence of the insect. A swarm of potato leafhoppers will often feed heavily on an area for a few days, then move as a group to a more favorable food source.

Management.   Control treatments (table 6–12) are usually impractical unless the leafhoppers persist in an area for a considerable time, and damage is evident. New seedlings are sometimes severely damaged, necessitating a control treatment.

6.4.4   Scale Insects and Mealybugs

Several species of scales infest turf, but the life cycles of these various bugs are almost alike. The rhodesgrass mealybug, also called rhodesgrass scale *(Antonina graminis)*, is found in the Gulf states and other southern states, including California, as a pest on rhodesgrass, bermudagrass, St. Augustinegrass, tall fescue, and centipedegrass. Bermudagrass scale *(Odonaspis ruthae)* attacks bermudagrass from Florida to California along the Gulf states.

Identification.   Adult scales are legless, small, immobile insects with an armored or waxy, shell-like protective cover. These tiny, globular or oval insects are easily overlooked when damaged grass areas are examined. Very young or immature scale insects are mobile. They settle on a grass leaf blade or stem, insert their needlelike mouthparts, suck plant juices, lose their legs, and begin to secrete the waxy covering.

The Rhodesgrass mealybug infests pasture grasses and also occurs in turf areas. The adults are globular, ⅛ in (3 mm) across, and dark purplish brown within masses of white, cottony filaments (figure 6–32). All adults are female and give birth to living young. The life cycle takes about 50 days to complete, and there are usually five generations annually.

Bermudagrass scales attach themselves to the nodes of bermudagrass stems. They are white when mature and ⅟₁₆ in (1.6 mm) long. Infested bermudagrass crowns have a moldy appearance. Infestations are more common in thatchy, shaded turf.

Ground pearls, a scale insect, attack the roots of centipedegrass, bermudagrass, and St. Augustinegrass (see sec. 6.2.4).

Importance.   Rhodesgrass scale attacks plant crowns, causing infested plants to wither and die (figure 6–33). Turf that is infested with bermudagrass scale grows slowly and deteriorates to a thin stand, which is then susceptible to disease attack and weed invasion.

Management.   Control suggestions for scales are given in table 6–12.

FIGURE 6–32   Rhodesgrass mealybug (scale).

FIGURE 6–33   Rhodesgrass scale injury to bermudagrass turf. (Courtesy Milwaukee Sewerage Commission.)

6.4.5 Spittlebugs

Two species of spittlebugs are found on turfgrass. The meadow spittlebug, *(Philaenus spumarius)* often makes its presence known by its spittle masses on clover in turf (figure 6–34) in the midwest and northeastern United States. The two-lined spittlebug *(Prosapia bicincta)* is found throughout the eastern United States, but in high numbers on turfgrasses in the southeast.

FIGURE 6–34   Spittlebug nymph and spittle.

Identification.   The meadow spittlebug is a small, wedge-shaped insect that resembles a leafhopper. The adults are heavy-bodied and about ¼ in (6 mm) long. They are usually mottled brown and cream colored, but vary from cream to almost black. They jump readily, making an audible thump. The small, yellow-to-white eggs are laid in rows on plant leaves and stems.

The nymphs or young cause the damage, if any. They are found behind leaf sheaths, in folded leaves, and on stems in masses of froth or spittle during the summer months. There is one generation each year. The meadow spittlebug overwinters in the egg stage.

The two-lined spittlebug is dark brown to black as an adult and ⅜ in (10 mm) long. The nymphs feed on bermudagrass, St. Augustinegrass, and centipedegrass within spittle masses.

Importance.   Neither species of spittlebug causes much damage, with some yellowing resulting from heavy feeding. Certainly feeding damage to clover and occasionally even weeds in the lawn by the meadow spittlebug is not an economic loss. Spittle masses are more of an aesthetic concern than damage from this insect's feeding.

Management.   Chemical control is rarely justified. If necessary, select one of the insecticides suggested in table 6–12.

6.4.6   Mites

Mites are microscopic, eight-legged relatives of insects. In a thorough examination of a piece of turf, several mite species may be present. Some are beneficial or predator mites that feed on pest mites. Many other mites, particularly most of the larger,

TABLE 6–7   MITE IDENTIFICATION

Pest	Color	Shape	Other traits
Eriophyid mites	Whitish	Cigar shaped	Very tiny
Clover mite	Blackish	Round	Front legs twice as long as other legs
Winter grain mite	Blackish	Round	Long front and rear legs
Banks grass mite	Greenish-yellow	Round	All legs of similar length
Chigger	Red	Round	All legs of similar length
Soil mites	White or blackish	Round	All legs of similar length

dark, hard-shelled ones, are scavengers and feed on decaying plant and animal material. Others are pests that penetrate leaf cells and consume the sap, causing severe chlorosis and browning of the leaves.

There are three eriophyid mites known to attack turfgrass. They are similar in appearance, the damage that they cause, and host specificity. Bermudagrass mite (*Eriophyes cynodoniensis*) attacks bermudagrass, zoysiagrass mite (*Eriophyes zoysiae*) attacks zoysiagrass, and buffalograss mite (*Eriophyes slyhuisi*) attacks buffalograss (table 6–7).

Identification.   Bermudagrass mites can cause considerable damage to bermudagrass lawns. These white, cigar-shaped mites are visible with a microscope. Bermudagrass mites, as with most mite species, multiply rapidly and cause severe damage in a short period. The females lay eggs under the leaf sheaths. Immature mites that hatch from eggs are six-legged rather than eight, as are the adults. The life cycle of the bermudagrass mite from egg to egg-laying adult is about 1 week. Zoysiagrass mite and buffalograss mite are similar in life cycle and appearance.

The clover mite (*Bryobia praetiosa*, figure 6–35) is more a nuisance pest inside and on the outside of homes rather than of turf. Although this mite has been known to cause significant injury to turf in home lawns, it feeds mainly on clovers and some other plants. Large populations can build up and invade homes. This migra-

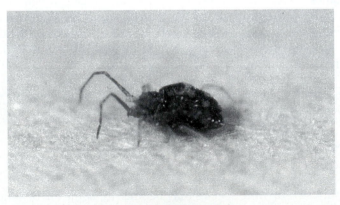

FIGURE 6–35   Clover mite. (Photo by D. Voegtlin.)

tion occurs mostly in late winter or spring but sometimes in the fall. There are several generations each year.

Clover mites resemble reddish-brown specks of dust crawling about on walls, windows, curtains, and other indoor surfaces. Under a hand lens they appear oval, eight-legged, and the front pair of reddish legs are more than twice the length of the other legs. When crushed on walls or curtains inside a home, clover mites leave a red stain. As a house pest, they are only a nuisance, but large numbers can build up on and around windows.

The winter grain mite *(Penthaleus major)* is a pest of small grains and some grasses. It is known as a pest of grass-seed fields in Oregon and also occurs in many other areas throughout the United States. It feeds at night, on cloudy days, and under snow cover on Kentucky bluegrass, ryegrass, fine fescues, bentgrass, and other grasses including crop species. It will also feed on legumes and other broad-leaved plants.

This mite is unique in that it is most numerous and damaging during the winter and early spring months. The eggs hatch in October. Young mites are reddish orange but become darker brown or black as they grow. The olive-black adults with eight reddish-orange legs emerge in November. The front and rear legs are longer than the middle two pair. The adults are small but can be easily observed with the naked eye.

On close inspection of damaged turf on winter days, winter grain mites can be seen in the thatch or on the damaged crowns. The mites increase to great numbers in late winter after the second generation eggs hatch. Several thousand per square foot is not unusual. Heavy mite populations diminish in April as eggs are laid on dead grass plants. The eggs do not hatch until late fall.

Banks grass mite *(Oligonychus pratensis)* attacks bluegrass, bermudagrass, St. Augustinegrass, and several other grasses including important crop species. They are similar in appearance to two-spotted spider mites in that they are greenish-yellow with two dark spots in the abdomen. They are barely visible to the unaided eye, but easily seen with magnification. Winter is spent as older nymphs and adult females and have several generations per year. They become numerous in hot, dry weather.

Importance. Bermudagrass mites suck juice from plants, causing them to turn yellow or brown. Leaves become twisted with rolled margins. Stem feeding causes shortening of internodes and a rosetted or tufted growth. Heavily infested plants turn brown and die. When mite damage is severe, the grass is thinned and soon invaded by weeds. Feeding and resulting damage is heaviest in late spring and summer. Zoysiagrass mite and buffalograss mite damage is similar to berumudagrass mite, except that they attack different grass species and are less likely to kill the turf.

Clover mites can cause localized turf damage when very numerous, such as along foundations, particularly to Kentucky bluegrass and perennial ryegrass. Damaged turf will appear whitish with close examination revealing whitish feeding streaks on the grass blades. Damage is more likely in dry areas.

Winter grain mite damage may be incorrectly diagnosed as winterkill or desiccation. Large turf areas will appear whitish to brown. Close examination will reveal whitish areas where the mites have been feeding near the top of the grass blade. Winter feeding is reduced during severely cold temperatures and unseasonably warm temperatures, but will resume when normal winter temperatures return.

Banks grass mite feeding causes small, yellow spots on the leaves. Heavy mite numbers result in yellowed, dead foliage, and is most likely during hot, dry weather on unwatered turf. Fine silk webbing is usually present at the base of attacked grass plants.

Management.   Bermudagrass mites are numerous in warm, dry conditions, so irrigation can reduce the amount of damage. Resistant bermudagrass varieties are available. There are predatory mites that feed on bermudagrass mites beneath the leaf sheath.

Keeping a zone of bare soil or nongrass plants about 18 to 24 in (46 to 61 cm) wide around the house foundation usually prevents clover mites from migrating into the home. This strip can be planted to flowers that do not attract clover mites. Irrigation can help reduce turf damage by clover mites.

Control measures applied against bermudagrass mites and winter grain mites are highly effective if the problem is correctly diagnosed. Suggested insecticides are listed in table 6–12.

6.5 MOUND- AND HOLE-PRODUCING INSECTS AND RELATIVES

Several uses of turfgrass are impacted by mounds of soil on the surface or even holes through the turf. A variety of insects and related animals cause these (table 6–8).

6.5.1 Ants

Identification.   Most species of ants in the United States are social insects that live in colonies or nests in which also live the egg-laying queens, the young or larvae, pupae, and many worker ants. The workers, all wingless sterile females, care for the colony as well as search for food and bring it to the nest. At one or more times during the growing season, ant colonies may produce winged males and females that fly about, mate, and have the ability to start new colonies. Each colony

TABLE 6–8   INSECTS AND RELATIVES THAT PRODUCE MOUNDS OR HOLES IN TURF

Animal	Size of hole	Other traits
Ants	⅛–¼ in (3–6 mm)	Loose soil around hole, ants numerous
Cicada killer	¾ in (19 mm)	Loose soil around hole, large yellow and black wasp
Cicadas	½ in (13 mm)	Chimney usually around hole
Yellowjackets	1–4 in (2.5–10 cm)	Numerous ½-in (13-mm) yellow and black wasps
Ground-nesting bees & wasps	1/4 in (6 mm)	Numerous bees or wasps present
Green June beetle	½ in (13 mm)	Loose soil around hole
Crayfish	1–2 in (2.5–5 cm)	Chimney up to 6 in (15 cm) usually around hole
Earthworm	⅛–¼ in (3–6 mm)	Mound of soil balls around hole

FIGURE 6–36   Little black ant.

of a particular species will usually do this at the same time. When an ant species that has many colonies in a turf area swarms, winged ants will seem to be everywhere for a day or two.

Some species of ants that frequent turfgrass areas and eventually construct anthills include the little black ant *(Monomorium minimum,* figure 6–36), the pavement ant *(Tetramorium caespitum),* and the thief ant *(Solenopsis molesta).* These ants all live in the soil and construct small hills. The worker ants can be observed constantly in search of food to bring back to the colony.

Other ant species may be in or near grass areas. Carpenter ants *(Camponotus spp.)* nest in dead trees, logs, and even structural wood in houses. Several common species are large, winged, black ants that often exceed ½ in (13 mm) in length and are commonly seen moving across turf. Winged males and females may swarm occasionally but do not nest in turfgrass soil.

The red imported fire ant *(Solenopsis invicta)* is an important turf pest, especially in the south from the Carolinas to Texas and north into Tennessee. The colony constructs honeycomb mounds containing up to half a million worker ants *(Color plate 29).* These mounds are found in pastures, roadsides, field borders, and forests. In urban areas they occur on playgrounds, golf courses, cemeteries, and in home lawns. The ants build mounds in many areas but prefer sunny sites and clay soils over sandy soils. In home lawns and other managed turfgrass areas, the mounds usually reach only the height of the grass. In unmowed situations, the mounds may be ½ to 2 ft (15 to 61 cm) tall.

Importance.   Ants are important to the ecology of turfgrass. These predators and scavengers feed on insects in the turf, particularly soft-bodied insects such as caterpillars. They are useful in removing dead insects and other debris. Their tunneling helps to aerate and incorporate organic matter into the soil.

Ants construct mounds or small hills of granulated soil which may smother out the surrounding grass. Grass may also be killed next to the mound as the soil around the grass roots dries out from the effects of the digging and burrowing. Active ant populations can travel under newly laid sod, promoting air movement beneath the sod that dries out its roots and keeps it from growing into the soil. Even small ant hills can interfere with golf course play, particularly on greens.

Imported fire ant stings are painful. When a person is stung, venom is injected. The site of a sting may only turn red, or a lesion may develop and lead to infection or scarring. Because of the potential for being stung, fewer people use parks and other turfgrass areas infested with fire ant mounds.

Management.   Ant control is temporary at best. When ant colonies are destroyed, new colonies will appear during the growing season as soon as the insecticide residue degrades. Thus, treatment of large turf areas is not recommended except when ants are numerous in areas where sod is going to be laid.

Individual, large ant colonies can be opened and the soil within the colony drenched with an insecticide recommended in table 6–12. Within a few weeks, a new colony will appear in the area. Hopefully it will not be a species that constructs a large, turf-choking nest.

The strategy for controlling the red imported fire ant is not its eradication but reduction of the population in the existing mounds. Fire ants increase their mound size in wet seasons to move above the moist area. Applied controls (table 6–12) are made to the mounds without disturbing the structure. Baits can also be used to kill the producing queen.

In many areas, local ant species compete favorably against red imported fire ants moving into the area and keep the fire ant populations low. Bacteria and fungi have been found that are effective against both the red imported fire ant and the parasitic phorid fly.

6.5.2   Cicada Killers

Identification.   Cicadas are fed upon by a black-and-yellow-marked wasp about 1½ in (38 mm) long called the cicada killer *(Sphecius speciosus,* figure 6–37). It is commonly seen buzzing and darting about lawns or gardens in late summer. As they fly about, you can hear their rapid wing beat. There may be a few or a dozen or more, depending on how well they prefer an area for nesting. These bright-colored wasps begin to appear in July and remain through September.

FIGURE 6–37   Cicada killer wasp.

Female cicada killers search for cicadas, struggle with, and sting them. The cicada is much larger than the wasp. The paralyzed cicada is carried, partially in flight, by the female wasp to previously prepared burrows in the soil. It is dragged into the burrow and placed in an enlarged cell at the bottom of the hole. The wasp lays an egg on the paralyzed cicada. The egg hatches in a few days into a larva that feeds on the cicada. It remains in the soil over the winter. The adults emerge in early to midsummer.

These wasps are expert and energetic diggers, throwing up a mound of soil near each burrow. These mounds may suddenly appear in the morning from digging performed the previous evening. The wasp prefers to dig in sandy soil on a slope or terrace. Often the wasps' burrows are under spreading shrubs in foundation plantings. The holes are about ¾ in (19 mm) in diameter. They extend 6 in (15 cm) or more into the soil and then branch out several times and extend another 6 to 8 in (15 to 20 cm) into the soil. Although formidable in appearance, these wasps are not vicious, but they can nevertheless inflict a severe sting if disturbed or annoyed.

Importance.   The cicada killer is a predator insect, but its digging and the soil mounds it creates can be a nuisance to high-quality turf.

These large wasps appear dangerous, and people are likely to avoid areas where they are present in large numbers. Male cicada killers, which cannot sting, establish territories and will fly up and look at anything that enters that area. The sight of a large wasp hovering a few inches from one's face can be very upsetting to many people. For this nuisance problem, control may be necessary.

Management.   Educating people about the activities and small likelihood of stings of cicada killers may reduce the need to control these insects. Soil insecticide drenches on infested areas will control these wasps, if needed (table 6–12).

6.5.3   Cicadas

Identification.   Cicadas are large, robust, winged insects that announce their presence with the daylong shrill singing of the males. The common dog-day or annual cicadas (*Tibicen* spp.) are a nuisance with their loud singing every year but cause very little, if any, damage.

Periodical cicadas (*Magicicada* spp.) are smaller than the annual cicada, and the body is darker (figure 6–38). The wings contain orange veins, and the eyes are red. There are six distinct species, but all are similar in appearance. Three species are usually present in broods occurring every 13 years in the southern half of the United States, and the other species in broods that emerge every 17 years in the northern half of the country. There are some 18 different broods, most found in the eastern half of the United States. Large numbers of cicada nymphs emerge from the soil at one time and change into adults after living in the soil as nymphs for almost 13 or 17 years. Adult cicadas begin emerging in late May.

In moist soil, the nymphs, before emerging, construct mud tubes or chimneys that extend up to 3 in (76 mm) above the ground level (figure 6–39). Great numbers of nymphs emerge at the same time. They crawl up tree trunks, posts, or other

FIGURE 6–38 Periodical cicada.

FIGURE 6–39 Periodical cicada nymph emergence holes or "chimneys."

objects. A few hours later, the adults emerge, leaving their many grayish-brown shed skins on posts, trees, and other upright objects. Cicada males announce their presence to the noiseless females by making a high-pitched sound. When many males are singing, the songs meld into a continuous trill. This sound is produced by vibrating membranes on the underside of the first abdominal segment.

After mating, the females lay eggs in rows in pockets that they cut in small branches and twigs of trees with their long, knifelike, egg-laying structure. The eggs hatch in 6 or 7 weeks; the newly hatched nymphs fall to the soil surface and burrow until they find tree roots on which to feed. With their sucking mouthparts, they immediately begin to suck sap from the root.

Importance.   Primary damage to turf by periodical cicadas occurs when the full-grown nymphs dig themselves out of the ground at the end of their 13- or 17-year cycle. If the number of nymphs that burrow into the soil and survive is large, the number of chimneys, burrows, or holes that suddenly appear under a tree may be quite striking—one for every square inch.

Management.   The nymphs of periodical cicadas are impossible to control, as they live at least a foot deep in the soil before emerging. The adults can be controlled (table 6–12) to prevent egg-laying damage to trees as well as to reduce the population of a generation 13 or 17 years later.

6.5.4   Ground-Nesting Bees and Wasps

Identification.   Ground-nesting bees (Andrenidae) and wasps (Sphecidae), other than the cicada killer described previously, may nest in burrows in the soil. The bees are also called mining or solitary bees. The adult bees are usually less than ½ in (13 mm) long and drab in color, commonly being brownish with dark brown or black bands (figure 6–40). Although most are capable of stinging, they very rarely do, even when their nests are disturbed. They feed on flower nectar and collect pollen and nectar to feed their white, legless larvae in the underground nest.

FIGURE 6–40   Ground-nesting bee.

Nest openings are usually about ¼ in (6 mm) in diameter and do not have loose soil around the opening. In the process, this foraging pollinates many species of plants.

Ground-nesting wasps provision their soil nests with adult insects such as flies, crickets, and grasshoppers. These are used as food for their larvae. The adult wasps live primarily on flower nectar, and they are important pollinators. These wasps are usually ¾ to 1½ in (19 to 38 mm) long depending on the species. Although capable of stinging, they rarely do so. Paralyzed insects are placed underground, and an egg is laid on it. The resulting larva feeds on the paralyzed prey. Nest openings are usually about ¼ in (6 mm) in diameter, and they are surrounded by loose soil.

A social wasp, the eastern yellowjacket *(Vespula maculifrons)* lives as a colony in an aerial nest in buildings or in the soil. Underground nests are constructed in old rodent burrows and commonly have a 2- to 3-in (51- to 76-mm) opening. The yellow and black, ½-in-long (6-mm) worker yellowjackets feed on ripening and overripe fruit, and forage in the summer for insects to feed their young. Individual nests can contain several thousand wasps by late summer. Similar species occur throughout the United States.

Importance.   Ground-nesting bees and wasps in a turf area may continually be constructing new burrows, or the insects may simply be a nuisance to turf owners or people using a public area such as a park or golf course.

Bees or wasps darting around through the air can be unnerving to many people, although most of these are very unlikely to sting. Nevertheless, treatment may need to be applied to ease their concerns. However, yellowjackets are likely to sting if their nest is disturbed. Serious injury or death can result from the multiple stings that are possible when large nests are disturbed in late summer. Some people are very sensitive to bee or wasp stings and can die from anaphylacticlike symptoms from even a small stinging incident.

Management.   Soil drenches, if needed, will reduce these pest insects (table 6–12). Avoid being stung by making control applications at dusk when these insects' activity is low.

6.5.5   Earthworms

Identification.   Earthworms live all year in the soil of many turfgrass areas. They are most evident in the spring and fall by their burrowing and feeding activity and the mounds, called casts or castings (figure 6–41), they leave behind. High populations of earthworms are usually associated with a moist, cool soil high in organic matter.

Importance.   Earthworms move to the soil surface during cool periods to pull leaves, stems, and other undecomposed debris from the soil surface. This feeding activity is a primary remover of thatch that accumulates on the soil surface. Large amounts of undecomposed material pass through earthworms and remain in the soil.

The small mounds or castings left by earthworms can be objectionable, if numerous, especially on a golf or bowling green or other fine-turf area. Castings on a

FIGURE 6–41   Earthworm and casting.

green that is topdressed with sand will contain sand which quickly dulls the greens mower blades. The benefits of earthworms have to be weighed against the objectionable, uneven soil surface caused by their castings.

Occasionally, earthworms will build up to such high numbers that the turf surface becomes spongylike and uneven from the very loose soil underneath created by this huge earthworm population. Digging into the soil will show it to be primarily consisting of earthworm castings. Turf in these areas may be healthy or can become sparse from too much air movement around its roots. Some turf areas become so soft from the loose soil that the wheels of mowers and other equipment sink into the soil and become stuck.

Management.   Most insecticides labeled for use on turf insects have no effect on the feeding activity of earthworms, but some do. The arsenicals, especially calcium arsenate, discourage earthworms from coming to the soil surface. Carbamate insecticides watered into the turf can reduce overly high earthworm populations to a small extent.

6.5.6   Crayfish

Identification.   Crayfish (Crustacea: cambaridae) look similar to small lobsters. They are only distantly related to insects. Crayfish have five pairs of legs with the first three pairs containing claws or pincers on the end. On the first pair of legs, these claws are greatly enlarged and are about 2 in long on adults of large species. They have one pair of long antennae that are about half as long as the body and another pair of short, branched antennae. The abdomen is long and slender. Large species of crayfish are commonly 6 to 8 in (15 to 20 cm) long as adults (figure 6–42).

FIGURE 6–42   Crayfish.

They breathe with gills that are attached to the base of the legs and are concealed under the carapace above the legs. As long as these gills remain moist, crayfish are able to breathe out of water. Most species of crayfish live only in streams and ponds and never venture onto land. A few species will construct burrows in low-lying areas in which the lower part of the burrow extends into the water table so that the animals are able to maintain their aquatic existence. These burrows commonly extend from 3 to 6 ft (0.9–1.8 m) deep with an enlarged area at the bottom where the crayfish lives most of the time. Some burrows have a side tunnel leading directly into a nearby pond or stream. Crayfish are scavengers and will feed on decaying plant and animal material, frequently leaving the burrow at night to forage.

Adult females carry their eggs on appendages under the abdomen. These eggs hatch into young crayfish that are similar in appearance to the adults except that they are smaller. The hatchling crayfish will be carried by the female for a few days before they drop off to fend for themselves. They live in streams or ponds when young, leaving those areas to live in burrows when about 1 year old. Most crayfish live about 3 years.

Importance.   Some species construct chimneys at the entrance to the burrow that may extend 6 in (15 cm) or more above the soil surface and be 2 or 3 in (5.1–7.6 cm) diameter (figure 6–43). These chimneys are balls of clay mud that bake in the sun and become very hard. Hitting them while mowing can dull mower blades and perhaps stop the engine. These chimneys' appearance can be an aesthetic problem as well.

Crayfish emerge from their burrows at night and wander on the turf. They are stepped on by people walking around at night. Barefoot walkers risk getting their toes pinched. When disturbed, crayfish quickly run backward and wave their large claws in defense. The claws are not dangerous; the pinch is unlikely to break the skin.

FIGURE 6–43   Crayfish chimney.

Management.   Control is best accomplished by improving drainage to eliminate the low-lying muddy areas that are attractive to crayfish. Although tiling may be expensive, it normally results in higher quality turfgrass and ornamental plants in the area than was possible earlier due to the poor drainage around their roots. Drainage along ponds and streams can be improved by reducing wide mud flats and excavating a more defined bank.

Solid wood or stone fences that extend several inches high have been used effectively to keep crayfish out of turf areas. Because crayfish are unable to fly or jump, an 8 in (20 cm) or higher fence should suffice.

Many golf courses and other professionally managed turf areas are solving the problem by turning such low-lying areas along ponds and streams into natural marshes by introducing native marsh plants. This not only solves the crayfish area maintenance problem, but also is usually popular with the public and clientele, thus improving public relations.

No pesticides are labeled for crayfish control. Any control achieved by killing the animals would be temporary at best. As long as the low-lying, wet habitat remains, the crayfish will return. The species that live in low-lying areas also live in ponds and streams, so bank areas will be rapidly repopulated. In addition, these species commonly travel overland on damp nights, making repopulation of more remote areas very likely.

6.6   INSECTS AND RELATIVES THAT DO LITTLE OR NO DAMAGE

While examining turf during pest scouting on other activities, the turf manager is likely to notice numerous insects and insect relatives other than pests. These animals feed on decaying plant material or otherwise function as part of the turf ecosystem (table 6–9).

TABLE 6–9   INSECTS AND RELATIVES FOUND IN TURF BUT DO LITTLE OR NO FEEDING DAMAGE

Animal	Length	Other traits
Sowbugs, pillbugs	¼–½ in (6–13 mm)	Oval, gray, with platelike segments
Millipedes	Usually 1–2 in (25–51 mm)	Hard shelled, many legs; animal coils when disturbed; slow moving
Centipedes	Usually 1–2 in (25–91 mm)	Many legs, fast moving
Earwigs	¾ in (19 mm)	Pair of "forceps" at rear of the body
Crickets	1–2 in (25–51 mm)	Enlarged hind legs for jumping
Ground beetles	¼–1 in (6–25 mm)	Oval, fast moving
Rove beetles	¼–¾ in (6–19 mm)	Slender, with short wing covers
Big-eyed bugs	⅛ in (3 mm)	Oval, fast-moving, bulging eyes
Springtails	1⁄16–⅛ in (1.6–3 mm)	Insects jump by means of a forked "tail"
Slugs	1–5 in (25–125 mm)	No legs, slimy body
Snails	1–3 in (25–75 mm)	No legs, slimy body with coiled shell

6.6.1   Sowbugs and Pillbugs

Identification.   Sowbugs *(Oniscus* spp.) and pillbugs *(Armadillium vulgare)* are not insects, being more closely related to crayfish. They are small, slate gray, hard-shelled, tubular-bodied, segmented animals about ¼ to ½ in (6 to 13 mm) long. The adults have seven pairs of legs. Because their segments are platelike in appearance, they superficially resemble armadillos.

Pillbugs *(Color plate 30)* are similar in appearance to sowbugs (figure 6–44), but have the ability to roll themselves into tight little balls when disturbed. Both animals live in damp areas such as under boards, stones, leaves, and other debris.

FIGURE 6–44   Sowbugs.

The development of sowbugs and pillbugs is quite similar. The eggs of both species develop in brood pouches. An average of two generations or broods are produced each year. Adult sowbugs and pillbugs live for about 2 years.

Importance.   Sowbugs and pillbugs feed primarily on decaying vegetable matter. Occasionally they feed on healthy plant tissue, including grasses. Sowbugs and pillbugs may invade structures in large numbers. They cannot remain in dry areas away from damp vegetation, litter, and stones, or they will desiccate and die.

Management.   The numbers of these animals can be reduced by removing their hiding places—leaves, rocks, boards, and other debris around the foundations of buildings. Chemical treatments are also available (table 6–12) but rarely needed, because these animals are primarily scavengers.

6.6.2   Millipedes and Centipedes

Identification.   Millipedes (Diplopoda), often called thousand-leggers, are not insects but elongated, wormlike animals with 30 or more pairs of legs. Milli-pedes curl up when disturbed (figure 6–45). The adults are 1 to 2 in (25 to 51 mm) long and colored brown, tan, or gray. They have two pairs of legs on each body seg-ment except the three segments behind the head.

Most millipedes are found in the damp areas under leaves, stones, boards, compost piles, or in the soil. They feed on decaying plant material. Sometimes they

FIGURE 6–45   Millipedes.

FIGURE 6–46   Centipede.

damage young plants as they feed on the roots. These animals may invade structures, causing considerable annoyance. Some species of millipedes eject a foul-smelling fluid through openings along the sides of their bodies.

Millipedes overwinter as adults in protected areas. They lay their eggs during the summer, often in nests in the soil or in damp areas without actually constructing a nest. Newly hatched young have three pairs of legs; more are added as the animal increases in size. Adult millipedes may live about 7 years.

Centipedes (Chilopoda) are elongated, flattened, wormlike animals with 15 or more pairs of legs, often referred to as hundred-leggers. They are 1 in (25 mm) or more in length with numerous body segments (figure 6–46). There is one pair of long legs for each body segment. The last pair of legs is longer than the others. A pair of long antennae project forward from the head.

Centipedes are predators that feed on small insects and spiders. If threatened, some centipedes may bite, producing swelling and some pain.

Centipedes overwinter as adults in protected areas and lay eggs during the summer. The eggs and hatching young are protected by the adults. Centipedes live approximately 6 years under favorable conditions.

Importance.   Millipedes are rarely a pest of living plants, preferring to feed on decaying plant material. As scavengers, they are helpful in recycling dead material into nutrients that plants can use. They can become a nuisance crawling in large numbers across sidewalks and up the walls of buildings at night and on damp mornings.

Centipedes prey on a wide range of insects and other small animals, probably feeding mainly on other scavenging creatures. As such, they are neither helpful nor harmful but are part of the natural ecology.

Management.   Some reduction in the population of millipedes and centipedes can be accomplished by changing the environment near turfgrass areas. Elimination of piles of organic matter, such as compost, grass clippings, and other damp debris, reduces their feeding and egg-laying sites. Removal of stones, boards, leaves, and other debris, together with the elimination of damp locations, also discourages a population buildup of both millipedes and centipedes.

Thick, mass plantings of shade-loving plants such as hostas, violets, and impatiens can provide the moisture that these animals need. Spacing plants farther apart to provide better air movement and drying under the plants helps reduce these animals' numbers. The decaying organic matter or the damp areas that they prefer may be on adjoining properties. Many new subdivisions are plagued by millipedes due to vacant or poorly managed areas near new homes and lawns.

If large populations of these animals are present and are a nuisance, they can be controlled on the turf, in waste areas, or near buildings. Chemical controls are listed in table 6–12, but these are much more effective against centipedes than millipedes.

6.6.3   Earwigs

Identification.   Earwigs are slender, beetlelike insects with short wings and a large pair of "forceps" on the rear of the body. They are dark reddish-brown and about ¾ in (19 mm) long. The most common species is the European earwig *(Forficula auricularia),* found in the northeastern and some western states (figure 6–47). Other species *(Euborellia annulipes* and *Labidura riparia)* are commonly found around homes in the southeastern United States.

The name "earwig" is derived from an old superstition that has these insects entering people's ears at night while they are asleep. This belief is totally untrue, as earwigs are completely harmless to human beings.

Earwigs are active primarily at night, hiding during the day in cracks and under boards, rocks, and in similar dark places. Most are scavengers, but occasionally they feed on ornamental plants, fruits, and vegetables. Large populations have been observed in piles of grass clippings. One earwig, *Labidura,* is an important predator of chinch bugs in south Florida. Earwigs overwinter as adults. The eggs

FIGURE 6–47   Earwig.

are laid in clusters within burrows beneath the soil surface. Young earwigs some-what resemble the adult.

Importance.   Earwigs can become very numerous. They tend to hide in cracks and crevices during the day and can be disturbing to people by being numerous in lawn-furniture crevices and other outdoor equipment.

When populations are large, earwigs are likely to feed on various flowers, particularly the blossoms of marigolds, roses, and daylilies. Control may be needed to reduce this feeding damage.

In general, earwigs are part of the ecology and serve an important role in their scavenging activities.

Management.   Control is rarely, if ever, necessary. An earwig population may build up over a 1- to 2-year period and then decrease dramatically. Effective controls when their numbers are a nuisance are listed in table 6–12. Do not use insecticides on the blooms of flowering plants such as marigolds that are commonly visited by pollinating insects.

6.6.4   Crickets

Identification.   Field crickets (*Gryllus* spp.) are brownish-to-black short-winged, hopping insects ¾ in (19 mm) long (figure 6–48). The chirping of male field crickets is a common sound to many people. Field crickets feed on some agricul-tural crops and, when numerous, also damage ripening garden vegetables. They spend the winter in the egg stage.

Egg hatch occurs in early summer, and young nymphs increase in size throughout the summer. Large adult crickets begin to be observed in August and are most noticeable with the cooler temperatures of autumn as they migrate and seek out sheltered areas, including homes. A single generation is completed each year in the northern United States. Sometimes great numbers of these black, jump-ing, chirping crickets can be found along the foundation of a house, garage, or other building.

FIGURE 6–48   Field Cricket.

FIGURE 6–49   Ground beetle.

Importance.   Crickets can be a nuisance to homeowners. Although they do not damage grass, they can be numerous in turf areas. These scavengers are useful in the role of helping break down dead plant material.

Management.   Applied control of crickets in turfgrass areas is unneeded, as they do not damage the grass; but the homeowner often desires to control them around the building foundation. Effective chemical controls are listed in table 6–12.

6.6.5   Ground Beetles

Identification.   Ground beetles (Carabidae) are oval to elongated beetles that feed on insects and other small animals. They are ¼ to ¾ in (6 to 19 mm) long, have long antennae, and are glossy black, brown, or green. They have long legs and move rapidly (figure 6–49). Adult ground beetles are good fliers and are strongly attracted to lights at night.

Larvae of ground beetles are white, elongated, and wormlike, with six legs and a brown head with large jaws. They live in the soil and feed on insects and other small animals. The larvae are frequently seen when they are flushed to the surface with irritants such as those used for sod webworms and cutworms.

Importance.   Although ground beetles feed on other insects, including pests, they do not appear to have a major impact on any important pests. They are, however, an important part of the turf ecology and should be accepted as such.

Management.   Because ground beetles are neither pests nor important predators of major turf pests, efforts either to reduce or increase their numbers are not needed.

6.6.6   Rove Beetles

Identification.   Rove beetles (Staphylinidae) are common in turf, and quite noticeable coming to the top when an irritant is used to scout for sod webworms, cutworms, or other insects. They are reddish-brown-to-brown or black, elongated beetles with short wing covers that extend for a short distance down the abdomen. They range from ⅛ to 1 in (3 to 25 mm) long, depending on the species. Species are either scavengers, feeding on decaying plant and animal material, or predators, feeding on insects and other small animals.

Importance.   Rove beetles are not pests of turfgrasses, and they are not responsible for any major turf problems. The predatory species do not appear to be a major control factor on any turfgrass pests. They are not of any economic importance. Rove beetles are just part of the ecology of most turfs.

Management.   Because rove beetles do not seem to affect turf health in any way, control efforts or efforts to increase their numbers are not warranted. They should be respected as part of the natural ecology of the turf and be allowed to exist in it.

6.6.7   Springtails

Identification.   Springtails (Collembola) are 1⁄16 to ⅛ in (1.6 to 3 mm) long, grayish to whitish insects that live in the soil. They have a forked structure on the back end of the body that, when released, throws them up into the air as high as 3 in. One type of springtail is elongate; the other is roundish. Springtails feed on decaying organic matter in the soil and are more common in moist soil. They can be very numerous.

Importance.   Springtails do not cause any damage because they do not feed on living plant material. Their feeding on decaying plant material helps recycle nutrients so that living plants can utilize them.

Management.   Springtails are rarely needed to be controlled because they cause no damage.

6.6.8   Big-Eyed Bugs

Identification.   Big-eyed bugs *(Geocoris* spp.) are similar in size to chinch bugs as adults (about ⅛ in or 3 mm long). They may be blackish or brownish, depending on the species, and they are identified by their bulging eyes. They move more quickly through the turf than do chinch bugs. Nymphs are similar in color and activity to the adults. Adults overwinter and emerge in the spring to feed and lay eggs. There are two to four generations produced per year.
Big-eyed bugs are primarily predators, feeding on chinch bugs, caterpillars, and other small soft-bodied insects. As is typical of many of the true bugs, they also feed on vegetation and have been associated with feeding on turfgrass. Big-eyed bugs apparently feed heavily in other areas besides turf, because they are relatively small in number most of the time. However, the use of irritants to scout for sod webworms, cutworms, and other insects will frequently flush out at least a few of these insects.

Importance.   Big-eyed bugs typically become numerous in turf when there are high numbers of chinch bugs. As with many predators, they build up in numbers after their prey's population gets large. They can be found in large numbers in turf as chinch bug numbers are declining, and they are probably at least partially responsible for the chinch bug reduction through their feeding on them.

Although not normally associated with turfgrass damage, there have been instances when this insect has been apparently responsible for declining turfgrass. When investigating turfgrass damage, be sure to look for evidence of a declining chinch bug infestation, but also be open to the possibility that damage could have been caused by big-eyed bugs.

Management.   Control is rarely needed for this insect's turf-feeding activities. In cases when control is necessary, insecticides listed in table 6–12 for chinch bugs should be effective against big-eyed bugs.

Too little is known of this insect's life history to be able to increase its numbers for predatory control of more severe turfgrass pests.

6.6.9   Slugs and Snails

Identification.   Slugs and snails are mollusks, not insects. They are more closely related to such aquatic animals as clams and oysters. Garden slugs are more common than snails and also a greater nuisance. Slugs are described as "snails without their shells." They are typically gray or spotted, elongate, soft-bodied animals that range from 1 in to more than 5 in (25 to 126 mm) long *(Color plate 31)*. They have a pair of antennaelike tentacles protruding from the front of the body, which is blunt, whereas the posterior end is tapered. Their lower body surface is a smooth foot that secretes a characteristic slimy mucous residue on plants or other objects over which they glide. This slimy trail is often the only visible evidence of their past activity.

Slug eggs are laid in clusters in hidden, moist locations. Most species overwinter in the egg stage. Slugs may live a year or more after hatching. Sometimes, in warmer regions, there is more than one generation each year.

Snails have bodies similar to slugs except that they are capable of retreating into their coiled shell that is carried upright above the body. They range from 1 to 3 in (25 to 76 mm) long. Their life cycle is similar to slugs.

Importance.   Slugs and snails are pests of flower beds, ornamentals, and seedling vegetable plants. They feed at night, eating holes in the foliage or crowns of plants. They can be especially numerous where the plants are heavily mulched.

When present in turf, they are likely either feeding on weeds in the turf or simply crossing the turf area in search of new food. Slugs and snails are not feeders on turfgrasses.

The slime trails that slugs and snails leave behind reflect the morning sunlight and can appear like silver ribbons of light before drying up. The presence of these trails can be disturbing to early-morning golfers and other turf users.

Management.   The numbers of slugs and snails can be reduced by removing their hiding places and egg-laying sites. In the fall, clear flower beds and vegetable gardens by picking up all debris, such as boards, dead leaves and stems, and other material. Reducing moisture is another effective way of reducing slug and snail populations. Space plants in beds farther apart to allow more airflow and drying action. Allow at least surface soil to dry between irrigations.

There is no effective insecticide that will control slugs or snails. Attractant bait formulations are available for use in and around flower and ornamental plantings and vegetable gardens. Metaldehyde has long been used as a chemical attractant and a stomach poison for slugs and snails. It is marketed as a pelleted bait. Shallow containers of stale beer are also attractive to slugs. The aldehyde odor attracts the slugs, causing them to crawl into the shallow dish of beer and drown. Although the number of slugs killed by beer is impressive, this technique usually does not significantly affect the slug population.

6.7 BITING, STINGING, AND NUISANCE PESTS

Several insects and insect relatives in turf are pests because they annoy or harm people trying to enjoy or otherwise utilize the turf area (table 6–10).

6.7.1 Fleas

Identification. Adult fleas (*Ctenocephalides* spp.), sometimes called sand fleas, are small, ⅛-in (3-mm), dark brown to nearly black insects that move swiftly among the hairs of animals (figure 6–50). They are capable of jumping as far as 1 ft. Their eggs are tiny, oval, and white. The hatching larvae are slender, whitish worms with light brown heads.

Fleas attack a wide variety of animals, including people. The eggs are laid among the hairs of the host animal but soon drop to the ground. The hatching larvae are scavengers and feed on a wide range of debris. This stage is harmless, but when the fleas become adults, they search for a host animal.

Importance. In turfgrass areas frequented by pets, both immature and adult fleas can build up in numbers to constantly reinfest pets running across the lawn. Occasionally, people who walk across an infested turf are bitten by fleas.

TABLE 6–10   INSECTS AND RELATIVES IN TURF THAT ANNOY, BITE, OR STING PEOPLE

Pest	Size	Other traits
Fleas	¹⁄₁₆–⅛ in (1.6–3 mm)	Insects jump; hard shelled, flat sided
Ticks	⅛–¼ in (3–6 mm)	Flattened, 8 legged, hard shelled
Chiggers	¹⁄₁₆ in (1.6 mm) or less	Roundish; adults are red
No-See-Ums	¹⁄₁₆–⅛ in (1.6–3 mm)	Dark-colored flies that bite
Mosquitoes	⅛–½ in (3–13 mm)	Slender, biting flies
Thrips	¹⁄₁₆ in (1.6 mm)	Slender
Spiders	¼–2 in (6–51 mm)	Eight legged, fast moving
Crane flies	½–1 in (13–25 mm)	Long, spindly legged with narrow, leathery wings

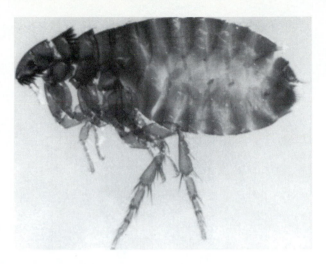

FIGURE 6–50   Flea.

Management.   Control of fleas is commonly directed at the infested pet by the use of systemic insecticides applied orally or to the skin. Flea collars, dusts, or sprays applied to the body of the animal are also effective. Dogs and cats are walking bait stations for fleas. As long as the animal is around, people are seldom attacked. If needed, there are effective controls (table 6–12) to apply to infested grass areas.

6.7.2   Ticks

Identification.   Ticks are close relatives of insects. Their primary food is the blood of warm-blooded animals. There are various species, but the one commonly associated with turfgrass areas is the American dog or wood tick *(Dermacentor variabilis,* figure 6–51). This tick is hard bodied, about ¼ in (6 mm) long and flattened, with a shield containing silver markings on its back. After feeding on one of several wild animal hosts, the female expands to about ½ in (13 mm) long and is only slightly flattened. Only the adult stage attacks larger animals, including people and dogs; the immature stages feed on smaller animals. After hatching, there are three stages: a small six-legged larva, an intermediate eight-legged nymph that resembles the adult, and the adult.

Importance.   Adult American dog or wood ticks are most active in early summer. They are especially numerous in brush or tall grass areas, waiting to attach themselves to passing animals. Ticks do not do well in short vegetation such as mowed lawns. Sites of tick infestations adjoin roadsides, fence rows, woods, brush, and other unmowed areas.

Management.   Avoiding tick-infested areas during early summer is the best prevention of ticks attaching themselves to the skin. Protective clothing such as long pants and boots helps prevent tick attacks. Mosquito repellents on exposed

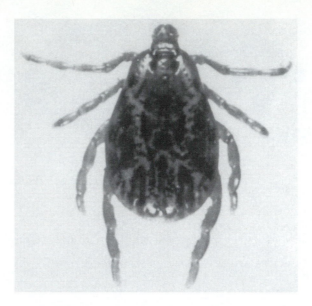

FIGURE 6–51   American dog tick.

parts of the body also discourage them. Tick-infested areas can be sprayed with an insecticide to control them (table 6–12).

6.7.3   Chiggers

Identification.   Chiggers (*Trombicula* spp.), also called redbugs, are mites that occur throughout the United States. There is more than one species, but their activities are similar. Chiggers pass the winter as adults near or below the soil surface. Eggs are laid in the spring, and the newly hatched larvae have six legs. Later they change to eight-legged nymphs and eventually to eight-legged adults. They overwinter in the soil. There is one generation per year.

Chiggers are only a pest to people in the six-legged larval stage (figure 6–52). They are almost invisible to the naked eye, being less than $\frac{1}{150}$ in (0.2 mm) in diameter. These animals are oval and a bright orange-yellow, and they move about quickly.

Importance.   Chiggers insert their mouthparts into the skin of human beings, usually in a skin pore or hair follicle. They are found especially in regions of the body with tight-fitting clothes. When fully fed, young mites drop off the host. Anyone who has been bitten by chiggers may not feel the bites for several hours after exposure.

Chiggers are found where vegetation is abundant, such as shaded areas, high grass or weeds, or fruit plantings. However, they may become serious pests of lawns, golf courses, and parks.

To check a suspected chigger-infested area, place a piece of black cardboard edgewise on the ground where chiggers may be. If present, tiny yellow or pink mites (chiggers) will soon move rapidly over the cardboard and gather along the upper edge.

FIGURE 6–52 Chigger.

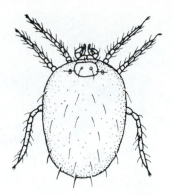

Management. Controlling chiggers should be directed at reducing nearby egg-laying sites. Wooded or grassy areas may be mowed to reduce their cover. Repellents applied to the body prevent chiggers from feeding. Chigger-infested areas can also be treated with a chemical (table 6–12).

6.7.4 No-See-Ums and Mosquitoes

Identification. No-see-ums, punkies, or sand flies are names given to a group of tiny, blood-sucking midges or flies (Ceratopogonidae) that can be a nuisance to lawn owners and turfgrass managers. The young, immature maggots develop in moist vegetation or water, such as creeks and ponds. The flies generally hide during the day and become active in the evening and early-morning hours. The surprising pain of the bite seems disproportionate to the tiny insect; hence the name "no-see-ums."

Mosquitoes (Culicidae) have habits similar to no-see-ums. They are larger flies, with the females doing the blood sucking (figure 6–53). The larvae or wrigglers need to live in standing water for about a week prior to becoming adults.

Importance. No-see-ums and mosquitoes are both nuisances due to their irritating bite. In addition, mosquitoes transmit several important diseases to human beings, including encephalitis, malaria, and yellow fever. Encephalitis diseases include St. Louis encephalitis, LaCrosse or California encephalitis, eastern and western equine encephalitis, and West Nile virus.

Management. Control of no-see-ums and mosquitoes can be partially or completely achieved by elimination of standing water in such places as eave troughs, old tires, tin cans, rain barrels, storm sewers, open garbage cans, and other waterholding containers. Floodwater mosquito larvae can develop in 2 weeks or less in temporarily flooded areas. Providing proper drainage or treating developing larval populations is needed to prevent outbreaks. Spraying shrubbery where the adults hide during the day reduces their numbers (table 6–12).

Aerosols or "flying insect" sprays in pressurized cans give a quick temporary knockdown of these pests at outdoor cookouts or parties. These are weak fliers, so

FIGURE 6–53   Mosquito.

a fan blowing across the area where people are gathered in the evening will help reduce the numbers of mosquitoes and no-see-ums. Effective insect repellents are also available for use on exposed skin.

6.7.5   Thrips

Identification.   Thrips are tiny, slender, black or yellow, brushy-winged insects up to ⅛ in (3 mm) long (figure 6–54). Oat bugs or oat lice are common names people use to refer to these nuisance pests. These minute insects move about quickly on grass blades or other plants.

Importance.   They feed on many species of plants. One species, the grass thrips *(Anaphothrips obscurus),* is almost specific to bluegrasses, though it has been found on other plants. Injury by thrips is done by their rasping mouthparts, which scratch the green tissue from the upper leaf surface. A damaged plant or turf area has a silver top or whitish appearance. Occasionally, thrips may crawl on a person and may even bite the skin surface.

Management.   If numerous, thrips can be controlled in turf areas (table 6–12).

6.7.6   Spiders

Identification.   Spiders (Araneae) can often be found near tall grass, flowers, shrubbery, and on or by buildings. All spiders are predaceous, feeding mostly on insects and other small animals. They have two body regions, a distinct abdomen and an anterior region containing the mouthparts and legs. The mouthparts contain two short leglike chelicerae followed by two longer, leglike pedipalps

FIGURE 6–54   Line drawing of a thrips.

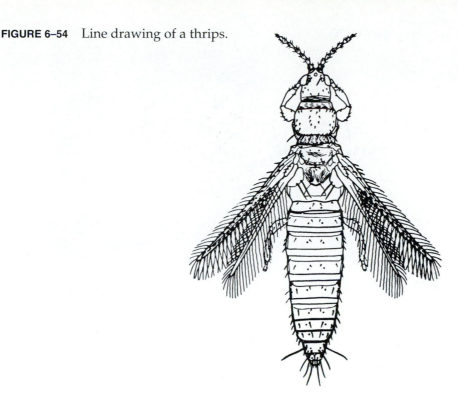

that the spiders use to manipulate their food. They have eight legs, but no antennae. Some species spin webs to trap their prey; others pounce on prey; still others lie in wait, hidden on plants and other sites.

Particularly in late summer and fall, sheet web spiders, also called funnel weavers, become numerous in turf areas. They become very obvious on foggy mornings where they appear white from light reflecting off of dew on the web. These webs are typically 6 to 12 in (15 to 31 cm) across in the form of a horizontal sheet laid across the turf. At one end the webbing is formed into a descending funnel in which the spider sits. Above the sheet web are several single silk strands into which insects fly, causing them to fall onto the web. If the insect is small enough, the spider runs out of its funnel to capture it. If the insect is too large, the spider remains in its tunnel and allows the insect to crawl off of the web.

Young spiders hatch from eggs and resemble their parents. They gradually undergo development to adults. Unusually high numbers of spiders often indicate the presence of other insects that provide their food.

Importance.   Spiders are disliked by many people, especially when found inside the home or on flowers. It is true that many species do bite, but only a few species are poisonous. These include the black widow (*Lactrodectus mactans*, figure 6–55) and brown recluse (*Loxosceles reclusa*, figure 6–56) spiders. Sheet web spiders and most other spiders are beneficial, playing an important role in keeping down populations of pest insects.

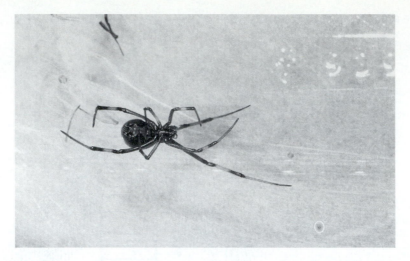

FIGURE 6–55 Black widow spider.

FIGURE 6–56 Brown recluse spider.

Management.　Spiders should not need to be controlled in turf areas. The few spiders that are poisonous are associated with buildings, animal burrows, or piles of debris. They are not likely to be encountered in turfgrass areas.

6.7.7　Crane Flies

Identification.　Crane flies, often called leatherjackets, resemble overgrown mosquitoes with extremely long legs (figure 6–57). They occur primarily in damp locations where vegetation is abundant. Larval crane flies are whitish-to-brownish,

FIGURE 6–57   Crane fly.

legless, wormlike, softbodied animals that live in bodies of water or very damp soil and feed on decaying vegetation.

Importance.   Some crane-fly species feed on living plants, including turfgrasses. No damage is done to turfgrass by crane-fly adults, but some larvae live in the soil and feed on living plants, including grass roots. This results in small, irregular brown patches in the turf. The size of crane flies and their similarity to oversized mosquitoes probably gives them an inflated nuisance rating.

The European crane fly *(Tipula paludosa)* occurs in the northwestern United States and can quickly damage lawns, putting greens, and other turf areas when maggot numbers are over 20 per square feet. The younger larvae feed on the roots, but the older larvae will come to the surface at night to feed on grass blades and stems. The larvae dry up and die when the relative humidity of the turf surface drops below 100%, making it unlikely that this insect will spread to other areas of the United States.

Management.   European crane-fly larvae are a problem on moist turf areas. Providing proper drainage and reducing irrigation will help reduce numbers of insects and associated damage. The insect-feeding nematode, *Steinernema carpocapsae,* is effective against the larvae. Chemical insecticide treatments are given in table 6–12.

6.8 *Mammal and Bird Pests*

Some mammals damage turf by their burrowing activities and may also leave mounds of soil. Other mammals and most birds damage turf while feeding on insect pests (table 6–11).

TABLE 6–11   MAMMALS AND BIRDS THAT PRODUCE MOUNDS OR HOLES IN TURF

Animal	Damage type	Size across	Other traits
Moles	Subsurface tunnels	3–4 in (8–10 cm)	6–12 in (15–30 cm) round mounds of loose soil
Pocket gophers	Subsurface tunnels	2–3 in (5–8 cm)	1–2 ft (30–60 cm) horseshoe-shaped mounds of loose soil
Voles	Surface runs	1–2 in (2.5–5 cm)	Long, winding runs
Ground squirrels	Burrow holes	2–3 in (5–8 cm)	No mound of soil
Raccoon	Sod pulled back	6–12 in (15–30 cm)	
Skunk	Sod pulled up	3–5 in (8–13 cm)	
Armadillo	Holes 3 in deep	3–5 in (8–13 cm)	
Birds	Holes through thatch	½ in (1.3 cm)	

6.8.1   Moles

There are several species of moles that can be a problem in turf areas in the United States. The most widespread species is the eastern mole (*Scalopus aquaticus*), which occurs over most of the United States east of the Rocky Mountains.

Identification.   Moles are 6- to 8-in-long (15- to 20-cm), short-legged, blackish, heavy-bodied mammals with elongated snouts, short tails, and broadened, shovel-like front feet (figure 6–58). They are better known by their subsurface tunnels, where a ridge of sod that is several inches wide is pushed up at least 2 in (5 cm) (figure 6–59). They also make deeper tunnels 6 to 10 in (15 to 25 cm) below the surface in which they rest and rear young. Large mounds of loose soil on the surface are created by the mole depositing soil dug from these deep tunnels.

Subsurface tunnels can be divided into feeding tunnels and main tunnels. Feeding tunnels tend to meander or wind back and forth. They commonly dead-end. Feeding tunnels are created by the mole when searching for food and are rarely reused. Main tunnels are straighter and connect other tunnels or mounds of loose soil. They are used almost daily by the mole not only to get to various parts of its territory, but

FIGURE 6–58   Eastern mole.

FIGURE 6–59   Mole runs or burrows. (Courtesy Milwaukee Sewerage Commission.)

also to trap earthworms. Earthworms entering these tunnels tend to turn and travel down the tunnel rather than cross into soil on the other side. Periodically, the mole goes down the tunnel and eats the earthworms. Moles feed primarily on earthworms, but grubs, other insects, insect relatives, and plant material make up the total diet.

Moles reproduce in the spring and apparently have only one litter per year. Much reproduction occurs in stable, natural areas where food is plentiful, such as in meadows, old fields, and wooded areas. Young moles leave the home range and are commonly the ones that become pests in managed turf. Moles normally have a home range of 1 to 3 acres, making it likely that mole tunnels in a residential lawn are being produced by only one mole.

Importance.   Moles are a recurring problem in turf areas next to meadows, forests, and other areas where their populations yearly produce offspring that invade those turf areas. Their subsurface tunnels can create mowing problems and cause turf dieback by exposing turfgrass roots to drying air, and they are aesthetically unpleasing. They do not feed heavily enough on white grubs to be useful in reducing grub numbers.

Management.   Moles feed heavily on earthworms, and, for that reason, the use of insecticides to kill grubs and other insects upon which they also feed is frequently unsuccessful in eliminating the moles.

Control moles by setting mole traps across main tunnels. There are two main kinds of traps. The harpoon trap is set by mashing a portion of tunnel and setting the trap across it so that the spring-loaded spikes are just penetrating the soil. This makes it unlikely that pets or children will be harmed by the trap. As the mole comes through the tunnel and pushes up the soil and trigger, the trap is sprung, killing the mole with the spikes. The choker trap sets across the tunnel and squeezes the mole to death as the trap is sprung. Main tunnels can be distinguished from feeding tunnels by the tips given above or by mashing portions of each tunnel and marking them off to the side with flags or other markers. Damage to main tunnels will be repaired within a couple of days.

Chemical repellents are available that show promise in keeping moles out of an area. Various repelling plants, such as castor bean, will be effective in repelling moles under some conditions. Commercial preparations containing castor bean oil are also available. These will be ineffective if the mole has no other areas to live. A young mole cannot return to the parents' home range, and if all turf areas are treated with repellent, the mole will live there anyway.

Other devices put vibrations into the soil. These may be effective for a period of time, but eventually the moles are likely to adapt to them and enter the turf area. Chemical toxicants that are supposed to poison the moles are usually not effective. Moles do not eat vegetable-based baits, and toxic gas escapes through the shallow subsurface tunnels.

6.8.2 Voles

Identification.   Voles are commonly known as field mice and meadow mice. Those that damage turf are usually the meadow vole *(Microtus pennsylvanicus)* and the prairie vole *(Microtus ochrogaster)*. They are blackish mice that have blunt faces and short tails. When fully grown, they range from 5 to 7 in (13 to 18 cm) long. They are different from house mice in that they are larger and heavier-bodied with blunt rather than pointed faces and short rather than long tails. They can reproduce year-round, but do so primarily in spring and fall.

Importance.   Voles live in a system of burrows several inches below ground. They also make open-topped surface runways that are about 2 in (5 cm) wide and extend through the thatch into the root zone of turf. They are completely free of plants and wind snakelike for several feet *(Color plate 32)*. Surface runways are more numerous where there is tall grass.

In maintained turf, surface runways are more common where there is heavy thatch and long periods of heavy snowcover which allow extensive runways to be built beneath it during the winter. Vole damage is also more common near areas planted to groundcovers or low-growing evergreens that provide cover for the vole population from predators such as cats, coyotes, hawks, and owls.

Voles will also feed on the bark of trees, particularly during the winter. Young fruit trees and other trees with thin bark can be severely damaged.

Management.   Turf damage can be reduced by reducing ground covers, low evergreens, and other cover. In most turf, the runways quickly grow over in the spring and this can be aided through seeding or sodding. Shrews are commonly seen in vole runways and blamed for their construction. Shrews are smaller than voles and house mice, usually 2 to 4 in (5 to 10 cm) long with pointed heads. They feed on small animals, including insects, earthworms, and voles.

The use of poisons or traps against voles is discouraged because birds or other nontarget animals are likely to get killed. If poisons or traps are used, they should be placed within belowground burrows or closed bait stations so that non-target animals are not likely to be affected.

6.8.3   Ground Squirrels

Identification.   Ground squirrels (*Spermophilus* spp.) are quick-moving, furtive mammals with bushy tails half as long as the body. They are tan or brown and may have stripes (figure 6–60). Their total full-grown length, including tail, is usually 10 to 12 in (25 to 30 cm).

Ground squirrels live underground in burrows that exit to the surface through holes that are about 2 in (5 cm) in diameter. The squirrels scatter excess soil over large areas so that there are no mounds next to the burrow openings. They are active above ground during the day. They feed on plants and insects. Ground squirrels feed through the warmer months, hibernating underground during the winter.

Importance.   Ground squirrels create problems in turf areas with their unsightly holes that can hinder play on golf courses. The animals will also feed on bulbs and roots of flowers and other ornamental plants.

FIGURE 6–60   Thirteen-lined ground squirrel.

Management.   Because the animals are not alone in an area but rather are part of an entire population, removing individuals rarely results in long-term control without constant removal. Individuals from nearby areas quickly move into a vacated habitat, and the reproduction of the population will increase to help replenish removed animals. Because the presence of an animal tends to keep out others of its own species, learning to live with the animals frequently results in a long-term answer to the problem.

Part of learning to live with the animals involves getting the burrow openings placed where they are less of a problem. Placing a cup of moth crystals or ammonia-soaked rags into the burrow and then closing the burrow opening with loose soil will cause the animal to move its burrow. These items frequently need to be put into the burrow opening more than a day to get the burrow moved. Repeating this process will eventually cause the animal to move its burrow out of an obvious problem area or away from plantings of damaged ornamental plants. This process will work with most burrowing mammals, including ground squirrels, chipmunks, woodchucks, and skunks.

6.8.4   Pocket Gophers

Identification.   Pocket gophers are brownish rodents that live in underground burrows. Several species (e.g., *Geomys, Thomomys)* occur in various parts of the United States, primarily west of the Mississippi River and also in Illinois, Alabama, Georgia, and Florida. Most species have elongated claws on their front feet and are 7 to 13 in (18 to 33 cm) in total length (figure 6–61). They can have two generations per year.

FIGURE 6–61   Plains pocket gopher. (Photo by M. Jeffords.)

Importance.   They feed on plant roots, bulbs, tubers, stems, and leaves as well as insects. They particularly like the bulbs and tubers of flowers, and they can cause heavy damage to flower beds. They construct subsurface tunnels to do much of this feeding, resulting in a ridge of soil being pushed to the surface. Burrows are also constructed several inches below the soil surface, with the loose soil piled into large mounds on the surface.

Management.   Traps placed in burrows are effective in controlling these mammals. Removing loose soil at the inside center of a crescent-shaped mound will usually uncover a burrow, although its entrance may be plugged with several inches of loose soil. Poisoned baits placed in either subsurface or deep burrows can also be effective. Fumigants can be forced into deep burrows, although the animal will frequently plug the tunnel with soil and escape harm.

6.8.5   Raccoons, Skunks, and Armadillos

These mammals become turf problems through their nighttime feeding on white grubs and other insects.

Identification.   The raccoon *(Procyon lotor)* occurs throughout the United States. It is grayish to brownish, stout-bodied, about 3 ft (91 cm) long, and has a black-ringed tail and black patches around its eyes. These mammals live in underground burrows, tree hollows, and rock crevices. They feed on a wide range of plant and animal material, including insects.

The striped skunk *(Mephitis mephitis)* occurs throughout the United States. It is 2 to 2½ ft (61 to 76 cm) long and completely black or black with a white stripe down its back. The spotted skunk *(Spilogale gracilis)* is smaller, 1 to 2 ft (30 to 61 cm) long, with a pattern of white spots and short white stripes. It occurs in the western United States. Both skunks live in burrows that are frequently dug underneath a log or rocks. They feed on fruits and nuts as well as insects and rodents.

The nine-banded armadillo *(Dasypus novemcinctus)* is found in the southern half of the United States as far west as Texas. These grayish mammals have elongated snouts and are covered with scaly plates. Including tail, they can be up to 3 ft (91 cm) long. They feed on insects and other small animals. Armadillos live in burrows that are about 6 in (15 cm) in diameter and located near trees, shrubs, or rocks.

Importance.   These mammals destroy turf in their efforts to feed on insects, particularly white grubs. Raccoons usually pull back large sheets of turf 6 to 18 in sq (15 to 46 cm sq) in their search for insects *(Color plate 33)*. Skunks will usually pull out 3- to 4-in (8- to 10-cm) pieces of turf that look similar to divots dislodged during golf *(Color plate 34)*. A single skunk may make up to 100 holes per night. Armadillos do more burrowing than raccoons or skunks, commonly digging down 3 in (8 cm) to find grubs or other insects.

Management.   These mammals cause turf damage in their search for grubs or other insects. Controlling these insects usually brings a halt to damage caused by such predators. However, damage can occur at white grub populations as low as 3 to 5 per square foot (30 cm sq), well below the 10 to 12 per square foot (30 cm sq) usually needed to cause turf damage.

6.8.6   Insect-Feeding Birds

Identification.   Various bird species, including starlings, cowbirds, robins, red-winged blackbirds, and rusty blackbirds, feed on insects in turf. Sod webworms, cutworms, armyworms, and other caterpillars are preferred, but they will also feed on white grubs. When feeding, large numbers are usually seen on the turf area.

Importance.   During feeding activities, birds punch ½-in (13-mm) holes into the thatch with bills in search for insects. These holes are usually not noticeable in most turf, but can be important in closely mowed sports turf such as golf greens. Heavy feeding on white grubs can cause large brownish areas where the loosely rooted sod can be torn up and damaged *(Color plate 35)*.

The presence of large numbers of birds on turf during the growing season is a good indication of a potentially damaging insect infestation. The wise turf manager looks to see what the birds are feeding upon. The presence of feeding flocks on winter-dormant turf usually does not indicate a problematic insect infestation. Overwintering sod webworms and other insects provide an important source of food for early migrating insect-feeding birds.

Management.   Damage from insect-feeding birds is best controlled by identifying and controlling the responsible insect pest. In addition, most of these birds are protected by law and cannot be harmed.

6.8.7   Canada Goose

Identification.   The Canada goose *(Branta canadensis)* is a large bird, up to 2 ft (61 cm) long, with a 5-ft (1.5-m) wingspread. It has a brownish body and a long black neck and head, with broad white cheeks. This bird is well known for its honking and flying in a V formation. It is capable of long flights of several hundred miles per day. This mobility allows it to range back and forth during the winter to avoid frozen water bodies and food-covering snow storms. It is attracted to bodies of water such as golf course water hazards, streams, and ponds.

Historically, large flocks overwinter in the southern states and migrate to reproduce during the summer in the northern United States and Canada. More recently, an increased amount of breeding occurs in central and southern areas of the United States, even though much of the population migrates. Pairs nest on the ground; from their eggs hatch brown goslings that follow the parents.

Importance.   Numerous large birds on fine-turf areas around office buildings and on golf courses can be quite a nuisance. These birds are also vegetarians and do feed on grass, occasionally damaging turf areas, particularly newly seeded areas. The main problem with these birds is their manure. Their fecal matter is 2 to 3 in (5 to 8 cm) long and about ½ in (13 mm) in diameter. Large amounts of manure can restrict golf play and the enjoyment of other turf areas.

Management.   Canada geese are a continual problem where ponds or other bodies of water are easily accessible by the birds. Because these birds are reluctant to cross even low fences, some turf managers have successfully managed the problem by surrounding ponds with 1-ft-high (30 cm) fences. A repellent has been developed from grapes, but it tends to be noticeable on the turf and sometimes causes an odor problem. It is sporadic in its effectiveness in repelling the geese.

In the past, molting, flightless geese have been rounded up under state conservation–authority supervision and shipped to other areas. The success of this bird's reproduction has been so high that few if any areas are still interested in these excess birds.

The use of dogs to run after and hassle the geese has met with considerable success. Goose control of this type is typically contracted from a company that owns and trains the border collies and other dogs. The dogs are brought out several times per week for part of a day to disturb and disperse any geese that are present. The goose-control company obtains the necessary permits from state conservation agencies, and its personnel monitor and manage the dogs so that neither the geese nor the property is harmed.

Another method used to reduce goose populations is to replace the eggs in the nest with artificial eggs or spin the eggs to kill the embryo inside. If the eggs are removed, the geese will lay another batch of eggs; however, they will continue to sit on eggs that do not hatch for a considerable period of time. As with most other birds, this bird is protected by law from harm except in legal ways during the proper hunting season or by special permit. Thus, disturbing the nest or eggs requires a special permit from the state conservation agency or U.S. Fish and Wildlife Service.

6.9   CONTROL OF TURFGRASS INSECTS

Insect control is an important part of turfgrass management. To be successful, the identification and some biological information about the pest insect is essential. The preceding text covered much of this information on each turf insect.

One insect specimen in a turf never causes damage to that area. Pest insects occur as populations of specimens of varying numbers, often including adults, young, and eggs. Low populations usually do not cause damage to the grass.

The aim of insect control is not to eradicate a pest population from a region, community, or turf area, but to suppress this damaging or potentially damaging population of a turf insect pest. Population suppression reduces the numbers below an economic threshold level capable of damaging the turf. Applied control (application of insecticide or changing maintenance practices) or natural control (climatic or biological factors) can be pursued. Applied control is by far the most common, but natural control plays a greater role than is realized. Many insects can successfully overwinter only in the southern states. Species of cutworms, armyworms, leafhoppers, and some aphids must migrate northward each year to be found in the north-central and northeastern regions of the United States. Predators, parasites, and diseases have a distinct limiting effect on populations of insects, especially aphids, caterpillars, and mites.

Applied control or suppression of a pest-insect population is often termed management of the insect, integrated pest management, or insect pest management. Proper insect pest management involves a series of decision-making steps to control insects rather than treat for turfgrass insects on a regular basis. The five basic steps are as follows:

1. *Identifying the insect and its population*
2. *Determining the potential for damage*
3. *Deciding what methods, if needed, to use to suppress the population*
4. *Taking action against the insect, if possible*
5. *Evaluating the results of the action*

These steps are well known to many experienced turf managers. The principles are not new, but incorporate many methods of reducing the pest-insect population or keeping it low into an overall, integrated system. These methods are combined into six types of control for easier understanding: legal or regulatory, genetic, cultural, biological, physical or mechanical, and chemical. Although they were discussed in general in chapter 2, they will be addressed here again as they relate to insect pest control.

6.9.1   Legal or Regulatory Control

Insect eggs or other stages are not easily transmitted on seed, so seed certification is a more appropriate control for other types of pests. However, sod is a turf product that can allow insects to be moved from one locale to another. State departments of agriculture personnel inspect sod, particularly any that will be shipped out of state, for evidence of insect pest infestation. This is done primarily by inspecting the sod field before it is cut. Inspections are made primarily for insect and other pests that are not found in the state receiving the sod, but high populations of common pests are likely to cause the sod to be held on a stop sale until the grower controls the pests and the sod passes inspection. Sod containing small numbers of pests may be restricted to states or intrastate areas that already have those pests.

6.9.2   Genetic Control

Genetic control of turfgrass insect pests is still used primarily in the selection of turfgrass species that are less susceptible to some insects than others. Of course, selected cultivars that will do well and thrive in a location are usually able to survive or thrive better after insect attack than others.

Host-plant resistance or reduced susceptibility is studied, and some differences are found; however, the differences in rate of attack or damage sustained are usually too small to be useful in turfgrass variety selection. A more likely source of host resistance will be genetically modified organisms (GMOs) that have had resistance genes implanted into them from distantly related sources. Research in that area on insect pest resistance looks promising.

6.9.3   Cultural Control

Maintaining healthy turf helps avoid severe insect problems. Although many insect pests tend to be attracted to the healthiest, most nutrient-laden grass, that is still the grass that is most likely to look better during and recover better after an insect pest infestation. A healthy stand of grass will be able to spread by runners or other means to fill in small areas lost to insect feeding.

When laying eggs, the adult beetles of white grubs appear to prefer turf that is moist. They tunnel into the soil to lay their eggs, so moist soil is softer and easier to move. White grub eggs will dry up and die in parched soil. The eggs will also die if the soil is very warm, a condition that is usually avoided in moist soil. Thus, allowing the soil and the turf to become dry in midsummer can result in fewer white grub problems later.

Alternatively, the common, marginally damaging grub population is large enough to cause the turf to wilt during dry periods, but is healthy after rains. Keeping the turf watered through the fall will allow the grass to grow new roots as fast as the grubs can eat them. In the spring, the heavier rainfall typical of spring weather, the increased turf growth of cool-season grasses with spring temperatures, and the shortened grub feeding time before pupation usually allows the turf to survive with no or only occasional irrigation. In this way, insecticide applications can be avoided if grub numbers are not too high.

Turf insect pests typically feed on portions of the grass plant, but rarely kill it outright. For instance, white grub larvae may eat the roots of the turf enough that the top turns brown for lack of water. If one controls the grubs and waters the turf, however, new roots typically emerge from the rhizomes and crowns and the turf will recover without the cost of seeding or sodding. Similarly, the turf may be brown from cutworms or sod webworms eating off the grass blades. If one controls the caterpillars and waters the turf, new shoots will emerge from the brown, but still alive, rhizomes and crowns of the grass plants and the turf will be restored.

6.9.4   Biological Control

Biological control methods show some promise in insect pest management, (for a review, see chapter 2). Some of the work involved with GMOs utilizes the implantation of endophytes into Kentucky bluegrass and other turfgrass species that do not naturally contain these protective organisms. Endophytes are numerous in the aboveground portions of the grass plants, making them useful in controlling sod webworms, chinch bugs, and other pests that feed on those parts of the plant. Endophytes are not numerous in the root system, so grubs and other root-feeding insects are not effectively controlled by them.

Parasitic wasps and flies can be important biological control agents of insect pests. Species of both parasitic wasps and flies that are effective against the Japanese beetle have been established in portions of the northeastern United States and are providing a useful level of control. Investigations are being made concerning their possible introduction into the midwest and other areas where the Japanese beetle has spread. Research into and introductions of a fly parasite of red imported fire ant are ongoing.

Beauvaria bassiana and other fungi have been demonstrated to control insect pests in controlled conditions. The main barrier to implementation is finding an effective method of applying these insect disease organisms so that they are reliably placed near host insects in a viable state under field conditions. Insect pathology research continues into bacteria, microsporidia, and other disease organisms of turfgrass pests with the aim of developing additional control options. Some of the more successful recent work has been on insecticidal nematodes.

Insecticidal Nematodes.   Insecticidal nematodes usually enter the mouth, anus, or spiracle (air tube opening) of an insect, although some can burrow through the insect's body wall. Within the insect, the nematode goes through the gut or air tube, releasing bacteria into the blood.

The released bacteria feed on and kill the insect within 2 days. The nematodes feed on the insect internal tissues as well as the bacteria, allowing the nematodes to grow, develop, mate, and produce infective juveniles within the dead insect. After the nematodes have developed for about three generations within the dead insect, the insect's body wall breaks down, releasing infective juveniles that carry the bacteria within their intestines to attack new insect hosts.

Insecticidal nematodes are either ambushers or cruisers. Ambushing nematodes wait for a host to come by and then attack. They are most effective in controlling active insects that live near the soil surface such as cutworms, armyworms, sod webworms, and other soil-living caterpillars and mole crickets. *Steinernema carpocapsae* and *S. scapterisci* are examples of ambushing insecticidal nematodes.

The infective juveniles of cruising nematodes search out their hosts by tunneling through the soil. Because of this, they are more effective against white grubs and other less active insects. They also control caterpillars and other more active pests. *Heterorhabditis bacteriophora* and *Steinernema glaseri* are examples of cruising nematodes.

Steinernema feltiae and *S. riobravis* are intermediate in activity between ambushers and cruisers. *S. riobravis* has been shown to be effective against mole crickets in the laboratory. It can also tolerate drier soils and needs higher soil temperatures than other insecticidal nematodes. It is most effective at temperatures above 95°F.

Insecticidal nematodes can be applied with most sprayers and irrigation systems. Infective juveniles are small enough to pass through most sprayer nozzle orifices and pumps; however, they are large enough to catch on pump and nozzle screens, so these screens should be removed before application. They typically need tank agitation to obtain the necessary oxygen to stay alive.

Because insecticidal nematodes are very sensitive to ultraviolet light and desiccation, application is normally made after 3 P.M. They should be applied to wet turf and then immediately irrigated with at least ½ in of water to help them move down into the soil.

The level of control is usually about 60 to 70% control in practical turf applications, which is usually sufficient to reduce pest numbers below damaging levels. Although insecticidal nematodes can survive in the soil through the growing season and may survive the winter, they are most effective when used as a biological insecticide. They should be applied when needed to control a pest population, as insufficient numbers are likely to survive from year to year to provide long-term control.

Insecticidal nematodes are expensive, perhaps costing 10 times more per unit area than conventional insecticides. With increased use and production, these costs should decrease. They also demand extra care in storage, length of storage, timing of application, field conditions, and application to avoid clogged equipment, but research may alleviate some of these problems. For the client that demands organic control, these nematodes can be a viable option.

6.9.5   Physical or Mechanical Control

Fences built to the ground have been suggested earlier for the exclusion of crayfish. Similarly, various types of fences can exclude raccoons, skunks, and armadillos that may tear up turf in their search for white grubs. Low fences near water bodies were also suggested to make adjacent turf areas less attractive to Canada geese. If Canada geese have to fly over even a 1-ft (30-cm) high fence to reach the water and escape a predator, they will tend to frequent areas where that effort is not necessary.

Moles can be effectively controlled in turf areas by either harpoon or choker traps placed over or in their tunnels. Proper placement and tunnel selection is crucial to successful control. Snap traps have been used to remove voles from turf areas, but are usually not recommended because one is apt to kill birds and other nontarget mammals as well, unless properly placed. Not only is this a moral concern, but most birds and many mammals also are federal- or state-protected species. Live traps are effective against voles, but they are noticeable and require daily maintenance and a release site or disposal method for the captured mice.

Similar problems arise with live trapping of larger animals such as raccoons and skunks. In addition, relocated mammals usually die because they are released into another animal's territory where it is difficult to find sufficient food, water, and shelter. These animals typically die slowly of starvation and/or exposure. It is considered to be more humane to destroy captured animals than to release them, which is difficult for many people to do. Removed mammals are usually replaced almost immediately, however, by another individual moving into the new, empty territory from the surrounding population.

6.9.6 Chemical Control

Insecticidal control should be viewed as a quick, effective, but short-term answer to a pest problem. Some pest problems occur sporadically and are difficult to predict. For instance, greenbugs can literally fall from the sky in some years to cause turf damage. Sod webworm may appear only on slopes and berms in dry years. In situations such as these, the use of an insecticide is usually the first and best choice.

However, some pests are perennial in turfgrass. Insecticides should not be used year after year to control the pests with little thought of why they are there and whether there is something that can be done to prevent the annual insecticide use. For instance, yearly chinch bug problems could be due to excess thatch that could be reduced by core aerification and a reevaluation of the fertilization program. White grubs in a football field may be greatly reduced by reducing irrigation in midsummer to make the area less attractive to egg-laying beetles. In this situation, the football coach would have to realize that summer practice will need to occur elsewhere and that the turf will look fine by football season in the fall. Similarly, many homeowners would trade drier grass in July for lush, untreated grass without grub injury in September and October. Not every homeowner would accept the trade-off, but many would if given a choice.

In certain situations, the annual application of insecticides cannot be avoided due to the use and purpose of the turfgrass. It is unlikely that a golf green will be able to escape insecticide application to control black cutworm. However, scouting with an irritant solution to confirm cutworm presence and population size will help reduce the number of sprays over calendar-mediated applications. Treatment of white grubs on golf course fairways will be necessary in dry years when the beetles are attracted to this irrigated turf. However, programs should allow the option of increased scouting for grub numbers and absence of insecticide application during rainy summers when the grubs are likely to be less numerous in these areas. This practice is possible only if the budget contains funds for an annual grub application even if it is not used.

Even with perennial pests, insecticide application can be reduced. A golf course may need to treat greens, tees, and fairways most years for white grubs, but roughs can be allowed to dry during midsummer to reduce egg-laying and subsequent grub populations. In many years, this will eliminate the need to treat those areas. In home lawns, grubs are rarely found under trees because the adult beetles tend to avoid laying eggs there. Alternatively, beetles are attracted to the edge of areas illuminated by street lights and to warmer areas near streets, sidewalks, and driveways for their egg laying. Scouting for grubs in various areas of the lawn can allow spot applications with less insecticide being applied.

The extra scouting and the individual turf area considerations take time, which costs money. In the past, clientele were more willing to pay for an insecticide application than for pest analysis. For many, that is still the case. However, much of the clientele is willing to pay for scouting and evaluation if it means a reduction in insecticide use. Locating these clientele can be done through marketing more than one program of turf management. The downside is that your personnel have to be knowledgeable enough to scout and identify pest problems. To retain these people may mean higher salaries or other amenities. There is also a greater chance of making a mistake, resulting in visible damage. The client must understand this possibility. As stated earlier, this program will not work for an entire client base, but it will work for some of it. If you do not provide this service to your clientele, evenually someone else will.

Evaluating the level of control is a means of fine-tuning any type of insect control program, but is especially important in a reduced pesticide-use program. Going back to the site an extra time to evaluate the level of control obtained or to evaluate whether a reduce pesticide-use program worked will cost money. Factor this cost into the fee charged to the customer or into the budget for the golf course.

A variety of insecticides are available to the turfgrass manager. Although some types of insecticides are gone from the marketplace or reduced in numbers, others have been developed to fill the needs. Arsenicals were used as stomach poisons on chewing insects more than 60 years ago. Organochlorines, also known as chlorinated hydrocarbons, such as chlordane, aldrin, dieldrin, and DDT, were used from the 1950s into the early 1970s, but these persistent chemicals have largely been replaced.

The federal Food Quality Protection Act of 1996 restricts the amount and type of exposure to pesticides more than previous legislation. If a pesticide or group of pesticides is used on many different crops and in many situations, its use is being reduced to lessen people's exposure, particularly that of children. This act has resulted in the elimination of homeowner packaging for diazinon and chlorpyrifos, leaving it available only in commercial-size quantities. Other pesticides have lost their federal registration and thus have been removed from the marketplace. This deregulation will continue at a rapid rate for several years and will likely result in some of the pesticides listed in table 6–12 being no longer labeled for that use.

Read the insecticide label to determine current legal uses, safety precautions, amounts, and timing of applications, and specific directions on how to apply the chemical for maximum effectiveness. Table 6–12 lists the common turfgrass insect pests and the suggested insecticides that are effective in reducing their populations. For further information on applying pesticides and precautions for their safe use, read chapter 7.

TABLE 6–12 CHEMICAL CONTROL OF TURFGRASS INSECTS

Insect	Suggested insecticide	Remarks
White grubs	Diazinon,* halofenozide, (Mach 2), imidicloprid (Merit),** trichlorfon (Dylox, Proxol)	Treat when small or at egg hatch; Merit and Mach 2 provide control throughout the growing season—allow 3 weeks for activation; drench with ½–1 in (1.3—2.5 cm) of water
Billbugs	Chlorpyrifos (Dursban), diazinon,* imidicloprid (Merit)**	Apply in spring for adult control; for larval control, drench into the soil in June or July with ½–1 in (1.3–2.5 cm) of water
Wireworms	Diazinon*	Drench into the soil with ½–1 in (1.3–2.5 cm) of water when worms are present
Ground pearls	No effective chemical control	Irrigate and fertilize the infested area to stimulate vigorous growth
Mole crickets	Acephate (Orthene), chlorpyrifos (Dursban), diazinon,* imidicloprid (Merit)**	Apply in early summer and repeat when damage appears later; drench into the soil with ½–1 in (1.3–2.5 cm) of water; apply Merit before or during peak egg hatch
Sod webworms	Bifenthrin (Talstar), carbaryl (Sevin), chlorpyrifos (Dursban), cyfluthrin (Tempo),* diazinon,* halofenozide (Mach 2), spinosad (Conserve), trichlorfon (Proxol, Dylox)	Apply as a spray or as granules when adults are present
Cutworms	Bifenthrin (Talstar), carbaryl (Sevin), chlorpyrifos (Dursban), deltamethrin (DeltaGard), diazinon,* halofenozide (Mach 2), spinosad (Conserve), trichlorfon (Proxol, Dylox)	Apply as a spray; repeat when damage reappears
Armyworms	Bifenthrin (Talstar), carbaryl (Sevin), chlorpyrifos (Dursban), diazinon,* halofenozide (Mach 2), spinosad (Conserve), trichlorfon (Proxol, Dylox)	Apply as a spray if many caterpillars are present and feeding on the grass
Grasshoppers	Carbaryl (Sevin), chlorpyrifos (Dursban)	Applied control is seldom necessary; if hoppers have migrated into turf areas and are feeding, apply as a spray
Lucerne moth	Carbaryl (Sevin)	Apply if damage is being done and larvae are present
Striped grass looper	Carbaryl (Sevin)	Apply if damage is being done and larvae are present
Frit flies	Chlorpyrifos (Dursban), diazinon*	Treat damaged area with spray

Pest	Insecticide	Remarks
Chinch bugs	Bifenthrin (Talstar), chlorpyrifos (Dursban), cyfluthrin (Tempo),* diazinon,* trichlorfon (Dylox, Proxol)	Apply as a spray, treatment may need repeating in 2 weeks
Aphids—greenbugs	Acephate (Orthene), chlorpyrifos (Dursban), insecticidal soap	Treat the infested area plus a 2–3 ft (61–91 cm) green border around the damaged area
Leafhoppers	Carbaryl (Sevin), diazinon*	Control is unnecessary unless many are present and damage is observed; apply as a spray if needed
Scale insects	Chlorpyrifos (Dursban), diazinon,* malathion	Apply as a spray and drench into the crowns
Spittlebugs	Acephate (Orthene), carbaryl (Sevin), diazinon*	Control is not needed unless spittle masses are objectionable; apply as a spray if needed
Mites	Diazinon,* chlorpyrifos (Dursban), malathion	Treat in the spring for the bermudagrass mite and repeat if necessary; treat for the winter grain mite, if needed, in late winter; apply the miticide as a spray
Ants	Chlorpyrifos (Dursban), diazinon*	Treat anthills or mounds by drenching insecticide into the nest in the soil
Cicada killer, ground-nesting wasps and bees	Carbaryl (Sevin), diazinon*	Chemical control is needed only if wasps and bees are a nuisance; drench the chemical into open holes in the soil at dusk or early evening
Periodical cicada	Carbaryl (Sevin)	Soil treatment is not effective; spray nearby shrubs and young trees if damage is occurring
Sowbugs and pillbugs	Chlorpyrifos (Dursban), diazinon*	Chemical control is seldom necessary; if necessary, treat the foundation of buildings
Millipedes and centipedes	Carbaryl (Sevin), diazinon*	Spray the turf area bordering the home or other buildings plus the foundation
Earwigs	Chlorpyrifos (Dursban), diazinon*	Spray areas with large numbers plus a border area around the home; apply only if earwigs are a problem
Crickets	Chlorpyrifos (Dursban), diazinon*	Spray the foundation plus a narrow area of grass around the home
Fleas	Carbaryl (Sevin), chlorpyrifos (Dursban)	Apply to turf as a supplement to dusting the pet and bedding
Ticks	Carbaryl (Sevin), chlorpyrifos (Dursban), diazinon*	Spray infested border areas of unmowed grass and shrubbery
Chiggers	Diazinon,* malathion	Spray infested turf and border areas; insect repellents can be used by persons entering infested areas
Thrips	Carbaryl (Sevin), diazinon*	Chemical control is rarely necessary; spray if numerous and damage is observed
Spiders	Chlorpyrifos (Dursban), diazinon*	Apply as a foundation spray to prevent entry; repeat in 4 weeks if needed
Crane flies	Chlorpyrifos (Dursban)	No control is usually necessary

*Diazinon and cyfluthrin (Tempo) are not labeled for use on golf courses and sod farms.
**Merit is not labeled for use on sod farms. Merit is limited to one application per year.

6.10 STUDY QUESTIONS

1. List five physical characteristics that separate insects from other animals.
2. Describe the life cycles of insects with gradual and those with complete metamorphosis.
3. What do white grub larvae look like?
4. Name the two ways that mole crickets damage turf.
5. How are insecticides applied to control root-feeding insects?
6. How do you identify sod webworm damage from other turf damage?
7. Describe the appearance of chinch bug nymphs and adults.
8. List the benefits that ants provide for turfgrass.
9. What problems do cicada killers cause for golfers and other turf users?
10. Describe the benefits that earthworms provide and the damage that they cause.
11. What can be done to control crayfish?
12. Are chiggers more common in mowed or unmowed areas?
13. Describe the difference between mounds made by moles and by pocket gophers.
14. List five types of turfgrass pest biological control agents.

6.11 SELECTED REFERENCES

Beard, J. B. *Turfgrass: Science and Culture.* Englewood Cliffs, NJ: Prentice Hall, 1973.

Brandenburg, R. L., and M. G. Villani. *Handbook of Turfgrass Insect Pests.* Lanham, MD: Entomological Society of America, 1995.

Hanson, A. A., and F. V. Juska (eds.). *Turfgrass Science.* Agronomy Monograph 14. Madison, WI: American Society of Agronomy, Inc., 1969.

Leslie, A. R. *Handbook of Integrated Pest Management for Turf and Ornamentals.* Boca Raton, FL: Lewis Publishers, 1994.

Metcalf, R. L., and R. A. Metcalf. *Destructive and Useful Insects.* New York: McGraw-Hill Book Company, 1993.

Niemczyk, H. D., and D. J. Shetlar. *Destructive Turf Insects.* Cleveland, OH: Lawn and Landscape Media Group, 2001.

Potter, D. A. *Destructive Turfgrass Insects: Biology, Diagnosis, and Control.* Chelsea, MI: Ann Arbor Press, 1998.

Shetlar, D. J., P. R. Heller, and P. D. Irish. *Turfgrass Insect and Mite Manual.* University Park, PA: Pennsylvania Turfgrass Council, Inc., 1983.

Swan, L. A., and C. S. Papp. *The Common Insects of North America.* New York: Harper & Row Publishers, Inc., 1972.

Turgeon, A. J. *Turfgrass Management.* Englewood Cliffs, NJ: Prentice Hall, 1991.

Vittum, P. J., M. G. Villani, and H. Tashiro. *Turfgrass Insects of the United States and Canada.* Ithaca, NY: Cornell University Press, 1999.

Watschke, T. L., P. H. Dernoeden, and D. J. Shetlar. *Managing Turfgrass Pests.* Boca Raton, FL: Lewis Publishers, 1995.

Application Equipment and the Safe Use of Pesticides*

CHAPTER

7

7.1 INTRODUCTION

As mentioned in earlier chapters, successful control of turfgrass weeds, diseases, insects, and other animal pests is based on an *early and accurate diagnosis* from knowledge of the pest's life cycle and dissemination, likely times and places of attack, parts of the turfgrass plant involved, and the cultural and chemical controls that have proven effective.

Pesticides should be applied *only* when the suggested management practices outlined earlier—proper planting, fertilization, watering, mowing, dethatching, and core aerification—are not providing the desired level of control. Turf-protection chemicals are generally formulated as liquid concentrates: solutions (S), flowable suspensions (F, FL, FLO, DF), emulsifiable concentrates (EC), soluble powders or concentrates (S, SC, SP), wettable powders (WP), granules (G), water and soluble dispersible granules (WDG or SDG), and water-soluble packets (WSP) that are dropped into the spray-tank water. Liquid concentrates and wettable powders are usually added to water and applied with a sprayer. Granular products are commonly applied with a drop (gravity) or rotary (centrifugal) fertilizer spreader.

The concentration of turf protection chemicals is usually expressed as a weight per unit volume or as a percent of the commercial product. Recommendations are usually given on the basis of 1000 sq ft (93 sq m) or an acre (0.4 hectare). For example, a 50% wettable powder (50 WP) contains 50% active ingredient (a.i.) and 50% inert material: emulsifying agent, carrier, diluent, and so on. If the manufacturer's recommended rate of application is 6 lb (2.42 kg) of a.i. per acre (0.4 hectare), then 12 lb of the commercially formulated product (50 WP) is needed to treat an acre. This is roughly equivalent to ¼ (4 oz) per 1000 sq ft (1 acre = 43,560 sq ft = 0.40 hectare).

Liquid formulations generally list the number of pounds of the active ingredient per gallon (lb a.i./gal) on the container label. If the concentration is 4 lb/gal, then

* Much of the following material is adapted from chapter 5 of *Application Equipment and Calibration in Illinois Pest Control Manual—Turfgrass* by R.E. Wolf, who kindly reviewed and edited this chapter. Dr. Wolf's assistance is greatly appreciated.

2 pints (1 quart, 4 cups, or 0.95 liter) are required per acre to supply 1 lb of active ingredient per acre (= 1.12 kg/ha).

Effective chemical control of any turfgrass pest involves using the *right* chemical at the *right* concentration at the *right* time in the *right* way. Earlier chapters have provided detailed information concerning suggested chemicals for use against each turfgrass weed, insect, disease, or animal pest (Tables 4.9, 5.5, and 6.12) as well as information on the best methods and timing of applications for optimal results. Additional precautions are given on pages 518–521 and 529–530 and should be read before applying any turf-protection chemical. The specific amount of material that should be applied to 1000 sq ft, acre, or hectare depends on the concentration of pesticide (the active ingredient or a.i.) in the preparation. For example, different manufacturers may sell the same insecticide in a half-dozen formulations where the percentage of a.i. may vary from 2 to 80%. All we can suggest is to read and follow *all* the manufacturer's directions as printed on the container label.

7.2   SAFE HANDLING OF PESTICIDES

Always read and adhere to the instructions and precautions outlined on pesticide labels in regard to the following:

1. Proper clothing to wear (e.g., gloves, boots, goggles, face mask, and approved protection clothing [figure 7–1])
2. Storage of pesticides in a locked, well-ventilated room or building away from people, especially young children and pets (If in doubt, consult OSHA [Occupational Safety and Health Administration] and FEPCA [Federal Environmental Pesticide Control Act] regulations in the United States.)
3. Mixing and application methods and procedures (see the section on general rules and principles)

7.2.1   What Application Equipment Should I Have?

At least three things should be considered when selecting equipment:

1. Consistent uniform distribution of chemicals is needed for efficient and maximum control.
2. The specific method of application is determined by the type of formulations you plan on using plus the size of the area or areas to be treated.
3. Your budget must be maintained.

Over 75% of all pesticide applications are made as liquid sprays. Liquid concentrates and wettable powders may be applied using a wide range of ground and air equipment satisfactory for spraying several thousand square feet a few times a year to hundreds of acres every 7 to 10 days throughout the growing season. Before purchasing any piece of equipment, consider these points. Does it handle and operate efficiently on sloping or uneven ground? Is it simple to fill and clean? Is it

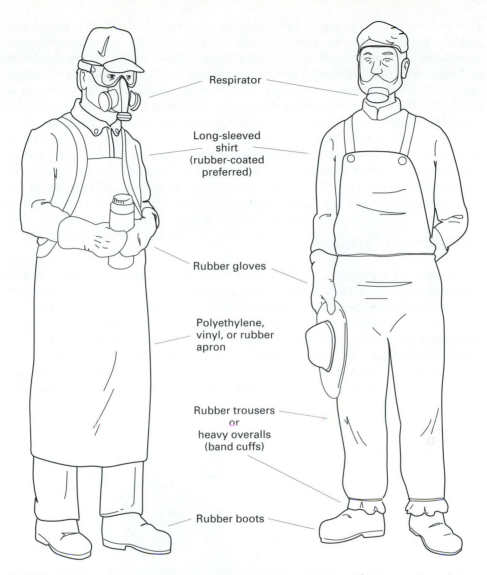

Respirator

Long-sleeved
shirt
(rubber-coated
preferred)

Rubber gloves

Polyethylene,
vinyl, or rubber
apron

Rubber trousers
or
heavy overalls
(band cuffs)

Rubber boots

FIGURE 7–1 Examples of protective clothing to wear when handling pesticide concentrates during mixing and loading or when using highly or moderately toxic materials that have DANGER—POISON or WARNING on the label. The pesticide label will indicate whether a respirator should be worn and other special precautions to take.

big enough to do the job without frequent refilling? Does it have a mechanical or hydraulic agitator? Is the manufacturer reliable? Are parts readily available? Is it well made? Does it have noncorroding parts? How long should it last if given proper cleaning and maintenance?

Many types and sizes of equipment—aerosol cans, hose-end and small compressed-air sprayers, small or large power sprayers, helicopters, and airplanes—are used to apply pesticides to lawn-type turf, athletic fields, golf courses, parks, cemeteries, and other turf areas. Despite the many possible variations and combinations of equipment, however, most pesticides are applied to turf with manually operated sprayers, power-operated sprayers with multi-nozzle booms, handheld spray guns, and granular applicators. Each sprayer or applicator has its distinct uses and features. Attention is given to selecting the proper type and size of nozzle needed for specific applications. In addition, information is provided on the use of electronics for applying pesticides, pesticide injection technology, and mixing pesticides. Chapter 8 thoroughly covers the basic steps of calibrating all types of ground equipment to ensure the proper application rate. Measuring ground speed and small land areas are given in appendices A and C, respectively.

7.2.2   Hose-End Sprayers

These sprayers are suitable for small turf areas, shrubs, and small trees. A hose proportioner is attached to a garden hose and operated by water pressure (figure 7–2). Concentrated pesticide added to a glass or polyethylene jar is metered out by suction. A jar of concentrate spray makes a specific number of gallons of dilute spray. Since accurate dosage control is difficult, the pesticide is often *not* applied at a uniform rate. Also, hose pressure may be insufficient to break up the spray into mist-size droplets that uniformly wet the grass blades and provide needed coverage. Adding an adjuvant (spreader-sticker or surfactant) to the spray mix may

FIGURE 7–2   Hose-end sprayer of 20-gal capacity made of shatterproof polyethylene plastic.

help. Formulations requiring agitation, such as wettable powders, must be frequently shaken. These formulations do not work as well as solutions, flowables, and emulsifiable concentrates. Other disadvantages include occasional drops in water pressure; and the need to drag the hose wherever you go, limiting the area you can reach and requiring you to kneel or stoop over to reach under shrubbery or flowers, where insects and mites occur and disease-causing organisms infect. The outstanding advantages, of course, are convenience and simplicity of operation. A steady hand and a constant walking speed are necessary for uniform application.

7.2.3   Manual Sprayers

Manual hand-operated, nonpowered sprayers such as compressed-air, knapsack, and backpack sprayers are extremely versatile and useful for a variety of applications, especially for spot treatment and treatment of areas unsuitable for larger units. They are relatively inexpensive, simple to operate, maneuverable, require minimum maintenance and easy to clean and store. Compressed air or small, rechargeable cylinders of carbon dioxide are used in most manual sprayers to apply pressure to the supply tank and force the spray liquid through a nozzle. Some backpack sprayers have small engines that can deliver 1 to 3 gal (3.8 to 11.4 l) per minute at pressures up to 300 lb per square inch (psi) or 2069 kPa. Adjustable spray guns are commonly used with these units, but spray booms are available on some models.

Most manual sprayers also have a number of accessories and special fittings available to meet practically every need. They are commonly equipped with a multipurpose nozzle that can be adjusted to produce everything from a fine spray to a coarse stream. It is very difficult to uniformly spray an area with this type of nozzle. Most turfgrass professionals prefer an even-fan-type nozzle. The spray distribution is uniform across the pattern, providing greater accuracy and better results. You can also vary the application volume by selecting a different-size nozzle tip with a known output, giving more accurate calibration and success.

Spray tanks made of rust-free material (stainless steel, brass, copper, or fiberglass) will outlast the cheaper tinplate, aluminum, or galvanized steel models. Lightweight polyethylene plastic sprayers are also widely available.

Compressed-air sprayers.   Compressed-air tank sprayers (figure 7–3) are available as handheld or backpack units. They are popular, low-priced, easy-to-operate, useful tools for a variety of jobs around the lawn, yard, and garden. Pressure for most sprayers is provided by a manually hand-operated air pump that fits into the top of the tank and supplies compressed air to force the spray liquid out of the tank and through a short hose to the spray gun. A valve at the end of the hose controls the flow of liquid. Agitation is provided by frequent shaking of the tank. Normal spraying pressure is between 20 lb (138 kPa) and 60 lb per square inch (414 kPa), maintained by occasional pumping. Because the pressure varies widely, manual sprayers have a tendency to apply in a nonuniform

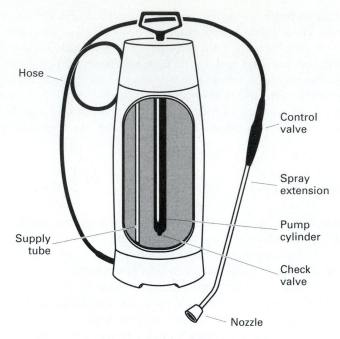

Hose

Control
valve

Spray
extension

Pump
cylinder

Supply
tube

Check
valve

Nozzle

FIGURE 7–3   A compressed-air sprayer is handy for spot spraying.

manner. A pressure-control valve should be used to help maintain a constant pressure. This allows the nozzle or nozzles to produce droplets of a fairly consistent size. If you do not have a pressure-control valve, equip the sprayer with a pressure gauge to monitor the pressure. The capacity of the tank ranges from 1 to 5 gal (3.8 to 19 l).

In some compressed-air sprayer units, a precharged cylinder of air or carbon dioxide supplies the needed pressure. These units include a pressure-regulating valve to maintain uniform spray pressure. Some models are mounted on wheels for easy portability. Pesticides may be applied through a spray gun or short boom with several nozzles.

Knapsack sprayers.   Knapsack or backpack sprayers (figure 7–4) that strap on the back are more expensive and heavier to carry than other sprayers but do a faster and better job of applying a uniform spray. The hose and nozzle are similar to those used on compressed-air sprayers. Pressure is maintained by a piston or diaphragm pump that is either operated by easy hand pumping, or a small, 2-cycle gasoline or by battery (12 volts DC) engine. An air chamber helps to smooth out pump pulsation. Tank capacity ranges from 2 to 6 gal (7.6 to 22.7 l), and pressures up to about 150 psi (1034 kPa) can be developed. Spray material in the tank is kept in suspension by a mechanical agitator or by bypassing part of the pumped solution back into the tank.

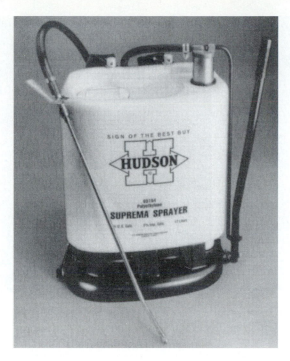

FIGURE 7–4   A knapsack sprayer is a convenient way to spray turf areas. (Courtesy Champion Sprayer Co.)

Wick applications.   These specialized applications are used for delivering water-soluble, postemergence herbicides. The herbicide flows by gravity from a reservoir to a sponge or rope wick. The wick is rubbed over the weed(s), leaving a thin film of herbicide. Wick applications are excellent for applying nonselective herbicides around or under trees and shrubs where drift to nearby plants may cause injury. These applications must be cleaned regularly.

7.2.4   Power Sprayers

Several types of power sprayers (figures 7–5, 7–6, and 7–7) are available that are capable of delivering 1 to 3 gal (3.8 to 11.4 l)/min at pressures up to 150 to 300 psi (1034 to 2068 kPa). You do not need anywhere near these pressures (at the nozzle) to spray turf: 20 to 60 psi (138 to 414 kPa) is all that is normally required. Adjustable spray guns or spray booms are used with these units. Uniformity of coverage is achieved by a constant speed, and a constant pressure is valve regulated.

Power sprayers for applying pesticides to turf areas use an auxillary gas, electric, or solar power source (portable or self-propelled) to develop the pressure

FIGURE 7–5   A small power sprayer equipped with a short spray boom. (Courtesy H. D. Hudson Manufacturing Co.)

FIGURE 7–6   Applying fungicide with a spray boom on a golf course. (Courtesy The Broyhill Co.)

required to meter and distribute the spray liquid. Handheld spray guns are not recommended for spraying turf areas. Spray pressures range from nearly 0 to over 500 psi (3448 kPa) or more, and application rates vary from less than 1 quart per 1000 sq ft to 100 or more gal per acre. Tractor-mounted, pull-type, and self-propelled sprayers are available in many models. Application rates can vary from 10 to 200 gal/acre (95 to 1900 l/ha).

All power sprayers used by turf applicators have several basic components: a pump, a tank, an agitation system, a flow-control assembly, and a distribution

FIGURE 7–7 Typical spray truck and the equipment widely used by lawn-care companies. (Courtesy The Broyhill Co.)

system (figure 7–8). Some newer systems also contain electronic components to help improve the accuracy of application. We suggest a large-wheeled, easily maneuverable model with extra-chemical-resistant hose and a spray boom or multinozzle, trigger-type spray gun.

Spray tanks. Common tank construction materials include stainless steel or heavy-gauge steel, fiberglass, and molded plastics. The material obviously should resist corrosion from any chemical applied.

Choose a tank design that keeps leftover spray mix to a minimum. This reduces waste and problems of disposal. Some tanks have a built-in sump that helps with disposal. All tanks have a drain plug at the lowest point.

Look for a tank that has a sight gauge visible from the operator's position or that is clearly marked so you can see the liquid level in the tank. Other considerations include the weight, cost, and ease of repair (table 7–1).

Sprayer pumps. The pump is the heart of a turf sprayer. It *must* deliver adequate flow and pressure. It should also handle all types of chemicals without rapid corrosion or wear. Both roller and centrifugal pumps are used in low-pressure sprayers for turf use. No matter which pump type you select, make sure it has enough capacity to supply all nozzles, provide adequate agitation, and offset pump wear.

Roller pumps (figure 7–8) operate efficiently with low initial cost and maintenance. The rollers are held in a slotted rotor that is revolving in an eccentric case. When the rollers pass the pump outlet, the spaces between and under the rollers expand and draw in liquid. When nearing the outlet, the spaces contract (due to the eccentric housing), forcing liquid out of the pump. Roller pumps are *not* well suited to abrasive materials (e.g., wettable powders and liquid fertilizers). Rubber rollers are slightly better than nylon rollers if abrasive materials are needed. To allow for normal wear, the pump's capacity should be 20 to 25% higher than originally necessary.

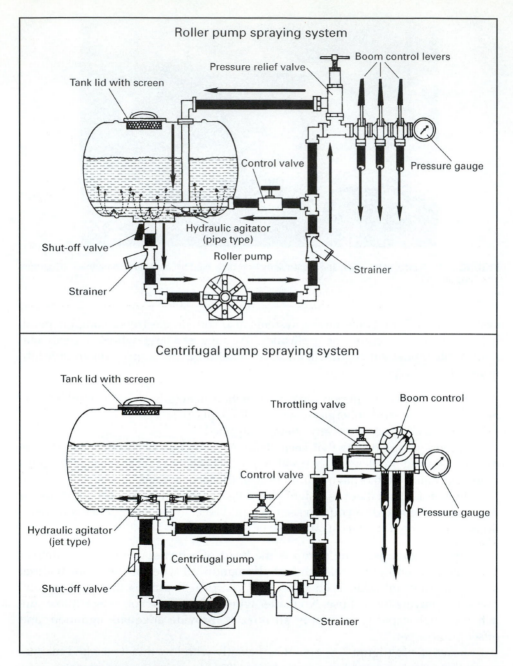

FIGURE 7–8   The principal types of low-pressure spraying systems used to apply pesticides and fertilizers to turfgrasses.

TABLE 7–1   TANK MATERIALS AND THEIR CHARACTERISTICS

Factors to consider	Molded plastic	Fiberglass	Stainless steel
Corrosion proof	Yes	Yes	Yes
Repairable by	Kits for some	Kit with resin	Welding
Weight	Light	Medium	Heavy
Sizes	10 to 1000+ gal	25 to 2000 gal	50 to 2000 gal
Liquid level visible to operator?	Yes	Some	No. Sight gauge is required
Cost	Low	Medium	High
Additional comments	• Not easy to repair • Sunlight may break down some tank materials • Mounting is critical	• Can break or crack under impact or with age • Mounting is critical	• Strong and durable but also most expensive

TABLE 7–2   BOOM-SPRAYER PUMPS FOR TURF USE

	Pump types	
	Roller	Centrifugal
Materials handled	Nonabrasive materials	Most
Purchase price	Low	Medium
Durability	Pressure decreases with wear	Long life
Pressure ranges	0 to 300 psi	0 to 100 psi
Operating speeds	300 to 1000 rpm	2000 to 4200 rpm
Flow rates in gallons per minute (gpm)	1 to 35	0 to 120
Advantages	• Low cost • Easy to service • Operates at PTO speeds • Medium volume • Easy to prime	• More easily handles all materials • High volume capacity • Long life span
Disadvantages	• Short life span when used with abrasive materials	• Low pressure • Not self-priming • A speedup drive or a high-speed hydraulic motor is required

Centrifugal pumps (figure 7–8) last a long time, even with abrasive materials, as the only moving parts are the shafts and the impellers. A high flow volume is produced, with working pressures of 30 to 40 psi (207 to 276 kPa). The impellers must operate between 2000 and 4200 revolutions per minute (rpm) to give high performance. Flow and pressure are created by centrifugal force from an impeller to the fluid. As liquid enters the center of the spinning impeller, centrifugal force throws the liquid into a spiral passage that leads to the outlet. Typical maximum working pressure for centrifugal pumps is 50 to 60 psi (345 to 414 kPa). Table 7–2 gives the principal characteristics of roller and centrifugal pumps for boom sprayers.

Agitation.   Spray-tank agitators are designed to maintain a uniform spray mix. Two types of agitation are common: hydraulic jet and mechanical.

1. *Hydraulic jet agitation*   (figure 7–8) uses a portion of the pump flow to create mixing action in the spray tank. The flow can be directed through a standard nozzle or a specially designed siphon nozzle that creates a venturi effect or vacuum that greatly increases the mixing action. Hydraulic jet agitation may use a sparger (a pipe or tube) with several discharge holes in the bottom of the tank. A sparger is most desirable in large and long tanks.

2. *Hydraulic agitation*   is produced by paddles or propellers in the bottom of the spray tank. The mixers are driven by the pump's power source or a 12-volt electric motor. Adjustable agitators, properly designed and operated, minimize foaming that may occur with vigorous agitation of some chemicals.

Pressure gauges.   The pressure gauge (figure 7–8) on the sprayer system should be designed to measure liquid pressure with a range of 1.4 to 2 times the maximum anticipated pressure. A damper installed between the gauge and the sprayer smooths pressure pulsations, making the gauge easier to read. Moving parts also last longer. Liquid-filled gauges are preferred over dry gauges, since a liquid-filled gauge dampens vibrations.

Boom-control valves.   Spray booms are commonly divided into sections with individual manual or electrical valves. The valves control the flow to the left, center, or right section of the boom, or any combination of these. The valves are useful when treating narrow turf areas or finishing a golf green.

Spray monitors and controls.   It is important for the operator to constantly monitor and control the spray operation. Numerous devices are available to aid sprayer operation (figure 7–9). Whether it is simply a monitor, spray-rate controller, or a more sophisticated computer system, more and more turf applicators are using sprayers equipped with electronic hardware and specially designed software to improve the accuracy of application. Electronic systems provide the versatility and "intelligence" to improve efficiency and make the application process more precise and automatic with less chance of damage to the environment while protecting the operator and those in contact with the turf.

The basic components of electronic control systems are a sensor to measure or detect a condition and a central processing unit (CPU) to translate the signal for display and activate a process. Sensors are the key to electronic control systems that continuously compute and monitor speed, flow, flow rate, pressure, clogged nozzles, and boom height. Monitors simply use the variables that determine gallons per minute (GPM)—ground or travel speed, flow and/or pressure, and spray width. These statistics are fed into a microcomputer with other information such as swath width. The console by the operator displays the pressure and travel speed. In addition, you can ask the monitor to compute and display such information as field capacity (acres per hour), gallons of spray mix left in the tank, acres covered, and distance traveled. Before the sensors will function, you must install and accurately calibrate them by following the directions in the installation and op-

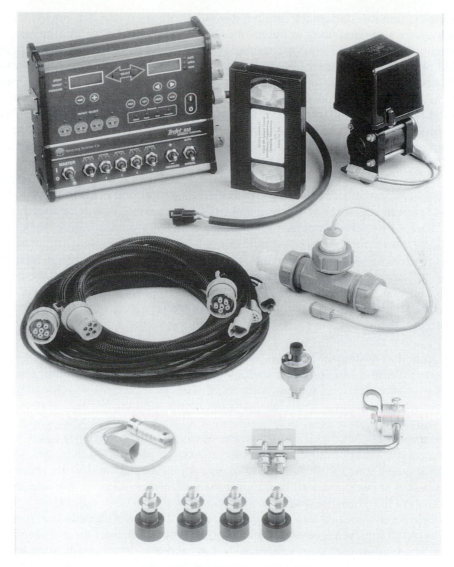

FIGURE 7–9 TeeJet Sprayer Control System. (Courtesy Spraying Systems Co.)

erator's manual. Once the sensors are calibrated, proper operation requires nothing more than flipping some switches and pushing some buttons. It is up to the operator to make the necessary adjustments needed to apply the desired gallons per acre (GPA) or minute (GPM).

A combination of the above electronic components constitutes a rate-controlling system that automatically adjusts application rates on the go. Rate controllers input the desired GPA and control the flow rate in GPM by activating a regulating valve (servo-valve) in the system to maintain the required GPM. As the speed sensor detects an increase or decrease in ground speed, the electronic control system calculates a new flow rate and automatically commands the servo-valve to adjust the application rate back to the original desired application rate. The new variable-rate systems use computers to determine the proper rate and control the amount of chemical applied. It is important to know that the limiting factor for precise application is your spray nozzles rather than the rate controller. With these units, GPA is controlled by changing nozzle pressure; it is critical to maintain the pressure within the recommended pressure range (for example, 10 to 50 psi for extended range flat-fan nozzles and 20 to 40 psi for RA Raindrop nozzles). Remember that pressure must increase four times to double the nozzle flow rate. Therefore, even with a rate controller, you must keep ground speed within a narrow range to maintain a quality spray pattern.

These same electronic components provide the operator the ability to detect any application malfunctions. Sensors located at critical points on the application system alert the operator to any problems that may occur (e.g., tractor speed is too variable for pressure compensation, a hose connection breaks, a line strainer or nozzle becomes plugged). The console either provides an audible warning or displays an error message. The system may also be capable of providing a percent application error by calculating the differences between the target rate and the actual application rate. Most spray controllers have boom sections that are individually monitored, so when a section is shut off, the controller automatically compensates to correct the application volume for the remaining sections.

Injection systems.   Efficient, safe use of inputs has always been the goal of the turf applicator. Due to public scrutiny of chemical use and regulations limiting the use of chemicals on turf, it is essential that technological developments to address environmental concerns are forthcoming. In addition, all users of pesticides are confronted with several potential hazards. Those who mix, load, apply, and handle pesticides have a risk of exposure, not only to themselves but also to the environment. Misapplication, spills, and unsafe application techniques are all potential major sources for contamination to people, wildlife, and water resources. Since pesticides will be a part of the pest management system for the foreseeable future, ways to reduce the risks caused by pesticides must be practiced.

The need to protect our environment from the above hazards has sparked several technological improvements in application equipment. Direct chemical injection is an important development with potential to help the application industry reduce the problems associated with applying chemicals. These systems can give a sprayer infinite rate control.

Direct chemical injection has potentially the greatest effect on the method of applying pesticides. With direct injection, the spray tank contains *only* the water or other carrier. Prior to exiting the nozzle, concentrated chemical formulations or specially blended materials are injected directly into the carrier spray lines at the desired rate depending on speed and other inputs. The type of mixing that occurs depends on whether the injection occurs before or after the carrier spray pump. Injection systems may be classified by the type of metering pump used. The systems use either a piston or cam metering pump to inject the chemical into the carrier. It is then combined in an inline mixer prior to spraying or through a series of peristaltic pumps that meter the chemical and inject it on the inlet side of the carrier spray pump.

The early direct-injection systems had several limitations, including a lag time for the chemical to reach the nozzles and improper mixing of the chemical before spraying. The units also had problems with wettable powder formulations. Many of the earlier troubles have been resolved. Improved metering-pump systems have reduced chemical lag time. Inline mixers have resulted in more uniform mixing. The addition of agitation to mix wettable powders now allows the use of a wide variety of chemical formulations. Direct-chemical-injection technology is more prominent in the agricultural-application industry and is available in ride-on, self-propelled, and skid-mounted sprayers. Control of injection with computers makes this technology well suited to adjusting rates on the go and for prescription applications. Rates can be accurately controlled to take advantage of site-specific needs requiring precise application. Online printers are also available to produce a permanent record of chemical use and job location.

Several application-equipment manufacturers offer direct-injection equipment systems as a factory option. Two companies that provide injection equipment to the turf industry are:

Raven Industries
Box 5107
Sioux Falls, SD 57117

Micro-Trak Systems, Inc.
P.O. Box 3699
Mankato, MN 56002

The injection systems are either included in the electronic controlling device or added on as a module to existing controlling devices.

Before making a decision about purchasing and installing a direct-injection system, ask the following questions.

1. How many chemicals will need to be injected at one time? Direct-injection systems are available to simultaneously apply multiple chemicals at a time. Each chemical requires a separate pump and returnable storage container. The operator can adjust the rate and type of chemical with on and off control at any time.

2. Will any of the injected chemicals need to be mixed? If so, special agitation devices may need to be purchased.

3. What is the chemical-carrying capacity for your needs? Most companies offer several tank sizes to meet those needs.

4. What is the range of application rates required for your products? You need to know this range so a pump of the proper size can be installed.

5. What types of carrier materials will be used?

Other factors also need to be considered when installing an injection system. Placement of the injection tank in reference to the injection pump, and the point of injection on the carrier line in reference to the last nozzle tip are important. The chemical-injection line should be kept as short as possible; the longer the line, the more potential waste during cleanup.

Another driving force behind much of this application technology is the development of sensors and the application of controllers. Spray controllers are being integrated into spray monitor systems. Electronic devices to control application rates have been widely used for years. Controllers (microprocessors) are designed to automatically compensate for changes in speed and application rates on the go. Some are computer based and work well with new application techniques such as direct injection and variable-rate application. Computers and controllers work together to place pesticide inputs in the precise position at the prescribed amount (figure 7–9). The turf applicator's ability to precisely place pesticides is an important factor in protecting the environment.

The acceptance of direct-injection technology has been spurred by environmental concerns, operator safety, federal and state regulations, and the development of new products that are effective at very low rates. Direct injection eliminates the need to tank-mix chemicals, thus eliminating pesticide compatibility problems. Cleanup of equipment is minimal; and, with no leftover solution, disposal of rinsates is not a major concern. If the chemicals are in returnable containers and handled in a closed system, the potential of operator exposure is greatly reduced. Due to the added precision and the ability to spot-spray only where pesticides are needed with the direct-injection process, a substantial savings to the turfgrass applicator or producer and the environment is also realized. The future success or failure in the pesticide-application industry largely rests on how well it manages the environmental issues.

Injection Technology and the Spray Technician.   At the beginning of the day, the technician turns a spigot to start filling the large spray tank with water. While the tank is filling, the water passes through a backflow preventer, which ensures that the water does not flow back into the water supply. The technician, based on scouting and perhaps also laboratory reports, decides which pesticides or other chemicals are needed to manage the problems for that day.

From a separate, locked storage building, the spray technician picks the small chemical-dispenser tanks to mount on the sprayer using quick-connect couplers that greatly reduce exposure. Many times, with these quick-connect couplers, the technician does not even smell the chemical as it is loaded. The sealed dispenser tanks hold 15 gal (57 l) or less of concentrated pesticide, enough for several applications. These tanks are filled and sealed at the factory. Their leak-resistant couplings fit only specific mating couplers to avoid the mixing of noncompatible chemicals. This closed delivery system practically eliminates human contact with chemicals.

The technician may choose a postemergent broadleaf herbicide, a fungicide or fungicide mixture, an adjuvant, and one or two different insecticides. The chemical tanks are then mounted on the sprayer, and their identification tags are inserted into the sprayer-control computer.

The tags on the individual chemical tanks contain encoded data that tell the computer what the pesticide is, the desired application rate, and physical information (e.g., viscosity) about the product. The computer uses this information to set the system components for the most efficient application rate for the particular chemical.

After the main tank is filled with water, the technician reviews and verifies the information stored in the sprayer-control computer. Where necessary, the setting generated by the computer from the encoded tags can be overridden so that the latest manufacturer's recommendation for the product can be used.

The sprayer is now ready to be driven over the turf area to be sprayed, applying the chemicals precisely as and where needed. All the chemicals in the dispenser tanks can be applied at the same time in some spots. In most spots, however, only a limited amount of one or two chemicals, based on lab or scouting reports, need to be applied. The technician can bypass the spray boom in these sites and just use a spray gun.

All spray application can be done in just one trip!

When spraying is completed, the technician simply flips a switch that starts the cleaning operation. Cleansers enter the spray system, followed by rinse water, and the small amount of rinsate is sprayed on a turf area, according to the label, where it cannot contaminate a water supply.

The technician then removes the chemical tanks and places them back in the pesticide storage building located some distance away from other buildings. The sprayer is then washed down on a wash pad and parked. Protective clothing and respirator (figure 7–1) are removed and washed, if necessary. The technician then takes a shower. Dressed in street clothes, the technician goes to the office to record the chemical usage for the day on the specific turf areas.

Advances in Injection Technology—Now and in the Future

1. The control microprocessor on the sprayer could interface with an office computer. This allows the technician to use the database and information files in the office computer to set up the program for the sprayer. The same system would also let the sprayer's computer download data from the spray-control computer back to the office computer for analysis and mapping after spraying. This keeps the database current and produces a record that may be required by state or federal regulations.

2. Satellites can be used on large turf areas such as golf courses to pinpoint the sprayer's exact location on the turf and relay this information to the onboard computer. The sprayer then knows exactly where it is for automatic application of pesticides and fertilizers only where needed. All the technician has to do is drive the sprayer! This technology, known as site-specific application, is based on premapping or on-the-go sensing.

Handheld spray guns. Boom sprayers provide excellent coverage and application efficiency, but in certain situations a handheld spray gun is the obvious choice (e.g., for spraying shrubs and trees and small turf areas with obstructions). Spray guns are available in a variety of styles and capacities. They range from those that produce a low flow rate with a wide, hollow, or solid conical spray pattern, to a flooding or showerhead nozzle pattern, to a high flow rate with a straight-stream spray pattern (figures 7–10 and 7–11). Spray guns are *not* usually recommended for spraying large turf areas such as industrial lawns, parks, or golf course greens; however, much of the commercial-lawn applications

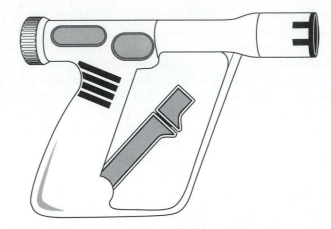

FIGURE 7–10   Handheld spray gun.

FIGURE 7–11   Spraying on a golf course with a handheld spray gun using a showerhead nozzle. Note the protective clothing worn by the operator. (Courtesy Spraying Systems Co.)

are currently being handled by turf guns equipped with showerhead nozzles (figures 7–7 and 7–11). If spray guns are used in this manner, four factors are critical for delivering the correct rate uniformly over the application area.

1. *The exact pressure must be monitored.*
2. *A proper walking speed or pace must be maintained.*
3. *A uniform hand/arm motion technique must be used.*
4. *A constant nozzle height and angle in reference to the ground must be maintained.*

FIGURE 7–12   The spray boom on this sprayer is enclosed within a fiberglass canopy and skirt to prevent drift to nontarget turf and other plants. Canopies, or spraying shields, are available to cover most spray booms.

If it is not possible to use a conventional sprayer with a boom, use a hand or walking boom with conventional nozzles (figure 7–12). When a handheld spray gun must be used because of rough or irregularly shaped areas, you should be aware of the difficulty in obtaining uniform spray coverage over the area. Make sure you also control drift. Pressure gauges at the pump and on the spray gun are recommended to indicate line-pressure loss and give some indication of output. Applying half the spray in one direction and the remaining half at right angles to the first half is another possibility to obtain more uniform coverage; but doubling the time raises the cost of application. To reduce operator fatigue and prevent accidental discharge, get a spray gun with a spring-loaded trigger (figure 7–10) that allows you to automatically lock the gun in the open position during spraying. Optional accessories include a low-volume adapter, an extension wand for spot spraying or touch-up work, and an inlet swivel that eliminates hose interference with the applicator.

Nozzle Types for Power Sprayers with Booms.   Regardless of the type of power sprayer, the proper selection of nozzle type and size is important in accurately applying pesticides to turf (figure 7–13). The nozzle determines how much spray is applied to a particular area, the uniformity of the application, the

Type of application			Nozzle type						
			Extended-range flat-fan	Even flat-fan	Turbo flat-fan	Flood and turbo flood	Drift-control flat-fan	Raindrop	Wide-angle full-cone
Preemergence herbicides		Broadcast	●		●	●	●	●	●
		Band		●					
Post-emergence herbicides	Contact	Band		●					
		Broadcast	●		●		●		
	Systemic	Band		●					
		Broadcast	●		●	●	●	●	●
Insecticides and fungicides	Incorporation		●		●	●	●	●	●
	Foliar banding			●					
	Broadcast		●		●	●	●		
Soil incorporation			●		●	●	●	●	●

FIGURE 7–13   Nozzle types for various types of application to turf.

coverage obtained on the sprayed grass surface, and the amount of drift. Each nozzle type has specific characteristics and capabilities and is designed for use under certain application conditions. You can minimize drift by selecting nozzles that give the largest drop size while still providing adequate coverage at the intended application rate and pressure. Although nozzles have been developed for almost every spray application, only a few types (extended-range flat-fan, flooding flat-fan, Turbo Flood, Turbo flat-fan, Raindrop, wide-angle full-cone, and drift-reduction preorifice flat-fan) are commonly used in spraying turfgrass (figures 7–13 and 7–14). When considering nozzles, look for the convenience of quick-attach, no-tools systems that allow you to change nozzles with a simple turn. ChemSaver diaphragm check valves will shut off nozzles below a certain pressure, thus preventing drips that waste chemicals, money, and may cause turf injury.

Extended-range flat-fan nozzles (figure 7–14 and table 7–3) are commonly used for broadcast soil and foliar applications when better coverage is required than can be obtained from the flooding flat-fan, Turbo Flood, or RA Raindrop nozzles. Extended-range flat-fan nozzles are available in both 80- and 110-degree fan

Extended range flat-fan

Flooding flat-fan (front)

(side)

Turbo-flood flat-fan (front)

(side)

Turbo flat-fan (front)

(side)

RA Raindrop®

Wide-angle full cone

Drift reduction preorifice flat-fan

FIGURE 7–14   Seven types of nozzles used on turf and their spray patterns.

TABLE 7–3   EXTENDED-RANGE FLAT-FAN NOZZLES

Manufacturer Delavan	Spraying systems	Liquid pressure (psi)	Capacity (GPM)
	XR8001	15	0.06
	XR11001	20	0.07
		30	0.09
		40	0.10
		60	0.12
80-1.5R	XR80015	15	0.09
110-1.5R	XR110015	20	0.11
		30	0.13
		40	0.15
		60	0.18
80-2R	XR8002	15	0.12
110-2R	XR11002	20	0.14
		30	0.17
		40	0.20
		60	0.25
80-3R	XR8003	15	0.18
110-3R	XR11003	20	0.21
		30	0.26
		40	0.30
		60	0.37
80-4R	XR8004	15	0.24
		20	0.28
		30	0.35
		40	0.40
		60	0.49
80-5R	XR8005	15	0.31
110-5R	XR11005	20	0.35
		30	0.43
		40	0.50
		60	0.61
80-6R	XR8006	15	0.37
110-6R	XR11006	20	0.42
		30	0.52
		40	0.60
		60	0.74
80-8R	XR8008	15	0.49
110-8R	XR11008	20	0.57
		30	0.69
		40	0.80
		60	0.98

TABLE 7–4   SPRAY ANGLES AND BOOM HEIGHTS

Spray angle (degrees)	Boom height, 20-in spacing (inches)
80	17 to 19
110	10 to 12

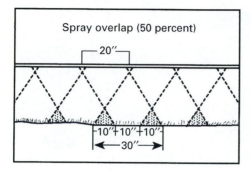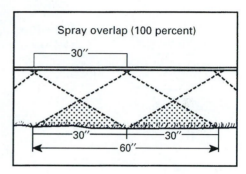

FIGURE 7–15   Two spray booms showing 50 and 100% overlap: flat-fan nozzles on the left boom, and flooding-fan nozzles on the right boom.

angles. The 80-degree fan nozzles are generally mounted on 20-in (51-cm) centers at a boom height of 10 to 19 in (25.5 to 127 cm). See table 7–4. The 110-degree nozzles are usually mounted on 30-in (76.2-cm) centers and at a boom height of 17 to 29 in (43.2 to 73.7 cm) or kept on 20-in centers and lowered to 10 to 12 in (25.5 to 30.5 cm). Regardless of the spacing and height, the spray patterns should overlap about 50% of the nozzle spacing for uniform spray distribution. The boom heights for 80- and 110-degree spray angles are given in table 7–4.

For soil and herbicide applications, keep the operating pressure between 10 to 30 psi. At these pressures, flat-fan nozzles produce medium-to-coarse droplets that are not as susceptible to drift. Smaller drops are produced at pressures from 30 to 60 psi, which increases the likelihood of drift. High pressures are a major concern at pressures above 40 psi and should be used only to apply foliar pesticides that must penetrate the grass canopy to provide maximum coverage.

Because the outer edges of the spray pattern have tapered or reduced volumes, it is necessary to overlap adjacent patterns along a boom to obtain uniform coverage. For maximum uniformity, the overlap should be about 40 to 50% of the nozzle spacing (figure 7–15). Foam markers (figure 7–16) are commonly used to help operators keep track of swath width overlap requirements on multiple passes.

Flooding flat-fan nozzles (figure 7–14 and table 7–5) produce a wide-angle, flat-fan pattern. They are used for applying herbicides and mixtures of herbicides and liquid fertilizers. The nozzle spacing should be 40 in (101.6 cm) or less.

FIGURE 7–16   Spraying turf. Note the white foam markers in the foreground dropped from the end of the boom, that help the operator keep track of swath width.

TABLE 7–5   FLOODING FLAT-FAN NOZZLES

Manufacturer Delavan	Spraying systems	Liquid pressure (psi)	Capacity (GPM)
—	TK-.5	10	0.05
		20	0.07
		30	0.08
		40	0.10
D-.75	TK-.75	10	0.075
		20	0.11
		30	0.13
		40	0.15
D-1	TK-1	10	0.10
		20	0.14
		30	0.17
		40	0.20
D-1.5	TK-1.5	10	0.15
		20	0.21
		30	0.26
		40	0.30

TABLE 7–5   FLOODING FLAT-FAN NOZZLES (continued)

Manufacturer Delavan	Spraying systems	Liquid pressure (psi)	Capacity (GPM)
D-2	TK-2	10	0.20
		20	0.28
		30	0.35
		40	0.40
D-2.5	TK-2.5	10	0.25
		20	0.35
		30	0.43
		40	0.50
D-3	TK-3	10	0.30
		20	0.42
		30	0.52
		40	0.60
D-4	TK-4	10	0.40
		20	0.57
		30	0.69
		40	0.80
D-5	TK-5	10	0.50
		20	0.71
		30	0.87
		40	1.00
D-7.5	TK-7.5	10	0.75
		20	1.10
		30	1.30
		40	1.50
D-10	TK-10	10	1.00
		20	1.40
		30	1.70
		40	2.00
D-15	TK-15	10	1.50
		20	2.10
		30	2.60
		40	3.00
D-20	TK-20	10	2.00
		20	2.80
		30	3.50
		40	4.00
D-30	TK-30	10	3.00
		20	4.20
		30	5.20
		40	6.00

These nozzles are most effective in reducing drift when they are operated within a pressure range of 8 to 25 psi (55.2 to 172.4 kPa). Pressure changes affect the width of the spray pattern more with the flooding flat-fan nozzle than with the extended-range flat-fan nozzle. In addition, the distribution pattern is usually not as uniform as that of the extended-range flat-fan tip. The best distribution is achieved when the nozzle is mounted at a height and angle to obtain at least double coverage or 100% overlap (figure 7–15). Uniformity of application depends upon the pressure, height, spacing, and orientation of the nozzles. Pressure directly affects droplet size, nozzle flow rate, spray angle, and pattern uniformity. At low pressures, flooding nozzles produce large spray drops; at high pressures, these nozzles actually produce smaller drops than flat-fan nozzles at an equivalent flow rate.

The spray distribution of flooding nozzles varies greatly with changes in pressure. At low pressures, flooding nozzles produce a fairly uniform pattern across the swath, but, at high pressures, the pattern becomes heavier in the center and tapers off toward the edges. The width of the spray pattern is also affected by pressure. To obtain an acceptable distribution pattern and overlap, operate flooding nozzles within a pressure range of 8 to 25 psi (55.2 to 172.4 kPa).

Nozzle height is critical in obtaining uniform application when using flood nozzles. The nozzle arrangement is recommended because it compensates best for pattern variations caused by changes in pressure and the height of the nozzles on the spray boom. In this arrangement, the nozzles are spaced on 30-in (76.2-cm) centers, and the height is adjusted to obtain double coverage or 100% overlap. With 100% overlap, the outer edges of the nozzle pattern reach the center of the pattern of the adjacent nozzles so that the entire surface receives spray from two nozzles. To achieve proper overlap, raise or lower the boom for each pressure and nozzle size. Automatic metering systems will cause unacceptable changes in overlap unless the nozzle pressure is maintained within the 8-to-25-psi (55.2-to-172.4-kPa) range.

Flooding flat-fan nozzles can be mounted vertically to spray backwards, horizontally to spray downwards, or at any angle in between. The position of the nozzles is not critical as long as double coverage is obtained. You can determine nozzle position by rotating the nozzle to the angle required to obtain double coverage at a convenient nozzle height. When the nozzle is mounted horizontally to spray downwards, heavy concentrations of spray tend to occur at the edges of the spray pattern. Rotating the nozzles 30 to 45 degrees from the horizontal will usually increase pattern uniformity over the recommended pressure range of 8 to 25 psi (55.2 to 172.4 kPa). For uniform distribution over a range of pressures, mount the nozzles to obtain double coverage at the lowest operating pressure. The boom heights for 80- to 110-degree spray angles are shown in table 7–4.

Turbo Flood nozzles (figure 7–14 and table 7–6) combine the precision and uniformity of extended-range flat spray tips with the clog-resistance and wide-angle pattern of flooding nozzles. The design of the Turbo Flood increases droplet size and distribution uniformity. The increased turbulence in the spray tip causes an improvement in pattern uniformity over existing flooding nozzles. At operating pressures of 8 to 25 psi (55.2 to 172.4 kPa), Turbo Flood nozzles produce

TABLE 7–6   TURBO FLOOD NOZZLES

Manufacturer spraying systems	Liquid pressure (psi)	Capacity (GPM)
TF-2	10	0.20
	20	0.28
	30	0.35
TF-2.5	10	0.25
	20	0.35
	30	0.43
TF-3	10	0.30
	20	0.42
	30	0.52
TF-4	10	0.40
	20	0.57
	30	0.69
TF-5	10	0.50
	20	0.71
	30	0.87
TF-7.5	10	0.75
	20	1.10
	30	1.30
TF-10	10	1.00
	20	1.40
	30	1.70

droplets that are 30 to 50% larger than standard flooding nozzles. The larger droplets reduce the number of driftable-size droplets in the spray pattern; thus, Turbo Flood nozzles work well in drift-sensitive applications. Turbo Flood nozzles, because of their improved pattern uniformity, probably need at least 50% overlap to obtain proper uniformity of application.

Turbo TeeJet flat-fan nozzles (figure 7–14 and table 7–7) are wide-angle, pre-orifice nozzles that create larger spray droplets across a wider pressure range (15 to 90 psi; 103.4 to 620.6 kPa) than comparable low-drift tips, thus reducing the amount of driftable particles. The design also creates a uniformly distributed spray pattern similar to the extended-range flat-fan and Turbo Flood nozzles. The unique design of the nozzles allows them to be mounted in a flat-fan nozzle body configuration. The wide spray angle allows for 30-in (76.2-cm) nozzle spacings and 50% overlap to achieve a uniform application along the boom width.

TABLE 7–7   TURBO TEEJET FLAT-FAN NOZZLES

Manufacturer spraying systems	Liquid pressure (psi)	Capacity (GPM)
TT11001	15	0.06
	20	0.07
	30	0.09
	35	0.09
	40	0.10
	50	0.11
	60	0.12
	70	0.13
	80	0.14
	90	0.15
TT110015	15	0.09
	20	0.11
	30	0.13
	35	0.14
	40	0.15
	50	0.17
	60	0.18
	70	0.20
	80	0.21
	90	0.23
TT11002	15	0.12
	20	0.14
	30	0.17
	35	0.19
	40	0.20
	50	0.22
	60	0.25
	70	0.27
	80	0.28
	90	0.30
TT11003	15	0.18
	20	0.21
	30	0.26
	35	0.28
	40	0.30
	50	0.34
	60	0.37
	70	0.40
	80	0.42
	90	0.45

TABLE 7–7   TURBO TEEJET FLAT-FAN NOZZLES (continued)

Manufacturer spraying systems	Liquid pressure (psi)	Capacity (GPM)
TT11004	15	0.24
	20	0.28
	30	0.35
	35	0.37
	40	0.40
	50	0.45
	60	0.49
	70	0.53
	80	0.57
	90	0.60
TT11005	15	0.31
	20	0.35
	30	0.43
	35	0.47
	40	0.50
	50	0.56
	60	0.61
	70	0.66
	80	0.71
	90	0.75

Raindrop nozzles (figure 7–14) are recommended when spray drift is a major concern. (See table 7–8 for ¼-in RA Raindrop nozzles.) When operated within a pressure range of 20 to 50 psi (139 to 344.8 kPa), these nozzles deliver a wide-angle, hollow-cone spray pattern and produce fewer smaller drops than flooding nozzles. For a uniform spray pattern, space the nozzles no more than 30 in (76.2 cm) apart and rotate 30 degrees from the vertical axis. RA Raindrop nozzles are best used with soil-applied herbicides, and they can replace traditional flood nozzles for increased control of drift. Although large droplets aid in drift control, they may result in less coverage than is required for foliar insecticides and fungicides. To ensure adequate coverage, take caution with RA Raindrop nozzles when applying certain pesticide products. Heavier application rates (50 to 75 gal per acre; 473.2 to 709.8 l per ha) will improve coverage in such cases. Be sure to adjust Raindrop nozzles to obtain double coverage or 100% overlap.

Wide-angle full-cone nozzles (figure 7–14 and table 7–8) produce large pesticide droplets over a wide range of pressures. The inline or straight-through nozzle design uses a counterrotating internal vane to create controlled turbulence. The design allows the formation of a 120-degree spray angle over a pressure range of 15 to 40 psi (103.4 to 275.8 kPa). This nozzle provides a uniform spray distribution pattern and requires only about 25% overlap, and is a good choice to reduce drift.

**TABLE 7–8   QUARTER-INCH RAINDROP
DRIFT-REDUCTION NOZZLE**

Manufacturer Delavan	Liquid pressure (psi)	Capacity (GPM)
RA-2	20	0.14
	30	0.17
	40	0.20
	50	0.22
RA-4	20	0.28
	30	0.35
	40	0.40
	50	0.45
RA-5	20	0.36
	30	0.44
	40	0.50
	50	0.56
RA-6	20	0.43
	30	0.52
	40	0.60
	50	0.67
RA-8	20	0.57
	30	0.70
	40	0.80
	50	0.90
RA-10	20	0.71
	30	0.87
	40	1.00
	50	1.10
RA-15	20	1.10
	30	1.30
	40	1.50
	50	1.70

Drift-reduction preorifice nozzles (figure 7–14 and tables 7–9, 7–10, and 7–11) produce an extended-range flat-fan pattern while lowering the exit pressure at the nozzle. The lowered exit pressure creates a larger droplet spectrum with fewer fine, driftable spray particles, thus minimizing the off-target movement of the spray pattern.

Two styles of drift-reduction flat-spray nozzles are currently available. The RF Raindrop flat-spray nozzles (table 7–9) are available with a 105- to 115-degree fan angle, and the Drift Guard flat-spray nozzles (tables 7–10 and 7–11) are available in both 80- and 110-degree fan angles. With a larger droplet size, drift-reduction preorifice nozzles can replace conventional flat-fan 80- and 110-degree tips in broadcast applications

TABLE 7–9   RF RAINDROP FLAT-SPRAY DRIFT-REDUCTION NOZZLES

Manufacturer Delavan	Liquid pressure (psi)	Capacity (GPM)
RF-1	20	0.071
	25	0.079
	30	0.087
	40	0.10
	50	0.11
	60	0.12
RF-1.5	20	0.11
	25	0.12
	30	0.13
	40	0.15
	50	0.17
	60	0.18
RF-2	20	0.14
	25	0.16
	30	0.17
	40	0.20
	50	0.22
	60	0.24
RF-3	20	0.21
	25	0.24
	30	0.26
	40	0.30
	50	0.34
	60	0.37
RF-4	20	0.28
	25	0.32
	30	0.35
	40	0.40
	50	0.45
	60	0.49
RF-5	20	0.35
	25	0.40
	30	0.43
	40	0.50
	50	0.56
	60	0.61
RF-6	20	0.42
	25	0.47
	30	0.52
	40	0.60
	50	0.67
	60	0.73

(continued)

TABLE 7–9   RF RAINDROP FLAT-SPRAY DRIFT-REDUCTION NOZZLES (continued)

Manufacturer Delavan	Liquid pressure (psi)	Capacity (GPM)
RF-8	20	0.57
	25	0.63
	30	0.69
	40	0.80
	50	0.89
	60	0.98

TABLE 7–10   WIDE-ANGLE FULL-CONE NOZZLES

Manufacturer spraying systems	Liquid pressure (psi)	Capacity (GPM)
FL-5	15	0.34
	20	0.38
	30	0.46
	40	0.52
FL-6.5	15	0.42
	20	0.48
	30	0.57
	40	0.65
FL-8	15	0.51
	20	0.58
	30	0.70
	40	0.79
FL-10	15	0.67
	20	0.76
	30	0.91
	40	1.00
FL-15	15	0.97
	20	1.11
	30	1.32
	40	1.50

TABLE 7–11   DRIFT GUARD FLAT-SPRAY DRIFT REDUCTION NOZZLE

Manufacturer Delavan	Liquid pressure (psi)	Capacity (GPM)
DG80015	30	0.13
DG110015	35	0.14
	40	0.15
	50	0.17
	60	0.18
DG8002	30	0.17
DG11002	35	0.19
	40	0.20
	50	0.23
	60	0.25
DG8003	30	0.26
DG11003	35	0.28
	40	0.30
	50	0.34
	60	0.37
DG8004	30	0.35
DG11004	35	0.37
	40	0.40
	50	0.45
	60	0.49
DG8005	30	0.43
DG11005	35	0.47
	40	0.50
	50	0.56
	60	0.61

where spray drift is a problem. The recommended pressure for drift-reduction preorifice nozzles is 30 to 60 psi (207 to 414 kPa). They require the same 50% overlap as the extended-range flat-spray tips. An alternative to the preorifice nozzle is a larger extended-range flat-fan nozzle operated at a lower pressure.

Nozzle Tip Materials.   Spray-nozzle assemblies consist of a body, cap, check valve, and nozzle tip. Various types of bodies and caps (including color-coded versions) and multiple nozzle bodies are available with threads as well as quick-attaching adapters. Nozzle tips are interchangeable in the nozzle cap and are available in a wide variety of materials, including hardened stainless steel, stainless steel, brass, ceramic, and various types of plastic (such as Kematal, Thermoplastic,

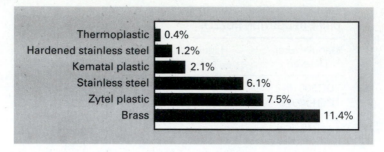

FIGURE 7–17 Percent increase in nozzle flow rate of flat-fan spray nozzles after a 40-hour wear test.

and Zytel). Hardened stainless steel and ceramic tips are the most wear-resistant materials, but they are also the most expensive. They have excellent wear resistance with either corrosive or abrasive materials. Plastic tips have good to excellent wear life, they are resistant to corrosion and abrasion, are very economical for applying pesticides, but swelling can occur with some chemicals and the nozzles are easily damaged when cleaned. Brass tips have been very common, but they wear rapidly when used in applying abrasive materials such as wettable powders, and they are corroded by some liquid fertilizers. Brass tips are economical for limited use, but other types should be considered for more extensive use (see graphs in figure 7–17). Severely worn nozzles can cause "hot spots" of turf injury or carryover and provide poor pest control. New spray-tip technology provides a wide range of tips that will spray better and longer than ever before.

Check a nozzle tip for wear by comparing it with an identical new tip using a graduated collection container, a stopwatch, and a pressure gauge mounted on the nozzle. A much easier way is to use a TeeJet Tip Tester, which fits over the nozzle and gives a direct readout of flow rate in gallons or liters per minute. A flow 10% over the rated value means the tips need replacing.

How to Reduce Spray-Tip Wear

1. When spraying abrasive or corrosive materials, use as low a pressure as possible and install more wear-resistant spray tips, such as ceramic or hardened stainless steel.
2. If spraying at high pressure (only 60 psi; 414 kPa maximum are needed on turf), use less abrasive materials with proper filters to avoid tip corrosion. Also use more wear-resistant tips.
3. If using small tips, apply as low a pressure as possible with more wear-resistant tips.
4. If using flat-fan or tapered orifice nozzles, use more wear-resistant tips and check all flat-fan nozzles frequently for wear.
5. Use an alternative design with a better wear life.

Nozzle Size Selection for Power Sprayers with Booms. The performance of any pesticide depends on the proper application of the correct amount of chemical. Most performance complaints about turfgrass pesticides relate directly to errors in dosage or to improper application. Again, choosing the correct type of nozzle is critical to making a proper application. In addition to the type of nozzle, the correct size of nozzle is equally important. Since the nozzle determines the amount of spray applied to a given area, you need to determine the size of tip needed to satisfy the parameters of the application.

Variables for Selecting the Proper-Size Nozzle Tips.   The size of the nozzle selected is based on three variables of calibration, regardless of the type of sprayer you use. The three variables affect the amount of spray mixture applied per 1000 sq ft, acre, or hectare:

1. The nozzle flow rate
2. The ground speed of the sprayer
3. The effective sprayed width per nozzle

To calibrate and operate your sprayer properly, you must know how each of the three variables (nozzle flow rate, ground speed, and sprayed width per nozzle) affects the sprayer output. (See chapter 8 for a complete discussion of calibration.)

Nozzle Flow Rate.   The flow rate through a nozzle varies with the size of the tip, the solution characteristics, and the nozzle pressure. You can increase the flow rate by installing a nozzle tip with a larger orifice or by increasing the pressure. Nozzle flow rate varies in proportion to the square root of the pressure. *Doubling the pressure will not double the flow rate. To double the flow rate, increase the pressure by a factor of four.* For example, to double the flow rate of a nozzle from 0.28 gallon per minute (GPM) at 20 psi (139 kPa) to 0.56 GPM, increase the pressure to 80 psi (4 × 20) or 551.6 kPa.

Pressure cannot be used to make major changes in the application rate, but it can correct minor changes due to nozzle wear. To obtain a uniform spray pattern and minimize drift, keep the operating pressure within the recommended range for each nozzle type. Remember, if you use check valves to prevent nozzle drip, then the pressure at the nozzle is 5 to 7 psi (34.5 to 48.3 kPa) lower than the boom pressure indicated on the pressure gauge. To protect against reduced application rates as a result of pressure loss, increase pressure and check the nozzle flow until it matches the desired amount.

Ground Speed.   The spray-application rate varies *inversely* with the ground speed. *Doubling the ground speed of the sprayer reduces the application rate by one-half.* For example, a sprayer applying 20 gal per acre (GPA) or 189.3 l per hectar at 3 miles per hour (mph) or 4.8 km/h would apply 10 GPA (94.6 l/ha) if the speed were increased to 6 mph (10.2 km/h) and the pressure remained constant.

Some power sprayers are equipped with control systems that maintain a constant spray-application rate (GPA) over a range of ground speeds. The pressure is changed to vary the nozzle flow rate according to changes in the ground speed. These systems require calibration at a set ground speed. When spraying turf, speed changes must be limited to those that maintain the nozzle pressure within its recommended range.

Sprayed Width Per Nozzle.   The effective width sprayed per nozzle also affects the spray-application rate. *Doubling the effective sprayed width per nozzle decreases the application rate by one-half.* For example, if you are applying 40 GPA (378.5 l/ha) with flat-fan nozzles on 20-in (50.8-cm) spacings and change to flooding nozzles with the same flow rate on 40-in (101.6-cm) spacings, then the application rate decreases from 40 to 20 GPA (378.5 to 189.3 l/ha).

The gallons of spray applied per acre can be determined using the following equation.

$$GPA = \frac{GPM \times 5940}{mph \times w}$$

$$or \ gallons \ per \ 1000 \ sq \ ft = \frac{GPM \times 136}{mph \times w}$$

where   GPA = output, gallons per acre
  GPM = output per nozzle, gallons per minute
  mph = ground speed, miles per hour
    w = effective sprayed width per nozzle, inches
  5940 = a constant used to convert gallons per minute, miles per hour, and inches to gallons per acre
    136 = a constant used to convert gallons per minute, miles per hour, and inches to gallons per 1000 sq ft

Problems of tank-mixing pesticides and fertilizers.   Combining chemicals to do several jobs with a single spray saves time and labor. In light of the many thousands of pesticides, soluble fertilizers, growth regulators, sticking-wetting agents or surfactants (see the section on adjuvants) and other chemicals that can be sprayed on turf, it is surprising that plant injury (or phytotoxicity) does not occur more often.

Mixing fertilizers and pesticides is common in the lawn-care industry to double up on visits and save time. But these mixes can be risky! A single tank mix might contain nitrogen, phosphorus, and potassium; a mixture of micronutrients; a pre- or postemergence herbicide; and even an insecticide or fungicide. Slow-release fertilizers, as liquids or powders, are available in sprayable forms. Golf course superintendents commonly spoon-feed their greens and tees with frequent, light applications of liquid fertilizer tank-mixed with one or more herbicides, fungicides, and insecticides. It is important to understand the problems that may arise from tank-mixing fertilizers and pesticides.

Monitoring the pH of tank mixes is critical, as most pesticides, especially insecticides, are sensitive to extremes in pH. Your pesticide supplier should be able to provide information about acceptable pH ranges for the pesticides you plan on using. Potassium sources in tank mixes generally influence pH much less than nitrogen and phosphate sources.

Most commonly used nitrogen fertilizers are alkaline. For example, the resultant solution when urea dissolves in water is a pH of 8.0 to 8.5 which, over time, may become even more alkaline. Most manufacturers of controlled-release liquid fertilizers stabilize and buffer these materials in a pH range of 9.0 to 10.5. Tank mixes with these fertilizer compounds have a pH near 9.0 and are difficult to acidify. Alkaline-sensitive pesticide breakdowns are likely to occur at a pH of 8.0 or above, resulting in compatibility problems.

Phosphate fertilizers tend to be acidic. For example, a commonly used liquid phosphate fertilizer (ammonium polyphosphate [10-34-0]) has a pH of 6.3 to 6.5 and is routinely highly buffered at this level. Some dry phosphate fertilizers are even more acidic. Phosphoric acid, for instance, can bring a tank mix below pH 3.0. If you use this acid, be sure that your liquid fertilizer supplier adjusts the fertilizer to a pH of 5.5 to 7.0 or 7.5.

When choosing a liquid fertilizer for tank-mixing, remember that ammonia, aqua-ammonia, and potassium hydroxide (0-0-30) are all highly alkaline. If one of these materials is combined with phosphoric acid, pH extremes may occur unless the ingredients are properly balanced and mixed. Talk with your supplier to better understand the components in liquid fertilizer mixes. You should specify an acceptable pH range for the finished product.

Fertilizer salts may cause problems by increasing the salt load (ionic strength) of the tank mix. Certain emulsifiable concentrates, liquid flowables, and other pesticide formulations cannot withstand the salt load, resulting in incompatibility. Such formulations break down and cause tank problems; further, they are phytotoxic. The biggest problem usually occurs with potassium fertilizers.

Iron sulfate is incompatible with amine formulations of certain phenoxy herbicides and can cause problems in the tank and on plants. At higher concentrations of iron, a cottage-cheese-like precipitate may form in the tank. The iron sulfate often causes the leaves of weeds to blacken, and the uptake of herbicide is reduced to where it is ineffective. You can avoid this problem by choosing an iron compound compatible with the herbicide you use in the tank mix.

Suppliers of fertilizers and pesticides should be able to provide information concerning the compatibility of their products in tank mixes. Talk with them about potential problems. They may know of successful uses by other applicators. Always read and follow all instructions and precautions listed on product labels. Pay special attention to any tank-mixing information. Thoroughly check out new fertilizer-pesticide tank-mix combinations before making extensive applications to turf.

Compatibility agents are designed to improve the stability of certain tank-mix combinations. The labels commonly provide specific information on compatibility. Manufacturers of these products (table 7–12) may also be good sources of information concerning specific tank-mix problems you may experience.

Over 60,000 proprietary pesticides have been formulated in various ways from some 1,200 basic active ingredients. The end products are used as bactericides, fungicides, herbicides, insecticides, miticides, nematicides, and so on. Sometimes, precipitates will form in the tank and screen filters; or spray nozzles will plug up, turf will be damaged, or pesticides will be made ineffective as a result of the indiscriminate mixing of two or more of these chemicals. An understanding of pesticide compatibilities is essential to avoid these problems.

Compatibility.   When pesticides are used in combination or as components of a mixture, they will either be compatible or incompatible.

Compatible refers to the reaction that occurs when two or more chemicals are mixed together without (1) impairing their toxicity to pests, (2) causing undesirable

**TABLE 7–12   SOME COMMON COMPATIBILITY AGENTS
AND THEIR MANUFACTURERS**

Manufacturer	Product(s)
Conklin	Kombind
Custom Chemicals	Merge
Helena	Induce pH, Blendex, Blendex VCH
Kalo	Complex
Loveland	Easy Mix, Li Combo, Unite
Miller	Spray-Aide
Plant Health Technologies	Comp-Ad, Compatibility Aid
Rhone Pôulenc	Rhodafac, Soprophor
Rohm and Haas	Latron Ag44M
Terra International	Riverside Combine
Western Nutrients	Hemiplus
Witco	Armix, Sponto

physical properties, or (3) making the combination more toxic to plants than the individual chemicals when used alone.

Incompatible refers to the reaction of pesticides that cannot be mixed safely without (1) impairing the effectiveness of one or more of the chemicals, (2) developing undesirable physical properties, or (3) causing plant injury.

There are two basic types of incompatibility: chemical and physical. It is possible to get one or both incompatibilities from the same mix. *Chemical incompatibility* involves the breakdown and loss of effectiveness of one or more products in the spray tank and possible formation of one or more new chemicals that are insoluble or phytotoxic. *Physical incompatibility* involves an unstable mixture that settles out, flocculates, foams excessively, or disperses poorly and reduces efficiency and causes the clogging of sprayer nozzles and screens. This type of incompatibility may result from the use of hard water.

A third type of incompatibility called *placement incompatibility* is not uncommon. This is where a recommended pesticide, at the manufacturer's suggested rate, is applied at the right time but to where it is not effective. For example, an insecticide to give effective control of grubs needs to get into the root zone where the grub larvae feed. If the insecticide is sprayed on the grass foliage and not washed or drenched into the root zone, it will not provide control of grubs (see the following section on general rules and principles).

General rules and principles

1. Follow all label instructions and precautions regarding uses, intervals, dosage, method of application (gallons of water to apply to 1000 sq ft or per acre), and all incompatibilities. Certain fungicides have special label instructions to avoid phytotoxicity and maintain efficiency. Practically all pesticide companies and experienced specialists recommend that chemicals should be applied separately and for a specific purpose.

2. If chemicals can be mixed, pour each separately into the spray tank with agitation (shaking). This frequently prevents settling out and plugging of screen filters and nozzles. A common practice is to make up a thin, uniform batter or slurry of the spray powder and dilute it with water. Some specialists suggest straining the spray solution through fine cheesecloth while it is being added to the tank.

3. Soluble fertilizers and trace elements can usually be added individually or mixed, provided that the amount will not exceed 1 oz of solid material per gallon of tank spray mix. (See the previous section on mixing fertilizers and pesticides.)

4. Apply spray solutions as soon after mixing as possible. The longer a spray combination remains in the tank, the greater the number of problems that can arise. Some labels specifically warn against premixing a certain number of hours or more before use due to possible breakdown.

5. Compatibility information is usually incomplete on most pesticide labels. For example, certain fungicide labels, such as chlorothalonil (Daconil) specifically warns against pesticide and/or fertilizer tank-mixing without the user's knowledge of compatibility. Fusetyl-Al (Aliette) should not be mixed with soluble fertilizers or other metal-containing pesticides (e.g., mancozeb [Fore, Dithane I/C]).

6. To determine the physical compatibility of a tank mix, check it yourself. Two steps are important: (1) Add all the tank-mix ingredients in correct proportions plus the proper volume of water to a 1-quart glass jar. Shake it briskly for 30 sec and let it stand for 30 min. Look for signs of incompatibility (e.g., precipitates that may settle to the bottom, oily droplets on the surface, or sediments that stick to the side of the jar). You may need to wait at least 24 hours for signs to appear. If the chemical mixture separates or settles out, it is unwise to use the mixture (figure 7–18). Regardless of the results in step 1, step 2 should be carried out if the material is at all sprayable.

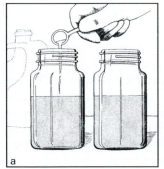  

FIGURE 7–18   Jar test for determining the physical compatibility of a tank mix. (a) Place a mixture of the precise dosage of pesticides in a quart jar. (b) Shake the jar briskly for 30 seconds. (c) Let the jar stand for 30 minutes. The mixture in the jar to the right has separated or settled out and probably should not be used as a tank mix.

In step 2, the mixture is applied to an out-of-the-way turf area, preferably during adverse conditions such as heat (above 85 to 90°F [29.5 to 32.2°C]) and moisture stress; overlap to determine phytotoxicity. A minimum of 48 hours should elapse before you can properly evaluate if injury has occurred.

7. If you are determined to apply a mixture of different pesticides, we suggest this order of adding them to the spray tank: wettable powders first, flowables second, solubles third, other powders fourth, adjuvants fifth, and emulsifiable concentrates last. Always remember that pesticides should be placed in a spray tank that has been filled with *clean* water while the agitator is running.

8. Use caution when mixing wettable powders with emulsifiable formulations or a soluble fertilizer. Wettable powders and emulsions both suspend in water. Depending on the concentration and water used, such mixtures may cause a breakdown of the emulsion, the formation of sludges, flocculation, and reduced efficiency. In most cases, the active ingredients are *not* at fault; the emulsifiers, solvents, fillers, and adjuvants are responsible. Sometimes, the products of a single company will be compatible, but the same active ingredients from two different companies will be incompatible. Yet many (or most) such formulations are compatible.

    Plant damage is most common and severe at high temperatures or under slow-drying conditions. Soluble fertilizers or nutrient mixes, such as zinc or iron sulfate and chelated compounds, added to a pesticide mix can destroy the emulsification or suspension of other chemicals. However, urea is compatible with most pesticides.

9. Do not mix strongly alkaline and acidic materials together. Strongly acidic or alkaline materials (sulfur, lime, lime sulfur, zinc sulfate and lime, ferrous sulfate, and ammonium sulfate) commonly cause acute compatibility problems. The container label should state if the product can be mixed with strongly acidic or alkaline materials; if unsure, check with your supplier.

10. The pH and chemical composition of the water may be important. City water that has been softened may be strongly alkaline (pH 8 to 9+). Hard water from deep wells may cause pesticides to precipitate. Water pumped from a stream or pond may contain chemical wastes. These factors have a great influence on how pesticides perform individually and in mixes.

11. Do not experiment with new combinations. If you must try out a new mix, apply it on a small area first. When possible, spray an out-of-the-way area with each product alone alongside a turf area sprayed with the test mixture. Apply these sprays several times at various strengths and different air temperatures. Check if the mix performs as well as when the materials are applied separately. Record your findings—do not trust your memory. Remember, no chemical company can possibly test all its products and those of its competitors in all possible combinations. The product you buy today may have different fillers, emulsifiers, solvents, or adjuvants than the same product bought a year or two ago.

12. As a general rule, insoluble and wettable powder fungicides do not produce chemical injury when used as recommended. Many formulations of soluble fungicides can be phytotoxic. They should be applied within their safety ranges and not combined with other chemicals unless the package label states otherwise.

13. Turfgrass in a low state of vigor is more easily injured by chemicals and mixes than vigorously growing grass. Plants may be predisposed to damage by winter or herbicide injury; drought; waterlogging of the soil; poor soil; disease or insect damage; or an imbalance of nitrogen, potassium, phosphorus, or other elements.

14. Do not mix a foliar fungicide, herbicide, or insecticide with fertilizers or other chemicals that require watering in. Incorrect placement of pesticides sometimes explains poor pest control. An example is a combination of a fungicide spray to control Helminthosporium leaf spot and an insecticide drench to control grubs applied at the same time.

15. Use caution when mixing chemicals that may be toxic by themselves with other materials. Such combinations commonly injure sensitive grasses at considerably lower concentrations than individual products in the mix.

16. Plant injury can sometimes be avoided by spraying when temperatures are between 40 and 80°F (4.4 and 26.6°C). Emulsifiable materials are more likely to cause injury than wettable powders.

17. Buy pesticides only in amounts you expect to consume in the current year. Different pesticides, especially in combination, vary greatly in their shelf life, particularly once the container has been opened. The same basic chemical, manufactured by different companies, may vary in stability due to differences in formulation. Storage temperature and humidity can also have important effects. Many pesticides last indefinitely if kept dry with the container sealed and not stored at or below 32°F (0°C).

18. Store chemicals only in their original, tightly closed containers so that contamination cannot occur. Clean sprayers, hose lines, nozzles, and mixing containers thoroughly after each use. Numerous cases of plant injury can be traced back to the careless contamination of insecticides, fungicides, fertilizers, or equipment with potent herbicides.

19. Be sure of the identity of the material. It may seem incredible, but whole towns have actually been fogged with the herbicide 2,4-D (instead of an insecticide) for mosquito control. Liability damages amounting to hundreds of thousands of dollars could have been avoided by simply reading the label—if the old container still had one.

    Commercial products that contain a mixture of ingredients are the result of many years of testing under a wide variety of environmental conditions. Many such mixes (tested by company representatives, the USDA or other federal agencies, state experiment stations, consultants, and extension or advisory-service specialists) are never sold because of reduced efficiency, short shelf life, or other problems. Yet many do-it-yourselfers with limited time and facilities think they can come up with an even better mix. The odds are not good.

20. Avoid irrigation and mowing of turfgrass treated with a pesticide for at least 24 hours after application.

21. If a large spill (>1 lb or 1 qt) occurs in storage, call your local hazardous materials team (dial 911) for cleanup instructions. Sand, "Kitty litter," a scoop, or large plastic bucket can be used for small spills (<1 lb or 1 qt).

Adjuvants.   When added to a pesticide or a plant-growth regulator, an adjuvant (or surfactant) reduces the surface tension between two unlike materials, such as a spray film and a solid surface. For example, by adding the proper surface-active agent to a tank mix, oil and water will mix and can be sprayed on plant surfaces. Adjuvants include activators; deflocculators; dispersants; compatibility agents; detergents; emulsifiers; foam and drift suppressants; spreading, sticking, and wetting agents; and buffering/acidifying agents. These materials are added to a spray mix to keep the pesticide in suspension, improve cohesiveness and dispersion of the spray, and increase the wetting (or coverage) of the foliage and thatch.

A *sticker* or *adhesive* is a material that, when added to a spray mix, improves the adherence (tenacity) of a chemical to a plant surface without increasing initial deposit. Commercial sticking agents are oily in consistency and increase the amount of suspended solids retained on plant surfaces by coating the particles in a resin or varnishlike film. (Most fungicides and insecticides already contain an adequate amount of sticker in the formulation.) Stickers may be measured in terms of resistance to wind and water, length of adherence, and mechanical or chemical action.

A *spreader-sticker* is an agent that combines the functions of both a spreader and a sticker.

A *wetting agent* is a material that, when added to a pesticide, lowers the interfacial tension between the liquid and a solid; in this case, a plant or soil surface. The effectiveness is measured by the increase in spread of a liquid over a solid surface and the ability of the spray film to make complete contact with it. When a wetting agent reduces surface tension, spreading naturally occurs.

Soil-wetting agents (such as Aquagro and Wet Foot) are commonly used to improve hydrophobic soils or localized dry or hot spots. The wetting agents reduce the surface tension that holds water droplets together, permitting better penetration of the thatch and soil and reducing surface runoff. Nutrients and water penetrate more deeply into the soil profile. Soil-wetting agents also help combat the formation of black layer. If layering or compaction due to small pore spaces is the problem in such soils, wetting agents may not be beneficial; in fact, they may actually be harmful.

A *spreader* or *film extender* (spreader-activator) is a substance that, when added to a pesticide mix, increases the area that a given volume of spray will cover and improves the contact between the pesticide and the plant surface. A spreading agent builds spray deposits and improves weathering ability. Most wettable powder fungicides and insecticides benefit from the addition of a spreader that is usually included in the formulation. Spreader/extenders lengthen the time the spray is effective by use of an ultraviolet screen, which slows the breakdown caused by the sun's rays.

An example of this involves the application of an adjuvant as a wetting agent, spreading the herbicide over the entire surface of a weed, a desirable characteristic when a contact herbicide is being applied with water in a low-gallonage rate. Without a wetting agent, the spray droplets will concentrate on small sections of the leaf surface, resulting in local tissue burning without killing the weed.

But when a *high* gallonage rate of the herbicide and water is being applied, a wetting agent used by itself can cause up to two-thirds of the spray to run off the plant. Under these circumstances, a specially compounded product is required that also contains a sticking agent to aid spray retention.

The foaming characteristics of an adjuvant (or surfactant) are important to its performance. When you reduce surface tension with a surfactant, you get foaming. When surface tension is too low, the spray tank fills up with foam, you get poor distribution of the pesticides, and you are left with a mess on your hands.

Buffering/acidifying agents are important for lowering the pH in water. A high pH (above about pH 7.5) can seriously shorten the period of time a pesticide is effective. Some pesticides tend to break down under alkaline conditions, so a pH range of 5 to 7 is best. Always check pesticide labels to see if the product breaks down in an alkaline pH.

Drift-control agents coagulate spray droplets into larger droplets that are less prone to drift. Add a drift-control agent to a pesticide if you expect windy conditions or a temperature inversion.

With increasing emphasis on safe application of pesticides, such factors as particle size, spray pattern, and pesticide drift have focused more on adjuvants to give ideal coverage for specific pesticides and plants. There are numerous terms that designate the surface-active components of a pesticide; these are often closely related and commonly perform two or more of the same functions.

Choosing an adjuvant is not simply a matter of finding something that works with one pesticide and using it with all your pesticides. The hardest part may be finding which adjuvant is best suited for a particular job. Otherwise you may be wasting money and risking discolored turf, damaged soil structure, ineffective spraying, or worse. Unless you have evaluated a spreader-sticker or other adjuvant, you have no way of knowing how it might affect the performance of a product or spray mixture. Although choosing an effective adjuvant to accompany a specific pesticide is no simple task, the container label should state if an adjuvant is needed and the brand or brands that should be used. Practically all turf-protection chemicals have one or more of these chemicals already in the spray mix. Certain labels specifically warn against the use of any surfactant with the product. If you are worried about possible turf injury, then, before tank-mixing an adjuvant, test the mix on a small, remote spot before applying to an entire area.

The manufacturer may recommend on the container label that you add a small amount of an adjuvant (spreader-sticker, surfactant, or other) to the spray solution. Recommendations are commonly given as percent concentration per volume (tank size). You will probably need to determine how much adjuvant to add per tankful. For instance, if you use an adjuvant at 0.5% concentration by volume, a few simple calculations are required to tell you how much adjuvant you should add to a 300-gal (1136-l) tank. Use one of the following methods to determine the amount.

Method 1.   Since 1% of 100 gal equals 1 gal (100 × 0.01), 0.5% of 100 gal equals ½ gal. You will need ½ gal per 100 gal or 1½ gal per 300 gal (½ × 3).

Method 2.   The recommendation is for 0.5% concentration by volume, and the volume of your tank is 300 gal. Multiply 0.5% (or 0.005) by the number of gallons (0.005 × 300 gal = 1.5 gal).

In a study of 70 adjuvants, 80% actually *reduced* the herbicide's effectiveness; 10% had no effect at all; and 10% increased the herbicide's efficiency. Many turfgrass specialists do *not* favor adding a surfactant to the spray mix unless it is specifically needed.

When selecting a product to add to a spray mix, consider such factors as the homogeneity of the surfactant concentrate or powder, its stability in storage, its ease of mixing in water, the effect of water hardness on the emulsion stability or dispersion, and the added cost.

Adjuvants and pesticides are sold separately, and adjuvants are not subject to EPA registration. The lack of authority to regulate performance standards has resulted in some misleading product claims and misunderstandings.

All commercial adjuvants should be mixed strictly according to label directions. Adding more adjuvant or surfactant than recommended may cause excessive runoff, resulting in a poor spray deposit and reduced pest control. In general, if the spray mix contains one or more pesticides produced or formulated by the same company, use an adjuvant sold or recommended by that company.

Whether a spray rolls off or sticks to a plant surface depends on the physical and chemical properties of the spray mixture and the properties of the surface itself. If the surface tension of the mixture is high or the plant surface is waxy, the spray droplets will roll off.

Sprayer maintenance and cleaning.   A sprayer is a long-term investment and should provide many years of satisfactory service. Carefully follow all manufacturer's instructions as printed in the sprayer service and maintenance manual regarding operation, maintenance, and lubrication.

Thoroughly check your sprayer before the start of your busy application season. Drain any antifreeze or water and check the pump for cracks or leaks. Test throttling valves, pressure gauges, hoses, and clamps for corrosion and leaks. Check nozzle gaskets for a tight fit. Clean line and tip strainers with fresh water and a soft-bristle brush. Identify nozzle needs for the upcoming season. New chemicals or application procedures may require different nozzle types, sizes, or pressure ranges. Replace worn nozzles *before* the spraying season and keep extra nozzles on hand to avoid unnecessary delays later. Your sprayer should be equipped to carry fresh water for rinsing gloves and tools and cleaning nozzle tips and screens.

Obtain appropriate safety protective gear including a long-sleeved shirt, long pants, sturdy shoes or boots, a chemical-resistant apron, gloves, and goggles or a face shield (see figure 7–1). Read label instructions regarding further safety precautions.

Be sure certain pumping and filling equipment has antibackflow devices, and check valves to prevent contaminating water supplies.

A regular maintenance program can identify many potential problems (such as worn or outdated parts, poor hose connections, hoses with kinks or cracks, and partially plugged strainers).

The spray tank should be in good shape and free of small cracks or other defects. It must be kept *clean*—free of rust, scale, dirt and sand, grass clippings, and other trash. Nothing wears out nozzle tips or a pump faster than sand particles and rust. Do not lay sprayer parts where they can pick up sand or dirt particles and grass clippings.

Everyone has had problems with strainers and spray nozzles clogging. Liquid concentrates, solubles, and flowables should present few problems. Clogging of wettable powders can often be prevented by mixing up a thin, smooth batter of water and spray powder before pouring the chemical into the spray tank through a fine screen or cheesecloth. Keep the agitator running. Be sure to install correctly sized line or tip strainers to prevent tip clogging.

Check and clean the strainers after use each day. Partially plugged strainers create a pressure drop and reduce the nozzle flow rate. Most power sprayers contain strainers: (1) at the tank's filter hatch, (2) on the suction hose to protect the pump, (3) in the line between the pump and the boom, (4) on supply lines to each boom section, and (5) in the nozzles (figure 7–8).

Self-cleaning strainers that use excess capacity to continually wash down the screens, moving contaminants and undissolved chemicals back to the tank, are available. A throttling device regulates the bypass flow from the strainers.

All types of sprayers should be thoroughly rinsed after *each* use.

1. Shake manual sprayers well before pumping a little pressure and flushing the discharge line with the nozzle cap off. Drain, pump some clean water through the system, and hang upside down. Lock the spray-control valve open for drainage. Where possible, take the pump out and hang separately.
2. For power sprayers, pump soapy water through the discharge system, followed by one or more rinses of clean water. Periodically remove the nozzles and strainers and wash with strong detergent; household ammonia; or trisodium phosphate, TSP (1 teaspoonful per gal; 1 quart to 25 gal or 1 l to 100 l) followed by hot, soapy water. Store dry in a place where the tank and discharge system will drain properly.

Before starting a new power sprayer, check all lubrication points. Flush out the system thoroughly before starting the pump to remove any metal chips or dirt accumulated during the manufacturing process. Then add water to the tank and operate at slow speed while checking the delivery system, control valves, pressure regulator, and other parts.

Occasionally during the season and definitely before winter storage, clean hand-operated and power sprayers more thoroughly. Clean, flush, and drain as outlined above, and then disconnect the hoses and soak nozzles, strainers, and screens in a solution of household ammonia or TSP. Scrub these with an old toothbrush or fine-bristle brush (*not* with anything metal) before rinsing with hot, soapy water. Replace nozzle discs with enlarged holes and other worn parts. Rinse hoses clean.

Before Winter Storage.   *Manual sprayers* should be taken apart and the pump cleaned, where possible. Add a few drops of light oil in the top of the pump cylinder to keep the plunger cup pliable and to lubricate the cylinder. For winter storage, leave the sprayer partially unassembled. Lightly oil all metal parts and wrap them in newspaper. When reassembling the sprayer, pump clean water through the open nozzle head to flush out the discharge line. Finally, pump to nearly full pressure and check hose and gaskets for leaks; maybe the shutoff valve needs a drop of oil. If the pump fails to develop full pressure, remove the plunger from the cylinder and replace the cup so that it seals tightly. It may need more lubrication or replacement.

Power sprayers should be thoroughly drained, flushed, and cleaned as outlined by the manufacturer. Pour a pint to 5 gal (0.47 to 19 l) of lightweight engine oil or radiator-rust inhibitor (depending on its size) into the tank, fill with water, and pump for 1 to 2 min. This provides a protective coating to the inside of the tank, pump, valves, and circulating system. Finally, drain and dry the sprayer completely and store where clean and dry. Remove the diaphragm in check valves and store them separately to ensure solid seals. To prevent corrosion, remove the nozzle tips and strainers, dry them, and store them in a can of light oil. The hoses should be stored unkinked inside in the dark where they will not freeze. When reassembling in the spring, check the hose or hoses carefully for deep cuts and cracks. Are the hose clamps tight and gaskets still pliable?

Corrosive fertilizers should not be used in certain sprayers. For example, liquid fertilizers are corrosive to copper, galvanized surfaces, brass, bronze, and steel. An ordinary sprayer can be ruined by using a liquid fertilizer just once. If you apply liquid fertilizers, use a sprayer made completely of stainless steel, molded plastic, or fiberglass.

7.2.5 Granular Applicators

Many turf-care applicators and homeowners use granular products as all or part of their pest control programs. Proper selection, care, calibration, and use of granular applicators can minimize costs and maximize the results obtained. Improper use can lower control of pests, cause injury (phytotoxicity) to the turf, increase costs, and damage the spreaders.

More and more pesticides are becoming available in granular form for turf use. Some come premixed with fertilizers or as mixtures to control a variety of pests. This is a fast, simple, and convenient way to apply chemicals to kill weeds, many soil insects, and nematodes. Some granular products, especially postemergence foliar herbicides, are less effective than their liquid counterparts. Granules, however, are *not* generally satisfactory for controlling disease-causing fungi that invade through the leaf blades unless you use a systemic fungicide. In addition, granular products take time to break down to release the active fungicide. Contact (protective) fungicides require the uniform coverage obtainable with a fine spray. Granular pesticides are generally safer formulations of pesticides than those available as liquids. The low percentages of active fungicide ingredient combined with the slow availability of the fungicide to affected turf make those products less desirable for disease control when treating diseases that are active.

Drop (gravity) and *rotary (centrifugal) spreaders* (figure 7–19) are widely available for applying granules to turf. Drop spreaders are usually more precise and deliver a more uniform pattern than rotary spreaders. Because the granules drop straight down, there is less chemical drift and less chemical spill on sidewalks, driveways, and roads. Some drop spreaders will not handle larger granules, however, and ground clearance in high-cut, wet, or uneven turf can be a problem. Because the edges of a drop-spreader pattern are sharp, any steering error will cause missed or doubled strips. Drop spreaders also usually require more effort to push than rotary spreaders.

Operating procedures Experienced turf-care operators are familiar with the proper use of granular applicators. New operators need to review the basic operating procedures. First, read the operator's manual and follow the manufacturer's

FIGURE 7–19   Drop or gravity spreader (left) and rotary or centrifugal spreader (right).

instructions carefully. Second, read the product label, and select the appropriate rate and pattern settings for specific conditions. Third, operate the spreader the long way of the turf area, but first mark off header strips across each end or around an irregular area (figure 7–20).

Header strips provide an area in which to turn around and realign the spreader. Always move the spreader at normal operating speed on the header strips, and then activate the spreader as it enters the untreated turf area. When you reach the other end, the spreader should be shut off while moving. Then stop and turn in the header strip. A spreader should *never* be open when stopped, because an excessive amount of product will be applied to a small area. In addition, the end turns should not be made with the spreader open, because the pattern will be very irregular while the spreader is turning.

Occasionally, it may be impossible to obtain a completely acceptable pattern with a rotary spreader, and streaking of the turf may result. A common solution to this problem is to reduce the setting by one-half and go over the area twice at right angles. Instead of averaging out the patterns, as is generally believed, this procedure usually changes the streaks into a diagonal checkerboard. It is better to reduce the setting and swath width by one-half and go back and forth in parallel swaths.

Do not operate your spreader backwards. When pulled backwards, most rotary spreaders deliver an unacceptable pattern, and drop spreaders will not maintain a constant application rate.

Finally, set and fill the spreader on a level, paved surface rather than on turf. If a spill occurs, a driveway or sidewalk is much easier than turf to sweep clean. Some rotary spreaders are provided with a means of shutting off one side of the pattern. This feature is desirable when edging a driveway, sidewalk, or other nonturf area.

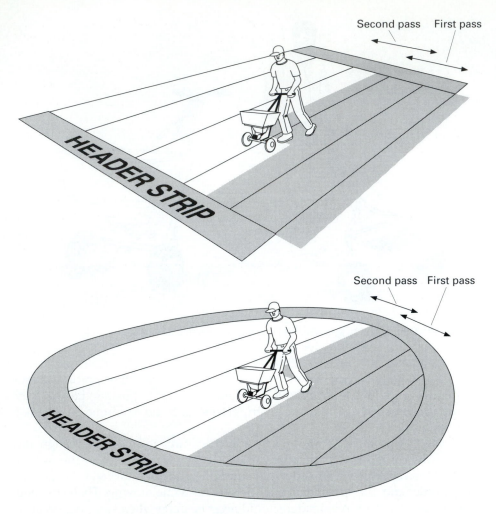

Second pass   First pass

HEADER STRIP

Second pass   First pass

HEADER STRIP

FIGURE 7–20 Operating drop and rotary spreaders. Header strips are first marked off at each end (above) or around an irregular area (below) before making a pass.

Uniform application.   There are two important aspects to the precision application of granular products. The first is the application rate (the amount of product applied in pounds per 1000 sq ft). Every turf product, whether a fertilizer or pesticide, is designed and recommended for application at a specific rate. Overapplication is costly, and it increases the risk of grass injury and may be illegal if label recommendations are exceeded. Underapplication may mean poor pest control.

The flow rate from granular applicators will not change in the same proportion as changes in speed. For example, doubling the speed will *not* double the flow rate. A constant ground speed is necessary to maintain a uniform application rate.

Uniform distribution of the product is as important as the application rate. For example, the pesticide label may indicate an application rate of 4 lb per 1000 sq ft, and the spreader may apply that amount; however, the pesticide may not be applied uniformly over the 1000 sq ft. It is very important to obtain uniform distribution of granules on turf, as even small differences in the application rate can result in obvious streaks or lack of control.

The pattern applied by a rotary spreader depends on impeller characteristics (height, angle, speed, shape, and roughness), ground speed, the drop point of the product on the impeller, the physical properties of the product, and environmental conditions (temperature, humidity, wind, and so on). Methods for adjusting the pattern include blocking off part of the metering port or ports on some units and moving the metering point or impeller on other units. If pattern skewing cannot be fully corrected by following the manufacturer's recommendations, try varying the speed or tilting the impeller. When a product is so light or heavy that skewing cannot be eliminated, it may be necessary to use a wider swath width on one side than the other.

Maintenance and care of spreaders. A spreader should provide many years of satisfactory service if the following maintenance techniques are practiced.

1. Carefully follow *all* the manufacturer's instructions as printed in the spreader service manual with regard to operation, maintenance, and lubrication.
2. Empty the spreader after each usage.
3. Rinse the outside and especially the inside of the hopper with a stream of water to remove any pesticide or fertilizer clinging to the surface of the spreader.
4. Allow the spreader to dry thoroughly.
5. Oil the bottom hopper surface of drop spreaders, plus the spring and inside of the control housing, axle bearing, and other lubrication points.
6. Store the spreader wide open.
7. Check the calibration periodically as outlined above.

7.3   SAFETY PRECAUTIONS

Following are other safety hints concerning the use of pesticides that may have been missed previously.

1. Reread the instructions, precautions, and warnings on the container label before opening. Use the product strictly according to package directions, on the *grasses* specified, in the *amounts* specified, at the *times* specified, and *only* when needed. Keep the container tightly closed except when preparing the mix.
2. Store all chemicals in their original containers—with a label securely attached—in a dry, locked, orderly, and well-ventilated cabinet or building outside the home. Place warning signs to indicate that pesticides are stored there. Never leave pesticides open to children, irresponsible or unauthorized adults, or pets. Do not store them near human food, animal feed, fertilizers, or other commonly used chemicals such as fuels, lubricants, and motor oil. A few minutes spent studying the directions and precautions on a product label will prevent misuse and needless accidents.

3. Do not breathe mists, vapors, or dusts of pesticides when mixing, handling, or applying. Do not allow smoking, eating, or drinking in or near pesticide handling, mixing, or storage areas. Mix and prepare pesticides in the open or in a well-ventilated place. Avoid spilling on the skin, shoes, or clothing. Immediately flush with warm, soapy water any part of the skin contacted. Promptly remove all contaminated clothing and launder before wearing. Bathe promptly after spraying.

4. When mixing, handling, or applying turf protection chemicals, wear full protective clothing when required. At minimum, wear a face mask to protect your eyes and unlined rubber gloves for your hands. Keep sleeves and trouser legs rolled down. Keep your collar buttoned, and wear a washable cap (see figure 7–1).

5. Wash hands and face thoroughly before eating, drinking, or smoking.

6. Cover birdbaths, pet dishes, and fish pools before applying any pesticide to turf. Avoid drift to swimming pools, fish ponds, streams, or other water supplies. Keep all chemicals away from herbs, fruits, and vegetables if these plants are not specifically mentioned on the label.

7. Do not mix emulsifiable concentrates (ECs) with wettable powder (WP) formulations unless the label states otherwise. Before mixing any two or more chemicals together, check the label for instructions.

8. Do not apply any spray when the temperature is above 80 to 85°F (16.6 to 29.5°C) or below 40°F (4.4°C), or injury to turf may result.

9. Properly dispose of leftover spray. First, check the label as well as local and state regulations for instructions. It may be possible to dilute and spray on another turf area. Use extra spray on a labeled crop, if possible. Never leave puddles on a hard surface that could attract pets or birds.

10. Promptly destroy empty or old pesticide containers so that they are not a hazard to people or pets; return reusable containers to the supplier. Wash out glass and metal containers before putting them in the trash can.

11. Do *not* contaminate a spreader used to apply insecticides and fungicides with herbicides, especially those used for broadleaf-weed control, such as 2,4-D, dicamba, or mecoprop (MCPP).

12. Keep the sprayer or granular applicator in good repair by following a regular maintenance program.

7.4   STUDY QUESTIONS

1. What are five common types of pesticide formulations?
2. Effective chemical control of any turfgrass pest involves applying the pesticide at the right time. What are the other four right conditions?
3. Before purchasing application equipment, what things should you consider?
4. What are the disadvantages of using a hose-end sprayer?
5. What are the main types of manual sprayers? What type of spray tank for a manual or power sprayer would you choose and why?

6. What are the five main components of a power sprayer?

7. What are the main disadvantages of a centrifugal pump in a power sprayer as compared with a roller pump?

8. What do electronic spray-monitor systems do?

9. Injection systems reduce unwanted pesticide exposure. How is this accomplished?

10. What does the spray technician currently do to start and finish a day of spraying? How will practices change in the future?

11. Four factors are critical when using a handheld spray gun to deliver the correct rate of pesticide uniformly over a turf area. What are these factors?

12. List the seven nozzle tips most commonly used to apply chemicals to turf. For what is each type of nozzle principally used?

13. What type of nozzle tip would you select when using abrasive materials (such as fertilizers)? What nozzle tips last the longest? How would you go about checking a nozzle tip for wear?

14. What are the three variables you need to determine when selecting the proper-size nozzle tips for a power sprayer? Why do you need to know these variables?

15. What are the three types of pesticide incompatibility? What do these involve? How would you go about solving these problems?

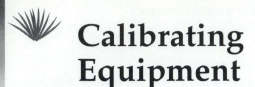

Calibrating Equipment

Proper calibration of spray and granular equipment is a necessity that requires only a few steps. It improves accuracy, maximizes the effectiveness of the fertilizer or pesticide, avoids turf injury, and helps to safeguard the environment. Inaccurate or sloppy calibration steps vary with each type of equipment.

8.1 CALIBRATION OF MANUAL SPRAYERS

Because manual sprayers are generally used to spray limited areas, the amount of spray volume should be determined on a small area such as 1000 sq ft. Most compressed-air sprayers do not have pressure gauges or pressure controls. The pressure in the tank will drop as the material is sprayed. This pressure drop can be partially overcome by (1) filling the tank to the fill line with spray material so that considerable air space remains above the spray mix; and (2) repressuring the tank at frequent intervals. If the sprayer has a pressure gauge, repressurize when the pressure drops approximately 10 lb per square inch (psi) from the initial reading. Equipping with a pressure-control valve will also help maintain a constant flow of material. When spraying, either hold the nozzle steady at a constant height and walk back and forth or swing the nozzle in a sweeping, overlapping motion. Always maintain a uniform walking speed during application. The fan pattern should be perpendicular to the direction of travel. Waving a flat-fan nozzle wand back and forth will result in nonuniform coverage. Handheld spray guns require considerable operator skill to evenly spray large areas. It is a good idea to practice on a paved area using water in the sprayer and then watching the drying pattern.

Use the following steps to ensure calibration.

Step 1. Measure and mark off an area of 1000 sq ft (for example, 20 ft × 50 ft). Practice spraying the area with water. For the most uniform application, always spray an area twice, each with half the spray, with the second application at right angles to the first.

Step 2. Once you are able to maintain a uniform spray, add a measured amount of water to the tank, spray the area in two directions, and then measure the amount of water remaining in the tank. Obviously, the difference between the amount in the tank before and after spraying is the amount used. For example, 2 gal added to the tank minus ½ gal remaining equals 1½ gal used per 1000 sq ft. Thus, the application rate for the above conditions is 1½ gal per 1000 sq ft.

TABLE 8–1 MANUAL SPRAYER CALIBRATION

Pints of spray caught in 1 min	Rate (gal/acre)
¼	5.0
⅜	7.4
½	10.0
⅝	12.5
¾	15.0
1	20.0
1¼	25.0
1½	30.0
1¾	35.0
2	40.0

You can also determine the time in seconds to spray this area in a normal manner. By this method, catch the spray from the nozzle (or nozzles) used in a quart jar marked in ounces for the time period you have determined (see table 8–1). You can calculate the rate per acre as follows: pints caught × 20 = gallons per acre.

8.2 CALIBRATION OF HANDHELD SPRAY GUNS

1. Check the flow rate of the spray gun by collecting spray in a bucket for 1 min.
2. Measure the time required to uniformly spray 1000 sq ft. Be sure to use the proper technique to maintain adequate coverage.
3. The application rate in gallons per 1000 sq ft is the time required to spray 1000 sq ft multiplied by the flow rate.
4. You can adjust the application rate by changing the pressure or application speed.

For example, if you collect 1.5 gal from the spray gun in 1 min, then to spray 1000 sq ft requires 2 min (1.5 gal × 2 min = 3.0 gal/1000 sq ft application rate).

8.3 CALIBRATION OF POWER SPRAYERS

There are many methods for calibrating power sprayers. Any technique for calibration that provides accurate and uniform application is acceptable. No single method is best for everyone. Sprayer calibration kits (figure 8–1) are available from nozzle manufacturers that contain all the formulas needed to calibrate your sprayer. You can also obtain tip testers for quick, in-season flow-rate checks, pattern checks, and specially coated papers that help you evaluate the uniformity of spray distribution patterns.

FIGURE 8–1   TeeJet Sprayer Calibration Kit. (Courtesy Spraying Systems Co.)

The gallons of spray applied per acre can be determined using the following equation:

$$\text{GPA} = \frac{\text{GPM} \times 5940}{\text{mph} \times w} \text{ or gallons per 1000 sp ft} = \frac{\text{GPM} \times 136}{\text{mph} \times w} \qquad (1)$$

where   GPA = output, gallons per acre
       GPM = output per nozzle, gallons per minute
       mph = ground speed, miles per hour
         w = effective sprayed width per nozzle, inches
      5940 = a constant used to convert gallons per minute, miles per hour, and inches to gallons per acre
       136 = a constant used to convert gallons per minute, miles per hour, and inches to gallons per 1000 sq ft

The following two sections use the variables in equation (1) to choose the proper-size nozzle tips for power sprayers. The steps may vary somewhat when choosing nozzles for a stationary boom versus a handheld one, but the variables from equation (1) remain constant. A thorough understanding of these variables allows you to calibrate most sprayers.

8.3.1   Selecting Proper-Size Nozzle Tips for Power Sprayers with Booms

The size of the nozzle tip depends on the application rate (GPA), ground speed (mph), and effective sprayed width (w) that you plan to use. Some manufacturers advertise gallons-per-acre nozzles, but this rating is useful only for standard conditions (usually 30 psi, 4 mph, and 20-in spacing). The GPA rating is useless if any one of your conditions varies from the standard.

An exact method for choosing the correct nozzle tip is to determine the gallons per minute (GPM) required for your conditions; then select nozzles that provide this flow rate when operated within the recommended pressure range. By following the five steps below, you can select the nozzles required for each application well ahead of the spraying season.

Step 1.   Select the spray-application rate that you want to use in gallons per acre (GPA) or in gallons per 1000 sq ft. Pesticide labels recommend ranges for various types of equipment and pesticide products. The spray-application rate is the number of gallons of carrier (water, fertilizer, and the like) and pesticide that you want to apply per acre or per 1000 sq ft.

Step 2.   Select or measure an appropriate ground speed in miles per hour (mph) according to turf conditions. Do not rely on speedometers as an accurate measure of speed. Slippage and variation in tire sizes can result in speedometer errors of 30% or more. If you do not know the actual ground speed, you can easily measure it (see the sections on measuring ground speed and tractor speed conversions in appendix B).

Step 3.   Determine the effective sprayed width per nozzle (w) in inches. For broadcast spraying, w equals the nozzle spacing.

Step 4.   Determine the flow rate required from each nozzle in gallons per minute (GPM) by using a nozzle catalog, nozzle tables, or the following equation.

$$\text{GPM} = \frac{\text{GPA} \times \text{mph} \times w}{5940} \text{ or GPM} = \frac{\text{gallons per 1000 sq ft} \times \text{mph} \times w}{136} \qquad (2)$$

where   GPM = gallons per minute of output required from each nozzle
GPA = gallons per acre from step 1
mph = miles per hour from step 2
w = inches sprayed per nozzle from step 3
5940 = a constant to convert gallons per minute, miles per hour, and inches to gallons per acre

Step 5.   Select a nozzle that will give the flow rate determined in step 4 when the nozzle is operated within the recommended pressure range. You should obtain a catalog that lists available nozzle tips. These catalogs may be obtained free of charge from equipment dealers or nozzle manufacturers (e.g., Spraying Systems Co., North Ave. at Schmale Road, P.O. Box 7900, Wheaton, IL 60189-7900 or Delavan Agricultural Products Operation, 20 Delavan Drive, Lexington, TN 38351); or check tables 7.3 to 7.11. If you decide to use the nozzles you already have, return to step 2 (see the section on determining the ground speed for nozzles installed on booms), and select a speed that allows you to operate within the recommended pressure range.

For example, if you want to broadcast a fungicide at 60 GPA (step 1) at a speed of 3.0 mph (step 2) using extended range flat-fan nozzles spaced 20 in apart on the boom (step 3), what size nozzle tip should you select? First, determine the required flow rate for each nozzle by using equation (2) in step 4:

$$\text{GPM} = \frac{\text{GPA} \times \text{mph} \times w}{5940} \text{ or GPM} = \frac{60 \times 3.0 \times 20}{5940} = 0.61$$

The nozzle you select must have a flow rate of 0.61 GPM when operated within the recommended pressure range of 20 to 30 psi. Table 7.3 shows the GPM at various pressures for several Delavan and Spraying Systems nozzles. For example, the Spraying Systems XR8008 and Delavan 80-8R nozzles have a rated output of 0.61 GPM at near 23 psi (step 5). Either of these nozzles would be suitable for this application.

The following three sections provide examples of alternative circulations that require the use of the previously described variables.

8.3.2   Selecting Proper-Size Nozzle Tips for Handheld Spray Booms

The size of the nozzle tip will depend on the application rate (gallons per 1000 sq ft), walking speed (minutes per 1000 sq ft), and effective spray width (the number of feet between the nozzles multiplied by the number of nozzles).

Choose the correct nozzle tip by determining the gallons per minute (GPM) required for your conditions; then select nozzles from a manufacturer's catalog that provide this flow rate when operated within the recommended pressure range. By following the steps outlined below, you can select the nozzles required for each application well ahead of the spraying season.

Step 1.   Determine the application rate in gallons per 1000 sq ft. Pesticide labels indicate recommended ranges for various types of equipment and pests. The spray application rate may be given in gallons of carrier (water, fertilizer, and so on) and pesticide applied per acre rather than in gallons per 1000 sq ft. Gallons per acre (GPA) can be converted to gallons per 1000 sq ft by the following equation:

$$\text{gallons per 1000 sq ft} = \frac{\text{GPA} \times 1000}{43.560} \qquad (3)$$

where 43.560 = number of square feet per acre

For example, if a rate of 50 GPA is recommended on the pesticide label, what is the application rate in gallons per 1000 sq ft?

$$\text{gallons per 1000 sq ft} = \frac{50 \times 1000}{43.560} = \frac{50.000}{43.560} = 1.15$$

As a rule of thumb, 1 gal per 1000 sq ft equals 43.56 GPA.

Step 2.   Determine the effective swath width in feet. For handheld booms, the swath width is the distance between the nozzles multiplied by the number of nozzles on the spray boom. For example, if your walking boom has three extended-range flat-fan nozzles spaced 20 in apart (20 in. equals 1.67 ft), then the effective swath width is 3 nozzles $\times$ 1.67 ft = 5 ft.

Step 3.   Measure the time in minutes required to spray 1000 sq ft. The time can easily be measured by using your swath width from step 2 and laying out a course measuring 1000 sq ft. To lay out a single-pass course that contains 1000 sq ft, use the following equation:

$$\text{distance of course} = \frac{1000 \text{ sq ft}}{\text{swath width in feet}} \tag{4}$$

For example, assume that your boom has an effective swath width of 5 ft. What length course is required for a 1000-sq-ft, single-pass course?

$$\text{distance of course} = \frac{1000 \text{ sq ft}}{5 \text{ ft}} = 200 \text{ ft}$$

Mark off a 200-ft course and time your walking speed, walking steadily along this course in both directions; then calculate your average time. For example, if one timing took 44 sec and a second took 46 sec, the average time would be 45 sec. To convert seconds to minutes, divide the average time by 60 sec per minute. In this example, the time required to spray a 200-ft course with your three-nozzle boom equals

$$\frac{45 \text{ sec}}{60 \text{ sec per min}} = 0.75 \text{ min}$$

Step 4.   Use the following equation to determine the flow rate required from each nozzle in gallons per minute (GPM).

$$\text{GPM} = \frac{\text{gallons per 1000 sq ft}}{\text{minutes per 1000 sq ft}} \tag{5}$$

For example, if the application rate is 1.15 gal per 1000 sq ft (step 1) and the time required to spray 1000 sq ft is 0.75 min (step 3), what is the required nozzle flow rate from the boom using equation (5)?

$$\text{GPM} = \frac{1.15}{0.75} = 1.5$$

Since there are three nozzles per boom, the required flow rate from the boom (1.5 GPM) would be divided by 3 (1.5 $\div$ 3 = 0.5). The required flow rate per nozzle would be 0.5 gal (or 2 quarts) per minute.

Step 5.   Select a nozzle from table 7.3 that will give the flow rate determined in step 4 when the nozzle is operated within the recommended pressure range.

Consider an example. Which extended-range flat-fan nozzle (table 7.3) would you select for your three-nozzle boom? The nozzle you select must have a flow rate of 0.5 GPM when operated within the recommended pressure range of 15 to 40 psi. Table 7.3 shows gallon-per-minute (GPM) rates at various pressures for several Delavan and Spraying Systems nozzles. For example, the Spraying Systems XR 8006 and Delavan 80-6R nozzles have a rated output of 0.5 GPM at near 29 psi. Either of these nozzles would be suitable for your boom.

Choosing the proper-size nozzle for a boom sprayer, whether stationary or handheld, is straightforward. However, at times you will need to use a sprayer already equipped with nozzles. In this situation, you may need to determine an appropriate ground speed to achieve the desired application rate; or you may need to determine the current application rate. The following sections discuss these two different situations.

Determining the ground speed for nozzles installed on booms.   If you plan to use a set of nozzles that are already on the sprayer, you will need to determine the nozzle flow rate so a ground speed can be calculated. This calculated speed allows you to achieve the proper application rate for the pesticide being used. Suppose that the pesticide you are applying requires a carrier rate of 60 gal per acre (GPA), and you are using RA Raindrop (RA-8, red) nozzles spaced on 30-in centers. Measure the flow rate in gallons per minute (GPM) from the existing nozzles at a pressure within the recommended range for that nozzle type. Using that flow rate, insert the measured GPM flow rate into the following formula. In this example, 0.70 GPM was collected at a pressure of 30 psi.

$$\text{mph} = \frac{\text{GPM} \times 5940}{\text{GPA} \times w \text{ (inches)}} = \frac{0.70 \times 5940}{60 \times 30 \text{ (inches)}} = 2.3 \tag{6}$$

The calculated speed is 2.3 mph. If you travel 2.3 mph using the RA-8 Raindrop nozzles on 30-in spacings at 30 psi, then the application rate will be the desired 60 GPA.

Determining the spray rate for nozzle tips installed on booms.   You may already have a set of nozzle tips in your boom, and you want to know the spray rate (GPA) when operating at a particular nozzle pressure and speed.

Add water to the spray tank and make a precalibration check to ensure that all spray components are working properly. Remember that the type, size, and fan angle of all nozzle tips must be the same, and the flow rate from each nozzle must be within 5% of the average flow rate from the other nozzles.

Step 1.   Operate the sprayer at the desired operating pressure. Use a quart jar (table 8–1) or other suitable container marked in ounces to collect the output from a nozzle for a measured length of time (such as 1 min). Check the other nozzles to determine the average number of ounces per minute of output from each nozzle.

Step 2.   Convert ounces per minute (OPM) of flow to gallons per minute (GPM) by dividing the OPM by 128 (the number of ounces in 1 gal).

Step 3.   Determine the spraying speed. For mounted boom sprayers, the speed in mph can easily be measured (see the section on measuring ground speed in appendix B). For hand-operated booms, lay out 1000 sq ft, and record the time required to cover the area uniformly.

Step 4.   Determine the sprayed width per nozzle (w) in inches. For broadcast spraying, w equals the nozzle spacing.

Step 5.   For mounted boom sprayers, use equation (1) to calculate the sprayer application rate in GPA:

$$\text{GPA} = \frac{\text{GPM} \times 5940}{\text{mph} \times w}$$

For instance, if the measured nozzle output is 54 ounces per minute (OPM), the measured ground speed is 6 mph, and the nozzle spacing (w) is 20 in, what is the sprayer application rate?

First, convert OPM to GPM (step 2) by dividing the OPM by 128 (the number of ounces in 1 gal):

$$\text{GPM} = \frac{54}{128} = 0.42$$

Using equation (1), calculate the application rate in GPA as follows:

$$\text{GPA} = \frac{42 \times 5940}{6 \times 20} = 20.8$$

For hand-operated walking booms, use the following equation to calculated the application rate.

$$\text{gallons per 1000 sq ft} = \text{GPM} \times \text{minutes per 1000 sq ft} \qquad (7)$$

Therefore, if the measured nozzle output is 0.6 GPM and 2 min are required to spray 1000 sq ft, the application rate equals 1.2 gal per 1000 sq ft (0.6 × 2).

The application rate (step 5) can be adjusted by changing the ground speed or nozzle pressure and recalibrating. Changes in nozzle pressure should be used *only* to make small changes in output. These changes must be within the recommended pressure range.

A final caution: Check your use patterns frequently. If you know the size of your turf areas (see appendix C) and how much spray should be applied, does this match your uses?

8.3.3   Calibration of Power Sprayers with Booms

Precalibration checking.   After making sure that your sprayer is clean, install the selected nozzle tips, partially fill the tank with clean water, and operate the sprayer at a pressure within the recommended range. Place a quart jar or other suitable

FIGURE 8–2 Nozzle flow device.

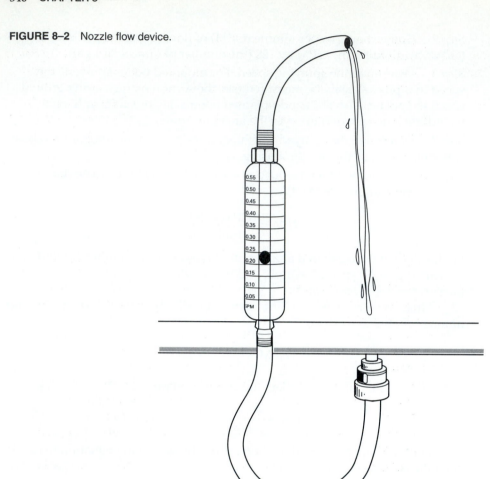

container under each nozzle (see table 8–1), or use a nozzle flow-rate device (figure 8–2). The nozzle flow-rate device will monitor nozzle flow in GPM. Check to see whether all the containers fill at about the same time or the flow rate measured from each nozzle is equal. Replace any nozzle that has an output of 5% more or less than the average of all the nozzles, an obviously different fan angle, or a nonuniform appearance in spray pattern.

To obtain uniform coverage, consider the spray angle, spacing, and height of the nozzle that you are using. The height must be readjusted for uniform coverage with various spray angles and nozzle spacing. Do *not* use nozzles with different spray angles on the same boom for broadcast spraying.

Worn or partially plugged nozzles and misalignment of nozzle tips will produce nonuniform patterns. Skips and uneven coverage also result if one end of the boom is allowed to raise or droop. An easy method for determining the exact nozzle height that will produce the most uniform coverage is to spray a warm, dry surface (such as a paved road or concrete drive) and observed the drying rate. Adjust the height to eliminate excess streaking.

Calibrating your sprayer.   After selecting and installing the proper nozzle tips, you are ready to complete the calibration of your sprayer. Check the calibration every few days during the season or when changing the pesticides being applied. New nozzles do *not* lessen the need to calibrate, because nozzles wear in and their flow rate increases most rapidly during the first few hours of use. Once you have learned the following calibration method, check application rates quickly and easily using the following five steps.

Step 1.   Determine the required flow rate for each nozzle in ounces per minute (OPM). To convert the gallons per minute (step 4) to OPM, use the following equation:

$$OPM = GPM \times 128 \tag{8}$$

where 128 = the number of ounces in 1 gallon. For example, if the required nozzle flow rate = 0.56 GPM, the required OPM = 0.56 × 128 = 71.7.

Step 2.   Collect the output from one of the nozzles in a container marked in ounces, or use the flow-rate device (figure 8–2) to measure flow in actual GPM. Adjust the pressure until the amount you collect is within plus or minus 5% of the desired number of ounces per minute (OPM) determined in step 1. If it is impossible to obtain the desired output within the recommended range of operating pressures, select larger or smaller nozzle tips, and recalibrate. It is important that spray nozzles be operated within the recommended pressure range. (The range of operating pressures is for pressure at the nozzle *tip*. Pressure losses in hoses or booms, nozzle check valves, and so on, may cause the pressure gauge to read much higher to obtain the proper flow rate.)

Step 3.   Determine the amount of pesticide needed for each tankful or for the acreage to be sprayed (see the section on mixing pesticides). Add the pesticide to a tank partially filled with a carrier (water, fertilizer, and the like); and then add more carrier to the desired level with continuous agitation.

Step 4.   Operate the sprayer at the measured ground speed and pressure as determined in step 2. You will be spraying at the application rate selected in step 1. After spraying an area of known size, check the liquid level in the tank to verify that the application rate is correct.

Step 5.   Check the nozzle flow rate frequenly. Adjust the pressure to compensate for small changes in nozzle output resulting from nozzle wear or variations in other spraying components. Replace the nozzle tips and recalibrate when the output has changed 10% or more from that of a new nozzle or when the pattern becomes uneven.

8.3.4   Mixing Pesticides

To determine the amount of pesticide to add to a spray tank, you need to know the recommended rate of pesticide, the capacity of the spray tank, and the calibrated output of the sprayer.

The recommended application rate of the pesticide is given on the container lable. The rate is usually indicated as ounces per 1000 sq ft or pounds per acre for wettable powders, and as pints, quarts, or gallons per 1000 sq ft or acre for liquids.

Dry Formulations

Example 1.   An insecticide recommendation calls for 2 lb off active ingredient (a.i.) per acre. You have purchased an 80% wettable powder (80 WP). Your sprayer has a 125-gal tank and is calibrated to apply 20 GPA. How much insecticide should you add to the spray tank?

Step 1.   Determine the number of acres you can spray with each tankful. Your sprayer has a 125-gal tank and is calibrated to apply 20 GPA:

$$\text{acres per tankful} = \frac{\text{tank capacity}}{\text{GPA}} = \frac{125 \text{ gal}}{20 \text{ GPA}} = 6.3 \text{ acres} \tag{9}$$

Step 2.   Determine the pounds of pesticide product needed per acre. Because only part of the material in the bag is an active ingredient, you have to add more than 2 lb of the product to each acre's worth of water in your tank. But how much more? The calculation is simple: Divide the percentage of active ingredient (80) into the total (100):

$$\text{pounds product per acre} = 2 \text{ lb per acre} \times \frac{100\%}{80\%} = 2.5 \text{ lb} \tag{10}$$

Step 3.   Determine the amount of pesticide to add to each tankful. With each tankful, you will cover 6.3 acres (step 1), and you want 2.5 lb of product per acre (step 2). Add 15.7 lb (6.3 acres × 2.5 lb per acre = 15.7 lb) of the 80 WP product to each tankful.

Pounds product per tank = 6.3 acres per tank × 2.5 lb product per acre = 15.7 lb.

Example 2.   An insecticide recommendation is 4 lb per acre. Your 4-gal compressed-air (pump-up) sprayer applies 1.25 gal per 1000 sq ft. How many ounces should you add to the spray tank?

Step 1.   Convert the recommended rate to ounces per 1000 sq ft by one of the following formulas:

$$\text{pounds/acre} \times 0.37 = \text{ounces/1000 sq ft for a dry material} \tag{11a}$$

$$\text{quarts/acre} \times 0.73 = \text{ounces/1000 sq ft for a liquid material} \tag{11b}$$

Therefore,

$$4 \text{ lb/acre} \times 0.37 = 1.48 \text{ ounces/1000 sq ft}$$

Step 2.   After you have converted the recommended rate to ounces per 1000 sq ft, you can find the amount of pesticide you should add to each tankful by using the following formula:

$$\text{ounces per tankful} = \frac{\text{gallons per tankful} \times \text{ounces per 1000 sq ft}}{\text{gallons applied per 1000 sq ft}}$$

$$= \frac{4 \text{ gal per tankful} \times 2.9 \text{ ounces per 1000 sq ft}}{5 \text{ gal applied per 1000 sq ft}} \qquad (12)$$

Ounces per tankful = 4.7.

Liquid Formulations

Example 1.   A preemergence crabgrass control is recommended at 1 lb of active ingredient (a.i.) per acre. You have purchased a product formulated at 4 lb per gallon. Your sprayer has a 150-gal tank and is calibrated at 15 gal per acre. How much product should you add to the spray tank?

Step 1.   Determine the number of acres that you can spray with each tankful using

$$\text{acres per tankful} = \frac{\text{tank capacity}}{\text{GPA}} = \frac{150 \text{ gal}}{15 \text{ GPA}} = 10 \text{ acres}$$

Step 2.   Determine the amount of product needed per acre by dividing the recommended active ingredient per acre by the concentration of the formulation.

$$\text{gallons per acre} = \frac{\text{pounds a.i. per acre}}{\text{pounds a.i. per gallon}}$$

$$= \frac{1 \text{ lb a.i. per acre}}{4 \text{ lb a.i. per gallon}} = \frac{1}{4} \text{ gal (1 quart)} \qquad (13)$$

One-fourth gallon (1 quart) of product is needed for each acre's worth of water in the tank to apply 1 lb of a.i. per acre.

Step 3.   Determine the amount of pesticide to add to each tankful. With each tankful, you will cover 10 acres (step 1), and you need ¼ gal or 1 quart of product per acre (step 2). Add 10 quarts (10 acres × 1 quart per acre of product to each tankful.

Gallons of product per tankful = 10 acres/tankful × ¼ gal product/acre = 2.5 gal (10 quart).

Example 2.   The insecticide recommendation calls for 1 gal (4 quarts) of product per acre. You have a 4-gal knapsack sprayer that has been calibrated to apply ½ gal per 1000 sq ft. How many ounces should you add to the spray tank?

Step 1.   Convert the recommended rate to ounces per 1000 sq ft. Use either pounds/acre × 0.37 = ounces/1000 sq ft for a dry material, or quarts/acre × 0.73 = ounces/1000 sq ft for a liquid material. Therefore, 4 quarts/acre × 0.73 = 2.9 ounces/1000 sq ft.

Step 2.   Determine the amount of pesticide to add to each tankful by using equation (12):

$$\text{ounces per tankful} = \frac{\text{gallons per tankful} \times \text{ounces per 1000 sq ft}}{\text{gallons applied per 1000 sq ft}}$$

$$= \frac{4 \text{ gal per tankful} \times 1.48 \text{ ounces per 1000 sq ft}}{1.25 \text{ gal applied per 1000 sq ft}}$$

8.4   CALIBRATION OF GRANULAR APPLICATION

Many suppliers recommend spreader settings and swath widths for their products and several products of their principal competitors. Use these *only* as initial guides for calibration runs prior to actual use. Every drop or rotary spreader should be calibrated for proper delivery rate with a particular product and operator due to variability in the flow characteristics of the product, the operator's walking speed, and environmental conditions. Calibration should be checked and corrected according to the manufacturer's directions at least once a week; more often if the spreader has received mechanical damage.

An easy method for checking the delivery rate of a pesticide (or fertilizer) spreader is to apply (spread) a weighed amount of product on a measured area (at least 1000 sq ft for a drop spreader and 5000 sq ft for a rotary spreader) and then weigh the product remaining in the spreader to determine the rate actually delivered.

To avoid contamination of the turf area during initial calibration, attach a catch pan or bag (most commercial drop spreaders come with a calibration pan) under the spreader and push the spreader a measured distance, usually on a driveway or sidewalk, at a uniform speed. Be sure to hang the container so there is no interference with the shutoff bar or rate-control linkage.

Check both the distribution pattern and the flow rate of a rotary spreader. The product manufacture may recommend a particular swath width, but verify this width before treating a large turf area. Check the pattern by laying out a row of shallow pans, with the same dimensions, in a row at right angles to the direction of travel in an open or paved area *not* on the turf. For commercial, push-type, rotary spreaders, the pans should be 1 to 2 in high with an area of about 1 sq ft, and they should be spaced on 1-ft centers. The row of pans should cover 1½ to 2 times the anticipated effective swath width. Add granules to the spreader, set it at the recommended setting for rate and pattern. Run the spreader at the speed to be used in active operation and make three passes in the same direction over the pans (figure 8–3). The material caught in the individual pans can then be weighed and a distribution pattern plotted (figure 8–4).

A simpler method for checking the distribution pattern is to pour the material from each pan into a small bottle. When the bottles are placed side by side in the proper order, a plot of the pattern can easily be seen. The pattern can be used to detect the correct skewing and determine the effective swath width (twice the distance out to a point where the rate is one-half the average rate at the center). For example, if the center three or four bottles have material 2 in deep and the bottles

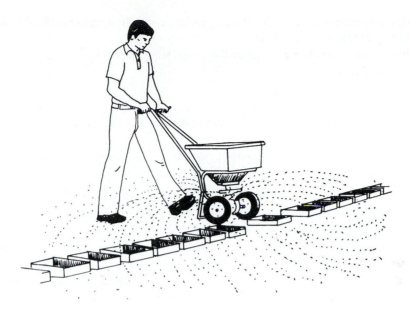

FIGURE 8–3 Pattern check pans. To make a quick pattern check when applying granules, lay out a series of shallow pans or cardboard boxes in a line perpendicular to the direction of travel, and make three passes over the containers.

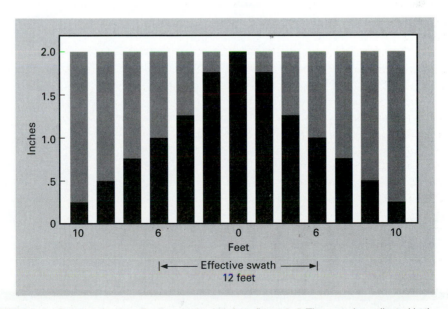

FIGURE 8–4 Determining the effective swath width from figure 8–3. The granules collected in the pans or boxes have been poured into a series of tall bottles. In this case, the effective swath width is 12 ft.

6 ft to the left and 6 ft to the right of the spreader centerline have material 1 in deep, the effective swath width is 12 ft.

For a conversion table for granular rates, see appendix A.

8.5 STUDY QUESTIONS

1. Outline how you would go about calibrating a manual sprayer to deliver 2 gal per 1000 sq ft and 40 gal per acre.
2. How does calibration of a handheld spray gun differ from a manual sprayer? What would you do to ensure applying 2 gal per 1000 sq ft?
3. What is the main advantage of using a sprayer calibration kit?
4. What equation would you use to calibrate a power sprayer to deliver 2 gal per 1000 sq ft and 40 gal per acre?
5. The constants 5940 and 136 are used in determining gallons per acre and 1000 sq ft, respectively. Why are the figures used?
6. How does the size of the nozzle tip used in power spray booms differ from selecting a nozzle tip for a handheld spray boom?
7. Why should the type, size, and fan angle of all nozzle tips on a spray boom be the same?
8. What do you need to know to determine the amount of pesticide to add to a spray tank?
9. When calibrating granular applicators why should you use at least 5000 sq ft for a rotary spreader and only 1000 sq ft for a drop spreader?
10. Describe how you would measure the distribution pattern of a rotary spreader.
11. What does a nozzle flow device do?

Appendix: Measurements and Conversions

WEIGHTS

1 gram (g) = 1000 milligrams = 0.035274 oz avoirdupois = 0.0022046 pound = 0.001 kilogram

1 ounce avoirdupois (oz av) = 0.911458 troy oz = 28.349527 grams = 0.0625 pound

1 ounce (troy) = 1.097 oz av = 31.10348 grams

1 kilogram (kg) = 1000 grams = 35.273957 oz av = 2.20462 pounds (av)

1 pound avoirdupois (lb av) = 16 oz av = 14.5833 troy oz = 453.5924 grams = 0.4535924 kilogram

1 ton (short, U.S.) = 2000 pounds (av) = 2430.56 troy pounds = 0.892857 long ton = 0.907185 metric ton = 907.18486 kilograms = 32,000 oz

1 ton (long, U.S.) = 2240 pounds = 2722.22 troy pounds = 1.120 short tons = 1.016 metric tons = 1016.04 kilograms

1 ton (metric) = 2204.6 pounds = 1.1023 short tons = 0.984 long ton = 1000 kilograms

SQUARE (OR SURFACE) MEASURES

1 square millimeter (sq mm) = 0.01 sq centimeter = 0.000001 sq meter = 0.00155 sq in.

1 square centimeter (sq cm) = 100 sq millimeters = 0.155 sq inch = 0.001076 sq foot

1 square inch (sq in) = 6.451626 sq centimeters = 0.0069444 sq ft

1 square foot (sq ft) = 144 sq inches = 0.111111 sq yard = 0.0929 sq meter = 0.003673 sq rod

1 square yard (sq yd) = 9 sq ft = 1296 sq inches = 0.83613 sq meter = 0.03306 sq rod

1 square meter (sq m) = 10.76387 sq ft = 1550 sq inches = 1.195985 sq yards = 0.039537 sq rod = 1 million sq millimeters = 10,000 sq centimeters

1 square rod (sq rod) = 30.25 sq yards = 25.29295 sq meters = 272.25 sq feet = 0.00625 sq acre = 0.0025293 hectare

1 acre = 43,560 sq feet = 4840 sq yards = 160 sq rods = 4046.873 sq meters = 0.404687 hectare = 0.0015625 sq mile = strip 8 ft wide and 1 mile long (approximate)

1 hectare = 2.471 acres = 395.367 sq rods = 10,000 sq meters = 0.01 sq kilometer = 0.0039 sq mile

1 square kilometer = 0.3861 sq mile = 247.1 acres = 100 hectares = 1 million sq meters

1 square mile = 640 acres = 1 section = 258.9998 hectares = 102,400 sq rods = 3,097,600 sq yards = 2.589998 sq kilometers

CUBIC MEASURES

1 cubic centimeter (cc) = 0.06102 cu inch = 1000 cu milliliters = 0.000001 cu meter

1 cubic inch (cu in) = 16.38716 cubic centimeters = 0.0005787 cu foot = 0.004329 gallon (U.S.)

1 cubic foot (cu ft) = 0.80356 bushel = 1728 cu inches = 0.037037 cu yard = 0.028317 cu meter = 7.4805 U.S. gallons = 6.229 British or Imperial gallons = 28.317 liters = 29.922 quarts (liquid) = 25.714 quarts (dry)

1 cubic foot of water = 62.43 pounds (1 pound of water = 27.68 cu inches = 0.1198 U.S. gallon = 0.01602 cu foot)

1 cubic foot (cu ft) of *dry* soil (approximate) = sandy (90 lb), loamy (80 lb), clay (75 lb)

1 bushel (bu) *dry* soil (approximate) = sandy (112 lb), loamy (100 lb), clay (94 lb)

1 cubic yard (cu yd) = 27 cu ft = 46,656 cu inches = 764.559 liters = 292 U.S. gallons = 168.176 British gallons = 1616 pints (liquid) = 807.9 quarts (liquid) = 21.694 bushels = 0.764559 cubic meter

1 cubic meter (cu m) = 1.30794 cu yard = 35.3144 cu feet = 28.3776 bushels = 264.173 gallons = 1056.7 quarts (liquid) = 2113.4 pints (liquid) = 61,023 cu inches

VOLUMES

Dry

1 quart (U.S.)* (67.2 cu inches = 2 pints = 1.1012 liters = 0.125 peck = 0.03125 bushel = 0.038889 cu foot = 67.2 cu inches

1 peck (U.S.) = 0.25 bushel = 2 gallons = 8 quarts = 16 pints = 32 cups = 8.80958 liters = 537.605 cu inches

1 bushel (U.S.) = 4 pecks = 32 quarts = 64 pints = 128 cups = 1.2445 cu feet = 35.2383 liters = 0.304785 barrel = approximately 1/20 cubic yard

---

* 1 pint or quart dry measure is about 16% larger than 1 pint or quart liquid measure.

Liquid

1 milliliter (ml) = 1 cubic centimeter (approximate) = 0.001 liter = 0.061 cu inch = 0.03815 fluid ounce

1 teaspoonful (tsp) = 5 milliliters = 0.17 fluid ounce

1 tablespoonful (Tbsp) = 3 teaspoons = 14.8 milliliters or cubic centimeters = ½ fluid ounce = 0.902 cu inch = 0.063 cup

1 fluid ounce (fl oz, U.S.) = 2 tablespoons = 0.125 cup = 0.0625 pint = 0.03125 quart = 0.00781 gallon = 29.573 millimeters = 1.80469 cu inches = 0.029573 liter

1 cup = 16 tablespoons = 8 fluid ounces = 0.5 pint = 236.6 cubic centimeters or milliliters

1 pint (U.S.)* = 16 fluid ounces = 32 tablespoons = 2 cups = 0.125 gallon = 473.167 milliliters = 1.04 pounds of water = 28.875 cu inches = 0.473167 liter = 0.01671 cu foot

1 liter = 2.1134 pints = 1.0567 liquid quarts (U.S.) = 0.9081 dry quart (U.S.) = 0.264178 gallon (U.S.) = 1000 milliliters or cubic centimeters = 33.8147 fluid ounces = 61.025 cu inches = 0.0353 cu foot = 0.028378 bushel

1 quart (U.S.)* = 2 pints = 4 cups = 32 fluid ounces = 57.749 cu inches = 64 tablespoons = 0.25 gallon = 0.946333 liter = 0.3342 cu foot

1 gallon (U.S.) = 4 quarts = 8 pints = 16 cups = 128 fluid ounces = 0.1337 cu foot = 0.83268 British or Imperial gallon = 3785.4 milliliters or cubic centimeters = 231 cu inches = 8.337 pounds water = 3.782 kilograms

1 barrel (U.S.) = 31.5 gallons = 7056 cu inches = 0.11924 cu meter

LINEAR (OR DISTANCE) MEASURES

1 millimeter (mm) = 0.1 centimeter = 0.01 decimeter = 0.001 meter = 1000 microns = 0.03937 inch (about ⅟₂₅ inch)

1 centimeter (cm) = 10 millimeters = 0.01 meter = 0.3937 inch = 0.03281 foot = 0.010936 yard (U.S.)

1 inch (in) = 25.4 millimeters = 2.54 centimeters = 0.0254 meter = 0.083333 foot = 0.027778 yard = 0.00505 rod

1 foot (ft) = 12 inches = 0.333 yard = 30.48 centimeters = 0.3048 meter

1 yard (yd) = 36 inches = 3 feet = 0.181818 rod = 0.9144 meter

1 meter (m) = 100 centimeters = 39.37 inches = 3.2808 feet = 1.09361 yards = 0.1988 rod = 0.001 kilometer

1 rod = 5.5 yards = 16.5 feet = 198 inches = 5.02921 meters = 0.003125 mile

1 kilometer (km) = 1000 meters = 3280.8 feet = 1093.6 yards = 0.62137 statute mile = 0.53961 nautical mile

1 mile (statute) = 5280 feet = 1760 yards = 1609.35 meters = 320 rods = 0.86836 nautical mile

---

* 1 pint or quart dry measure is about 16% larger than 1 pint or quart liquid measure.

DILUTIONS

1 part per million (ppm) = 1 milligram per liter or kilogram = 0.0001% = 0.013 ounce by weight in 100 gallons = 0.379 gram in 100 gallons = 1 inch in nearly 16 miles = 2 crystals of sugar in a pound = 1 ounce of salt in 62,500 pounds of sugar = 1 minute of time in about 2 years = a 1-gram needle in a 1-ton haystack = 1 ounce of sand in 34¼ tons of cement = 1 ounce of dye in 7530 gallons of water = 1 pound in 500 tons = 1 penny of $10,000

1 part per billion (ppb) = 1 inch in nearly 16,000 miles = 1 drop in 20,000 gallons = 1 ounce in 753 million gallons = 2 crystals of sugar in 1000 pounds

1 percent (%) = 10,000 parts per million = 10 grams per liter = 1.28 ounces by weight per gallon = 8.336 pounds per 100 gallons

MISCELLANEOUS WEIGHTS AND MEASURES

1 micron = 0.00039 inch

1 acre inch of water = 27,154 gallons = 624.23 gallons per 1000 sq feet

1 pound per cubic foot = 0.26 gram per cubic inch

1 gram per cubic inch = 3.78 pounds per cubic foot

WEIGHTS PER UNIT AREA

1 ounce per square foot = 2722.5 pounds per acre

1 ounce per square yard = 302.5 pounds per acre

1 ounce per 100 square feet = 27.2 pounds per acre

1 ounce per 1000 square feet = 2.72 pounds per acre

1 pound per 100 square feet = 435.6 pounds per acre

1 pound per 1000 square feet = 43.56 pounds per acre

1 pound per acre = 1 ounce per 2733 square feet (0.37 oz/1000 sq ft) = 0.0104 g per sq ft = 1.12 kilograms per hectare

100 pounds per acre = 2.5 pounds per 1000 square feet = 1.04 g per sq ft

5 gallons per acre = 1 pint per 1000 square feet = 0.43 ml per square foot

100 gallons per acre = 2.5 gallons per 1000 square feet = 1 quart per 100 square feet = 935 liters per hectare

1 quart per 100 gallons (approximate) = 10 ml per gallon

1 pound per gallon = 120 grams per liter

1 kilogram per 100 square meters = 2.05 pounds per 1000 square feet = 1 kilogram per hectare = 89 pounds per acre

1 kilogram per hectare = 0.0205 pound per 1000 square feet = 0.1 kilogram per 100 square meters = 0.89 pound per acre

CONVERTING TEMPERATURE FROM FAHRENHEIT TO CELSIUS (CENTIGRADE), AND VICE VERSA

To convert from Fahrenheit to Celsius: subtract 32 from the Fahrenheit reading, multiply by 5, and divide the product by 9. *Example:* $131°F - 32 = 99 \times 5 = 495$; $495 \div 9 = 55°C$.

To convert from Celsius to Fahrenheit: multiply the Celsius reading by 9, divide the product by 5, and add 32. *Example:* $25°C \times 9 = 225 \div 5 = 45$; $45 + 32 = 77°F$.

EQUIVALENT VOLUMES (LIQUID) FOR COMMON MEASURES

Measuring unit used	tsp	Tbsp	cup	pint	cc	liter
1 teaspoonful	1.00	0.33	0.021	0.010	4.9	0.0049
1 tablespoonful	3.00	1.00	0.663	0.031	14.8	0.0148
1 fluid ounce	6.00	2.00	0.125	0.062	29.6	0.0296
1 cup	48.00	16.00	1.000	0.500	236.6	0.2366
1 pint	96.00	32.00	2.000	1.000	473.2	0.4732
1 quart	192.00	64.00	4.000	2.000	946.3	0.9463
1 gallon	768.00	256.00	16.000	8.000	3,785.3	3.7853
1 liter	202.88	67.63	4.328	2.164	1,000.0	1.0000
1 milliliter (cc)	0.20	0.068	0.0042	0.0021	1.0	0.0010

Number of units to fill measure in the first column

*M*ETRIC-*E*NGLISH *C*ONVERSION *F*ACTORS

To change	To	Multiply by
Inches	Centimeters	2.54
Centimeters	Inches	0.3937
Feet	Meters	0.3048
Meters	Inches	39.37
Square inches	Square centimeters	6.452
Square centimeters	Square inches	0.155
Square yards	Square meters	0.836
Square meters	Square yards	1.196
Cubic yards	Cubic meters	0.765
Cubic meters	Cubic yards	1.308
Cubic inches	Cubic centimeters	16.387
Cubic centimeters	Cubic inches	0.061
Cubic centimeters	Fluid ounces	0.034
Fluid ounces	Cubic centimeters	29.57
Quarts	Liters	0.946
Liters	Quarts	1.057
Grams	Ounces (avoirdupois)	0.0352
Ounces (avoirdupois)	Grams	28.349
Pounds (avoirdupois)	Kilograms	0.454
Kilograms	Pounds (avoirdupois)	2.2046
Pounds (apothecary)	Kilograms	0.373
Kilograms	Pounds (apothecary)	2.205
Ounces (apothecary)	Grams	31.103
Grams (sq ft)	Pounds (acre)	96

Appendix: Calculating Land Areas

Rectangle, square.   The most common shape of a turf area is that of a rectangle or square. A rectangle is a four-sided figure (figure B–1) with opposite sides parallel in which adjacent sides make angles of 90 degrees with each other. A square is a rectangle with all sides equal. To compute the area of a rectangle, multiply the length by the width, as follows (assume in figure B–1 a length of 125 ft and a width of 85 ft):

$$area = 125 \times 85 = 10,880 \text{ sq ft}$$

$$area \text{ in acres} = \frac{10,880}{43,560} = 0.25 \text{ acre}$$

Four-sided figure with two sides parallel.   The area is found by multiplying the average length of the parallel sides (P_1, P_2), by the perpendicular or shortest distance between them (h). See figure B–2.

$$area = \frac{P_1 + P_2}{2} \times h$$

Assuming that P_1 measures 280 ft, P_2 measures 350 ft, and h measures 180 ft:

$$area = \frac{280 + 350}{2} \times 180$$

$$= 43,700 \text{ sq ft}$$

$$area \text{ in acres} = \frac{43,700}{43,560} = 1,003 \text{ acres}$$

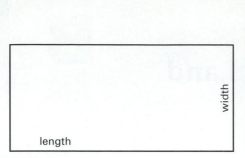

length
width

FIGURE B–1

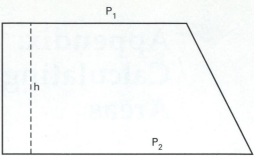

P_1

h

P_2

FIGURE B–2

Right triangle.   The area of a right triangle (figure B–3) is

$$\frac{\text{base} \times \text{altitude}}{2}$$

Assume in figure B–3 a base of 375 ft and an altitude of 450 ft.

$$\text{area} = \frac{375 \times 450}{2} \times 84{,}375 \text{ sq ft}$$

$$\text{area in acres} = \frac{84{,}375}{43{,}560} = 1.937 \text{ acres}$$

Triangle.   The area of a triangle (figure B–4) is

$$\frac{\text{base} \times \text{altitude}}{2}$$

If the base is 500 ft and the height is 350 ft, then

$$\text{area} = \frac{500 \times 350}{2} \times 87{,}500 \text{ sq ft}$$

or

$$\frac{87{,}500}{43{,}560} = 2.008 \text{ acres}$$

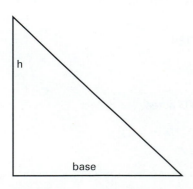

h

base

FIGURE B–3

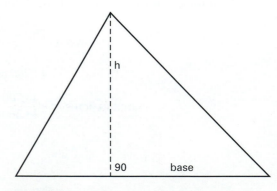

h

90   base

FIGURE B–4

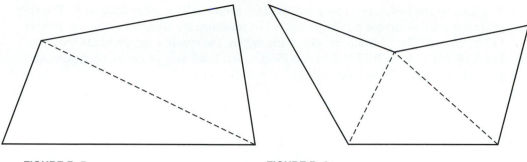

FIGURE B–5   **FIGURE B–6**

Four-sided area; no sides parallel.   This type of figure should be divided into two triangles by a diagonal (figure B–5), and then the area of each triangle is computed. The area sum of both triangles will be the area of the figure.

Any number of sides.   Any turf area bounded by more than four sides may be divided into a series of triangles by diagonal lines (figure B–6) and the area of each triangle computed as before. The sum of the triangular areas is the area of the figure.

Irregularly shaped area.   The simplest way to measure an odd-shaped area (figure B–7) is to reduce it to simple geometrical forms, then work out the different areas. The area of the triangle is the base multiplied by h divided by 2.

$$\text{area} = \frac{120 \times 45}{2} \times 2700 \text{ sq ft}$$

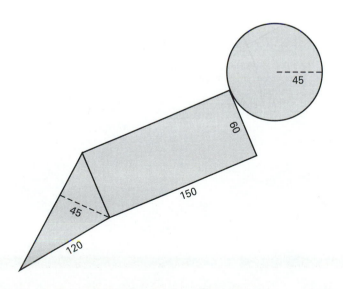

FIGURE B–7

The area of the rectangle is length $\times$ width. Area $= 150 \times 60 = 9000$ sq ft. The remaining area is almost a perfect circle. To measure its area, measure the radius (45 ft) and use the formula πr^2 (45×45), where the symbol π represents the number 3.14. ($45 \times 45 = 2025 \times 3.14 = 6358.5$ sq ft). Total area of circle, rectangle, and triangle $= 2700 = 9000 + 6358.5/43{,}560 = 0.415$ acre

Appendix: Land-Grant Institutions and Agricultural Experiment Stations in the United States

C

For help in diagnosing and controlling turfgrass problems, write to the extension horticulturist or extension agronomist (turfgrass specialist) to answer questions on turfgrass management and weeds, the extension entomologist (insects, mites), and the extension plant pathologist (diseases). Write to the specialist at the college of agriculture of your land-grant university or to your state experiment station. Turfgrass bulletins, circulars, fact sheets, and spray schedules are often available free. Write to the bulletin room or mailing clerk, college of agriculture, at your state university.

ALABAMA: Auburn University, Auburn, AL 36849.

ALASKA: University of Alaska, Fairbanks, AK 99775; Experiment Station, Palmer, AK 99645.

ARIZONA: University of Arizona, Tucson, AZ 85721; University of Arizona Research Station, Route 1, Box 587, Yuma, AZ 85365.

ARKANSAS: University of Arkansas, Fayetteville, AR 72701; Cooperative Extension Service, P.O. Box 391, Little Rock, AR 72203; Southeast Research Center, P.O. Box 3508, Monticello, AR 71655.

CALIFORNIA: University of California, Berkeley, CA 94720; Riverside, CA 92521; Davis, CA 95616; Kearney Agricultural Center, 9240 S. Riverbend Ave., Parlier, CA 92648.

COLORADO: Colorado State University, Fort Collins, CO 80523.

CONNECTICUT: University of Connecticut, Storrs, CT 06268; Connecticut Agricultural Experiment Station, P.O. Box 1106, New Haven, CT 06512.

DELAWARE: University of Delaware, Newark, DE 19711.

DISTRICT OF COLUMBIA: University of the District of Columbia, Cooperative Extension Service, Washington, DC 20002.

FLORIDA: University of Florida, Gainesville, FL 32611; Agricultural Research & Education Center, 18905 SW 280 St., Homestead, FL 33031; Agricultural Research & Education Center, 3205 SW 70 Ave., Ft. Lauderdale, FL 33314; Agricultural Research & Education Center, P.O. Box Drawer A, Belle Glade, FL 33430; North Florida Research Center, Route 3, Box 4370, Quincy, FL 32351; SW Florida Research and Education Center, P.O. Drawer 5127, Immakalee, FL 33934.

GEORGIA: University of Georgia, Athens, GA 30602; Agricultural Experiment Station, Experiment, GA 30212; Rural Development Center, P.O. Box 1209, Tifton, GA 31793; University of Georgia, Griffin, GA 30223.

GUAM: University of Guam, Cooperative Extension Service, Box EK, Agana, GU 96910.

HAWAII: University of Hawaii, Honolulu, HI 96822; Hilo, HI 96720.

IDAHO: University of Idaho, Extension Service, Boise, ID 83702; Agricultural Experiment Station, Moscow, ID 83843; Research & Education Center, Kimberly, ID 83341; Research & Extension Center, Parma, ID 83660; Research & Extension Center, Caldwell, ID 83605; University of Idaho, 1330 Filer Ave. E, Twin Falls, ID 83301.

ILLINOIS: University of Illinois, Urbana, IL 61801.

INDIANA: Purdue University, West Lafayette, IN 47907.

IOWA: Iowa State University, Ames, IA 50011.

KANSAS: Kansas State University, Manhattan, KS 66506.

KENTUCKY: University of Kentucky, Lexington, KY 40546; Research & Education Center, P.O. Box 469, Princeton, KY 42445.

LOUISIANA: Louisiana State University, University Station, Baton Rouge, LA 70803.

MAINE: University of Maine, Orono, ME 04473; Aroostook County Extension Office, Houlton Road, Presque Isle, ME 04769.

MARYLAND: University of Maryland, College Park, MD 20742; Vegetable Research Farm, Quantico Road, Route 5, Salisbury, MD 21801.

MASSACHUSETTS: University of Massachusetts, Amherst, MA 01003.

MICHIGAN: Michigan State University, East Lansing, MI 48824.

MINNESOTA: University of Minnesota, St. Paul, MN 55108.

MISSISSIPPI: Mississippi State University, Mississippi State, MS 39762.

MISSOURI: University of Missouri, Columbia, MO 65211; Delta Center, Box 160, Portageville, MO 63873.

MONTANA: Montana State University, Bozeman, MT 59717.

NEBRASKA: University of Nebraska, East Campus, Lincoln, NE 68583; Panhandle Station, 4502 Ave. 1, Scottsbluff, NE 69361; South Central Station, Box 66, Clay Center, NE 65833.

NEVADA: University of Nevada, Reno, NV 89557.

NEW HAMPSHIRE: University of New Hampshire, Durham, NH 03824.

NEW JERSEY: Rutgers, The State University of New Jersey, New Brunswick, NJ 08903; Rutgers Research & Development, R.R. #5, Bridgeton, NJ 08302.

NEW MEXICO: New Mexico State University, Las Cruces, NM 88003.

NEW YORK: Cornell University, Ithaca, NY 14853; Agricultural Experiment Station, Geneva, NY 14456; Highland Fruit Laboratory, Highland, NY 12528; Long Island Horticultural Research Laboratory, Riverhead, NY 11901.

NORTH CAROLINA: North Carolina State University, Raleigh, NC 27695; Mountain Horticulture Crops Research Station, Fletcher, NC 28732.

NORTH DAKOTA: North Dakota State University, Fargo, ND 58105.

OHIO: The Ohio State University, Columbus, OH 43210; Ohio Agricultural Research & Development Center, 1680 Madison Ave., Wooster, OH 44691.

OKLAHOMA: Oklahoma State University, Stillwater, OK 74078.

OREGON: Oregon State University, Corvallis, OR 97331.

PENNSYLVANIA: The Pennsylvania State University, University Park, PA 16802; Fruit Research and Extension Center, P.O. Box 309, Biglerville, PA 17307.

PUERTO RICO: University of Puerto Rico, Rio Piedras, PR 00927.

RHODE ISLAND: University of Rhode Island, Kingston, RI 02881.

SOUTH CAROLINA: Clemson University, Clemson, SC 29634; Edisto Experiment Station, Box 247, Blackville, SC 29817; Pee Dee Experiment Station, Box 5809, Florence, SC 29502; Sandhill Experiment Station, Box 528, Elgin, SC 29045.

SOUTH DAKOTA: South Dakota State University, Brookings, SD 57007.

TENNESSEE: University of Tennessee, P.O. Box 1071, Knoxville, TN 37901; Jackson, TN 38301; P.O. Box 110019, Nashville, TN 37222.

TEXAS: Texas A & M University, College Station, TX 77843; Agricultural Experiment Station, Lubbock, TX 79401; 2415 East Highway 83, Weslaco, TX 78596; Uvalde, TX 78801; Fort Stockton, TX 79735; Stephenville, TX 76401; 17360 Coit Road, Dallas, TX 75252; Overton, TX 77684.

UTAH: Utah State University, Logan, UT 84322.

VERMONT: University of Vermont, Burlington, VT 05405.

VIRGINIA: Virginia Polytechnic Institute and State University, Blacksburg, VA 24061; Virginia Truck Experiment Station, Norfolk, VA 23501; Piedmont Fruit Research Laboratory, Charlottesville, VA 22903; Winchester Fruit Research Laboratory, Winchester, VA 22601; Southern Piedmont Research Center & Continuing Education, P.O. Box 448, Blackstone, VA 23824; Cooperative Extension Service, Virginia State University, Petersburg, VA 23803; Eastern Virginia Research Station, Warsaw, VA 22572; TR & CEC, Holland, VA 23437.

VIRGIN ISLANDS: College of Virgin Islands, Cooperative Extension Service, P.O. Box L, Kingshill, St. Croix, VI 00850.

WASHINGTON: Washington State University, Pullman, WA 99163; Western Washington Research & Extension Center, Puyallup, WA 98371; Irrigated Agricultural Research & Extension Center, Prosser, WA 99350.

WEST VIRGINIA: West Virginia University, Morgantown, WV 26506; Fruit Experiment Station, Kearneysville, WV 25430.

WISCONSIN: University of Wisconsin, Madison, WI 53706; Peninsula Branch Experiment Station, Sturgeon Bay, WI 54235.

WYOMING: University of Wyoming, University Station, Laramie, WY 82071.

Appendix: Key to Turfgrass Weeds

The identification of a turfgrass weed from a mature, intact specimen can be both time-consuming and laborious. When one or more essential plant parts are under-developed or missing from the plant, the task becomes even more difficult. To sim-plify identification, the structures normally found on weeds in mowed turf are utilized together with floral or seedhead parts for identification process.

The key is organized to present pairs of possible identifying structures. After you have a representative weed available for identification, select the statement that best describes or applies to the plant you wish to identify. Your choice directs you to the next appropriate subordinate pair of statements. Again, select between the two statements. Proper choices direct you until a choice ends with the identifi-cation of any weed included in the key.

Main Key

1. Plant not appearing grasslike, leaves sometimes wider than long, leaf veins generally branched   *Broadleaf key*
1. Plant appears grasslike, leaves longer than wide, leaf veins generally parallel   2
   2. Leaves long, narrow, in groups of two; stems, if present, are flat, round, or hollow
                                                     Grass key
   2. Leaves long, narrow, in groups of two or three; stems triangular or round   3
      3. Leaves long, slender, somewhat thickened with a white underground bulb, with garlic odor or taste   4
         4. Narrow, nearly round stem, hollow (at least toward base); aboveground bulb-lets present in clusters at top of stem or greenish-to-white flowers
                                                    Wild garlic
         4. Narrow, flat leaves arising from base of plant only; few flowers or aboveground bulblets   *Wild onion*
      3. Leaves long, narrow, thin, and grasslike, in whorls of three around a triangular stem   4
         4. Grasslike, three-sided stem, corms present, closed sheath, ⅛ to ⅔ in wide, sheath approximately as wide as mature stem; flowers yellow-brown in nar-row spikelets in an umbel-like inflorescence   *Yellow nutsedge*

561

4. Grasslike, three-sided stem, corms absent, closed sheath that is generally shorter than stem when mature; flowers dark brown or purple
Purple nutsedge

Broadleaf Key

1. Leaves compound   2
  2. Leaf shape palmate   3
    3. Leaves opposite; leaves glabrous; stems or petioles hairy; petals yellow; flower regular, perfect; inflorescence arranged in cyme   *Yellow woodsorrel*
    3. Leaves alternate   4
      4. Leaflet margins entire or serrate   5
        5. Leaflet margins serrate; leaves glabrous; stems and petioles with fine hairs; petals white; flower regular, perfect; single flowers in leaf axils
Wild strawberry
        5. Leaflet margins entire   6
          6. Leaflets attached by petiole   7
            7. Leaves, stems, or petioles glabrous or without hair; petals yellow; flower irregular, imperfect; cluster of flowers in leaf axils
Little burclover
            7. Leaves, stems, or petioles hairy; petals purple; flower irregular, perfect; flowers single or arranged in raceme   *Common lespedeza*
          6. Leaflets sessile   7
            7. Petals white to pink; leaves, stems, or petioles glabrous or without hair; flower irregular, perfect; cluster of flowers in terminal inflorescence   *White clover*
            7. Petals yellow; leaves, stems, or petioles glabrous or without hair; flower irregular, perfect; cluster of flowers in terminal inflorescence
Hop clover
      4. Leaflet margins lobed   5
        5. Leaflets sessile; leaves, stems, or petioles glabrous or without hair; petals red; flower irregular, perfect; cluster of flowers in terminal inflorescence
Crimson clover
        5. Leaflets attached by petiole   6
          6. Stems or petioles glabrous or without hair; leaves hairy; petals yellow; flower regular, perfect; flowers arranged in panicle   *Silvery cinquefoil*
          6. Stems or petioles hairy   7
            7. Flowers green; leaves, stems, or petioles hairy; flower regular, imperfect; flowers in panicles arising from leaf axils   *Parsley-piert*
            7. Disk flowers yellow, ray flowers white; leaves, stems, or petioles hairy; flower regular, perfect; single terminal flowers
Creeping buttercup
  2. Leaf shape pinnate   3
    3. Leaf margins lobed or serrate   4
      4. Leaflet margins lobed; leaves alternate, petioled; leaves, stems, or petioles glabrous or without hair; flower regular, perfect; flowers arranged in panicle
Poison oak

      4. Leaflet margins serrate   5

        5. Leaves glabrous or without hair; stems or petioles glabrous; leaves alternate; petals yellow; flower irregular, perfect; flowers axillary in spike or spikelike arrangement   *Black medic*

        5. Leaves hairy; stems or petioles glabrous, petals green to yellow; flower regular, perfect; flower panicles arranged in axils   *Mugwort*

   3. Leaflet margins entire   4

      4. Leaves opposite, hairy; spiny seedpods; petals yellow; flower regular, perfect; single flowers in leaf axils   *Puncturevine*

      4. Leaves alternate   5

        5. Stems woody; flowers green and regular; flowers in panicle rising from leaf axils   *Poison ivy*

        5. Stems herbaceous; flowers purple and irregular; flowers arranged in raceme   *Hairy vetch*

1. Leaves simple   2

  2. Leaves with entire or finely serrate edges   3

    3. Leaf arrow-shaped with distinct pointed or rounded lobes   4

      4. Leaves in whorls, attached by petioles; leaves, stems, or petioles glabrous or without hair; petals red; flower irregular, imperfect; flowers in inflorescence arranged in raceme   *Red sorrel*

      4. Leaves alternate; leaves, stems, or petioles glabrous or without hair
                                           Field bindweed

    3. Leaf not arrow shaped   4

      4. Leaves lanceolate; longer than wide   5

        5. Leaf margins entire   6

          6. Leaves in whorls   7

           7. Leaves sessile; leaves hairy, stems with spines; petals white; flower regular, perfect; usually single flowers in leaf axils   *Bedstraw*

           7. Leaves attached by petioles   8

              8. Flowers white; leaves, stems, or petioles glabrous or without hair; flower regular, perfect; single flowers in leaf axils
                                           Carpetweed

              8. Flowers green or reddish brown, regular, imperfect; leaves, stems, or petioles glabrous or without hair; flowers arranged in spike or spikelike   *Buckhorn plantain*

          6. Leaves not in whorls   7

           7. Leaves alternate   8

              8. Leaves glabrous or without hair   9

                9. Flowers arranged in raceme; basal leaves in whorl; flower regular, perfect   *Curly dock*

                9. Single flowers in leaf axils; petals white; flower regular, perfect   *Prostrate knotweed*

              8. Leaves hairy, stems, or petioles glabrous; petals green; flower regular, perfect; single flowers to panicles arranged in axils or in inflorescence   *Kochia*

           7. Leaves opposite   8

8. Leaves attached by petiole; leaves, stems, or petioles glabrous or without hair; petals white; flower regular, perfect; inflorescence arranged in cyme   *Little starwort*

8. Leaves sessile   9

9. Single flowers in leaf axils   10

10. Sepals four; petals purple; leaves, stems or petioles hairy; flower regular, perfect   *Poorjoe*

10. Sepals two; petals white to purple; leaves, stems or petioles hairy; flower regular, perfect   *Virginia buttonweed*

10. Petals yellow; leaves, stems, or petioles hairy; flower regular, perfect   *Mouseear hawkweed*

9. Flowers in inflorescence; petals white; leaves, stems, or petioles hairy; flower regular, perfect; flowers arranged in panicle   *Mouseear chickweed*

5. Leaf margins serrate to lobed   6

6. Leaf margins serrate; leaves glabrous; stems or petioles hairy; petals white; flower regular, perfect; flowers arranged in raceme   *Shepherdspurse*

6. Leaf margins lobed   7

7. Stems glabrous; petals yellow; flower regular, perfect; single flowers in terminal inflorescence   *Spotted catsear*

7. Stems covered with spines   8

8. Flower regular, imperfect; petals purple; single flowers in terminal inflorescence   *Canada thistle*

8. Flower regular, perfect; petals red to purple; single flowers in terminal inflorescence   *Bull thistle*

4. Leaf cordate or ovate; wider than long   5

5. Leaves cordate   6

6. Leaves opposite; leaves, stems, or petioles hairy; petals purple; flower irregular, perfect; inflorescence arranged in cyme   *Catnip*

6. Leaves alternate   7

7. Petals purple; leaves, stems, or petioles glabrous or without hair; flower regular, perfect; flowers arranged in spike or spikelike   *Creeping bellflower*

7. Petals green; leaves, stems, or petioles glabrous or without hair; flower regular, perfect; flowers axillary in spike or spikelike arrangements   *Common lambsquarters*

5. Leaves ovate   6

6. Leaf margins entire   7

7. Leaves in a basal whorl; leaves, stems, or petioles glabrous or without hair; petals green; flower regular, imperfect; flowers arranged in spike or spikelike shape   *Broadleaf plantain*

7. Leaves not in a basal whorl   8

8. Leaves alternate   9

9. Flowers imperfect; leaves, stems, or petioles glabrous or without hair; petals green; single flowers in leaf axils   *Prostrate pigweed*

        9. Flowers perfect   10

           10. Flowers yellow; leaves, stems, or petioles glabrous or without hair; single flowers in leaf axils
Common purslane

           10. Flowers white to green; stolons rooting at nodes; leaves, stems, or petioles glabrous or without hair
Dichondra

     8. Leaves opposite   9

        9. Leaves hairy   10

           10. Petals purple; leaves, stems, or petioles hairy; flower irregular, perfect; inflorescence arranged in cyme
Healall

           10. Petals white; leaves hairy, stems or petioles glabrous; flower regular, imperfect; single flowers in leaf axils
Prostrate spurge

        9. Leaves glabrous   10

           10. Flowers imperfect; leaves, stems, or petioles glabrous or without hair; petals white to red; single flowers in leaf axils *Spotted spurge*

           10. Flowers perfect   11

              11. Petals white; leaves, stems, or petioles glabrous or without hair; single terminal flowers
Common chickweed

              11. Petals yellow; leaves, stems, or petioles glabrous or without hair; single flowers in leaf axils
Moneywort

  6. Leaf margins serrate to lobed   7

    7. Leaf margins lobed   8

     8. Leaves alternate; leaves, stems, or petioles glabrous or without hair; disk flowers yellow, ray flowers white; flower regular, perfect; single terminal flowers *Oxeye daisy*

     8. Leaves opposite   9

        9. Annual growth habit without rhizomes; leaves hairy, stems or petioles glabrous; petals purple; flower irregular, perfect; cluster of flowers in whorl in leaf axils *Henbit*

        9. Perennial growth habit with rhizomes or runners; leaves hairy, stems or petioles glabrous; petals purple; flower irregular, perfect; cluster of flowers in leaf axils *Ground ivy*

    7. Leaf margins serrate   8

     8. Leaves alternate   9

        9. Stems hairy; petals white to violet; flower regular, perfect; single flowers in leaf axils *Common mallow*

        9. Stems glabrous   10

           10. Flowers irregular, petals purple; leaves, stems, or petioles glabrous or without hair; single flowers in leaf axils
Violets

10. Flowers regular, petals white to yellow   11

  11. Flowers single in inflorescence; leaves glabrous to hairy, stems or petioles glabrous; disk flowers yellow, ray flowers white   *English daisy*

  11. Flowers in clusters; petals white   *Lawn pennywort*

8. Leaves opposite   9

  9. Leaves glabrous; petals purple; flower irregular, perfect; inflorescence arranged in cyme   *Florida betony*

  9. Leaves hairy; petals purple; flower irregular, perfect; cluster of flowers in whorl in leaf axils   *Purple deadnettle*

2. Leaves with coarsely serrate to deeply or double-lobed edges   3

  3. Leaf margin serrate or sawtoothed; leaves alternate; leaves, stems, or petioles glabrous or without hair; petals white; flower regular, perfect; flowers arranged in raceme   *Field pennycress*

  3. Leaf margin lobed   4

    4. Leaves pinnately lobed   5

      5. Leaves in a basal whorl   6

        6. Leaf lobes rounded   7

          7. Stems or petioles glabrous or without hair; petals yellow; flower regular, perfect; flowers arranged in raceme   *Yellow rocket*

          7. Stems or petioles hairy; petals white to green; flower regular, perfect; flowers arranged in raceme   *Field pepperweed*

        6. Leaf lobes pointed   7

          7. Petals blue; leaves, stems, or petioles glabrous or without hair; flower regular, perfect; single flowers in leaf axils   *Chicory*

          7. Petals yellow; leaves, stems, or petioles glabrous or without hair; flower regular, perfect; single flower head in terminal inflorescence   *Dandelion*

      5. Leaves not in a basal whorl   6

        6. Leaves opposite; leaves, stems, or petioles hairy; petals purple; flower irregular, perfect; flowers arranged in spike or spikelike shape   *Prostrate vervain*

        6. Leaves alternate   7

          7. Stems glabrous; leaves with spines; petals yellow; flower regular, perfect; flowers arranged in panicle   *Annual sowthistle*

          7. Stems hairy or with spines   8

            8. Stems with spines; petals yellow; flower regular, perfect; cluster of flowers   *Buffalobur*

            8. Stems hairy   9

              9. Leaves glabrous; petals white; flower regular, perfect; flowers arranged in raceme in leaf axils   *Swinecress*

              9. Leaves hairy; petals yellow; flower regular, perfect; flowers arranged in panicle   *Wild mustard*

    4. Leaves palmately lobed; radiating from a central point or double and deeply dissected   5

      5. Leaves double and deeply dissected   6

6. Leaves arranged in a basal whorl 7

    7. Stems or petioles with hair; hair present on either leaf surface; petals red to purple; flower regular, perfect; inflorescence in terminal umbels
Redstem filaree

    7. Stems or petioles glabrous; upper leaves alternate; petals white; flower regular, perfect; flowers arranged in raceme
Hairy bittercress

6. Leaves not arranged in a basal whorl 7

    7. Lower leaves alternate 8

        8. Upper leaves alternate 9

            9. Leaf margins lobed; leaves, stems, or petioles glabrous or without hair; petals white to yellow; flower regular, perfect; single terminal flowers   *Mayweed chamomile*

            9. Leaf margins serrate or pointed; leaves, stems, and petioles hairy; petals white to yellow; flower irregular, perfect; inflorescence in terminal umbels   *Common yarrow*

        8. Upper leaves opposite; leaves, stems, or petioles hairy; petals white to blue; flower regular, perfect; single flowers in leaf axils
Waterpod

    7. Leaves opposite 8

        8. Flowers single; leaves, stems, or petioles hairy; petals green; flower regular, perfect   *Lawn burweed*

        8. Flowers arranged in an umbel; leaves, stems, and petioles hairy; petals white; flower regular, perfect   *Wild carrot*

5. Leaves palmately lobed; leaves in whorls; leaves, stems, or petioles glabrous or without hair; petals pink to purple; flower regular, perfect; single flowers in terminal inflorescence   *Carolina geranium*

Grass Key

1. Seedheads not present, immature or no longer intact 2

  2. Leaves folded in bud shoot 3

    3. Auricle short or rudimentary with tips not touching; sheath compressed, margins overlapping; collar broad, divided; leaf blade from 5 to 10 mm wide, bunch-type growth habit   *Perennial ryegrass*

    3. Auricle absent or not developed 4

      4. Ligule truncate or flat; short membranous; sheath compressed, margins overlapping; collar broad, continuous; leaf blade greater than 10 mm wide; both rhizomes (weak) and stolons present   *Bahiagrass*

      4. Ligule not truncate or flat 5

        5. Ligules ciliate or a ring of hairs 6

          6. Sheath hairy, compressed, margins overlapping; collar narrow, continuous; blade tip flattened, both rhizomes (weak) and stolons present
Kikuyugrass

          6. Sheath glabrous, compressed, margins overlapping; collar narrow, continuous; blade tip pointed, both rhizomes and stolons present
Bermudagrass

  5. Ligules not ciliate or a ring of hairs   6
    6. Ligule acute, membranous; sheath compressed, margins overlapping; collar narrow, continuous; leaf blade less than 5 mm wide; bunch-type growth habit   *Annual bluegrass*
    6. Ligule membranous and toothed   7
      7. Sheath margins overlapping; collar broad, continuous; leaf blade from 5 to 10 mm wide; bunch-type growth habit   *Goosegrass*
      7. Sheath fused or closed, not easily removed from stem; collar broad, continuous; leaf blade greater than 10 mm wide; bunch-type growth habit   *Orchardgrass*
2. Leaves rolled in the bud shoot   3
  3. Auricle clawlike; with tips touching or wrapping around the sheath   4
    4. Rhizomes absent; sheath round and split with margins overlapping; collar broad, continuous; leaf blade greater than 10 mm wide; bunch-type growth habit   *Italian ryegrass*
    4. Rhizomes present; sheath round and split with margins overlapping; collar broad, continuous; leaf blade from 5 to 10 mm wide   *Quackgrass*
  3. Auricle absent or not developed   4
    4. Ligule absent or not developed   5
      5. Leaf blade greater than 10 mm wide; sheath compressed, margins split but not overlapping; collar broad, continuous; bunch-type growth habit   *Barnyardgrass*
      5. Leaf blade from 5 to 10 mm wide; sheath compressed, margins split but not overlapping; collar broad, continuous; bunch-type growth habit   *Junglerice*
    4. Ligule present, membranous or ciliate   5
      5. Ligule ciliate or a ring of hairs   6
        6. Sheath round and split with margins overlapping; collar broad, continuous; leaf blade from 5 to 10 mm wide; both rhizomes and stolons present   *Zoysiagrass*
        6. Sheath compressed, split with margins overlapping   7
          7. Collar broad, continuous   8
            8. Leaf blade greater than 10 mm wide; bunch-type growth habit   *Yellow foxtail*
            8. Leaf blade from 5 to 10 mm wide; bunch-type growth habit   *Field sandbur*
          7. Collar narrow or narrow to broad, continuous   8
            8. Leaf blade from 5 to 10 mm wide; bunch-type growth habit   *Witchgrass*
            8. Leaf blade less than 5 mm wide; bunch-type growth habit   *Stinkgrass*
      5. Ligule membranous   6
        6. Ligule truncate or flat   7
          7. Sheath compressed, split with margins overlapping   8
            8. Both rhizomes and stolons present; collar broad, continuous; leaf blade greater than 10 mm wide   *Bahiagrass*

    8. Bunch-type growth habit; collar broad, continuous; leaf blade greater than 10 mm wide   *Smooth crabgrass*

  7. Sheath round   8

    8. Sheath round, margins split but not overlapping; collar broad, continuous; leaf blade greater than 10 mm; bunch-type growth habit   *Tall fescue*

    8. Sheath round, fused or closed, not easily removed from stem; collar broad, divided; leaf blade greater than 10 mm wide; rhizomes present   *Smooth brome*

6. Ligule not truncate or flat   7

  7. Ligule round   8

    8. Sheath compressed, split with margins overlapping   9

      9. Collar broad, continuous; leaf blade less than 5 mm wide; stolons present   *Nimblewill*

      9. Collar narrow, divided; leaf blade less than 5 mm wide; bunch-type growth habit or rhizomatous   *Velvetgrass*

    8. Sheath round, split with margins overlapping   9

      9. Collar narrow, continuous; leaf blade less than 5 mm wide; stolons present   *Creeping bentgrass*

      9. Collar broad, divided; leaf blade more than 10 mm wide; rhizomes present   *Johnsongrass*

  7. Ligule not round   8

    8. Ligule acuminate; sheath round, margins split but not overlapping; collar broad, continuous; leaf blade greater than 10 mm; rhizomes present   *Dallisgrass*

    8. Ligule not acuminate   9

      9. Ligule acute to toothed   10

        10. Sheath round, split with margins overlapping; collar broad, divided; leaf blade greater than 5 mm wide; rhizomes present   *Redtop*

        10. Sheath compressed, split with margins overlapping; collar broad, continuous; leaf blade greater than 10 mm wide; bunch-type growth habit   *Large crabgrass*

      9. Ligule toothed   10

        10. Leaf blade 5 to 10 mm wide or less   11

          11. Sheath compressed, split with margins overlapping; collar narrow, divided; bunch-type growth habit   *Foxtail barley*

          11. Sheath round, fused or closed, not easily removed from stem; collar narrow, divided; bunch-type growth habit   *Downy brome*

        10. Leaf blade greater than 10 mm wide   11

          11. Sheath fused or closed, not easily removed from stem; collar broad, divided; bunch-type growth habit   *Rescuegrass*

          11. Sheath round   12

         12. Sheath split with margins overlapping; collar narrow, continuous; bunch-type growth habit *Timothy*

         12. Sheath margins split but not overlapping; collar broad, continuous; bunch-type growth habit *Wild oats*

1. Seedheads or flowers present   2

  2. Florets encased in hard, spiny burs   *Field sandbur*

  2. Florets not encased in hard, spiny burs   3

    3. Glumes longer than terminal floret enclosing entire spikelet   4

      4. Florets disarticulating or breaking away from inflorescence above the glumes   5

        5. Lemma not ending in an awnlike structure   6

          6. Collar broad, divided; leaf blade greater than 5 mm wide; rhizomes present   *Redtop*

          6. Collar narrow, continuous; leaf blade less than 5 mm wide; stolons present   *Creeping bentgrass*

        5. Lemma with bifid awn or split tip   *Wild oats*

      4. Florets disarticulating or breaking away from the plant below the glumes   5

        5. One floret per spikelet in a spikelike arrangement   *Zoysiagrass*

        5. Two florets per spikelet in a panicle arrangement   *Velvetgrass*

    3. Glumes shorter than or equal to terminal floret   4

      4. Florets disarticulating or breaking away from inflorescence above the glumes   5

        5. Lemma ending in an awnlike structure   6

          6. Lemma with bifid awn or split tip   *Downy brome*

          6. Lemma with awn short to long, arising from tip   7

            7. One flower or fertile floret per spikelet   8

              8. Inflorescence narrow or closed panicle   *Nimblewill*

              8. Inflorescence spike or spikelike shape   9

                9. Sheath compressed   *Foxtail barley*

                9. Sheath round   *Timothy*

            7. More than one floret per spikelet   8

              8. Ten to 20 florets per spikelet   *Italian ryegrass*

              8. Four to 6 florets per spikelet   9

                9. Inflorescence a spike   *Quackgrass*

                9. Inflorescence short branching panicle   *Orchardgrass*

        5. Lemma not ending in an awnlike structure   6

          6. One fertile floret or flower per spikelet   7

            7. Inflorescence a whorl of one-sided spikes   *Bermudagrass*

            7. Inflorescence in panicle or paniclelike arrangement either opened or closed   8

              8. Inflorescence partially or totally enclosed in sheath   *Annual dropseed*

              8. Inflorescence extended beyond sheath   *Smutgrass*

          6. More than one fertile floret per spikelet   7

7. Two fertile florets per spikelet   8

    8. Leaf blade 2 to 4 mm wide   *Alkaligrass*

    8. Leaf blade greater than 10 mm wide   *Yellow foxtail*

7. More than two fertile florets per spikelet   8

    8. Inflorescence in panicle or paniclelike arrangement either opened or closed   9

        9. Ligules ciliate or ring of hairs   *Stinkgrass*

        9. Ligules not ciliate or ring of hairs   10

          10. Ligule truncate or flat   11

            11. Sheath round, margins split but not overlapping; collar broad, continuous; leaf blade greater than 10 mm; bunch-type growth habit   *Tall fescue*

            11. Sheath round, fused or closed, not easily removed from stem; collar broad, divided; leaf blade greater than 10 mm wide; rhizomes present   *Smooth brome*

          10. Ligule truncate or flat   11

            11. Ligule acute, membranous, sheath compressed, margins overlapping; collar narrow, continuous; leaf blade less than 5 mm wide; bunch-type growth habit   *Annual bluegrass*

            11. Ligule toothed, sheath fused or closed, not easily removed from stem; collar broad, divided; bunch-type growth habit   *Rescuegrass*

    8. Inflorescence not in panicle or paniclelike arrangement either opened or closed   9

        9. Inflorescence a whorl of spikelike racemes   *Goosegrass*

        9. Inflorescence a true spike   *Perennial ryegrass*

4. Florets disarticulating or breaking away from the plant below the glumes 5

  5. Lemma of at least one floret ending in an awnlike structure; spikelets in pairs   *Johnsongrass*

  5. Lemma not ending in an awnlike structure   6

    6. More than one floret per spikelet   7

      7. Inflorescence a panicle; both rhizomes and stolons present   *Kikuyugrass*

      7. Inflorescence a raceme; bunch-type growth habit   *Hardgrass*

    6. One flower or fertile floret per spikelet   7

      7. Inflorescence raceme or near raceme   8

        8. Ligule absent or not developed   *Junglerice*

        8. Ligule present   9

          9. Rhizomes present   10

            10. Ligule truncate or flat; stolons present   *Bahiagrass*

            10. Ligule acuminate; rhizomes often poorly defined; stolons absent   *Dallisgrass*

9. Bunch-type growth habit, rhizomes and stolons absent
    Fringeleaf paspalum

7. Inflorescence not raceme or near raceme   8

8. Inflorescence spike or spikelike shape   9

9. Ligule truncate or flat   10

10. Sheath compressed, split with margins overlapping
     Smooth crabgrass

10. Sheath round, split with margins overlapping
     Cogongrass

9. Ligule toothed to acute   *Large crabgrass*

8. Inflorescence not spike or spikelike shape   9

9. Inflorescence in panicle or paniclelike arrangement either
    opened or closed   10

10. Ligule absent   *Barnyardgrass*

10. Ligules ciliate or ring of hairs   11

11. Rhizomes present   *Torpedograss*

11. Rhizomes not present; bunch-type growth habit 12

12. Sheath without hair   *Fall panicum*

12. Sheath hairy   *Witchgrass*

GLOSSARY

A

A horizon. The surface layer of soil of varying thickness or a mineral soil having maximum organic matter accumulation, maximum biological activity, and/or eluviation by water of materials such as iron and aluminum oxides and silicate clays.

Abiotic. Of or pertaining to the nonliving; inanimate.

Absorption. Entrance of a pesticide (or other substance) into the body or system of a human, animal, plant, or microorganism.

Acaricide (miticide). A chemical or physical agent that kills or inhibits the growth of mites and ticks.

Acervulus (pl. acervuli, adj. acervular). Erumpent, saucer-shaped, cushionlike, asexual fruit body of a fungus bearing conidiophores, conidia, and sometimes setae.

Achene. A small, dry, thin-walled fruit, such as that of a buttercup and dandelion, that does not split open when ripe.

Acid soil. A soil with a pH value of less than 7.0; more technically, a soil having a preponderance of hydrogen ions over hydroxy ions in solution.

Acre. Unit of land measure = 160 rods = 43,560 sq ft = 0.40469 hectare = 4,046.87 sq m.

Acropetal. Upward from the base toward the apex. *See* Basipetal.

Activated sludge. Material containing a very large and active microbial population produced in one method of sewage disposal by vigorous aeration of sewage.

Activator. A material added to a pesticide that directly or indirectly increases its toxicity.

Active ingredient (a.i.). The amount of actual chemical pesticide in a formulation that is toxic or otherwise inhibits the pest. The active ingredient may be listed by its common name, chemical name, or both. The concentration of the active ingredient may be given as a percentage of active ingredient or as pounds of active ingredient in the formulation.

Acuminate. Tapering to a slender point.

Acute. Forming an acute (less than 90-degree) angle or point.

Acute toxicity. A single or limited exposure to a pesticide that occurs from immediately after to within several hours after an exposure to a pesticide; it may result in injury, illness, or death; the relative amount of poisoning that occurs following a single dose or exposure to a substance such as a pesticide. *See* Chronic toxicity.

Adherence.   The property of a substance (such as a pesticide) to adhere or stick to a given surface; retention.

Adhesion.   The molecular force of attraction in the area of contact between unlike bodies (e.g., a spray or dust material and a plant surface) that acts to hold them together.

Adhesive.   In pesticides, a material that increases the chemical's sticking power to the treated area; sticker, spreader.

Adjuvant.   Any substance that is either in a pesticide formulation or is added to the spray tank to enhance the pesticide's physical or chemical properties. Adjuvants include wetting agents and emulsifiers, spreaders or surfactants, stickers, penetrants, drift-control additives, defoaming agents, invert emulsifiers, spray colorants, and buffers.

Adsorption.   The binding of a substance to the outer surface of a particle (e.g., a pesticide to organic matter or clay in the soil).

Aeciospore.   Dehiscent, dikaryotic (N + N) spore produced in a cup-shaped aecium of a rust fungus.

Aecium (pl. aecia).   Cuplike fruiting body (sorus) of rust fungi.

Aeration, mechanical.   *See* Cultivation.

Aeration, soil.   Process by which water in soil pores is replaced by air from the atmosphere.

Aerobic.   Relating to the growth of organisms in the presence of molecular oxygen.

Aerosol.   A fine mist produced when a pesticide dissolved in a liquid is released as a gas into the air from a pressurized container.

Aesthetic injury level.   A level of disease severity or the number of pests that might cause enough damage to the appearance of a turf or other ornamental planting to warrant the cost of control.

Agar.   A dried mixture of certain red seaweeds, used for preparing semisolid growth media on which fungi or other microorganisms are grown.

Agitation.   The continuous mixing of a pesticide or other chemical in suspension to prevent it from settling or separating out.

Airborne.   Transported from place to place through the air.

Air pollutants.   The most common air pollutants that affect plants are ozone, sulfur dioxide, nitrogen oxides, fluorides, and particulate deposits (e.g., smoke, heavy metals, and photochemical smog).

Alate.   Winged; a form in the life cycle of certain insects (e.g., aphids). *See* Apterous.

Algicide.   A pesticide used to control algae.

Alkaline soil.   A soil with a pH value greater than 7.0.

Allergic effect.   The reaction that a particularly sensitive person or animal has to a substance such as pollen, dust, pesticide group, or all pesticides.

Allelopathy (adj. allelopathic).   Ability of one species of plant to inhibit or prevent the growth of the same or other species of plants through the excretion of toxic substances.

Alluvial soil.   A recently developed soil from deposited soil material that exhibits essentially no horizon development or modifications.

Alternate.   Leaves arranged singly at different heights and on different sides of the stem.

Alternate host.   One of two different types of plants on which some fungi (e.g., heteroecious rusts) must develop to complete their life cycle; host of lesser economic importance.

Alternative host.   One of a parasite's several hosts; alternative hosts are not required for completion of the developmental cycle of the parasite. Weeds and wild plants are often alternative hosts of certain pathogens and pests.

Amendment, physical.   Any substance added to the soil for the purpose of altering its physical condition (e.g., sand, calcined clay, peat moss, lime, sulfur, gypsum, and rubber crumbs).

Amenity turf.   Higher-cut, cool-season turf including home and industrial lawns, athletic fields, parks, and golf course fairways; turf that increases physical or aesthetic comfort.

Amorphous.   Without definite form, shape, or organized structure.

Anaerobic.   Relating to biological processes that occur in the absence of molecular oxygen.

Anamorph.   Asexual or imperfect state in the life cycle of a fungus. *See* Teleomorph.

Anionic surfactant.   A surface-active additive to a pesticide with a negative surface charge. They perform better in cold, soft water. Most wetting agents are anionic surfactants.

Annual plant.   A plant that completes its life cycle from seed and dies within a year or less. Examples include common chickweed, crabgrass, and goosegrass. *See also* Summer annual, Winter annual.

Antagonism.   The action of the combination of two or more pesticides that reduces the effectiveness of one or all of the pesticide components; the phenomenon of one microorganism producing toxic metabolic products that kill, injure, or inhibit the growth of some other microorganisms in close proximity.

Anterior.   Toward the front or head (as opposed to posterior).

Anther.   The saclike, pollen-bearing portion (stamen) of a flower.

Antheridium.   Male sex organ found in some fungi (e.g., *Pythium*).

Anthracnose.   A leaf, stem, flower, and/or fruit disease with characteristic limited, sunken, necrotic lesions (often zonate and with a distinct margin), caused by a fungus producing asexual spores in an acervulus, often with setae.

Antibiosis.   Antagonistic association between two organisms, or between one organism and a metabolic product of another organism, to the detriment of one of the organisms.

Antibiotic.   Damaging to life; especially a chemical produced by one microorganism that, in very small amounts, inhibits or kills other living microorganisms.

Antidesiccant.   Chemical compound that, when applied to a plant, minimizes water loss through transpiration.

Antidote.   A substance used to counteract the effects of a poison.

Antimicrobial.   Inhibiting microorganism by biological or chemical agents.

Antioxidant.   A substance capable of chemically protecting other substances against oxidation or spoilage.

Apex (pl. apices, adj. apical).   Tip or end; uppermost point; part of root or shoot containing apical meristem.

Aphid.   A small, piercing-sucking insect of the family Aphididae (order Homoptera) that produces honeydew and, in large populations, injures plants. Aphids may serve as a plant virus vector.

Apical growth.   Growth that occurs at the tip or apex of an organ.

Apical meristem.   Terminal growing point of a root or shoot; a mass of undifferentiated cells capable of division.

Application rate.   Amount of a fertilizer, pesticide, and so on, applied to a given area.

Applicator.   The person (in an organization) who has the ultimate legal responsibility for all pesticide purchasing, storage, handling, and use.

Apomixis (adj. apomictic).   The asexual (vegetative) production of grass seedlings in the usual sexual structures of the flower but without the mingling and segregation of chromosomes (genetic recombination). Seedling characteristics are the same as those of the maternal parent.

Apterous.   Wingless stage in the life cycle of certain insects (e.g., aphids). *See* Alate.

Aqueous.   Indicating the presence of water in a solution.

Aqueous suspension.   A water suspension of minute droplets of a liquid pesticide surrounded by oil.

Arid.   Dry.

Arthroconidium.   A jointed conidium of more than one cell that becomes separated; formed by hyphal fragmentation.

Arthropod.   Invertebrate animals with articulated bodies and limbs. Insects, arachnids, and crustaceans are arthropods.

Ascocarp, ascoma (pl. ascomata).   Sexual fruit body (ascus-bearing organ) of a fungus in the ascomycete group.

Ascomycete (adj. ascomycetous).   Fungus that produces sexual spores (ascospores) within a tubular or oval ascus (sac). The mycelium has cross walls (septate). These fungi often also produce spores called conidia.

Ascospore.   A sexually produced spore formed inside an ascus.

Ascus (pl. asci).   Typically a saclike structure within an ascocarp in which ascospores are borne.

Aseptate.   Without cross walls.

Aseptic.   Free of microorganisms.

Asexual.   Vegetative; without sex organs, gametes, or sexual spores; imperfect.

Atomize.   To reduce a liquid to fine droplets by passing it under pressure through a suitable nozzle or by applying drops to a spinning disc.

Attractant.   A substance that attracts a pest (such as an insect) to a bait or trap. Generally used with a killing agent or trap.

Auricle.   Claw- or earlike structure in grasses that occurs in pairs at the base of the leaf blade or at the apex of the leaf sheath.

Auxin.   A natural growth-regulating hormone in plants that controls cell elongation. Auxin is required for cell division.

Available nutrient.   That portion of a molecule or compound in the soil that can be readily absorbed, assimilated, and utilized by growing plants.

Available water.   That portion of water in a soil that can be absorbed by plant roots; soil moisture held in the soil between field capacity (F.C.) and permanent wilting percentage (P.W.P.). Available water = F.C. – P.W.P.

Avicide.   A pesticide used to prevent, destroy, repel, or attract birds.

Avirulent.   Unable to cause disease; nonpathogenic.

Awn.   Bristlelike structure at the apex of the outer bract of some grass flowers.

Axenic or gnotobiotic culture.   Growth of a single species of organism without another organism being present; germ free; pure culture; uncontaminated.

Axil.   Upper angle formed between a leaf (or spikelet) and the stem axis.

B

B horizon.   A soil layer of varying thickness (usually beneath the A horizon) characterized by an accumulation of silicate clays, iron and aluminum oxides, and humus, alone or in combination and/or a blocky or prismatic structure.

Back-siphoning.   The movement of a liquid by gravity suction back into the source from which it was drawn. Refers to the movement of a pesticide solution from the spray tank back into a well or other water source while the tank is filling. Prevent back-siphoning by maintaining an air gap between the end of the hose and the level of liquid when filling a spray tank.

Bactericide.   A pesticide (often an antibiotic) used to prevent, destroy, or repel bacterial infection.

Bacterium (pl. bacteria).   A microscopic (0.25–2 microns), generally one-celled, nongreen organism (prokaryote) that lacks a nuclear membrane and reproduces by dividing in half (binary fission). Some bacteria attack plants, causing blights, cankers, crown gall, leaf spots, rots, wilts, and other plant diseases.

Bar.   Unit of pressure used to express water potential (1 atmosphere = 1.013 bar; equal to 1 million dynes/cm^2).

Basal application.   A treatment applied to the stems or trunks of plants at or just above the ground line.

Basal whorl.   Group of leaves attached at the same point at the base of a plant.

Basidiocarp, basidioma.   Sexual fruit body of a fungus in the basidiomycete group on or in which basidia and basidiospores develop.

Basidiomycete (adj. basidiomycetous).   Fungus characterized by septate mycelium, often with clamp connections, that forms sexual spores (basidiospores) on a club-shaped basidium.

Basidiospore.   Haploid sexual spore produced on a basidium; also called sporidium in rust and smut fungi.

Basidium (pl. basidia, adj. basidial).   Short, often club-shaped haploid cell or organ produced by basidiomycetous fungi on which basidiospores are borne.

Basipetal.   From the apex down toward the base.

Bed knife.   The fixed blade of a reel mower against which the rotating reel blades cut with a shearing action.

Beetle.   Any insect of the order Coleoptera characterized by thickened outer wings, chewing mouthparts, and complete metamorphosis.

Bench setting.   The height at which the bed knife of a mower is set above a firm, level surface.

Biennial plant.   A plant that lives for more than 1 year but not more than 2 years. The first year it produces leaves in a low-growing cluster (rosette) and stores food; the second year it produces fruits and seeds.

Bifed.   Two cleft or dividing into two points; forked.

Big roll.   Sod harvested as a piece 21–48 in (53–122 cm) wide by 150 ft (45 m) long.

Binucleate.   Having two nuclei per cell.

Bioassay.   The qualitative or quantitative determination of a substance by the systematic measurement of the response of living organisms when compared with the measurement of the response to a standard or standard series of tests.

Biocide.   A compound that is toxic to all forms of life.

Biodegradable.   Materials readily decomposed by microorganisms such as bacteria and fungi.

Biological management or control.   Techniques for managing (or controlling) pests by means of living organisms (e.g., predators, parasites, disease-producing organisms, competitive microorganisms) and other natural components of the environment that reduce pest populations to economically acceptable levels.

Biomass.   Volume (quantity, weight, and the like) of organisms (or living material) in a particular environment (such as turfgrass, weeds, bacteria, fungi, nematodes, and insects in soil); sometimes extended to the quantity of organic matter in a material (e.g., domestic refuse).

Biotechnology.   The use of genetically engineered microorganisms and/or modern techniques and processes with biological systems for industrial production; also, technology concerned with machines in relation to human needs.

Biotic.   Of or pertaining to living organisms; animate.

Biotype.   Subspecies group of organisms that differ from other biotypes in one or a few biochemical, physiological, or behavioral properties; a group of individuals with a like genetic makeup; a subdivision of a pathologic race; a geographical isolate or a clone of the same vector species. *See* Physiological race.

Black layer.   A dark layer in the soil primarily of high-sand golf greens or greens topdressed with sand. The layer forms when excess sulfur or sulfur derivatives, under anaerobic conditions, are reduced to the sulfide form (as hydrogen sulfide), which combines with metal ions (such as iron, magnesium, or manganese) in the soil to form the corresponding sulfide (iron sulfide). This sulfide forms a dark precipitate; the black layer.

Blade.   The flat or expanded part of a grass leaf located above the sheath; the portion of leaf that grows away from the stem.

Bleached.   White to straw colored; used to describe areas of necrotic tissue.

Blend.   A combination of two or more cultivars of a single turfgrass species having similar properties, often differing in resistance to one or more pathogens.

Blight.   General and rapid killing of leaves and stems from a single point of infection, resembling severe heat injury.

Blotch.   A large, irregularly shaped, superficial, straw-colored-to-brown discoloration.

Boom.   Several or more nozzles joined together by sections of pipe or tubing to apply pesticide or fertilizer over a wide area at one time.

Boot.   Sheath of the uppermost leaf that encloses the head of a grass plant.

Borer.   Insect or insect larva that forms tunnels within stems of grass and other plants.

Bract.   A reduced leaf associated with a flower or inflorescence; modified leaf from the axil of which a flower arises.

Brand.   The name, number, trademark, or designation applied to a substance (such as a pesticide) by the manufacturer, distributor, importer, or vendor.

Broadcast application.   Application of a pesticide or fertilizer in spray, dust, or granule form made uniformly over an entire turfgrass area.

Broadleaf or broadleaved.   Any plant with a flat leaf and netted veins other than a grasslike plant.

Buffer.   A substance that is so strongly and thoroughly acid or alkaline that it greatly resists change to another pH. Also a mixture of an acid and a base in a solution capable of maintaining hydrogen-ion concentration and thereby avoiding rapid changes in acidity or alkalinity of a solution.

Bug.   Any insect of the order Hemiptera characterized in part by piercing-sucking mouthparts, a triangular scutellum, two pairs of wings, and gradual metamorphosis. It is incorrect to apply this term to all insects or creatures that resemble them.

Bulb.   A short, flattened, usually globose or disc-shaped underground perennial storage organ consisting of a stem axis and numerous overlapping scales; essentially a subterranean bud.

Bulbil.   Small sclerotiumlike survival structure of fungi lacking a distinct rind layer.

Bulblet.   A little bulb; usually applied to the bulblike structures produced by some plants in the axils of leaves or replacing the flowers.

Bunch-type growth.   Plant development at or near the soil surface without production of rhizomes or stolons.

C

C horizon.   A soil layer beneath the B layer that is relatively unaffected by biological activity and pedogenesis and is lacking in properties diagnostic of an A or B horizon; parent material from which upper horizons develop.

Calcareous, calciferous.   Rich in calcium carbonate (lime) or lime compounds; chalk-white; chalky.

Calcined clay.   Clay minerals, such as montmorillonite and attapulgite, that have been fired at high temperatures to obtain absorbent, stable, granular particles that resist breakdown into smaller particles; used as amendments in soil modification.

Calyx (pl. calyxes).   Outermost whorl of flower parts; sepals collectively.

Canister respirator.   A respirator with a large filter that is contained in a canister and worn on the worker's belt or chest when in the presence of toxic chemicals (such as pesticides) for an extended period of time.

Canopy.   Expanded leafy top of a plant or plants.

Capillary water.   Water retained in the spaces among and on the surfaces of soil particles after drainage with a tension greater than 60 cm of water.

Carbamate insecticide.   A synthetic compound (salt or ester) derived from carbamic acid. Carbamate insecticides are contact killers with relatively short-lived effects. Examples are carbaryl, carbofuran, bendiocarb, and propoxur.

Carbohydrate.   A compound (foodstuff) of carbon, hydrogen, and oxygen (CH_2O) with the last two frequently in a 2-to-1 ratio; various chemical compounds such as sugars, starches, and celluloses.

Carbon-nitrogen ratio.   The ratio of the weight of organic carbon to the weight of total nitrogen in a soil or organic material.

Carcinogen (adj. carcinogenic).   A substance or agent capable of producing or inciting cancer.

Cardinal temperatures.   The minimum, optimum, and maximum temperatures for growth or germination of an organism.

Carrier.   Inert gas, liquid, or solid material (such as dust, clay, oil, water, and air) used to dilute or suspend a pesticide during its application and provide for its more uniform dispersal; plant, animal or human harboring an infectious disease agent (such as a virus or bacterium) but not showing marked symptoms. A carrier plant can be a source of infection to others. An insect contaminated externally with a plant pathogen (such as a bacterium, fungus, virus, or nematode) is sometimes called a carrier. *See* Vector.

Cartridge respirator.   A respirator with a filter cartridge that attaches to the mask. Always wear when pesticides are present.

Caryopsis.   Dry, indehiscent, one-seeded fruit with a thin pericarp fused to the seed coat; the "seed" (grain) or fruit of grasses.

Castings, earthworm (wormcasts).   The roundish fecal pellets of earthworms deposited on the turf surface or in the burrow; forms a relatively stable soil granule that can be objectionable on closely mowed turf.

Catenate; catenulate.   In chains or rows.

Caterpillar.   The wormlike stage (larva) of a moth or butterfly.

Cation exchange capacity (base-exchange capacity).   A measure of the total amount of exchangeable cations that a soil can hold; expressed in mEq/100 g of soil at pH 7.

Cationic detergent.   A detergent having positively charged surface ions.

Causal agent.   The organism or abiotic factor that produces a given disease or injury.

Cell.   The structural and functional unit of all plant and animal life. The essential feature of a cell is its living protoplasm limited by a membrane. It is surrounded by a wall.

Cellulose.   A complex polysaccharide of hundreds of beta-glucose molecules (polymers) linked in an unbranched chain that makes up to 40 to 55% by weight of plant cells.

Cereal. Grass grown primarily for its edible seed (e.g., barley, oats, rice, and wheat).

Certification. Seeds, plant parts or plant material, the progeny of foundation, registered or certified seed or other plant material, produced and sold under inspection control to maintain genetic identity and purity, freedom from harmful diseases, insect and mite pests, and weed seeds. It is approved and certified by an official certifying agency.

Chaff. Nonseed portion of a mature grass head.

Chelating (or sequestering agent). A large, organic compound, such as sodium EDTA, which attracts and tightly binds with specific bivalent (divalent) and trivalent metallic cations. Often used to correct nutrient deficiencies, to inhibit biological interactions that require bivalent ions and for other purposes.

Chemical control. Control of pests usually by applying one or more pesticides. It may be an integral part of an IPM program.

Chemical name. The name that indicates the chemical composition and/or chemical structure of a compound (such as a pesticide).

Chemosterilant. A chemical used to sterilize insects.

Chemotaxis. A movement or growth of organisms in response to a chemical stimulus.

Chimney. A cylindrical structure consisting of balls of mud that surrounds the opening of a crayfish burrow. It may be several inches high.

Chlamydospore. A thick- or double-walled asexual fungal spore formed directly from a hyphal or conidial cell that functions as a resistant or overwintering stage.

Chlorinated hydrocarbon insecticide. A synthetic pesticide that contains hydrogen, carbon, oxygen, and chlorine. These compounds are persistent and kill insects mainly by contact action. They are insoluble in water and are decomposed by alkaline materials and high temperatures. Examples are DDT, aldrin, chlordane, dieldrin, heptachlor, lindane, toxaphene, and methoxychlor. Most of these materials are not sold at present in industrial countries because of environmental concerns.

Chlorophyll (adj. **chlorophyllous**). Green, light-sensitive pigments found chiefly in the chloroplasts of leaves and other green parts of higher plants, which absorb the light energy used in the process of photosynthesis.

Chloroplast. Specialized cytoplasmic organelle (plastid) in plant cells that contains chlorophyll and is the site of photosynthesis.

Chlorosis (adj. **chlorotic**). Paling, yellowing, or whitening of normally green tissue characterized by the partial to complete destruction of chlorophyll. May be due to a virus, the lack of or unavailability of some element (such as iron, manganese, zinc, nitrogen, boron, and magnesium), lack of oxygen in a waterlogged soil, alkali injury, or some other factor.

Cholinesterase (acetylcholinesterase). An enzyme capable of influencing the rate of hydrolysis of acetylcholine, limiting the activity of nerve impulses.

Cholinesterase test. A blood test to determine the amount of poisoning that is occurring from some types of insecticides (such as organophosphate or carbamate). These tests are based on the cholinesterase level in blood plasma or red blood cells.

Chromosomes.   Self-replicating, strandlike bodies within cell nuclei in eukaryotes, formed from chromatin and bearing an aggregate of genes (hereditary determiners); a store of genetic information composed of protein and DNA. One chromosome of each pair is inherited from each parent. In bacteria, the entire genome is contained within one double-stranded, circular DNA molecule.

Chronic.   Slow-developing, persistent, or recurring symptoms that appear over a long period of time; an infection that lingers, often without symptoms; pertaining to a condition that is of long duration. *See* Acute.

Chronic toxicity.   Poisoning by a prolonged exposure to a substance such as a pesticide from small, repeated dosages over a period of time that may result in injury or death.

Ciliate, ciliolate.   Fringed or edged with fine hairs; having protoplasmic filaments (cilia) similar to flagella but shorter and more numerous on the cell.

Cilium (pl. cilia).   Hairlike, protoplasmic appendage like a flagellum that propels certain types of unicellular organisms through water. *See* Flagellum.

Circulative.   Pertaining to viruses that accumulate within or pass through the gut of insect vectors before being transmitted to plants. This type of transmission is often called persistent.

Cirrhum, cirrus (pl. cirrhi or cirri).   A curled, tendril-like mass of exuded spores held together by a slimy matrix as it issues from an ostiole (opening); also termed a spore horn.

Clamp connections.   Bridges around the septa of a hypha; taken as evidence that the fungus belongs to the basidiomycetes.

Clay.   Complex, colloidal, inorganic fraction of soil consisting largely of aluminosilicates; clay particles are usually negatively charged and absorb to positively charged ions; soil particles less than 0.002 mm in equivalent diameter; soil material containing less than 40% clay, less than 45% sand, and less than 40% silt.

Claypan.   A compact, slowly permeable layer of varying thickness and depth in the subsoil having a much higher clay content than the overlying material. Claypans are usually hard when dry and plastic and sticky when wet.

Cleistocarp, cleistothecium (pl. cleistothecia, adj. cleistocarpous).   Closed, usually spherical ascocarp that ruptures at maturity to release its ascospores; survival structure; typical of powdery mildew fungi.

Clippings.   The tips or upper ends of grass blades and, in some cases, stems removed by mowing.

Clod.   A compact, coherent mass of soil produced artificially, usually by tillage operations, especially when performed on soils either too wet or too dry.

Coalesce, coalescent.   Growing or joined together into one body or spot; to overlap; to merge.

Cocoon.   A silken case constructed by an insect larva (such as a caterpillar) to protect the pupal stage.

Cohesion.   Holding together; a force holding a solid or liquid together, owing to attraction between like molecules.

Cold-water-insoluble nitrogen (CWIN).   Fertilizer nitrogen not soluble in water at 77 (25°C).

Cold-water-soluble nitrogen (CWSN).   Fertilizer nitrogen soluble in water at 77°F (25°C).

Coleoptile.   Protective sheath of an embryonic shoot (the first true leaf of a grass seedling). It protects the plumule as it emerges through the soil.

Collar.   A light-colored band at the junction of the leaf blade and sheath of grasses on the outside of the leaf.

Colloid.   State in which a finely divided solid remains suspended in a liquid. Colloidal systems are usually more stable than emulsions or suspensions, and they are usually electrically charged and turbid; of soil, organic and inorganic matter with very small particle size and a correspondingly large surface area per unit of mass.

Colorant.   A paintlike material, usually a dye or pigment, applied to (a) brown warm-season turfgrasses that are in winter dormancy; (b) brown cool-season turfgrasses that are in summer dormancy; or (c) turfs that have been discolored by environmental stress, turfgrass pests, or the abuses of human beings. The purpose of a colorant is to maintain a uniform, favorable green appearance.

Commercial pesticide applicator.   A person licensed (certified) to apply pesticides to land or commodities.

Common (coined) name.   An official generic chemical name given to the active ingredient of a pesticide by a recognized committee.

Compatible chemicals.   Chemicals that can be mixed together without decreasing their effectiveness against the intended pests or becoming phytotoxic.

Complete development or metamorphosis.   The life cycle that passes through the four stages of change, including egg, larva, pupa, and adult. Examples include beetles, butterflies, moths, mosquitoes, bees, ants, and wasps. *See* Incomplete development.

Complete fertilizer.   Any fertilizer containing the three basic elements often deficient in soil for optimum plant growth: nitrogen (N), phosphorus (P), and potassium (K).

Complete flower.   One possessing all floral parts (i.e., sepals, petals, stamens, and pistils) all attached to a receptacle.

Compost.   A mixture of organic residues and soil that has been piled, mixed, moistened, and allowed to decompose biologically. Mineral fertilizers are sometimes added.

Compound leaf.   Leaf blade composed of two or more distinct leaflets.

Concentration.   The amount of actual pesticide or active ingredient (a.i.) contained in a formulation or mixture, usually expressed as percent or pounds per gallon.

Conidiogenous cell.   Any fungus cell that directly produces one or more conidia.

Conidiophore (pl. conidiophores).   A specialized, simple or branched hyphal cell or group of cells bearing cells that produce conidia.

Conidium, conidiospore (pl. conidia).   Any asexual, nonmotile spore (except sporangiospores or chlamydospores) that develops externally or is liberated from a conidiogenous cell.

Conjugate.   Joined; in twos.

Contact pesticide.   A pesticide that kills on contact. *See* Protectant pesticide.

Contagious.   Spreading from one to another.

Containment system.   Procedures and facilities established to contain pesticide inadvertently released during mixing, loading, handling, or an emergency.

Contaminated, contamination.   The pollution of water, soil, organisms, or any non-target part of the environment, as with pesticides; bearing, or intermixed with, a pathogen as spores, seeds, or fungi in soil; entry of undesirable organisms into some material or object; (of cultures) not pure.

Cool-season turfgrass.   Turfgrass species that has optimum growth at temperatures between 60 to 75°F (15.5 to 24°C).

Cordate.   Heart-shaped; usually used to describe leaves with a pair of rounded basal leaves.

Core aerification.   A method of turf cultivation in which soil cores are removed by hollow tines or spoons to control soil compaction and to aid in the penetration, distribution, and movement of pesticides, water, and air through soil.

Corm.   A short, enlarged, solid, often globose, fleshy base of an underground stem for reproduction or food storage, and containing undeveloped buds.

Corolla.   The petals, collectively, of a flower.

Cortex (adj. cortical).   The primary, soft, mostly parenchymatous tissue between the epidermis and phloem in culms and the pericycle in roots.

Cotyledon.   The seed leaf; one in the monocotyledonae (including grasses), two in the dicotyledonae; primary embryonic leaf; a food-digesting and food-storing part of an embryo.

Coverage.   The degree of spread or distribution of a chemical over a discontinuous surface such as leaves, thatch, and seeds; the number and uniformity of pesticide particles per square inch of target area.

Crawler.   Newly hatched or first-stage insect that is still able to crawl.

Crop tolerance.   The ability of a turfgrass to endure treatment with a chemical or injurious environmental factor with minor adverse effect.

Crown.   Compacted series of nodes from which shoots and adventitious roots arise at the base of the culm (stem) of grasses.

Crumb.   A soft, porous, irregular, natural unit of structure from 1 to 5 mm on diameter.

Culm.   Jointed stem of a grass plant often hollow except at the swollen nodes.

Cultivar (cv.).   A cultivated variety; an assemblage of closely related plants of common origin within a species that differs from other cultivars in certain specific details (such as form, color, flower, fruit, leaf width or disease resistance) that, when reproduced sexually or asexually, retain their distinguishing features. *See* Variety.

Cultivation turf.   Working of the soil and/or thatch without destruction of the turf, as by coring, slicing, spiking, or some other means growing, forking, shattering, and water injection.

Cultural control.   Control of pests by improving plant health in order to make the plant better able to compete with or tolerate pests.

Cultural intensity.   The amount of time, money, and effort (mowing fertilization, irrigation, cultivation, and pest control) required to maintain a particular turfgrass.

Cuticle (adj. **cuticular).**   A thin, waxy, protective, discontinuous layer (of platelets) over the outer wall of epidermal cells, consisting mostly of wax and cutin; outer sheath or membrane of a nematode.

Cutin.   A clear or transparent waxy material related to cellulose, very impermeable to water, that comprises the inner layer of the cuticle of plants.

Cutting height.   Of a mower, the distance between the plane of travel as defined by lower surface of the wheel, the roller, or the skid and the parallel plane of the mower blade.

Cyme.   A type of inflorescence with a broad, more-or-less flat-topped, determinate flower cluster with the central flower at the apex being the first to mature and open.

Cyst.   In nematodes, the egg-containing carcass or oxidized cuticle of a dead adult female of certain genera.

Cytoplasm.   All living substance (protoplasm) of a cell except the nucleus; consists of a complex protein matrix or gel including essential membranes and cellular organelles.

D

Damage threshold level.   The lowest pest population density at which significant damage will likely occur.

Damping off, damped-off.   Rapid decay and death of seeds in the soil or young seedlings before or after emergence. Most evident in young seedlings that suddenly wilt, topple over, and die from rot at the stem base.

Decay.   Disintegration or decomposition of plant tissue or other substrates by bacteria, fungi, and possibly other microorganisms. *See* Rot.

Deciduous woody plant.   A plant that sheds its leaves at the end of each growing season.

Decline.   Reduced vigor of perennial plantings (e.g., turf) as a result of chronic symptoms of a disease or disorder; the gradual reduction in health and vigor of a plant or planting that is in the process of slowly dying.

Decompose, decomposition.   Degradation into simpler compounds; rotting of colonized plant tissue, usually by microorganisms.

Defoliate.   To remove the leaf tissue of a plant by insect feeding or other means, leaving no more than the midrib and other major leaf veins.

Deflocculating agent.   A material added to a suspension of a pesticide to prevent settling.

Degradation.   The process by which a chemical compound (such as a pesticide) is reduced to a less complex compound or compounds by chemical or microbial breakdown.

Degree-days.   One degree of deviation on a single day of the daily mean temperature from a given standard temperature; $5°$ above or below the standard temperature would be recorded as five degree-days. The more rapid or slower development of plants as well as pests is largely determined by the number of degree-days.

Delayed effect.   Illness or injury that occurs long after single or repeated exposures to a toxic chemical (such as a pesticide). *See* Chronic effect.

Deposit or spray residue.   The quantity of a pesticide deposited per unit area of plant or other surface at any given time. *Initial deposit* is the amount deposited initially.

Dermal.   Of or pertaining to the skin.

Dermal exposure.   Exposure to a substance (such as a pesticide) through skin contact.

Dermal toxicity.   Ability of a compound when absorbed through the skin of animals and humans to produce symptoms of poisoning.

Desiccant.   A substance (such as a pesticide) that accelerates drying of plant or other tissues.

Desiccate, desiccation.   Drying out; removal of water.

Desorption.   The release of pesticide molecules or ions from particle surfaces to which they had been adsorbed. *See* Adsorption.

Dethatching.   Removing excessive thatch from a turfgrass area by hand raking or with various types of machinery equipped with vertical knives or tines.

Deuteromycete.   One of a subdivision of fungi with no known sexual states. Also called imperfect fungi or Fungi Imperfecti.

Diagnosis (pl. **diagnoses).**   Investigation or analysis of the nature and cause of a plant problem.

Dichotomous key.   Identification method generally with two choices to consider at any point leading to the next choice and, eventually, the name or identification.

Dicotyledonae.   A flowering plant (dicot) having two cotyledons (seed leaves) in an embryo in contrast to monocotyledons (such as grasses). *See* Monocotyledonae.

Dieback (v. **die back).**   Progressive death of leaves, stems, or roots, generally starting at the tip. Dieback may be due to stem, crown, or root rot; insect feeding; nematodes winter injury; deficiency or excess of moisture or nutrients; some other factor; or a stress complex.

Dikaryon (adj. **dikaryotic).**   A fungus cell having two genetically distinct, haploid (sexually compatible) nuclei.

Diluent.   An essentially inert or nonreacting gas, liquid, or solid material used to dilute the concentration of the active ingredient in a pesticide formulation to the desired concentration. *See* Carrier.

Diploid.   Having a double set of chromosomes per cell.

Directed application.   Application of a pesticide to a specific area (such as weeds or soil) or to the leaves, stems, and thatch of turfgrass plants.

Disarticulation.   Breaking or falling away from the plant.

Disease.   In plants, any disturbance that interferes with normal growth and development (such as structure and function), economic value, or aesthetic quality, and leads to development of symptoms; a continuously, often progressively affected condition that interferes with the normal activity of the plant's cells or organs or causes abnormality. Injury, in contrast, results from momentary damage. *See* Disorder, Injury.

Disease complex.   Disease resulting from the combined or sequential actions of two or more biotic or abiotic agents; often manifested by a greater-than-normal array of symptoms.

Disease cycle.   The sequence of events involved in disease development, including the stages of development of the pathogen and the effect of the disease on the host; the chain of events that occurs between the time of infection and the final expression of disease.

Disease range.   The geographic distribution of a disease.

Disease severity.   *See* Severity.

Disease tolerance.   In plants, the ability of a turfgrass to survive and produce satisfactorily at a level of infection that causes economic loss to other cultivars or types of the same turfgrass plant.

Disease triangle.   The simultaneous occurrence of a virulent pathogen, a susceptible host, and a favorable environment that allows an infectious disease to develop.

Disinfectant.   A chemical or physical agent that kills pathogenic organisms (fungi, bacteria, nematodes) *after* they have entered plant tissues.

Disinfestant.   A chemical or physical agent that removes, kills, or inactivates disease-causing organisms *before* they can cause infection. It may be applied on the surface of a seed or other plant part, on tools, or in the soil.

Disk flower.   One of a number of small tubular flowers composing the disk of certain composite plants such as sunflower and chrysanthemum. *See* Ray flower.

Disorder.   A harmful, nonpathogenic deviation from normal growth; abiotic disease. *See* Disease.

Dispersing agent.   A material that reduces the cohesive attraction between like particles, as in a pesticide.

Dissected.   Deeply cut into numerous segments, as a leaf; toothed with fine serrations; lobed, a partial division of an organ such as a leaf. The term generally applies to a division that extends less than halfway to the base of the leaf.

Dissemination (dispersal).   The transfer (spread) or transport of pests or infectious material (inoculum) from its origin by wind, water, humans, insects and other animals, machinery, or other means.

DMI fungicides.   Demethylation-inhibiting fungicides that prevent demethylation processes from occurring when fungi attempt to synthesize ergosterol from squaline for cell-membrane formation. They include cyproconazole, fenarimol, mycobutanil, propiconazole, and triadimefon; also called sterol-inhibiting fungicides (SIS) or sterol-biosynthesis inhibitory fungicides (SBIS).

DNA (deoxyribonucleic acid).   Occurs in nuclei of plant and animal cells and is a molecule composed of repeating subunits of nucleotides containing deoxyribose (a 5-carbon sugar), phosphoric acid, and four nitrogenous bases. Every inherited characteristic has its origin in the genetic code of an individual's complement of DNA.

Dormant, dormancy.   Not actively growing due to internal causes; resting; being in a state of reduced physiological activity. In turfgrass, a condition wherein the grass grows slowly if at all and commonly turns brownish. In insects, a condition wherein the insect does not eat and grows slowly if at all.

Dose, dosage.   Quantity of a substance applied per unit of plant, soil, or other surface; also called rate.

Drift (in relation to pesticides).   The movement by air of pesticide particles outside the intended target area during or shortly after application. *Particle drift deposit* is the deposition of pesticide particles outside the intended target area. *Airborne drift* is the dispersion of pesticide particles to the atmosphere by air transport and diffusion. *Vapor drift* is the dispersion of pesticide fumes to the atmosphere and areas surrounding the target area during and following application.

Drought.   An extended period of dry weather that usually causes stress to turf.

Dry flowable (DF).   A dry pesticide formulation of uniformly sized microgranules from an aqueous suspension of a wettable powder.

Dust.   In pesticides, a dry, finely ground formulation usually containing 4 to 10% of the active ingredient mixed with an inert material such as talc, clay, powdered nut hulls, volcanic ash, and similar materials.

E

Echinulate.   Having spines or other sharp projections.

Ecology.   The science dealing with the relationships between organisms and their environment.

Economic injury level.   The lowest population density of pests that will cause economic damage; the point at which the cost of pest control equals the amount of revenue loss caused by the pest.

Economic threshold level.   The density of pests at which control measures should be implemented to prevent an increasing pest population from reaching the economic injury level; point at which a control measure is justified to prevent a pest population from reaching the economic injury level.

Ecosystem.   The community of animals, plants, and other organisms and the relationship of these with each other and the chemical and physical environment.

Ectoparasite.   A parasite that feeds on the outside of a turfgrass plant. Many nematodes are ectoparasites. *See* Endoparasite.

Ectotrophic.   Describing a fungus that makes substantial growth primarily over the surface of roots. *See* Runner hyphae.

ED_{50}.   The *e*ffective *d*ose that gives a 50% response in a population.

Edaphic.   Of or pertaining to soil as it affects plant growth.

Eelworms.   Nematodes.

Emulsifiable concentrate (EC, E).   A liquid formulation of a pesticide containing an emulsifier and a high concentration of active ingredient dissolved in one or more solvents. Designed to be diluted, usually with water, to the desired strength before application. When added to water, a milky mixture is formed that is a suspension of active ingredient and emulsified solvent in the water.

Emulsification.   The breaking up of large particles or liquids into minute ones that remain suspended in another liquid, forming an emulsion where both liquids are immiscible (e.g., oil droplets in water).

Emulsifier (emulsifying agent).   A surface-active agent that facilitates the suspension of minute droplets of one liquid in another to form a stable emulsion.

Endodermis.   A single layer of living cells with thick walls and no intercellular spaces that surrounds the vascular tissues of stems and roots; the innermost layer of the cortex that is most conspicuous in roots.

Endogenous.   Produced, living, or undergoing development within.

Endoparasite.   A parasite that enters and feeds within its host. Some nematodes are endoparasites. *See* Ectoparasite.

Endophyte (adj. **endophytic).**   Generally, a fungus developing and living within turfgrass plants (especially fescues and ryegrasses) that produces toxins that kill or inhibit the growth of certain insects and other pests that eat it.

Entomology.   The science or study of insects.

Environment.   Our surroundings (soil, water, the atmosphere, plants, animals, humans, and all other organisms) make up the environment.

Environmental Protection Agency (U.S. EPA).   The federal agency in the United States responsible for reviewing product data, registering pesticides, setting tolerances, and enforcing laws related to pesticide use, pollution, disposal, groundwater protection, endangered species protection, and so on. Each state agency is responsible for matters of pesticide use that might result in adverse environmental effects such as groundwater contamination, spills, or disposal.

Enzyme.   A complex, high-molecular-weight protein produced in living cells by protoplasm that catalyzes a specific biochemical reaction but does not enter into the reaction itself.

Epidemic.   A rapidly developing and widespread outbreak of an infectious disease of humans in a community; used loosely for plants and animals. *See* Epiphytotic, Epizootic.

Epidemiology.   Study of the factors influencing the initiation, development, and spread of infectious causes of disease; the study of epidemics (epiphytotics and epizootics).

Epidermis (adj. **epidermal).**   The outermost layer of cells on all plant parts but absent from the root cap and apical meristem; in nematodes, the outer cylinder of cells consisting of a single layer of epithelium that secretes the cuticle (hypodermis).

Epiphyte.   An organism (such as a bacterium) growing on the surface of a plant part, from which it gains physical and nutritional support without causing disease.

Epiphytotic.   The sudden, widespread, and destructive outbreak of a plant disease, usually over large areas; increase of disease with time in a plant population. *See* Epidemic.

Epizootic.   An epidemic among animals.

Eradicant.   A fungicide or other chemical that kills a pathogen after infection has been established.

Erosion.   The wearing away (movement) of surface soil particles by wind, moving water, or other means.

Erumpent.   Bursting or erupting through the host surface in the course of development; bursting forth.

Escape (n).   A plant or section of turf that missed treatment or failed to respond to treatment; a susceptible plant that avoids a pest through some character of the plant or its location.

Establishment number.   The number on a pesticide label that indicates the facility where the product was produced.

Etiolated, etiolation.   Blanching or yellowing of tissue (absence of chlorophyll); elongating of stems (spindliness), and failure of normal leaf development caused by reduced light.

Etiology.   The science of the causes or origins of disease, together with the relations of the causal factors to the host.

Eukaryote (adj. **eukaryotic).**   An organism having membrane-limited organelles and nuclei that divide by mitosis; all organisms except bacteria and blue-green algae. *See* Prokaryote.

Evapotranspiration.   Total loss of moisture through the combined processes of evaporation from the soil surface and transpiration from plants, for a given area during a specified period of time.

Exoskeleton.   An external, usually hard, supportive covering of an insect (molt) that is shed and reformed several times as the insect grows.

Exotic.   An organism introduced from another country or area; not indigenous.

Explant.   To take living cells or tissues from a plant and place them in an artificial medium for tissue culture.

Exposure.   Oral, dermal, or respiratory contact with a substance, such as a pesticide.

Extracellular.   Outside the cell wall.

Extrude (n. **extrusion,** v. **extruded).**   To push out; emit to the outside.

Exudate.   Substance, often a liquid ooze or slime, discharged or secreted from diseased, injured, or healthy plant tissue; metabolic by-products discharged from roots into surrounding soil.

F

Facultative parasite.   A normally saprophytic organism that can live as a parasite under certain conditions and be cultured on laboratory media. *See* Parasite, Obligate parasite.

Facultative saprophyte or saprobe.   A normally parasitic organism that, under proper conditions, can live saprophytically during part of its life cycle.

Fairway.   The large central area of a golf hole between the tee and the green. This area is mowed shorter and maintained better than adjoining turf areas.

Fairy ring.   A ring of mushrooms or puffballs on the ground representing the periphery of mycelial growth of a basidiomycetous fungus; the distinct zone of greener grass associated with a ring of mushrooms.

Fastidious organism.   An organism that is difficult to isolate or cultivate on ordinary culture media.

Fertilizer.   Any organic or inorganic material of natural or synthetic origin containing nutrients (elements) essential to the growth of plants. To be labeled and sold as such, it must be state-licensed, and an analysis must be printed on the container label.

Field capacity (field moisture capacity).   The maximum amount of water the soil can hold against the force of gravity.

Filament (adj. filamentous).   Threadlike.

Filler.   A diluent in powdered form.

Fission.   Transverse splitting in two of bacterial cells (binary fission); asexual form of cell division.

Flagellum (pl. flagella, adj. flagellar).   A long; delicate; flexible; hair-, whip-, or tinsel-like structure on certain bacterial cells or zoospores of the lower fungi that enables them to swim through a liquid; a flagellum is functionally similar to a cilium but longer. *See* Cilium.

Flail mower.   A mower that cuts turf by high-speed impact of inverted T-blades rotating in a vertical cutting plane relative to the turf surface.

Fleck.   A small or minute (0.2–2 mm), white-to-tan lesion, often translucent and visible through a leaf.

Flexuous.   Having turns, bends, or windings alternately in opposite directions; capable of bending.

Flock.   Cottonlike tuft.

Floret.   A grass flower enclosed by a lemma and palea.

Flowable (F).   A finely ground, wettable powder formulation of a pesticide that is sold as a thick suspension in a liquid. Flowables require only moderate agitation and seldom clog spray nozzles.

Foliar, foliose.   Pertaining to leaves; leaflike.

Foliar application.   Application of a pesticide, fertilizer, or other chemical to above-ground parts of turfgrass plants.

Foliar burn.   Injury to shoot tissue caused by dehydration due to contact with high concentrations of chemicals, such as certain fertilizers or pesticides.

Forma specialis (pl. *formae speciales;* abbr. *f. sp.* or *ff. sp.*).   Special form; a biotype (or group of biotypes) of a species of pathogen that differs from others in the ability to infect selected genera or species of susceptible plants.

Forage.   Green or dehydrated vegetation used as food by livestock; hay, pasture, and silage.

Formulation.   The form in which a pesticide is offered for sale (as emulsifiable concentrate [EC], solution [S], flowable [F] or liquid [L], wettable powder [WP], dry flowable [DF], soluble powder [SP], water-dispersible granule [WDG], granule [G], dust [D], oil solution [OS], ultralow volume concentrate [ULV], microencapsulated [ME or M], aerosol [A], poisonous bait [B], fumigant [F], and so on. Includes both the active and inert ingredients.

Frass.   Fecal matter and small clipped bits of vegetation produced by insects.

Friable.   Easily crumbled into small aggregates or powder when handled, as with soil.

Fruit body, fruiting body.   A complex, multicellular fungal structure (fructification) that contains or bears spores.

Fumigant.   A vapor-active (volatile) pesticide used to kill disease-causing organisms, insects, nematodes, weeds, and other pests; a gaseous or readily volatilizable disinfectant or disinfestant that destroys organisms by vapor action in an enclosed area or under plastic laid on the soil.

Fumigation.   The application of a fumigant, applied in an enclosed area or usually under plastic laid on the soil, for disinfesting an area.

Fungicide (adj. **fungicidal).**   A chemical or physical agent used to prevent, destroy, or repel fungal infections; a type of pesticide.

Fungistat (adj. **fungistatic).**   A chemical or physical agent that inhibits the growth of fungi but does not cause death; a type of pesticide.

Fungus (pl. **fungi).**   A single- to many-celled, eukaryotic organism (thallophyte) that lacks chlorophyll, usually with a chitinous wall; reproduces by sexual and/or asexual spores and commonly produces mycelium; a nonchlorophyllous organism, the vegetative body (thallus) of which consists of threadlike filaments (hyphae) usually aggregated into branched systems (mycelia). Some fungi attack plants and cause plant diseases such as rusts, smuts, mildews, wilts, leaf spots, and patch diseases.

G

Gene.   Smallest functioning unit of genetic material on a chromosome; the fundamental unit of inheritance; the base triplets of the DNA molecule in a chromosome that determines or conditions one or more hereditary characteristics.

General-use pesticide.   A pesticide that will not harm the applicator or the environment to an unreasonable degree when used according to directions. It can be purchased over the counter and used by persons on areas they own or lease.

Genetic engineering or genetic manipulation.   Alteration of the genetic composition of a living cell by various tissue-culture procedures (transformation, protoplast fusion, and so on) so that the cell can produce more or different chemicals or perform completely new functions.

Genetically modified organism (GMO).   Any living organism into which foreign nuclear material was initially inserted by an artifactual method.

Genus (pl. **genera).**   A taxonomic category ranking above species and below family that includes one or more closely related (structurally or phylogenetically) and definable groups of plants comprising one or more species; the genus or generic name is the first word in a Latin binomial. *See* Species.

Germinate, germination.   To begin growth of a seed, spore, sclerotium, zygote, or other reproductive body starting with imbibition of water.

Germ tube.   Hyphal thread resulting from an outgrowth of the spore wall and/or cytoplasm during generation of a fungus spore, bulbil, or sclerotium. The hyphal threads elongate, branch, and become a new fungus body (*mycelium*).

Girdle.   To encircle.

Glabrous.   Smooth, without hairs or scales; lacking pubescence.

Globose.   Nearly spherical, globe shaped.

Glumes.   A pair of sterile, chaffy bracts, usually present at the base of a grass spikelet.

Grade.   To establish elevations and contours prior to planting.

Gramineae.   Former taxonomic family name for grasses; now the Poaceae.

Graminicole, graminicolous.   Growing on grass.

Gram-positive or gram-negative.   Bacteria staining violet or pink-red, respectively, following treatment with Gram's stain.

Granules, granular (G).   A ready-to-use formulation in which the pesticide is attached to particles of an inert carrier such as clay or ground corncobs. Granules are generally in the size range of about 15- to 40-mesh.

Grass.   Any plant that is a member of the family Poaceae (Gramineae).

Green.   The closely mowed and highly maintained area of grass on a golf course hole.

Ground cover.   Low-growing, nongrass plants used to cover the soil in areas where mowing is impractical or turf will not grow.

Groundwater.   The water stored in the saturated portions of soil and porous rock formations beneath the earth's surface.

Growth or plant regulator.   *See* Plant-growth regulator (PGR).

Guttation.   Exudation of water and solutes from hydathodes of plants.

H

Half-life.   The length of time required for microbial and chemical processes to degrade a fertilizer, pesticide, or other chemical to one-half its original concentration.

Haploid.   Having a single complete set of unpaired chromosomes in each cell nucleus.

Hardpan.   An impervious layer of soil or rock in the lower A horizon or upper B horizon that prevents downward drainage of water and root growth.

Haustorium (pl. haustoria).   A specialized, simple or branched hypha of a fungus that penetrates a host plant, makes intimate contact with the host cell's protoplast, and absorbs food. Powdery mildews, downy mildews, and rusts produce haustoria.

Hazard.   The likelihood that an injury will result from the use of a substance (such as a pesticide) in a prescribed quantity and manner. A hazard constitutes both toxicity and exposure.

Heaving.   The partial lifting of turfgrass plants out of the ground, frequently breaking their roots, due to alternate freezing and thawing of the surface soil during the winter.

Hectare (ha).   A land area in the metric system equal to 2.471 acres = 395.367 rods = 10,000 sq meters = 0.01 sq kilometer = 0.0039 sq mile.

Herbaceous.   A vascular plant with mostly soft, succulent, nonwoody stems (e.g., annuals, biennials, and perennials) that normally die back to the ground in the winter.

Herbage.   The succulent leaves and stems of herbaceous plants.

Herbicide.   Any chemical or physical agent used to prevent or control plants that grow where they are not wanted; a weed or grass killer; a type of pesticide.

Heteroecious (n. heteroecism).   Requiring two unrelated host plants for completing the life cycle, as in many rust fungi.

Heterokaryon.   Cell or mycelium with two or more nuclei of different constitution.

Heterothallism (adj. **heterothallic).**   Self-sterile; a sexual condition in which an individual produces only one kind of gamete; having sexes separated in different mycelia. *See* Homothallism.

Heterozygous.   Having mixed heridary factors not a pure line or breed; having different alleles at various loci.

Hilum.   A dot, mark, or scar, especially on a spore at the point of attachment to a conidiophore or sterigma; scar on a seed coat, marking the place of attachment of the seed stalk to the seed.

Homonym.   A name of an organism that must not be used because of an earlier name used in a different sense; that is, the names are the same, but the types are different.

Homothallism (adj. **homothallic).**   Self-fertile; the condition in which sexual reproduction can occur without the interaction of two different thalli; compatible male and female gametes on the same mycelium.

Homozygous.   Having paired identical genes present in the same cell or organism; a pure breed or line. *See* Heterozygous.

Honeydew.   A sweet, sticky secretion given off by aphids and other sucking insects, in which sooty mold fungi grow.

Hormone.   A naturally occurring or synthetic organic compound produced in one part of a plant and used in minute quantities to induce a growth response in another part; frequently referring particularly to auxins. *See* Plant-growth regulator.

Host.   Any plant or animal that is invaded by a parasite and from which the parasite obtains part or all of its food and on which it reproduces.

Hot-water-insoluble nitrogen (HWIN).   Fertilizer nitrogen not soluble in hot water at 212°F (100°C). Used to determine the activity index of ureaforms. *See* Nitrogen activity index.

Humidity, relative.   The weight of water vapor in a given quantity of air as compared with the total weight of water vapor that the air is capable of holding at a given temperature and expressed as a percentage.

Humus.   The organic fraction of soil in which decomposition by microorganisms is so far advanced that its original form cannot be distinguished.

Hyaline.   Transparent or nearly so; translucent; clear or colorless.

Hybrid (v. **hybridize).**   The offspring of two individuals of different genetic character; the crossing of two individuals differing in one or more heritable characteristics; also used for the offspring resulting from a cross between two species.

Hydathode.   Special epidermal gland or pore structure in leaves and stems at the ends of vascular tissue through which water of guttation is exuded.

Hydrocarbon poisoning.   Poisoning caused by oil-based materials such as fuel oil, diesel, gasoline, kerosene, and tar.

Hydrogen-ion concentration.   *See* pH.

Hydrolysis, hydrolyze.   Enzymatic disintegration of a large molecule into smaller component ones; to undergo or cause to undergo a chemical reaction with ions of water.

Hydroseed.   To seed in a water mixture by pumping through a nozzle that sprays the mixture onto a seedbed. The water mixture may also contain amendments such as fertilizer and certain mulches.

Hymenium (pl. **hymenia).**   Spore-bearing layer of fungal fruit body consisting of asci or basidia.

Hypersensitive.   A response of plants to invasion by certain pathogens, that includes rapid dying of local tissue, which prevents further advance of the pathogen; especially the reaction of a rust fungus or other obligate parasite.

Hypha (pl. **hyphae,** adj. **hyphal).**   Single, tubular, largely microscopic filament that constitutes the basic vegetative unit of structure and function of most fungi. Hyphae may be single-celled or divided into multiple cells by cross walls (septa).

Hyphopodium (pl. **hyphopodia).**   Short, lateral, hyphal branch or appressorium (one or two cells) of an epiphytic fungus, usually flattened or lobed, specialized for host attachment or penetration; characteristic of *Gaeumannomyces*. A pale spot in the center often indicates the point of origin of a narrow filament that penetrates the host turfgrass cell.

Hypocotyl.   That portion of a developing embryo or seedling just below where the cotyledon is attached; the portion between the cotyledon and the radicle. The root meristem is situated at the tip of the hypocotyl.

I

Immune, immunity.   Not affected by or responsive to disease; exempt from infection due to its inherent properties (e.g., tough outer wall, hairiness, nature of natural openings, waxy coating, and thick cuticle).

Imperfect.   The asexual portion of a fungal life cycle. *See* Anamorph and Teleomorph.

Imperfect flower.   One lacking either stamens or pistils; unisexual.

Imperfect fungus.   An asexually reproducing fungus that may or may not produce sexual spores. *See* Deuteromycete.

Incidence of disease.   Number of individuals affected by disease as distinguished from disease severity. *See* Disease severity.

Incomplete development.   The life cycle of an insect that passes through the three stages of change (metamorphosis), including egg, nymph, and adult. Examples include grasshoppers, thrips, leafhoppers, aphids, and scale insects. *See* Complete development.

Incorporate.   To work or blend a fertilizer, pesticide, or other chemical completely into the soil.

Inert ingredients.   Inactive components of a pesticide formulation (such as water, sugar, dust, wetting and spreading agents, carrier, conditioning agents, emulsifiers, and propellants) that are not toxic to the target pest.

Infect (n. **infection).**   The dynamic process or act of entering (invasion, penetration) and establishing a parasitic (often pathogenic) relationship with a host plant. Once infection has been effected, colonization may begin.

Infection court.   Site in or on a host plant where infection can occur.

Infection cushion.   An aggregation of hyphae amassed on a surface of a host, which serves as a support for one or more infection pegs that penetrates the host cell wall.

Infection peg.   A small hyphal protrusion that penetrates the host cell wall.

Infectious disease.   A disease caused by a pathogen that multiplies and can be transmitted from plant to plant. *See* Noninfectious disease.

Infest, infested (n. infestation).   Containing or covered with large numbers of pests (insects, mites, nematodes, bacteria, fungi, weeds, and so on). Refers to the presence of pests in the environment, not to infection of hosts.

Infiltration.   Movement of water into the soil.

Inflorescence.   The flowering portion of a grass shoot, including the spikelets and any supporting axis or branch system.

Inhalation exposure.   Exposure resulting from breathing in pesticide vapors, dust, or spray particles.

Inhibition.   Prevention of growth or multiplication of microorganisms.

Injury.   Momentary (transitory) damage to a plant by a biotic, physical, or chemical agent. *See* Disease.

Inoculate (n. inoculation).   Introduction of a pathogen at the site of infection of a host plant (the infection court).

Inoculum (pl. inocula).   The pathogen or its parts (such as fungus spores, mycelium, bacterial cells, nematodes, virus particles) that come in contact with a grass host and cause disease.

Insect.   In the adult stage, true insects usually have wings, six walking legs, and three body divisions (head, thorax, and abdomen). The term is sometimes loosely used to describe numerous small invertebrate (no backbone) animals such as spiders, mites, ticks, centipedes, and millipedes. *See* Insect relative.

Insect relative.   One of the group of invertebrate (no backbone) animals resembling a true insect. The adult stage, however, does not have wings, six walking legs, and three body divisions. Examples include mites, ticks, spiders, centipedes, millipedes, sowbugs and pillbugs, and crayfish.

Insect vector.   An insect that transmits a disease-inducing organism or agent. Many aphids transmit plant viruses (see figure 5.6).

Insecticide.   A physical or chemical agent used to prevent, destroy, repel, or attract undesirable insects and certain insect relatives; a type of pesticide.

Instar.   Growth phase of an insect or nematode between successive molts.

Integrated pest management (IPM).   An organized program in which the best management methods available (including chemical, physical, cultural, biological, and regulatory) are used to keep pest populations below the economic and/or aesthetic injury level while avoiding adverse effects on humans, wildlife, and the environment.

Intercellular.   Between or among cells.

Interseeding.   Seeding into an established turf.

Intracellular.   Inside or through a cell or cells.

In vitro.   Biological reactions in glass or in an artificial environment; in culture; outside the living host, as opposed to in vivo.

In vivo.   Biological reactions taking place in the host; used in contrast to in vitro.

Involucre.   A circle of bracts surrounding a flower cluster or single flower.

IPM.   *See* Integrated pest management.

Irrigation.   Applying water to turfgrass or soil other than by natural rainfall.

Isolate.   To separate a culturable microorganism from an infected plant and grow it in the absence of other organisms.

J

Juvenile.   An immature nematode or insect. This term is usually applied to animal forms in which there is little dimorphism between immatures and adults.

K

Koch's postulates.   Four rules to be followed to prove the pathogenicity of a microorganism. The rules are (a) consistent association of a suspected causal agent with a disease syndrome; (b) isolation of the infectious agent, which is grown in pure culture; (c) reproduction of the disease syndrome after transmitting the infectious agent to healthy plants; and (d) reisolation of the infectious agent from the host (which should be identical to that found originally).

L

Label.   Printed information affixed to the pesticide container or wrapper by the manufacturer listing the contents, directions for use, and precautions. A pesticide label must be approved and registered by a federal and state agency (e.g., the U.S. Environmental Protection Agency [EPA] and the State Department of Agriculture).

Labeling.   All printed information about a pesticide provided by the manufacturer. It includes the label and any supplemental literature such as leaflets, flyers, and pamphlets supplied through the dealer.

Labile.   Easily destroyed; unstable.

Lacerate.   As if deeply split or finely torn.

Lanceolate.   Lance- or spear-shaped; oblong and tapering to a point.

Larva (pl. larvae).   Juvenile or immature stage that is distinctly different from the adult; immature or worm stages between the embryo or egg and the adult, as in insects and nematodes.

Larvacide.   An insecticide used to kill larvae of insects.

Latent.   Present but not manifested or visible; dormant, delayed.

Lateral shoot.   A shoot originating from a vegetative bud in the axil of a leaf or from the node of a stem, rhizome, or stolon.

Layering, soil.   Undesirable stratification within the A horizon of a soil profile; can be due to construction design, topdressing with different textured materials, inadequate onsite mixing of soil constituents, or blowing and washing of sand on other soil constituents.

LC$_{50}$.   Median *l*ethal *c*oncentration; amount of a substance (such as a pesticide) in water or air required to kill 50% of a test animal population in a 24-hour period. Usually expressed in parts per million (ppm) or micrograms per liter (mg/l). The lower the LC$_{50}$ value, the more poisonous the substance. It is often used as a measure of acute inhalation toxicity.

LD.   Lethal dose; number of pests required to cause death to a given species of plant or animal.

LD$_{50}$.   The milligrams of toxicant per kilogram body weight of an organism (mg/kg), administered orally or dermally, calculated to be lethal to 50% of the organisms in a specific test situation. If a chemical has an LD$_{50}$ of 50 mg/kg, it is more toxic than one having an LD$_{50}$ of 500 mg/kg. LD$_{50}$ is often used as a measure of acute oral or dermal toxicity.

Leach, leaching (adj. **leached).**   Removal of soluble nutrients, pesticides, and other materials through the soil profile by usually downward flowing or percolating water from deep irrigation or rain.

Leaf.   A lateral outgrowth of a stem produced in definite succession from the stem apex.

Leaf axis.   Upper angle between a leaf and the culm (stem) from which it grows.

Leaf bud.   Bud that develops into a leaf shoot and does not produce flowers.

Leafhopper.   Active insects, family Cicadellidae (order Homoptera), with sucking mouthparts; often a vector for mycoplasmas and viruses; may also cause direct plant injury during feeding.

Leaflet.   One of the several blades of a compound leaf.

Leaf margins.   Edge of leaf or leaflets.

Leaf spot.   Self-limiting necrotic area (lesion) on a leaf.

Lemma.   Lowermost of the two bracts enclosing the flower of a grass floret.

Lesion.   Well-marked, localized, often sunken area of diseased or disordered tissue; a wound.

Lethal.   Capable of causing death.

Life cycle, life history.   The series of stages in the growth and development of a plant, animal, or microorganism that occurs until the reappearance of the first stage.

Liquid concentrate (LC).   *See* Emulsifiable concentrate.

Ligule.   A membranous or hairlike appendage in grass leaves on the upper and inner side of the leaf blade where the leaf joins the sheath.

Lime.   A material containing the carbonates, oxides, and/or hydroxides of calcium and/or magnesium. Lime is used to raise the soil pH and neutralize soil acidity.

Loam.   A mellow, textural class of soil composed of about equal parts silt and sand and less than 20% clay.

Localized dry spot.   A dry spot of sod and soil amid normal, moist turf that resists rewetting by normal irrigation and rainfall; associated with a number of factors including thatch, soil fungal activity, shallow soil over buried material, or elevated sites in the terrain. Localized dry spots are most common in high-sand-content soils.

Localized infection.   An infection (such as a leaf spot) confined to a particular area and not spreading throughout the grass host.

Low-volume (LV) spray.   A concentrate spray applied to uniformly cover the crop being treated, but not as a full-cover spray to the point of run-off.

M

Macroconidium (pl. macroconidia).   Long or large conidium of a fungus relative to microconidia.

Macronutrients.   Essential chemical elements required in relatively large quantities for the growth of plants (such as nitrogen [N], phosphorus [P], and potassium [K]), usually in amounts greater than 1 ppm.

Maggot.   The immature stage or larva of a fly.

Manual transmission.   Spread or introduction of inoculum to infection courts by hand manipulation. *See* Mechanical transmission.

Masked symptoms.   Symptoms that are absent under certain environmental conditions but appear under others.

Mat.   Thatch intermixed with mineral matter that develops between the zone of green vegetation and the original soil surface; commonly associated with golf greens that have been topdressed.

Matrix.   Basal or foundation substance from which an object is formed or in which it is embedded.

Mechanical control.   Control of pests by physically eliminating them, such as by hand pulling, mowing, and aquatic harvesting of weeds.

Mechanical transmission or inoculation.   Artificial spread or introduction of inoculum (virus, bacterium, or fungus) to an infection court (especially a wound) by hand manipulation (such as mowing), accompanied by physical disruption of the host tissue.

Medulla (adj. medullary).   Central part of an organ or tissue, as in sclerotium; the part of fungal sporocarps composed mainly or entirely of longitudinal hyphal.

Melting-out.   A disease of turfgrasses (primarily Kentucky bluegrass) caused by the fungus *Drechslera poae*. It is characterized by the appearance of dark leaf spots, often followed by the killing of irregular areas of turf in warm-to-hot weather.

Membranaceous, membranous.   Membranelike; resembling a thin skin or parchment; thin and compact, but pliant.

Meristem (adj. meristematic).   Undifferentiated plant tissue that functions principally in cell division and is thus responsible for the first phase of growth.

Mesh.   Standard screens that separate solid particles into size ranges. The mesh is stated in the number of openings per linear inch. The finest practical screen is the 325-mesh, which has openings 44 microns in diameter and over 105,000 openings per square inch. A common range of granular formulations is in the 15–30 range. Particles that pass a 60-mesh screen are considered dusts.

Mesophyll.   Chlorophyllous parenchyma tissue in leaves between the two epidermal layers.

Mesostemic pesticide. One that has a high affinity with the plant surface and is absorbed by the waxy layers of the plant. It redistributes itself at the plant surface by superficial vapor movement and redistribution. It penetrates plant tissues, and has translaminar activity but is not systemic within the vascular system.

Metabolism. The sum total of the physical and chemical processes (physiological activities) occurring in the body of a living organism.

Metabolite. Any chemical component of a reaction series or process; a product of metabolism.

mg/kg. Used to express the amount of a substance (such as a pesticide) in milligrams per kilogram of animal body weight to produce a toxic or desired effect (1,000,000 mg = 1 kg = 2.20462 lb avoirdupois).

Microconidium (pl. **microconidia).** Small conidium of a fungus relative to macroconidia; microspore. *See* Macroconidium.

Microflora. Composite of microscopic plants and certain other microorganisms at a site.

Micronutrients. Essential chemical elements required in only minute amounts (less than 1 ppm) for the growth of plants, such as boron, chlorine, copper, iron, manganese, molybdenum, and zinc; trace elements. Micronutrients are often toxic when necessary amounts are exceeded.

Microorganism. A microscopic organism (microbe) such as a bacterium, fungus, nematode, or protozoan.

Microscopic. Too small to be seen except with the aid of a microscope.

Midrib. The central, thickened vein of a leaf extending from the stem to the leaf tip.

Miscible. Capable of being mixed and remaining mixed under normal conditions.

Mite. Minute animals ¼₄ to ½₂ in long of the families Tetranchidae and Eriophyidae. Mites have no evident body divisions, and they are six-legged as larvae and eight-legged as adults.

Miticide. *See* Acaricide.

Mixture. A combination of two or more turfgrass species planted together.

MLO. *See* Mycoplasmalike organism, Phytoplasma.

Mold. Any microfungus with conspicuous, profuse, or woolly superficial growth (mycelium and/or spore masses) on various substrates; often saprophytic. Molds grow commonly on damp or decaying matter and on the surface of plant tissues.

Mollicute. One of a group of prokaryotic microorganisms lacking a cell wall and bounded by flexuous membranes, such as MLO.

Molluscicide. A pesticide used to control mollusks (slugs and snails).

Molt. To shed periodically an outer layer, as with the exoskeleton of an insect or nematode.

Monilioid cells. Rounded fungal cells.

Monoclinous. Having the gametangium and oogonium originating from the same hypha.

Monocotyledoneae, monocots (adj. **monocotyledonous).** The subclass of flowering plants, including the grasses, the embryo of which has one cotyledon (seed leaf).

Monocyclic.   Having one cycle; no secondary infection.

Monogenic.   Containing or controlled by one gene. In nematodes, producing offspring of only one sex.

Moribund.   Being in a dying state.

Morphology.   Study or science of the form, structure, and development of organisms.

Mosaic.   Disease symptom characterized by a patchy mottling of the foliage or by variegated patterns of light green to yellow that form a mosaic; symptomatic of many viral infections.

Mottle, mottling.   Disease symptom comprising an irregular pattern of light and dark areas; often symptomatic of viral diseases.

Mowing frequency.   The number of times a turfgrass community is mowed per week, month, or growing season. (The reciprocal of mowing frequency is mowing interval—the time elapsed between successive mowings.)

Mowing height.   The distance above the ground surface at which the turfgrass is cut during mowing.

Mowing pattern.   The patterns of back-and-forth travel recommended for mowing turf. Patterns may be changed regularly to distribute wear and compaction, to avoid creating "grain," and to create visually aesthetic effects, especially for spectator sports.

Muck (soil).   Highly decomposed black soil in which the original plant parts are not recognizable; similar to peat soil, but having a higher percentage of organic matter.

Mulch.   Any nonliving material (e.g., straw, sawdust, dry leaves, and plastic film) spread over the soil surface to protect it.

Multigenic.   Containing or controlled by a number of genes.

Multinucleate.   Having more than one nucleus per cell.

Multiple infection.   Invaded by more than one parasite.

Multiseptate.   Having a number of septa or partitions.

Muriform.   Having both cross (transverse) and longitudinal septa.

Mutagenesis.   The production of mutations, often caused by chemical and physical agents.

Mutagenic.   Pertaining to or causing mutagenesis.

Mutation, mutant.   An abrupt heritable or genetic change in a gene or an increase in chromosome number.

Mutualism.   A form of symbiosis in which two or more organisms of different species are living together for the benefit of both or all. *See* Symbiosis.

Mycelia sterilia.   Members of the Deuteromycetes (Fungi Imperfecti) for which spores, except for chlamydospores, are not present.

Mycelium (pl. mycelia).   The strands, group, or mass of interwoven filaments (hyphae) that comprises the vegetative body (thallus) of a true fungus. The mycelia of fungi show great variation in appearance and structure.

Mycoplasma.   A genus of prokaryotic, parasitic bacteria of the class Mollicutes, measuring 0.1 to 1 micron, that lacks a rigid cell wall and is highly variable in shape (pleomorphic). Mycoplasmas induce viruslike symptoms in infected plants.

Mycoplasmalike organism (MLO).   Bacterium with apparent features of a mycoplasma, which has not been proven to be a mycoplasma. Found in phloem and phloem parenchyma of diseased plants and assumed to be the cause of the disease. Now referred to as *phytoplasma*. A phytoplasmal disease of bermudagrass is white leaf.

Mycorrhiza (pl. mycorrhizae or mycorrhizas, adj. mycorrhizal).   A usually intimate, symbiotic association of the mycelium of a typically nonpathogenic or weakly pathogenic fungus with the roots of a higher plant, that may aid in the uptake of certain nutrients by the plant host.

Myxamoeba (pl. myxamoebae).   The zoospore state (swarm cell) that succeeds an amoebalike state (as in slime molds); an amoeboid cell.

Myxomycete.   The true slime molds; a class of primitive organisms characterized by an amoeboid, multinucleate, motile mass of protoplasm.

N

Necrosis (pl. necroses, adj. necrotic).   Localized or general death and disintegration of plant cells or plant parts, usually resulting in tissue turning brown or black due to oxidation of phenolics. Commonly a symptom of disease or injury.

Nematicide.   A chemical or physical agent that prevents, destroys, or repels nematodes; a type of pesticide.

Nematodes.   Generally microscopic, unsegmented roundworms that usually live free in moist soil, water, or decaying matter, or as parasites in plants and animals.

Neutral soil.   A soil in which the surface layer is neither acidic nor alkaline in reaction. *See* pH.

Nitrate reduction.   The reduction of nitrates to nitrites or ammonia.

Nitrification.   Formation of nitrates and nitrites from ammonia by soil microorganisms.

Nitrogen activity index (AI).   Applied to ureaformaldehyde compounds and mixtures containing such compounds. The AI is the percentage of cold-water-insoluble nitrogen (CWIN) that is soluble in hot water. AI = % CWIN – % HWIN ÷ % CWIN × 100.

Node.   The swollen joint of a culm or stem; the meristematic tissue from which a leaf root or stolons, rhizomes or flowers may develop.

Noninfectious disease.   A disease (or disorder) that is caused by unfavorable growing conditions (e.g., extremes in weather, air or soil pollutants, nutrient deficiencies or excesses, and toxic chemicals) and not by a pathogen. It cannot be transmitted from plant to plant. *See* Infectious disease.

Nonpersistent.   Dissipating, said of viruses that are infectious within an insect vector for short periods and are transmissible without a latent period and without prior multiplication and translocation within the vector.

Nonselective pesticide.   A pesticide effective against a wide range of plants, animals, or other organisms, both desirable and undesirable.

Nonseptate.   Without cross walls.

Nontarget organism.   A plant or animal other than the one against which a pesticide is applied.

Nozzle.   A device for metering and dispensing a spray solution.

Nucleus (pl. **nuclei).**   The dense, usually spherical or ovoid, protoplasmic body present in most living cells of plants and animals and essential in all synthetic, developmental, and reproductive activities of a cell; contains the chromosomes, which transmit hereditary characteristics.

Nursery, turfgrass.   A place where turfgrasses are vegetatively propagated for increase and planting as stolons or sprigs or where sod is grown for later transplanting by sodding or plugging. Sometimes also used for experimentation.

Nutlet.   A small nut or nutlike fruit.

Nutrients, plant.   Essential elements available to plants through soil, air, and water that are utilized in metabolism and growth.

Nymph.   The immature stage (resembling an adult) of an insect that passes through three stages (egg, nymph, and adult) in its development. Nymphs are smaller than adults; they have no wings and cannot reproduce.

O

Obligate.   Necessary; essential; obliged.

Obligate parasite.   An organism or agent that can grow and multiply in nature only on or in living tissue and that cannot be cultured on an artificial medium. Examples include rusts, powdery mildews, viruses, and plant-parasitic nematodes. *See* Parasite, Facultative parasite.

Obligate saprophyte or saprobe.   An organism that can develop only on or in dead organic matter. These organisms are important in the decomposition of thatch and the breakdown of organic matter in the soil.

Off-site mixing.   The mixing of soil and amendments during root-zone modification at a place other than the site where they are to be used.

Off-target movement.   The movement of a pesticide or fertilizer out of the area intended for application.

Oncogenic.   A substance that causes tumor formations.

Oogonium, oogone, oocyst (pl. **oogonia).**   One-celled female sex organ (gametangium) containing one or more eggs (gametes or oospheres) of the fungal group oomycetes (such as *Pythium*).

Oomycete.   Typically aquatic, saprobic or parasitic fungus that produces oogonia, antheridia, sexual oospores, and motile asexual zoospores.

Oospore.   A thick-walled, resting fungus spore in the oomycetes that develops from a fertilized oosphere or by parthenogenesis.

Opposite leaves.   Those paired opposite each other on a stem at a node.

Operator.   A person who may apply pesticides only under the direct supervision of the sponsoring applicator and only to areas covered by the sponsoring applicator's license.

Oral exposure.   Exposure to a substance (such as a pesticide) that occurs through the mouth.

Oral toxicity.   Toxic effects brought about by the passage of a material into the body through the mouth.

Organ.   One of the major parts of a plant body—leaf, stem, root. An organ is composed of various cells and tissues and adapted to perform certain specific functions.

Organelle.   Membrane-delimited structure or body within a cell (such as Golgi apparatus, mitochondria, chloroplasts) that has a specialized function.

Organic matter.   Any plant or animal material capable of being decomposed and resynthesized (in the soil).

Organic phosphorus insecticide.   A synthetic compound derived from phosphoric acid. Organic phosphorus insecticides are primarily contact killers with relatively short-lived effects. They are decomposed by water, pH extremes, high temperatures, and microorganisms. Examples are malathion, diazinon, chlorpyrifos, Aspon, phorate, isofenphos, trichlorfon, dimethoate, and fenthion.

Orifice.   An opening in a nozzle tip, duster, or granular applicator through which the spray, dust, or granules flow.

Osmosis (adj. osmotic).   Diffusion of fluids (usually water) through a differentially permeable membrane from the side of the higher concentration of water to the side of the lower concentration of water.

Ostiole, ostiolum.   Small, more-or-less circular opening in a fruit body (spermogonium or pycnium, perithecium, or pycnidium) through which spores are discharged.

Ovate, oviform, ovoid.   Shaped like a hen's egg, with one end narrower than the other.

Overlap, percent.   The proportion of treated turf area that receives pesticide or fertilizer from two or more nozzles or adjoining swaths. The percent of overlap can be calculated by subtracting the nozzle spacing from the individual pattern width, and then dividing the nozzle spacing and multiplying the result by 100.

Overseason.   To live over from one planting season to the next.

Oviposit.   To deposit or lay eggs; in insects, with an ovipositor.

Ovipositor.   Egg-laying organ of a female insect, located at the end of the abdomen.

Overseeding.   The sowing of grass seed over an established turfgrass area, resulting in filling in sparse areas of the turf with additional turfgrass plants. In transitional and southern regions, overseeding is usually achieved with a temporary turfgrass to provide green, active grass growth during dormancy of the original turf, which is usually a warm-season turfgrass.

Overwinter.   To survive the winter period.

Ovicide.   A pesticide used to kill the eggs of insects.

Ovoid.   Egg shaped.

Ovule.   A rudimentary seed; an enclosed structure, consisting of a female gametophyte (egg cell), nucellus, and one or two integuments. After fertilization, the ovule becomes a seed.

Oxidation-reduction reaction.   A chemical reaction in which one substance *oxidizes* (loses electrons, or loses hydrogen ions and their associated electrons, or combines with oxygen) and a second substance *reduces* (gains electrons, or gains hydrogen ions and their associated electrons, or loses oxygen).

Ozone (O₃). A highly reactive form of oxygen (a photochemical oxidant) that in relatively high concentrations may injure plants, humans, and animals. A common air pollutant in areas with heavy vehicular traffic.

P

Palea. Uppermost and smaller of the two bracts enclosing the flower of a grass floret.

Palmate leaf. A compound leaf having three or more lobes or leaflets radiating from a common center; shaped like a hand with the fingers spread.

Palmately veined. Type of net venation in which the main veins of a leaf blade branch out from the apex of the petiole like the fingers of a hand.

PAN (peroxyacyl nitrates). Toxic air pollutants produced by photochemical reactions in daylight air originating from the exhaust of internal combustion engines, which are injurious to plants.

Panicle. A loosely branched, open-flower cluster common in the grass family (Poaceae); a compound raceme.

Pappus. A tuft of bristles or similar structure surmounting the achene of certain plants (such as dandelions and thistles).

Parasite. An organism (such as a fungus, bacterium, mollicute (phytoplasma, spiroplasma), nematode, or insect), virus, or viroid living in close association with, another living organism (host) and obtaining food from it. *See* Facultative parasite, Facultative saprophytes, Obligate parasites, Pathogen.

Parasitic insect. An insect that lives in or on the body of another insect.

Parenchyma. Physiologically active, soft tissue of higher plants composed of thin-walled, often isodiametric cells that commonly store food or perform other functions, retain meristematic potential, and have intercellular spaces between them.

Parthenogenesis (adj. parthenogenetic). A type of asexual reproduction. Development of an egg (female gamete) into a new individual without fertilization by a sperm (male gamete).

Parts per million (ppm). The number of parts by weight or volume of a given compound in 1 million parts of the final mixture; given in milligrams per liter; 1 ppm = 1 mg/l.

Pasteurization. Freeing a medium of selected pathogenic microorganisms in soil, or other propagating media, using heat. The treatment does not materially change the natural characteristics of the substance treated.

Patch. Distinctly delimited, somewhat circular area of turfgrass plants in which most or all are affected by disease.

Pathogen. An organism or agent (such as a fungus, bacterium, mollicute (phytoplasma, spiroplasma), nematode, virus or viroid) capable of causing disease in a particular host (suscept) or range of hosts. Most pathogens are parasites, but there are exceptions.

Pathogenesis. The sequence of processes in disease development, from the initial contact between a pathogen and its host to the final reaction in the host; production and development of disease.

Pathogenicity. The ability of a pathogen to cause (incite) disease; the state or condition of being pathogenic.

Pathovar (pv).   A type of subspecies; a strain or group of strains of a bacterial species differentiated by pathogenicity in one or more hosts (species or cultivars); pathotype.

Peat.   An unconsolidated soil mass, high in semicarbonized organic materials, consisting of partially decomposed plant tissue formed in water of marshes, bogs, or swamps, usually under conditions of high acidity.

Pedicel, pedicle.   Small, slender stalk of a single grass floret.

Peduncle.   Stalk or main stem of an inflorescence, part of an inflorescence, or a fructification; stalk.

Penetrant.   A chemical that, when applied to the surface of a plant, passes into underlying tissue in quantities that are toxic to the target organism. There are fungicides that are local penetrants (such as iprodione), which remain in place and are not translocated; acropetal penetrants (such as fenarimol and thiophanate materials), which move only upward in the plant after penetration; and systemic penetrants (such as fosetyl aluminum), which are translocated uniformly throughout the plant after penetration.

Percolation.   The usually downward movement of water through the soil profile.

Perennial plant.   A plant that lives 2 years or more.

Perfect flower.   One having both stamens and pistils (carpels); a hermaphroditic flower.

Perfect state.   State in the life cycle of a fungus in which sexual spores are formed after nuclear fusion or by parthenogenesis; sexual state or teleomorph; capability for sexual reproduction. *See* Teleomorph.

Perianth.   The external envelope of a flower, consisting of sepals and petals (calyx and corolla).

Pericycle.   The layer of cells in a stem or root located between the endodermis and the vascular cylinder. Branch roots arise from the pericycle.

Perithecium (pl. perithecia).   Flasklike to more or less globose, thin- or thick-walled ascocarp with an ostiolelike opening, asci, and ascospores.

Permanent wilting percentage.   The amount of water in the soil at the time plants become permanently wilted.

Permeability.   The relative ease by which water moves through soil.

Persistent, persistence.   The condition of a pesticide that retains its pesticidally active form for some time after application. Persistence is usually measured by half-life or months of pesticidal activity. *See* Residual. Also pertains to circulatory viruses that are infectious within insect vectors for long periods.

Pest.   Any plant or animal (insect, mite, rodent, nematode, fungus, weed, and the like) injurious to the health of a beneficial plant, plant product, animal, human, or environment.

Pesticide.   A substance or mixture of substances used to destroy, prevent, inhibit, repel, or attract any animal, plant, or plant pathogen considered a pest (except microorganisms living in or on the body of humans and animals), and any substance intended for use as a plant regulator, chemical defoliant, or desiccant. Examples are insecticides, miticides or acaricides, herbicides, fungicides, nematicides, disinfectants, bactericides, and rodenticides.

Pesticide dealer. A person who distributes registered pesticides.

Petal. Portion of a flower surrounding the stamen and pistil; sometimes colorful or showy.

Petiole. The stalk that attaches a leaf blade to a stem.

pH. A numerical measure of the acidity or hydrogen ion activity within the range of 0 to 14. A pH of 7 indicates neutrality; values above 7 are increasingly basic (alkaline), while those below 7 are increasingly acidic. The scale is logarithmic [pH = $-\log (H+)$], so that one pH unit change is equal to a tenfold change in the hydrogen ion concentration.

Pheromone. A chemical used to affect the reproductive or aggregation behavior of animals, especially insect pests.

Phloem. Food-conducting tissues in plants; complex vascular tissue consisting of sieve tubes, companion cells, fibers, and phloem parenchyma.

Photodegradation. The breakdown of a chemical (such as a pesticide) by means of sunlight.

Photosynthesis. The complex fundamental process by which carbohydrate food (sugar) is produced from carbon dioxide, water, and light energy in chlorophyll-containing plants.

Phycomycete. An outdated class of fungi (now treated as Chromista, Chytridomycota, and Zygomyco) where mycelium has no cross walls.

Physiogenic (physiological) disease. A disease (or disorder) produced by some unfavorable genetic, physical, or environmental factor. *See* Noninfectious disease.

Physiological form or race. Subdivision within a species that differs from other physiological forms or races in behavior or other characteristics but not in morphology. Differences may occur in virulence, symptom expression, biochemical and physiological properties, or host range. *See* Biotype.

Phytoplasma. *See* Mycoplasmalike organism.

Phytotoxic. Harmful (injurious or poisonous) to plants; usually describing a chemical.

Phytotoxicity, acute. The immediate dramatic impact of a chemical on the well-being of a plant or plant part.

Phytotoxicity, chronic. The continuous low impact of a chemical on the well-being of a plant or plant part.

Pinnate. Having branches, lobes, leaflets, or veins; arranged in a featherlike manner; attached or arranged on two sides of a stem.

Piscicide. A pesticide used to control fish.

Pistil. The seed-producing organ in a seed plant, typically consisting of an ovary and one or more stigmas and styles.

Plant disease. *See* Disease.

Plant-growth regulator (PGR). A substance that affects plants or plant parts through physiological rather than physical action by accelerating or retarding growth, prolonging or breaking a dormant condition, promoting decay, or inducing other physiological changes.

Plasmid.　Generally a small, covalently closed, circular piece of nonchromosomal, double-stranded DNA found in certain bacteria and fungi that carries generally nonessential genetic information and is self-replicating. (Some plasmids are linear molecules.) Plasmids are used in recombinant DNA experiments as acceptors of foreign DNA.

Plasmodium (pl. plasmodia, adj. plasmodial).　Naked, amoeboid mass of protoplasm, generally reticulate, without cell walls and containing multiple nuclei, resulting from the fusion of uninucleate amoeboid cells. Plasmodia move and feed in amoeboid fashion. They are produced by slime molds. *See* Myxomycete.

Pleomorphic.　Able to assume various shapes and perhaps sizes; having more than one independent form or spore stage in the life cycle; polymorphic. Examples include mycoplasma and spiroplasmas.

Plug.　A small, usually round piece of turfgrass with adhering soil used in vegetative propagation; to propagate turfgrasses vegetatively by means of small pieces of sod.

Poison.　A substance that, when absorbed or taken orally, can cause illness, death, retardation of growth, or shortening of life; also called biocide.

Poison Resource Center.　A designated hospital that provides information on the treatment of poisonings by pesticides and other substances 24 hours a day via a toll-free telephone number.

Polar.　At the ends or poles.

Polycyclic.　A disease in which many cycles occur in one year or one growing season, resulting in many secondary infections. *See* Monocyclic.

Polygenic.　A character controlled by three or more genes. *See* Multigenic.

Polyhedral, polyhedron.　A spheroidal particle or crystal having many sides or plane faces (as in certain viruses).

Porosity.　That percentage of the total bulk volume of soil not filled by solid particles.

Postemergence.　The period after the appearance of a specified weed or turfgrass; a material (such as a pesticide or fertilizer) applied after emergence of the grass or weed.

Powdery mildew.　Fungus (or disease) that forms a superficial white mantle of mycelium on the surface of grass leaves, usually in the shade.

Power raking.　Using a machine to mechanically remove excessive thatch from a turfgrass area.

ppm (parts per million).　A method of expressing the concentration of chemicals, proportions, or the relative content of one substance in another. One part per million is 1 lb in 500 tons; 1 ounce in 8000 gal; 1 in in about 16 mi.

Predaceous or predacious fungi.　Fungi that parasitize amoebae, nematodes, and other small terrestrial or aquatic animals.

Predator.　An animal (including insects and nematodes) that preys, destroys, or devours other animals.

Predispose.　To make prone to infection and disease.

Preemergence.　The period before a turfgrass or weed emerges from the soil; an application of a material (such as a fertilizer or pesticide) before emergence of the grass or weed.

Preplanting.   Any time that falls before the crop is planted; application of chemicals (such as fertilizer or pesticide) before turfgrass is planted.

Pressure rinsing.   A method of removing most of the pesticide residue from a container by flushing it with water or other wash materials under force.

Preventive control.   Control of pests by inhibiting the entry and spread of the pest into an area.

Primary host.   Plant on which the sexual forms of an aphid mate and lay eggs to overwinter. *See* Secondary host.

Primary inoculum.   Propagules or vegetative structures of a pathogen, usually from an overwintering source, that causes initial rather than secondary outbreaks of disease. *See* Secondary inoculum.

Private pesticide applicator.   A person certified to use or supervise the use of a restricted-use pesticide on property owned, leased, or rented by himself/herself or by his/her employer on more than two neighbors' farms, as exchange labor, for the purpose of producing an agricultural commodity primarily intended for sale, consumption, propagation, or other use by humans or animals.

Profile (soil).   A vertical section of the soil that extends through all the horizons and into its parent material.

Progeny.   The young or offspring.

Prokaryote (adj. **prokaryotic).**   An organism (such as bacteria and blue-green algae) lacking membrane-limited nuclei or organelles and not exhibiting mitosis; organisms with a genome of a simple, circular, double-stranded DNA free in the cytoplasm and not in organelles. *See* Eukaryote.

Promycelium (pl. **promycelia).**   Basidium of the rusts and smuts; initial short and short-lived hypha produced upon teliospore germination.

Propagule.   Any part of an organism capable of initiating independent growth when separated from the parent body.

Propellent.   A gas or liquid used in a pressurized pesticide product to expel the contents.

Protectant pesticide.   A pesticide applied to a plant or animal prior to the appearance or occurrence of the pest to prevent or inhibit infection or injury by the pest. *See* Contact pesticide.

Protection.   Placement of a chemical or physical barrier between the pest and the host that prevents infection or injury.

Protein.   Complex, high-molecular-weight, organic, nitrogenous substance (polymer compound) composed of amino acids joined by peptide bonds.

Protoplasm.   Living material within a cell in which all vital functions of nutrition, secretion, growth, and reproduction depend; essential semifluid, viscous, translucent colloid of all plant and animal cells.

Pseudothecium (pl. **pseudothecia).**   A perithecium like ascocarp with a dispersed rather than an organized hymenium.

psi.   Pressure; measured in pounds per square inch.

Public pesticide applicator.   An applicator who works for a governmental organization supported by tax money that requires its employees to use pesticides only in the maintenance of its property.

Pupa (pl. pupae).   The resting, nonfeeding stage of an insect that passes through four stages (egg, larva, pupa, adult) in its development.

Pupate.   To be inactive, as with certain insects in a cocoon or case, before ultimate maturation into an adult.

Pure live seed (PLS).   Percentage of the content of a seed lot that is pure and viable.

Pustule.   A small, blisterlike or pimplelike, frequently erumpent elevation (spot or sorus) from which erupts a fruiting structure of a fungus (such as rust or smut) that produces spores. *See* Sorus.

pv.   *See* Pathovar.

Pycnidium (pl. pycnidia, adj. **pycnidial).**   An asexual, closed, usually cup- or flask-shaped-to-globose, ostiolate, thin-walled, brown fruiting structure (or variously shaped cavity) of certain fungi (*Ascochyta, Septoria*) that is speck-sized and contains conidia. Commonly found in diseased tissue.

Pycniospore.   Haploid, sexually derived spore (spermatium) formed in a pycnium (spermogonium).

Pycnium (pl. pycnia).   Older term now replaced by *spermogonium.* Haploid, flask-shaped structure that contains spermatia (pycniospores) and filaments (receptive hyphae); produced by rust fungi.

Pyrethrin.   Also known as pyrethrum. An insecticide that, at low levels, excites the nervous system of insects, causing affected insects to leave the area where the insecticide was placed.

Q

Quarantine.   State and/or federal legislative control of the transport, import, export and/or sale of plants or plant parts, usually to prevent spread of pathogens, insects, mites, weeds, or other pests; holding of imported plants or plant parts in isolation for a period to ensure their freedom from diseases and pests.

Quiescent, quiescence.   Dormant; quiet; a rest period caused by external conditions unfavorable to germination or growth.

R

Race (or strain).   A subgroup or biotype of pathogens within a species, variety, or pathovar distinguished by behavior (differences in virulence, symptom expression, or, to some extent, in host range) but not by morphology; different in virulence to cultivars of the same host species; a genetically and often geographically distinct mating group within a species. *See* Physiologic race.

Raceme.   Type of inflorescence in which the main axis is elongated and unbranched; the flowers are borne on pedicels of roughly equal length.

Rachilla.   Internal axis of a grass spikelet.

Rachis.   Elongated main axis of a grass inflorescence (spike or raceme).

Radicle.   The part of the embryonic axis that becomes the primary root; the root primordium of an embryo; the first part of the embryo to start growth during seed germination.

Raster.   The pattern of hairs or setae on the underside of the last abdominal segment on white grub larvae that is used in identification.

Rate.   The amount of active ingredient of a pesticide applied per unit area (usually 1000 sq ft); also called dose and dosage.

Ray flower.   One of the marginal florets surrounding the disk of tubular flowers in the flower heads of certain composite plants such as aster and marigold. *See* Disk flower.

Receptacle (pl. receptacla).   The more or less expanded terminal portion of the stem on which flower parts are borne; enlarged upper end of a pedicel or peduncle to which flower parts are attached.

Receptive hypha.   Specialized hypha that protrudes from the top of a spermogonium (pycnium) and functions in sexual reproduction.

Recuperative potential.   The ability of a turfgrass to recover from injury through vegetative growth.

Reentry or restricted-entry time interval.   The period of time following a pesticide application during which worker entry to the treated area is restricted without wearing protective clothing.

Reestablishment turf.   Rebuilding a lawn or other turf area by complete removal of any existing turf, followed by site preparation and planting; does not encompare rebuilding.

Registration number.   A number on the label that indicates that a product has been registered (in the United States by the federal EPA).

Relative humidity (RH).   The ratio of the quantity of water vapor present in the atmosphere to the quantity that would saturate it at the same temperature.

Renovation turf.   The complete reconstruction of a turf area, including the removal of existing turf; cultivation and other seedbed preparation; and complete seeding, sprigging, plugging, or sodding of the area with turfgrass.

Repeated exposure.   Low-level constant or intermittent exposure to one or more pesticides or other poisons over a long period of time.

Repellent.   A pesticide used to repel rather than kill an animal pest.

Replicate.   Repetition of an experiment or procedure at the same time and place (one of several identical experiments, procedures, or samples); formation of new virus particles within an infected host cell.

Reproduction, sexual.   Development of new plants by seeds (except in apomixis); the sexual cycle of a fungus or other microorganism.

Reproduction, vegetative (vegetative propagation).   Reproduction by other than sexually produced seed, including grafting, cutting, layering, and apomixis; the asexual cycle of fungi or other microorganisms.

Resident bacteria.   Pathogenic and nonpathogenic bacteria that persist and multiply on surfaces of leaves, fruit, and expanding buds of plants without causing disease. *See* Epiphyte.

Residual.   The property of a pesticide to persist after application in amounts sufficient to kill pests for several days to several weeks or even longer.

Residue.   The amount of pesticide present following application.

Resistance.   The inherent ability of an organism (host plant) to overcome, retard, or avoid, completely or to some degree, the activity (infection) of a pathogen or other damaging factor; the ability of a pest to withstand exposure to certain pesticides. A plant may be slightly, moderately, or highly resistant.

Resistant plant.   Plant strains, cultivars, or varieties that are not as heavily attacked by pests or better able to survive pest attack.

Resistant pest species.   Pests that survive relatively high rates of pesticide application.

Respiration.   A series of enzymatic reactions (oxidation processes) whereby a living organism produces energy for cellular activities, usually by oxidizing carbohydrate and releasing carbon dioxide.

Respirator.   A device that filters materials and fumes out of the air before it is breathed.

Respiratory toxicity.   Intake of a substance (such as a pesticide) through nasal, oral, and throat passages into the lungs.

Resting.   Temporarily dormant, usually thick-walled spore or sclerotium capable of later germination and initiation of infection after a resting period, frequently overwinter; resistant to extremes in temperature and moisture.

Restricted-entry interval.   *See* Reentry or restricted-entry time interval.

Restricted-use pesticide.   A pesticide that could harm the environment or the applicator, even when used as directed. A restricted-use pesticide is for retail sale and use only by certified applicators for persons under their direct supervision and only for those uses covered by the certified applicator's certification.

Retention.   Ability of a plant or other surface to hold a pesticide.

RH.   *See* Relative humidity.

Rhizome (adj. **rhizomatous).**   A usually horizontal, jointed, commonly underground stem of grasses that forms both roots and leafy shoots at its nodes; often enlarged by food storage; may originate from the main stem or from tillers.

Rhizosphere.   The soil microenvironment immediately surrounding (within 5 mm) a living root; the microflora is frequently richer and different from that of soil away from the root.

Rind.   The firm outer layer of a rhizomorph, sclerotium, or other organ; cortex.

Rod.   A bacterial cell with a straight central axis that is longer than the diameter of the cross section of the cell.

Rodent.   A kind of mammal. Rodents feed mostly on plants and grains, although some occasionally eat meat. Examples are rats, mice, and squirrels.

Rodenticide.   A pesticide used for the control of rats, mice, and other rodents.

Root hair.   Threadlike, single-celled outgrowth from a root epidermal cell, through which water, nutrients, and other substances are absorbed into a plant.

Root knot.   A nematode-caused disease characterized by round-to-irregular galls (knots) on the roots, caused by *Meloidogyne* species. Most common in sandy soils, where it attacks over 2000 kinds of plants.

Root system.   The total mass of roots of a single plant.

Root zone.   The layer of soil or growth medium in which most of the turfgrass roots are active; usually defined by the depth from the surface.

Rosette.   A basal, circular cluster of leaves or other organs not separated by evident internodal stem elongation.

Rot.   State of decomposition and putrefaction; the softening, discoloration, and often disintegration of plant tissue by enzymes produced by fungal or bacterial infection. *See* Decay.

Rough.   The area of a golf course adjacent to the fairway that is mowed at a higher height and not as highly managed.

Runner hyphae.   Thickened hyphal strands mostly on roots that may spread a fungus from one plant to another. *See* Ectotrophic.

Runoff.   Pesticide or other material that is carried away from an area by the flow of surface water. Also used to describe the rate of application to a surface—spray to runoff.

Rust.   A disease caused by one of the rust fungi (Uredinales in the basidiomycetes) or the fungus itself; a disease giving a rusty appearance (sign) to a plant.

S

Safener.   A material added to a pesticide that reduces the phytotoxicity of another chemical.

Safety.   The practical certainty that injury will not result from the use of a substance in a proposed quantity or manner.

Saline soil.   A soil containing sufficient soluble salts to impair turfgrass growth.

Sand.   A soil particle between 0.05 and 2.0 mm in diameter.

Sanitation, sanitize.   Destruction (removal) of infected and infested plants or plant parts; elimination of disease inoculum and insect vectors; decontamination of tools, equipment, containers, hands, and so on; cultural methods of disease control that reduce inoculum.

Saprobe, saprophyte (adj. saprophytic, saprogenic, saprogenous, saprophilous, saprobic).   An organism that obtains its food from dead organic matter commonly causing its decay; a necrophyte on dead material that is not part of a living host. *See* Parasite.

Saranex.   A waterproof material applied to Tyvek clothing that provides additional protection from dermal pesticide exposure.

Scald, turf.   A condition that develops when a turfgrass collapses and turns brown after sudden exposure from intense sunlight, hot water, or heat.

Scalp.   To remove an excessive quantity of functioning, green turfgrass leaves at any one mowing, resulting in a shabby, brown appearance caused by exposed crowns, stolons, dead leaves, and even bare soil.

Scientific method.   An approach to a problem that consists of stating the problem, establishing one or more hypotheses as solutions to the problem, testing these hypotheses by experimentation or observation, and accepting or rejecting the hypotheses.

Sclerotium (pl. sclerotia).   A small, compact, often hard interwoven mass of dormant fungus hyphae, usually more or less spherical or flat and dark in color with differentiated rind and medulla. A vegetative resting body of a fungus composed of usually thick-walled, special-size hyphal cells with or without the addition of host tissue or soil. The structure may remain dormant in soil, plant refuse, or seed for long periods and is capable of surviving under unfavorable environmental conditions.

Scorch.   Sudden browning (burning) and death (necrosis) of leaf tissue due to infection or unfavorable environmental conditions found typically at leaf margins or tips.

Scouting.   The process of systematically searching for pests in an area, identifying them, mapping their location, estimating their numbers, and determining their damage potential.

Scum.   The layer of algae on the soil surface of thinned turfs; drying can produce a somewhat impervious layer that can impair subsequent shoot emergence.

Scutellum, scutellate (pl. scutella, adj. scutellar).   The single cotyledon of a grass embryo; the rudimentary leaflike structure at the base of the first node of the embryonic culm of a grass plant; triangular structure on the back of true bugs.

Secondary cycle.   Of plant disease, any cycle initiated by inoculum generated during the same season.

Secondary host.   The host plant on which the asexual aphid form occurs; also used to indicate alternative hosts for pathogens. Do *not* confuse with alternate host, which involves two hosts required for completion of the life cycle of a fungus.

Secondary inoculum.   Inoculum resulting from primary infections or other secondary infections in the same season. *See* Primary inoculum.

Secondary organism (pathogen).   Organism that multiplies in already diseased tissue; not the primary pathogen.

Sedentary (sessile).   Remaining in a fixed location; stationary.

Sedge.   A grasslike plant with triangular stems that spreads by rhizomes and overwinters as a tuber.

Seed.   The mature ovule of a flowering plant containing an embryo, sometimes an endosperm, and a seed coat.

Seed blend.   A combination of seeds of two or more cultivars of the same turfgrass species.

Seed (certified).   Progeny of foundation, registered, or other certified seed that is so handled as to maintain satisfactory genetic identity and/or purity and that has been approved and certified by a certifying agency.

Seedbed.   Soil that has been prepared for planting seeds, plugs, sprigs, sod, and the like.

Seedborne disease.   Disease in which inoculum is borne on seed.

Seed coat.   Testa or outer covering of a true seed.

Seed disinfectant or disinfestant.   A chemical that destroys certain disease-causing organisms carried in (disinfectant) or on (disinfectant or disinfestant) the seed. Neither is necessarily a seed protectant.

Seedhead.   Floral development; in the case of grasses, usually a fruiting cluster or spike; inflorescence.

Seed mixture.   A combination of seeds of two or more turfgrass species.

Seed treatment.   A chemical or nonchemical treatment used to control pathogen propagules associated with true seeds.

Selection.   The process of isolating and preserving certain individuals or characters from a group of individuals or characters.

Selectivity.   The ability of a pesticide to kill some pests and not others without harming related desired plants or animals.

Self-fertile, self-fertility.   Capable of fertilization and producing viable seed after self-pollination; a condition in which sexual reproduction occurs as a result of the fusion of eggs and sperms produced by the same individual.

Semiarid.   Climate in which evaporation exceeds precipitation; a transition zone between true desert and a humid climate. The annual precipitation is usually between 10 and 20 in (250 and 500 mm).

Senesce, senescence (adj. **senescent).**   Decline or degeneration of tissues as with maturation or a physiological aging process; often hastened by environmental stress, disease, or insect attack; growing old.

Sensitize.   To cause a person's body to react abnormally when exposed to a poison (such as a pesticide). Symptoms include asthma, shock, and skin and eye irritation.

Sepals.   A division of the outer floral envelope or calyx; the modified leaflike structures, usually green, collectively forming the calyx.

Septum (pl. **septa,** adj. **septate).**   A cross wall in a hyphal filament; divided by partitions or septa; having septa.

Sequential application.   An application in which different pesticides are applied at different times during the growing season.

Serrate.   Having edges with sharp notches or teeth, like a saw blade.

Sessile.   A structure such as a leaf, leaflet, flower, floret, or fruit that is attached directly to the axis without a stem (stalk), petiole, or pedicel.

Seta (pl. **setae).**   Stiff, hair- or bristlelike sterile fungus appendage, a modified hypha; usually erect, deep yellow to black, and thick walled.

Settling.   A lowering of the soil surface resulting from a decrease in the volume of a soil previously loosened by tillage; occurs naturally and can be accelerated mechanically by tamping, rolling, cultipacking, or watering.

Severity.   The measure of damage done by a plant disease as distinguished from disease incidence, which is a measure of the number of individuals affected.

Shade-tolerant.   Turfgrass cultivar that can grow in reduced sunlight.

Sheath, leaf.   The lower tubular part of a grass leaf that clasps the culm (stem) and envelops new leaves; a membranous cover.

Shoot density.   The relative number of shoots per unit area.

Sieve tube.   A series of thin-walled phloem cells forming a long cellular tube through which food materials are transported.

Sign.   Any indication of disease on a turfgrass plant from direct visibility of the pathogen or its parts and products (spores, mycelium, sclerotia, exudate, fruit bodies, rhizomorphs, ooze). *See* Symptom.

Silt.   A class of textural soil having particles between 0.05 and 0.002 mm in diameter.

Silvicide.   A pesticide used for the control of trees and other woody vegetation.

Single-cycle disease.   A disease whose causal agent (or its progeny) does not infect additional host plants during the current growing season. *See* Monocyclic.

Skeletonize.   The removal, usually by an insect, of all tissue in an area of a leaf except for one surface and the veins.

Slicing.   The use of a machine with knives that make slits through the turf into the underlying soil for the purpose of cultivation or to insert seeds, fertilizer, or pesticides.

Slime molds.   Primitive organisms, the plasmodium of which flows over low-lying vegetation like an amoeba; also the superficial diseases caused by these organisms on low-growing plants. *See* Myxomycete.

Slit trench drain.   A narrow trench, usually 2 to 4 in (5 to 10 cm) wide, backfilled to the surface with a porous material such as sand, gravel, or crushed rock; used to intercept surface or lateral subsurface drainage water.

Slowly available fertilizer.   Designates a rate of dissolution less than that obtained for completely water-soluble fertilizers; may involve compounds that dissolve slowly, materials that must be decomposed microbiologically, or soluble compounds coated with substances highly impermeable to water. Used interchangeably with delayed release, controlled release, controlled availability, slow acting, and metered release.

Slurry.   A thick suspension of a finely divided material in a liquid paste. Generally used for treating seeds.

Smut.   A disease caused by a smut fungus or the fungus itself; it is characterized by masses of dark brown or black, dusty or greasy teliospores that generally accumulate in black, powdery sori.

Snow molds.   Fungal diseases that commonly develop under snow and as snow is melting.

Sod.   Turf harvested as a thin layer composed of a living, dense population of grass plants with shortened or shallow roots; organic matter; a variety of microbes, small plants and animals; in soil or a growth medium, and in some cases artificial materials used to increase tensile strength.

Sodding.   Planting turf by means of sod.

Soil application.   The application of chemicals made primarily to the soil surface.

Soil fumigant.   A compound that kills most organisms (e.g., bacteria, fungi, insects, nematodes, plants, seeds) in soil by vapor action, usually permitting replanting soon after.

Soil incorporation.   Mixing of pesticides, fertilizers, or other substances in the soil to increase their effectiveness and in some cases to prevent their loss by volatility or photodecomposition.

Soil inhabitant.   A microorganism that is usually strongly competitive with other normal microflora of the soil and which often survives many years as a saprophyte in the complete absence of suitable hosts. Examples include species of the fungi *Fusarium, Pythium,* and *Rhizoctonia.*

Soil injection.   Mechanical placement of a chemical (such as a pesticide) beneath the soil surface with a minimum of mixing or stirring.

Soil invader.   A microorganism (e.g., most leaf-spotting fungi) that is poorly competitive with normal soil microflora and seldom survives over one or two years in the soil in the complete absence of suitable hosts.

Soil mix.   A prepared mixture used as a growth medium for turfgrass.

Soil modification.   Alteration of soil characteristics by adding soil amendments; commonly used to improve phyiscal conditions of the soil.

Soil organic matter.   The organic fraction of the soil that includes plant and animal residues at various stages of decomposition, cells and tissues of soil organisms, and substances synthesized by the soil population.

Soil probe.   A cylindrical soil sampling tool with a cutting edge at the lower end.

Soil reaction.   *See* pH.

Soil salinity.   The amount of soluble salts in a soil, expressed as parts per million (ppm), millimho/cm, or other convenient ratios.

Soil sterilant.   A chemical (pesticide) that, when present in the soil, completely prevents the growth of plants and microorganisms.

Soil sterilization.   Treating soil by heat or chemicals to kill living organisms.

Soil texture.   The relative percentages of sand, silt, and clay in a soil.

Soil tilth.   The physical condition of soil as related to its ease of tillage, fitness as a seedbed, and suitability for plant growth.

Solubility.   The maximum amount of a gas, liquid, or solid that will dissolve in a liquid. The solubility varies with temperature; hence it is usually expressed at a standard temperature of 77°F (25°C).

Soluble powder (SP).   A finely ground powder formulation that dissolves and forms a solution in water or other liquid.

Soluble salts.   Dissolved salts (anions and cations) in the soil that can become toxic to roots when exceeding certain levels.

Solute.   A dissolved substance; the dispersed phase of a solution.

Solution.   A homogeneous mixture; the individual molecules of the dissolved substance (the solute) are uniformly dispersed as individual molecules among the molecules of the solvent.

Solvent.   A substance, usually a liquid, that can dissolve other substances (solutes); the continuous phase of a solution.

Sooty molds.   Fungi with dark hyphae that grow on the excreta (honeydew) secreted by insects (e.g., aphids), forming a dense, superficial, sooty coating on foliage, stems, and fruit.

Sorus (pl. sori).   Compact fruiting structure, especially the erumpent spore mass of rust and smut fungi. *See* Pustule.

Spawn.   Mycelium.

Species (pl. species).   A group of closely related individuals of the same ancestry, resembling one another in certain inherited characteristics of structure and behavior, and relative stability in nature; the individuals of a species ordinarily interbreed freely and maintain themselves and their characteristics in nature; one kind of plant or animal life subordinate to a genus but above a race, strain, or variety. A genus name followed by sp. means that the particular species is undetermined; spp. after a genus name means that several species are grouped together without being named individually. *See* Genus, Race.

Spermatium (pl. spermatia).   Nonmotile, uninucleate, haploid ($\div$ or $-$) sex cell; haploid male gamete, variously termed as pycniospore or microconidium.

Spermogonium, spermagonium (pl. spermogonia).   Flask-shaped, walled fungus structure (male sex organ) of a rust fungus in which spermatia are produced; sometimes called a pycnium.

Sphagnum peat.   Dead and partially decomposed sphagnum moss that characteristically grows in bogs that is mined and used to loosen soil, improving air and water movement.

Spike.   Type of inflorescence in which the spikelets are directly attached to the main axis with no connecting stem (pedicel).

Spikelet.   The basic unit of the grass inflorescence, consisting of two glumes and one or more florets; a small spike.

Spiking.   A method of turf cultivation in which solid tines or flat, pointed blades penetrate the turf and soil surface.

Spiroplasma.   A genus of small, pleomorphic, prokaryotic bacteria that lack rigid cell walls; flexuous, often helical mycoplasma present in the phloem of diseased plants. There are also spiroplasmas that live on plant surfaces.

Split application.   An application in which the same chemical (such as a fertilizer or pesticide) is applied at different times during a single growing season.

Spongy mesophyll.   Tissue composed of loosely arranged, chlorophyll-bearing parenchyma cells of diverse form found just inside the upper and lower epidermis of a grass leaf.

Spoon, coring.   A method of turf cultivation by which curved, hollow, spoonlike tines remove small soil cores and leave a hole or cavity in the sod.

Spora, air spora.   The population (spore flora) of airborne particles of plant or animal origin.

Sporangiophore.   A differentiated hypha (a sporophore or stalk) bearing one or more sporangia.

Sporangium (pl. sporangia).   Commonly a sac- or flasklike fungal structure, the contents of which are usually converted into an indefinite number of asexual spores (zoospores, sporangiospores).

Spore.   A microscopic, one- to many-celled reproductive body of a fungus or lower plant that detaches, germinates, and can develop into a new plant; may be sexually or asexually produced and of a wide variety of sizes, shapes, colors, and origins; (of bacteria) more correctly *endospore:* a resting phase, a resistant body within the main part of the vegetative microbial cell.

Spore ball.   Unit of dispersal comprised of a firmly aggregated group of smut spores; spores and sterile cells joined closely together of varying structure.

Spore print.   A deposit of basidiospores obtained from a basidiocarp (e.g., a mushroom) that is made to fall on a sheet of paper below.

Sporidium (pl. sporidia).   Basidiospore of rusts, smuts, and other basidiomycetes.

Sporocarp.   Organ or fruit body in or on which spores are borne.

Sporodochium (pl. sporodochia).   Superficial, erumpent, cushion-shaped (pulvinate), asexual fruit body (stroma) bearing closely packed conidiophores and conidia.

Sporophore.   Any structure on which spores are borne.

Sporulate, sporulation.   To form or produce spores.

Spot.   A definite, localized, chlorotic or necrotic, circular to oval area on leaves or other plant parts.

Spot treatment.   Application of a chemical (such as a fertilizer or pesticide) to a restricted or small area.

Spray deposit rate.   The amount of spray liquid deposited per acre, hectare, or 1000 sq ft. *Mean spray deposit rate* is the average amount of deposit over the entire spray swath. *Effective spray deposit rate* is the mean deposit from center to center of adjoining swaths.

Spray drift.   The movement of airborne spray particles from the intended target area.

Sprayed width per nozzle.   The equivalent width sprayed by a single nozzle. For broadcast spraying, the spray width per nozzle is the nozzle spacing.

Spreader.   Material added to a spray preparation to make the spray droplets spread out. This increases the area covered and improves contact between the chemical and the plant or other treated surface; also termed *film extender.*

Sprig.   A single turfgrass stem (stolon, rhizome, tiller, or combination), usually with some attached roots and leaves, that is used in vegetative propagation.

Sprigging.   Vegetative planting by placing stolons, rhizomes, or tillers in furrows or small holes.

Sprigging, broadcast.   Vegetative planting by broadcasting stolons over a prepared soil and, in most cases, covering by topdressing or press rolling.

Spring green-up.   The initial seasonal appearance of green shoots as spring temperature and moisture conditions become favorable for chlorophyll synthesis, thus breaking winter dormancy.

ssp.   Subspecies.

Stage.   A phase in the life cycle.

Stamen (adj. staminate).   The male or pollen-producing structure of a flower, consisting of one or more anthers (pollen-bearing portions) borne on a stalk or filament.

Stand.   The number of established individual plants or shoots per unit area.

Stele (adj. stelar).   The vascular cylinder of a plant, including xylem; phloem; and (when present) pith, pericycle, and interfascicular parenchyma.

Sterilant.   A chemical that makes pests unable to reproduce, or one that destroys all living organisms in a substrate (such as soil).

Sterile.   Infertile; devoid of living (reproducible) microorganisms; uncontaminated; nonsporing.

Sterilization.   The elimination of living cells or organisms by means of heat, chemicals, light, or some other source, from soil, containers, and so forth; the killing of all forms of life.

Sticker.   A material added to a spray or dust to improve adherence (tenacity) to plant or other surface rather than to increase initial deposit.

Stigma.   The feathery portion of the pistil, usually the apex, that receives the pollen and upon which pollen grains germinate; also, a darkened area in the upper margin near the end of the wings of certain insects.

Stipe.   Stalk or stem of a basidiocarp (such as a mushroom).

Stippling.   Series of small dots or specks in which chlorophyll is absent.

Stipule.   Small leaflike structure or appendage, usually paired, found at the base of leaf petioles in many species of plants.

Stolon.   A slender jointed horizontal stem (or shoot) that grows horizontally above the soil surface and is capable of developing leaves, roots, and stems at its tip or at nodes may originate extravaginally from the main stem of tillers. It is used to propagate certain grasses.

Stoloniferous.   Bearing or developing stolons.

Stoma or stomate (pl. stomata or stomates).   A microscopic, regulated opening (pore) in the epidermis of a leaf or stem controlled by two guard cells, through which gases and water vapor are exchanged; sometimes serves as a point of entry for pathogens.

Strain.   An organism or group of similar organisms that differ in minor aspects from other organisms of the same species or variety; biotype; race; form; isolate.

Streak.   An elongated, necrotic lesion, usually with irregular sides, along vascular bundles in leaves or stems of grasses.

Stress.   The condition of plants that are unable to absorb enough water to replace that lost by transpiration. This may result in wilting, cessation of growth, or death of the plant or plant parts.

Striate (n. striations).   Marked with minute parallel or radiating lines, grooves, furrows, projections, or ridges.

Stripe.   Elongated necrosis of tissue between vascular bundles in leaves or stems of grasses.

Stroma (pl. stromata, adj. stromatic, stromatal).   Compact mass or matrix of specialized vegetative hyphae (with or without host tissue or substratum), in or on which fruit bodies (reproductive structures) and/or spores are produced.

Style.   Stalklike structure between the stigma and ovary in the pistil of most flowers through which pollen tubes grow toward the ovule.

Stylet.   Relatively long, pointed, stiff, slender, hollow feeding organ in the mouth portion of plant-parasitic nematodes and some insects (such as aphids) for piercing and withdrawing nutrients from plant cells.

Subglobose.   Not uniformly spherical.

Subgrade.   The surface grade of a turf site prior to the addition of topsoil.

Submicroscopic.   Too small to be seen with a light (compound) microscope.

Substrate, substratum (pl. substrata).   Surface or medium on or in which a microorganism is growing (attached) or living and from which it may get its nourishment.

Succulent.   Plant or plant part with tender, juicy, or watery tissues.

Summer annual.   An annual plant that germinates in the spring, flowers, produces seed in mid- to late summer, and dies in the fall.

Summer dormancy.   The cessation of growth and subsequent death of shoots of perennial plants due to heat and/or moisture stress.

Supplement (syn. adjuvant, axillary spray material).   A material added to a pesticide to improve its physical or chemical properties, including the spreader, sticker, safener, and wetting agent but not the diluent. *See* Adjuvant.

Supplied-air respirator.   A respirator in which the air a person breathes is contained in a tank so that the worker does not breathe air containing toxic chemicals.

Suppressive soils.   Soils in which certain diseases are suppressed due to the presence of microorganisms antagonistic to the pathogen or pathogens.

Surface water.   Lakes, streams, ponds, rivers, creeks, or any other aboveground body of water.

Surface tension.   The property of liquids in which the exposed surface tends to contract to assume the smallest possible area, resulting in the formation of spherical drops on plants, bodies of insects, and other surfaces.

Surfactant.   A monomolecular, surface-active compound that increases the emulsifying, dispersing, spreading, wetting, penetrating, or other surface-modifying properties of a pesticide formulation. It reduces surface tension between two unlike materials such as oil and water. Surfactants are used in agricultural sprays as wetters, stickers, emulsifiers, penetrants, or other agents.

Suscept.   Any living organism attacked or susceptible to a given disease, pathogen, or toxin; an abbreviated term for susceptible plant or susceptible species. *See* Host.

Susceptible, susceptibility.   Not immune; lacking resistance the inability of a plant to overcome disease; prone to infection. *See* Tolerance.

Susceptible species.   Pests readily killed by relatively low rates of pesticide application.

Suspension.   Very small solid particles mixed with, but not dissolved in, a liquid, such as soil particles in water; a solid form of a pesticide (of finely divided particles) suspended in a liquid, solid, or gas.

Sward.   Turf; carpet of grass.

Swarm spore, swarmer.   Zoospore, as in *Pythium* spp.

Swath, effective width.   The center-to-center distance between overlapping broadcast applications.

Symbiosis, symbiont (pl. symbioses, adj. symbiotic).   Two or more dissimilar organisms (symbionts) living together in close association; usually applied to cases where the relationship is mutually beneficial. *See* Mutualism.

Symptom.   The visible expression of a turfgrass plant's reaction to the activities of a pathogen reaction by a person to pesticide or other poisoning. *See* Sign.

Syndrome.   Total effects produced in a plant by one disease, whether all at once or successively.

Synergism (adj. synergistic).   The mutual association (living together) of two or more organisms or environmental factors acting at one time and eliciting a host response that one alone could not make; the magnitude of host response to concurrent pathogens exceeds the sum of the separate responses to each pathogen. Also, the action of two pesticides that produces a more than additive effect when the pesticides are used together than when they are used individually.

Synergist.   Any substance that, when mixed with a pesticide, increases its efficiency or toxic effects; it may or may not have pesticidal properties of its own.

Synergistic.   Pertaining to a host response to concurrent pathogens that exceeds the sum of the separate responses to each pathogen.

Synonym or (syn.).   In biology, another name for an organism (species or group), especially a later or illegitimate name; a rejected scientific name, other than a homonym. *See* Homonym.

Synthesis.   The coming together of two or more substances to form a new material.

Syringe.   To spray turf with small amounts of water; usually on a hot, dry, windy day to reduce water loss (transpiration).

Systemic.   Applied to chemicals and pathogens (or single infections) that generally spread internally throughout the plant body (away from the original site of entry) as opposed to remaining localized. *See* Penetrant.

Systemic (or translocated) pesticide.   A pesticide that is absorbed by treated plants or animals and translocated (moved) from the site of uptake to most tissues. *See* Penetrant.

T

Tank mix.   A mixture of two or more chemicals (pesticides, fertilizers) in a spray tank at the time of application. A tank mix should be used with caution until assured that the ingredients are compatible.

Taproot.   The primary, elongated, deeply descending root of a plant from which the secondary or lateral roots branch.

Target pest (or species).   The organism for which a pesticide application is intended (as opposed to a nontarget species).

Taxonomy.   Science or scientific art dealing with the systematic describing, naming, and classification of organisms based on their natural relationships.

Tee.   The grassed area of a golf course on which a golf ball is placed and from which the ball is initially hit down the fairway toward the hole. Due to high usage, this area requires a high level of maintenance.

Teleomorph, teleomorphosis.   The sexual or perfect state of a fungus. *See* Anamorph.

Teliospore.   Thick-walled resting or overwintering spore produced by rust and smut fungi. It germinates to form a promycelium (basidium) in which meiosis occurs.

Telium (pl. telia).   Sorus of a rust fungus producing teliospores.

Tenacity (adherence).   The resistance of a pesticide deposit to resist weathering as measured by retention.

Tendril.   Long, slender, coiling mass of fungus spores.

Teratogen.   A chemical, exposure to which is liable to produce malformations, monstrosities, or serious deviations from the normal in an animal embryo or fetus.

Testa.   The outer coat of a true seed developed from the integuments of an ovule.

Texture.   In soil, the relative proportions of the various sized groups of individual soil grains in a mass of soil; refers to the proportions of sand, silt, and clay in a given amount of soil.

Thallus (pl. thalli).   Vegetative body of a fungus; a relatively simple organism that lacks true stems, roots, and leaves; characteristic of algae, fungi, liverworts, and the like.

Thatch.   A tightly intermingled layer of living and undecomposed or partially decomposed plant litter (from long-term accumulation of dead grass roots, crowns, rhizomes, and stolons) that develops between the layer of green vegetation and the soil surface.

Thatch control.   The process of preventing excessive thatch accumulation by cultural manipulation and/or reducing excess thatch from a turf by either mechanical or biological means.

Threshold.   The number of pests that need to be present to warrant control.

Thorax.   The middle of the three major divisions of an insect's body; between the head and abdomen.

Thrips.   Small, slender insects (order Thysanoptera), usually with four long, narrow wings. Thrips have rasping-sucking mouthparts and an abdomen of 10 segments.

Tiling.   The use of porous piping in the soil to collect water and carry it away to reduce the moisture content of the soil.

Tiller.   A lateral shoot, culm, or stem, usually erect, arising from a crown bud; common in grasses.

Tilth.   State of soil aggregation or consistency; good tilth implies porous, friable texture. *See* Soil tilth.

Tip burn, tip necrosis.   A whitening of the leaf tip; may be caused by internal water stress, wind desiccation, or salt.

Tissue.   A group of cells, usually of similar structure and function.

Tissue analysis.   Analysis of leaf tissues for major and minor elements.

Tolerance, tolerant.   The ability of a plant to endure an infectious or noninfectious disease; adverse conditions; insect, mite, other pest; or chemical injury without serious damage or yield loss; of pesticides, the amount deemed safe and legally permitted on an agricultural product entering commercial channels and usually measured in parts per million (ppm). The tolerance in the United States is set by the federal Environmental Protection Association (EPA).

Topdressing.   A prepared soil amendment added to the turf surface and usually incorporated into the root zone by raking, matting, and/or irrigating; materials such as fertilizer or compost are applied to the soil surface while plants are growing.

Topical application.   Application to the top or upper surface of the plant.

Top soil. The surface layer of soil of varying thickness. *See* A horizon.

Toxicant. A poisonous or toxic substance.

Toxicity. The relative capacity of a substance (such as a pesticide) to interfere adversely with the processes of a living organism; quality, state, or degree of being poisonous.

Toxin. A poison produced by a plant or animal.

Tracheid. Thick-walled, elongated xylem cells that conduct water and solutes from roots to shoots.

Trade name (syn. brand name, trademark). The proprietary name given to a product by the manufacturer or formulator to distinguish and identify it.

Transfer mixing system. Pesticide-handling process that is a closed system, thus eliminating exposure to workers during mixing operations.

Transitional climatic zone. The suboptimal zone between temperate and subtropical climates.

Translocated pesticide. *See* Systemic pesticide, Penetrant.

Translocation. Distribution of a chemical from the point of absorption (plant leaves, stems, roots) to other leaves, buds, and root tips; movement of water, minerals, food, pathogens, wastes, and the like within or throughout a plant.

Translucent. Transmitting light without being transparent.

Transmission. The transfer or spread of virus or another pathogen from plant to plant or from one plant generation to another.

Transpiration. Diffusion and loss of water vapor from aerial parts of plants, chiefly through stomata in the leaves.

Transverse. Crosswise.

Trenching. Physical separation of soil in a vertical plane to sever grafted roots between trees or prevent roots from entering high-maintenance turf, such as golf greens.

Triple rinsing. A method of removing most of the pesticide residue from a container by flushing it three times.

Tropism. Movement or curvature in a plant due to an external stimulus that determines the direction of movement.

Truncate. Ending abruptly as though the end was cut off squarely.

Tuber. A short, fleshy, much enlarged, mostly underground stem borne at the end of a rhizome, having numerous buds.

Turbid. Cloudy; not clear.

Turf. A covering of closely mowed dense vegetation, usually a turfgrass, growing intimately with an upper soil stratum of intermingled roots and stems.

Turfgrass. A grass species or mixture of grass species which is maintained as a turf.

Turfgrass community. An aggregation of individual turfgrass plants that have mutual relationships with the environment as well as among individual plants.

Turfgrass culture. The complete cultural practices involved in growing turfgrasses for purposes such as a lawns, greens, sports fields, and roadsides.

Turfgrass management.   The development of turf standards and goals which are achieved by planting and direct labor, capitol, and equipment, with the objective of manipulating cultural practices to achieve those standards and goals.

Turfgrass quality.   The degree to which a turf confirms to an agreed standard of uniformity, density, texture, growth habit, smoothness, and color, as judged by objective visual assessment.

Turgid, turgidity.   Cells or tissues swollen, distended, plump, or rigid due to internal water pressure.

Turgor, turgor pressure.   Inflation of a plant cell by its fluid contents; pressure within a cell resulting from the absorption of water into the vacuole and the imbibition of water by the protoplasm. Lack of turgor pressure causes plants to wilt.

Tyvek.   A spun-bonded, olefin, nonwoven, comfortable, lightweight fabric worn to provide protection against toxic pesticides.

U

Ubiquitous.   Constantly encountered; widespread.

ULV (ultralow volume).   Sprays applied at a total volume of ½ gal (1.891 kg) or less per acre (0.4047 ha) or applied in undiluted form. *See* Low-volume (LV) spray.

Umbel.   A type of inflorescence whereby flowers are borne at the end of a common stalk, forming a more or less flattened or rounded cluster; can be compound with subsets of umbels.

Undulate.   Wavy, not flat or uniformly curved.

Unicellular.   Having one cell; referring to an organism, the entire body of which consists of a single cell.

Uninucleate.   Having one nucleus per cell.

Unisexual.   An organism that produces eggs or sperms but not both; or a flower that bears stamens or pistils, but not both.

Ureaformaldehyde.   A synthetic, slowly soluble nitrogen fertilizer consisting mainly of methylene urea polymers of different lengths and solubilities; formed by reacting urea and formaldehyde.

Uredinium, uredium (pl. uredinia, uredia).   Fruit body (sorus) of a rust fungus that produces urediospores (urediniospores, uredospores) formed after the aecium and before the telium in the life cycle.

Urediospore, uredospore, or urediniospore.   Binucleate, dikaryotic, asexual, one-celled, repeating or summer spore of rust fungi.

V

Vacuole.   The relatively clear, bubblelike, membrane-lined space in the cytoplasm of a plant cell filled with a watery solution of sugars and various other plant products and by-products.

Variety, botanical (var.).   A subdivision within a species, the members of which, in minor details, are distinct in form, color, flower, and fruit from other similar groups of individuals; group of plants of common origin within a species. *See* Cultivar.

Variety, cultivated.   *See* Cultivar.

Vascular.   Pertaining to conductive (xylem and phloem) tissue. A vascular parasite grows and/or moves in these tissues.

Vascular bundle or strand.   A distinct group of elongated conducting cells in a stem or leaf, consisting of primary xylem for water conduction and primary phloem for food conduction.

Vector.   A living organism (e.g., an insect, mite, bird, higher animal, nematode, parasitic plant, or human) able to carry or transmit a pathogen (virus, bacterium, fungus, nematode) and disseminate disease; in genetic engineering, a vector or cloning vehicle is a self-replicating DNA molecule, such as a plasmid or virus, used to introduce a fragment of foreign DNA into a host cell. *See* Transmission.

Vegetation.   The plants that cover an area or region.

Vegetative.   Relating to growth functions that do not involve reproductive functions.

Vegetative propagation.   Asexual propagation using pieces of vegetation, such as sprigs or sod pieces.

Vein.   A vascular strand of xylem and phloem in a leaf, petal, or other plant part.

Venation.   The arrangement of veins in a leaf blade.

Verdant.   A grass-green color.

Vermicular, vermiculate, vermiform.   Worm-shaped, thickened and bent in places; having the motion of a worm; marked with irregular lines like worm tracks.

Vernation.   The arrangement of the folded leaves in a bud.

Vertical cutting.   Involves a mechanical device having vertically rotating blades that cut into the face of a turf for the purpose of controlling thatch or grain.

Vertical mower.   A mechanical device, the vertically rotating knives of which cut into the face of a turf for the purpose of reducing thatch, grain, and surface compaction.

Viable, viability.   State of being alive; capable of growth and development.

Vicid.   Slimy, sticky, mucilaginous, glutinous, lubricous, viscous.

Virion.   Complete virus particle consisting of nucleic acid and a protein shell; the infectious unit of a virus.

Virulence (adj. **virulent).**   Degree or measure of pathogenicity of a given pathogen; relative capacity to cause disease; highly pathogenic.

Viruliferous.   Containing or carrying a virus; term applied particularly to virus-laden insect and nematode vectors capable of transmitting the virus.

Virus.   A submicroscopic, filterable agent that causes disease and multiplies only in living plant or animal cells; an intercellular parasite consisting of a core of one or more infectious nucleic acid molecules (RNA or DNA) usually surrounded by a protein coat. Viruses are capable of producing flower breaking, growth malformations, mosaics and mottles, ringspots, and other plant diseases.

Volatile, volatility.   Evaporating or vaporizing readily at ordinary temperatures or exposure to air; rate of evaporation of a pesticide.

Volatilization.   The conversion of a substance from a liquid to a gaseous or vapor state.

W

Warm-season turfgrass. Turfgrass that has optimum growth at temperatures between 80 to 95°F (26.6 and 32°C) and usually dormant during cold weather.

Water-dispersable granules (WDG). *See* Dry flowable (DF).

Waterlogged soil. Without soil aeration due to excess water and a lack of or poor soil drainage.

Watersoaked. Disease symptom of plants, giving a wet, dark, and usually sunken and translucent appearance.

Water table. The upper boundary of the groundwater or that level below which the soil is saturated with water.

Wear. The collective direct or indirect effects of traffic on turf; is distinct from the indirect effects of traffic caused by soil compaction.

Weed. Any plant growing in a place where it is not wanted.

Wettable powder (WP). A powder formulation of a pesticide containing a wetting agent that causes the powder to form a suspension in water.

Wetting agent. A compound added to a spray preparation that reduces surface (interfacial) tension and causes a liquid to contact plant or other surfaces more thoroughly, usually a surfactant.

Wet wilt. Wilting of turf in the presence of free soil water when evapotranspiration exceeds the ability of roots to take up water.

Whorl, whorled. Three or more leaves, flowers, twigs, hyphae, or other plant parts arranged in a circle and radiating from one point.

Wilt, wilting. Lack of turgor (rigidity) or drooping of plant parts from lack of water (inadequate water supply or excessive transpiration); a vascular, crown, or root disease that interrupts the normal uptake and distribution of water by a plant.

Wind burn. Death and browning of tissues most commonly on the uppermost leaves of semidormant grasses caused by atmospheric desiccation. *See* Winter desiccation.

Winter annual. An annual plant that usually germinates in the fall, lives over winter, and completes growth, flowers, and produces seed in mid- to late spring, and dies in the summer.

Winter desiccation. The death or partial death of leaves or plants by drying during winter dormancy.

Winterkill. Any injury to turfgrass plants that occurs during the winter period.

Winter overseeding. Seeding cool-season turfgrasses over warm-season turfgrasses at or near their start of winter dormancy; practiced in subtropical climates to provide green, growing turf during the winter period when the warm-season species are brown and dormant.

Wound. An injury to a plant in which the surface is cut, scraped, torn, or otherwise broken.

X

Xerophile, xerophyte. A plant growing in soil with a scanty water supply, or in a soil where water is absorbed with difficulty.

Xylem. The complex water- and mineral-conducting tissue of vascular plants.

Y

Yellowing. To make or render tissue yellow that once was green.

Yellows. A plant disease or symptom characterized by yellowing and stunting of the host plant or affected parts, usually caused by phytoplasmas or deficiencies and imbalances of one or more essential elements.

Z

Zonate, zonation. Marked with concentric zones (bands or lines) of different colors and/or textures; targetlike; any leaf or stem development appearing in concentric rings.

Zoosporangium (pl. zoosporangia). A usually thin-walled sac (sporangium) that contains or produces spores endogenously.

Zoospore. An asexually produced fungus spore bearing flagella and capable of independent movement in water; found especially in the lower fungi (such as *Pythium*); a swarm spore or sporangiospore; swimming spore.

Zygote. A diploid cell or fertilized egg resulting from the union of two haploid gametes; the beginning of a new organism in sexual reproduction.

INDEX

Illustrations are indicated by page numbers in **bold**

16. White grub

17. Japanese beetle

18. Annual white grub adult

19. Grub damage pattern

20. Black turfgrass ataenius -
    adult, pupa, larva

21. Billbug damage

22. Mole cricket damage

23. Sod webworm damage

24. Bluegrass webworm adult

25. Black cutworm larva

26. Chinch bug damage

27. Greenbug damage

28. Greenbug damage

29. Red imported fire ant mound

30. Pillbugs